25th Anniversary of Molecules—Recent Advances in Analytical Chemistry

25th Anniversary of Molecules—Recent Advances in Analytical Chemistry

Editors

Gavino Sanna
Stefan Tsakovski

MDPI • Basel • Beijing • Wuhan • Barcelona • Belgrade • Manchester • Tokyo • Cluj • Tianjin

Editors

Gavino Sanna
Department of Chemical,
Physical, Mathematical and
Natural Sciences
Sassari University
Sassari
Italy

Stefan Tsakovski
Department of Analytical
Chemistry Faculty of
Chemistry and Pharmacy
University of Sofia St.
Kliment Ohridski
Sofia
Bulgaria

Editorial Office
MDPI
St. Alban-Anlage 66
4052 Basel, Switzerland

This is a reprint of articles from the Special Issue published online in the open access journal *Molecules* (ISSN 1420-3049) (available at: www.mdpi.com/journal/molecules/special_issues/25th_anniversary_analytical).

For citation purposes, cite each article independently as indicated on the article page online and as indicated below:

LastName, A.A.; LastName, B.B.; LastName, C.C. Article Title. *Journal Name* **Year**, *Volume Number*, Page Range.

ISBN 978-3-0365-6945-1 (Hbk)
ISBN 978-3-0365-6944-4 (PDF)

© 2023 by the authors. Articles in this book are Open Access and distributed under the Creative Commons Attribution (CC BY) license, which allows users to download, copy and build upon published articles, as long as the author and publisher are properly credited, which ensures maximum dissemination and a wider impact of our publications.
The book as a whole is distributed by MDPI under the terms and conditions of the Creative Commons license CC BY-NC-ND.

Contents

About the Editors vii

Preface to "25th Anniversary of Molecules—Recent Advances in Analytical Chemistry" ix

Natalia Wilkosz, Michał Czaja, Sara Seweryn, Katarzyna Skirlińska-Nosek, Marek Szymonski and Ewelina Lipiec et al.
Molecular Spectroscopic Markers of Abnormal Protein Aggregation
Reprinted from: *Molecules* **2020**, *25*, 2498, doi:10.3390/molecules25112498 1

Paulina Grocholska and Remigiusz Bachor
Trends in the Hydrogen−Deuterium Exchange at the Carbon Centers. Preparation of Internal Standards for Quantitative Analysis by LC-MS
Reprinted from: *Molecules* **2021**, *26*, 2989, doi:10.3390/molecules26102989 27

Shenggang Wang, Yue Huang and Xiangming Guan
Fluorescent Probes for Live Cell Thiol Detection
Reprinted from: *Molecules* **2021**, *26*, 3575, doi:10.3390/molecules26123575 53

Laurynas Jarukas, Olga Mykhailenko, Juste Baranauskaite, Mindaugas Marksa and Liudas Ivanauskas
Investigation of Organic Acids in Saffron Stigmas (*Crocus sativus* L.) Extract by Derivatization Method and Determination by GC/MS
Reprinted from: *Molecules* **2020**, *25*, 3427, doi:10.3390/molecules25153427 115

Michael K. Appenteng, Ritter Krueger, Mitch C. Johnson, Harrison Ingold, Richard Bell and Andrew L. Thomas et al.
Cyanogenic Glycoside Analysis in American Elderberry
Reprinted from: *Molecules* **2021**, *26*, 1384, doi:10.3390/molecules26051384 125

Brahmam Kapalavavi, Ninad Doctor, Baohong Zhang and Yu Yang
Subcritical Water Extraction of *Salvia miltiorrhiza*
Reprinted from: *Molecules* **2021**, *26*, 1634, doi:10.3390/molecules26061634 143

Nadia Spano, Sara Bortolu, Margherita Addis, Ilaria Langasco, Andrea Mara and Maria I. Pilo et al.
An Analytical Protocol for the Differentiation and the Potentiometric Determination of Fluorine-Containing Fractions in Bovine Milk
Reprinted from: *Molecules* **2023**, *28*, 1349, doi:10.3390/molecules28031349 155

Yanyu Ren, Xiumin Shi, Pengcheng Xia, Shuang Li, Mingyang Lv and Yunxin Wang et al.
In Situ Raman Investigation of TiO_2 Nanotube Array-Based Ultraviolet Photodetectors: Effects of Nanotube Length
Reprinted from: *Molecules* **2020**, *25*, 1854, doi:10.3390/molecules25081854 169

Marin Senila, Oana Cadar and Ion Miu
Development and Validation of a Spectrometric Method for Cd and Pb Determination in Zeolites and Safety Evaluation
Reprinted from: *Molecules* **2020**, *25*, 2591, doi:10.3390/molecules25112591 181

Fangqun Gan, Ke Wu, Fei Ma and Changwen Du
In Situ Determination of Nitrate in Water Using Fourier Transform Mid-Infrared Attenuated Total Reflectance Spectroscopy Coupled with Deconvolution Algorithm
Reprinted from: *Molecules* **2020**, *25*, 5838, doi:10.3390/molecules25245838 195

Neil Yohan Musadji and Claude Geffroy-Rodier
Simple Derivatization–Gas Chromatography–Mass Spectrometry for Fatty Acids Profiling in
Soil Dissolved Organic Matter
Reprinted from: *Molecules* **2020**, *25*, 5278, doi:10.3390/molecules25225278 205

Xindan Liu, Ying Zhang, Menghua Wu, Zhiguo Ma and Hui Cao
The Potential Transformation Mechanisms of the Marker Components of Schizonepetae Spica
and Its Charred Product
Reprinted from: *Molecules* **2020**, *25*, 3749, doi:10.3390/molecules25163749 213

Lulu Wang, Mengxin Gao, Zhipeng Liu, Shuang Chen and Yan Xu
Three Extraction Methods in Combination with GC×GC-TOFMS for the Detailed Investigation
of Volatiles in Chinese Herbaceous Aroma-Type Baijiu
Reprinted from: *Molecules* **2020**, *25*, 4429, doi:10.3390/molecules25194429 229

Xun Jia, Lucas J. Osborn and Zeneng Wang
Simultaneous Measurement of Urinary Trimethylamine (TMA) and Trimethylamine *N*-Oxide
(TMAO) by Liquid Chromatography–Mass Spectrometry
Reprinted from: *Molecules* **2020**, *25*, 1862, doi:10.3390/molecules25081862 247

Ghada Rashad Ibrahim, Iltaf Shah, Salah Gariballa, Javed Yasin, James Barker and Syed
Salman Ashraf
Significantly Elevated Levels of Plasma Nicotinamide, Pyridoxal, and Pyridoxamine Phosphate
Levels in Obese Emirati Population: A Cross-Sectional Study
Reprinted from: *Molecules* **2020**, *25*, 3932, doi:10.3390/molecules25173932 263

Sundus M. Sallabi, Aishah Alhmoudi, Manal Alshekaili and Iltaf Shah
Determination of Vitamin B3 Vitamer (Nicotinamide) and Vitamin B6 Vitamers in Human Hair
Using LC-MS/MS
Reprinted from: *Molecules* **2021**, *26*, 4487, doi:10.3390/molecules26154487 277

About the Editors

Gavino Sanna

Gavino Sanna received the master's degree in chemistry from the University of Sassari (Sardinia, Italy). After spending five years at the EniChem Research Centre in Porto Torres, Italy, he became Assistant Professor of Analytical Chemistry at the Department of Chemistry at the University of Sassari. He is currently Associate Professor of Analytical Chemistry in the Department of Chemical, Physical, Mathematical, and Natural Sciences at the University of Sassari. In 2018, he received the scientific qualification (called Abilitazione Scientifica Nazionale in Italian) as a full professor in analytical chemistry. He has published more than 100 papers, and several books, book chapters, and patents. His general interest in research is focused on assessing, validating and applying on real samples of instrumental methods for measuring organic and inorganic analytes in food, materials, environmental, and biological matrices. His most representative research lines are the determination and speciation of organic and inorganic minority analytes of interest for the characterization and valorization of traditional Sardinian foodstuffs (e.g., dairy, seafood, and beehive products), the role of irrigation methods in the bioaccumulation of health-threatening elements in rice and on electroanalytical studies on the interaction between metal ions and biologic ligands. He is also published contributions on food safety, on environmental analysis and monitoring, on material sciences and on biomedical analysis.

Stefan Tsakovski

Stefan Tsakovski received his Ph.D. degree from Sofia University "St. Kliment Ohriski" in 1999 in the field of Analytical Chemistry (Chemometrics). He is currently working as a full-time professor and head of the Chemometrics and Environmetrics Unit in the Department of Analytical Chemistry at the Sofia University "St. Kliment Ohridski". His research interests include a number of areas related to data mining of information derived from chemical experiments, epidemiological studies and environmental monitoring. He is a co-author of two book chapters and over 100 publications in peer-reviewed journals.

Preface to "25th Anniversary of Molecules—Recent Advances in Analytical Chemistry"

Analytical chemistry between the 20th and 21st centuries is almost entirely instrumental analytical chemistry. The use of analytical instruments in experimental studies related to molecular biology, medicine, geology, food science, materials science, environmental science, etc., is crucial for gathering important information for the understanding of the investigated phenomena. Nowadays, the widespread usage of a large number of analytical techniques used for analysis of variety of analytes in different matrices raises the question of the correctness of the obtained analytical data. This volume, which gathers 16 contributions (13 research articles and 3 reviews) covering different areas of analytical chemistry, aims to be a small but qualified contribution representing a sample of the potential applications of modern methods of instrumental analytical chemistry to solve real problems in multiple research fields. The selected scientific papers point out that a comprehensive validation is key for obtaining reliable analytical information from the implemented complex analytical procedures. Many of these contributions deal with the development and application of chromatographic methods interfaced with mass spectrometry, but also spectroscopic, electrochemical and separation methods to real matrices. We really hope that this volume will be of interest, not only for analytical chemists but also for scientists working in areas where trustworthy analytical data are needed.

Gavino Sanna and Stefan Tsakovski
Editors

Review

Molecular Spectroscopic Markers of Abnormal Protein Aggregation

Natalia Wilkosz, Michał Czaja, Sara Seweryn, Katarzyna Skirlińska-Nosek, Marek Szymonski, Ewelina Lipiec * and Kamila Sofińska *

M. Smoluchowski Institute of Physics, Jagiellonian University, 30-348 Kraków, Poland;
natalia.szydlowska@uj.edu.pl (N.W.); michalandrzej.czaja@student.uj.edu.pl (M.C.); saras@poczta.onet.pl (S.S.); katarzyna.skirlinska@gmail.com (K.S.-N.); ufszymon@cyf-kr.edu.pl (M.S.)
* Correspondence: ewelina.lipiec@uj.edu.pl (E.L.); kamila.sofinska@uj.edu.pl (K.S.);
  Tel.: +48-12-664-45-86 (E.L. & K.S.)

Academic Editor: Gavino Sanna
Received: 28 April 2020; Accepted: 25 May 2020; Published: 27 May 2020

Abstract: Abnormal protein aggregation has been intensively studied for over 40 years and broadly discussed in the literature due to its significant role in neurodegenerative diseases etiology. Structural reorganization and conformational changes of the secondary structure upon the aggregation determine aggregation pathways and cytotoxicity of the aggregates, and therefore, numerous analytical techniques are employed for a deep investigation into the secondary structure of abnormal protein aggregates. Molecular spectroscopies, including Raman and infrared ones, are routinely applied in such studies. Recently, the nanoscale spatial resolution of tip-enhanced Raman and infrared nanospectroscopies, as well as the high sensitivity of the surface-enhanced Raman spectroscopy, have brought new insights into our knowledge of abnormal protein aggregation. In this review, we order and summarize all nano- and micro-spectroscopic marker bands related to abnormal aggregation. Each part presents the physical principles of each particular spectroscopic technique listed above and a concise description of all spectral markers detected with these techniques in the spectra of neurodegenerative proteins and their model systems. Finally, a section concerning the application of multivariate data analysis for extraction of the spectral marker bands is included.

Keywords: abnormal protein aggregation; secondary structure; amyloids; neurodegenerative diseases; multivariate data analysis; principal component analysis (PCA); hierarchical cluster analysis (HCA); molecular spectroscopy; nanospectroscopy

1. Introduction

Following the increase of global average life expectancy, the world population has experienced an upsurge in the number of individuals suffering from deadly and debilitating neurodegenerative diseases, such as Alzheimer's, Parkinson's, and Huntington's disorders [1–3]. Recent estimations imply that over 50 million people worldwide are living with dementia, and the cost of treatment exceed already 1% of the global GDP [4]. Neurodegenerative diseases not only affect individuals and their families, but also pose a crippling burden on the healthcare systems. Despite intense scientific efforts all over the world, dementia is still incurable. The neurodegenerative processes at the heart of those diseases are caused by an abnormal aggregation of highly cytotoxic, structurally pathological proteins called amyloids [1–3]. A detailed investigation into the secondary structures of neurodegenerative proteins and model proteins, that undergo fibrillation processes, is crucial for an understanding of the etiology of neurodegenerative diseases, which will allow for the development of successful treatment regimes.

Abnormal aggregation of proteins such as β-amyloid, tau, α-synuclein and polyglutamine-containing proteins is known to be related to the pathology of neurodegenerative diseases including Alzheimer's disease (AD), Parkinson's disease (PD) and Huntington's disease (HD) [5–7]. Protein aggregates are toxic and lead to neuronal death in different brain regions depending on the disease [8,9]. The etiology of neurodegenerative diseases has not been explained yet. A lot of scientists working on this issue have suggested various mechanisms of protein misfolding and their aggregation. However, a complete understanding of this issue still requires further research at the molecular and cellular level.

The first important factor that affects protein aggregation in neurodegenerative diseases is the protein structure. In particular, primary and secondary structures are critical factors affecting the physical and chemical properties of proteins/peptides and their three-dimensional conformation (tertiary structure). In the primary structure, the position and number of different characteristic amino acid residues can accelerate or slow down the aggregation process. In general, the number of hydrophobic amino acids in proteins is proportional to the aggregation tendency [6]. Mutations are another factor that affects the protein structure and are considered to have an impact on the aggregation process by changing protein solubility and stability [10].

Neurodegenerative diseases are associated with the occurrence of amyloid plaques, which are specific for each disease. AD pathogenesis is considered to be related to tau protein. Tau protein in its native state has important physiological functions, such as microtubule stabilization [11,12]. To perform their functions, tau proteins must be phosphorylated at a normal, physiological level. The hyperphosphorylation causes the loss of their biological activity due to conformational changes and abnormal aggregation of the tau protein [13].

Amyloid β (Aβ) is a small peptide with a mass of 4–4.4 kDa [14]. It is the main component of amyloid deposits found in AD, mainly in the cortex, hippocampus, forebrain, and brain stem [8]. The native Aβ occurs in neurons, astrocytes, neuroblastoma cells, hepatoma cells, fibroblasts, and platelets [15]. Its functions are associated with inflammatory and antioxidant activity. Aβ also affects the regulation of cholesterol transport and the activation of the kinase enzyme. Native Aβ protein has mainly disordered structure but several research reveals the presence of local regions displaying a secondary structure [16,17]. The relative β-sheet content increases with the ongoing aggregation associated with the AD [18].

α-synuclein (α-syn) is associated with Parkinson's disease. The mass of the α-synuclein monomer is ca.14 kDa. The name "α-synuclein" comes from its synaptic and nuclear location. α-Syn regulates dopamine neurotransmission by modulation of vesicular dopamine storage. This protein interacts with tubulin. It also has a molecular protective activity in the folding of SNARE (soluble N-ethylmaleimide-sensitive-factor attachment protein receptor) proteins [19]. Native α-syn is soluble in the cytoplasm in contrast to its abnormal aggregates, which form Lewy bodies. The secondary structure of the α-syn aggregates is complex and consists of β-sheet, β-turns, α-helics/disordered conformations [20].

One of polyglutamine-containing, neurodegenerative proteins is a Huntington protein (Htt). It has a large mass of 350 kDa and consists of 3144 amino acids. Native protein is highly expressed in the peripheral tissues and the brain. This protein is involved in endocytosis, vesicle trafficking, cellular signal transduction, and membrane recycling. In the brain, abnormally aggregated Htt protein damages cells and it is toxic, forming pathological aggregates [21,22].

Monitoring the aggregation process with spectroscopic methods involves tracking the changes in the protein secondary structure. To correctly interpret the progress of the amyloid aggregation based on the spectroscopic data, it is necessary to understand the course of the aggregation process itself. The abnormal protein aggregation is considered to be related to protein misfolding leading to the exposure of hydrophobic amino acid residues or with the change of the protein net charge [8,23]. The exact aggregation pathway of amyloid proteins depends on the secondary structure of aggregating molecules. However, the main principles of the amyloid aggregation process can be simplified to a three-step fibril formation process consisting of the lag-phase, the elongation phase and the plateau phase (Figure 1) [8,24,25]. Within the lag phase, also called a nucleation phase, the soluble oligomeric intermediates are formed. The elongation phase (fibril growth phase) includes the formation of other "aggregation-prone" intermediates known as protofibrils. A self-template growth mechanism leads to their rapid growth until the formation of insoluble mature fibrils [8,26]. The plateau stationary phase is reached when the content of mature fibrils is at a constant maximum level. There are also theories describing the fragmentation of fibrils during the aggregation process changing the kinetics of mature fibril formation [26,27].

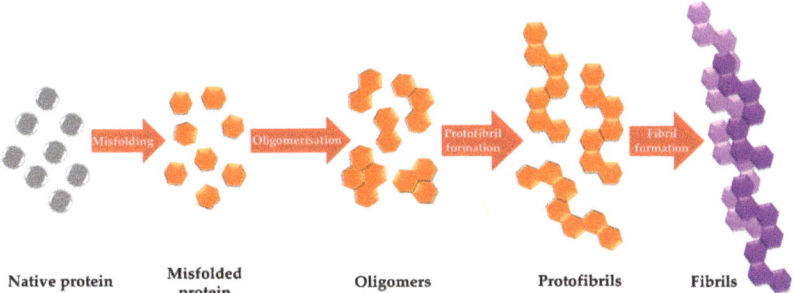

Figure 1. The schematic representation of successive species formation in amyloid aggregation process.

Mature amyloid fibrils, regardless of the peptide/protein they are made of, possess universal characteristic structural features. The morphology of fibrils is represented by a unique cross-β structure consisting of hydrogen-bonded β-sheets arranged perpendicularly along the fibril axis. The fibril diameter is usually 6–13 nm and is formed from 2–8 protofilaments twisted around each other [8,24,26].

The amyloid aggregation process and its subsequent steps can be followed with vibrational spectroscopy techniques. Spectra of proteins have already been described well. Individual bands are attributed to vibrations of various chemical groups present in the peptide chain. Amide bands A and B as well as amide bands I-VII (wavenumbers and assigned vibrations are shown in Table 1) can be observed in the Raman and infrared spectra of peptides and proteins. Molecular spectroscopy techniques are sensitive to the changes of the protein secondary structure. Each of the intermediate species of amyloid aggregation, as well as fibrils, are represented by a different ratio of individual secondary structures. In general, the β-sheet content increases with the progress of the aggregation process [28,29]. Celej et al. [28] registered the decrease in random coil and/or helical structure content (absorbance at 1660–1650 cm^{-1}), for α-synuclein aggregation, from ca. 60% in monomers to 52% in oligomers and 27% in fibrils while the β-sheet contribution increased from 8% for monomers, 26% in oligomers up to 51% in fibrils. In this case, the β-turn content (at 1670 cm^{-1}) was ca. 22% for oligomers and fibrils. The different ratio of secondary structures for fibrils formed from various peptide variants is considered to be a strong premise for different aggregation pathways [30–32].

This review is devoted to the description and organization of all molecular spectroscopic markers of abnormal protein aggregation. Due to comprehensive presentation of the infrared, Raman TERS, SERS, nanoFTIR and AFM-IR spectral changes related to the secondary structure modifications of neurodegenerative proteins and model peptides, researchers can easily verify their results, find an interpretation of the observed spectral changes and compare them with results obtained so far for various amyloids systems with all known molecular spectroscopic techniques. In addition, the usefulness of multivariate data analysis in extraction of spectroscopic markers of the amyloids aggregation is presented.

Table 1. Characteristic Raman and infrared bands for proteins [33,34]

Amide Bands	Wavenumber [cm^{-1}]	Assigned Vibration [1]
A	3500	ν(N-H)
B	3100	ν(N-H)
I	1700–1600	80% ν(C=O), 10% ν(C-N), 10% δ(N-H)
II	1580–1480	40% ν(C-N), 60% δ(N-H)
III	1300–1230	30% ν(C-N), 30% δ(N-H), 10% ν(CH$_3$-C), 10% δ(O=C-N), 20% other
IV	770–625	40% δ(O=C-N), 60% other
V	800–640	γ(N-H)
VI	600–540	γ(C=O)
VII	200	skeletal mode

[1] ν—stretching, δ—bending, γ—out-of-plane bending

2. Infrared Spectroscopy Studies of Abnormal Aggregation of Proteins/Peptides

Infrared spectroscopy (IR), along with Raman spectroscopy, is one of the vibration spectroscopy techniques. The leading phenomenon underlying IR spectroscopy is the absorption of electromagnetic radiation in the infrared range by molecules, exciting vibrations and oscillations of the functional groups occurring in the studied analyte [35,36]. Each chemical bond undergoes various types of vibrations (i.e., stretching, bending, twisting, rocking or wagging motions), whose energies correspond to the IR region including far, mid, and near infrared. IR spectroscopy is selective for vibrations leading to the change of the molecule dipole moment and these vibrations are detected in the IR spectrum [35,37]. Due to the chemical sensitivity, as well as the susceptibility to intra- and intermolecular effects (which affects the vibration frequency and the bonds polarity), IR spectroscopy is very common in the science of biological systems. Although the IR spectra of biomolecules are very complex and consist of several overlapping bands (resulting in a loss of some information), they are still extremely useful for tracking structural changes of biological molecules. For example, it is possible to follow each process that affects the molecule geometry. In the case of proteins, IR spectroscopy makes it possible to analyze modifications in the protein secondary structure which, according to Barth [35], is probably the most popular application of this technique. The IR spectrum of proteins or peptides contains a few characteristic bands carrying the information about the secondary structure, known as the amide I (1700–1600 cm^{-1}), amide II (~1550 cm^{-1}) and amide III (1400–1200 cm^{-1}) bands. The amide I region is especially sensitive to the structure of the protein backbone and is the most useful in neurodegenerative peptide aggregation research.

For studying biological systems, in addition to the transmission FTIR spectroscopy (Fourier transform infrared spectroscopy), the attenuated total reflectance (ATR) technique seems to be advantageous. In an ATR, a sample is placed on a crystal with a high refractive index. The light is reflected once or several times on the crystal–sample interface, which helps to increase the sensitivity allowing to measure thin films. The evanescent wave, perpendicularly oriented to the crystal–sample interface, arises and penetrates into the sample, where it can be absorbed. Due to this phenomenon, the light reaching the detector contains information about the sample structure [35,38]. ATR is especially helpful in biological molecules science, where researchers are struggling with small amounts of the studied material (even less than 100 ng) and low sample concentrations (in the range of μM) [26].

Historically, the interpretation of amyloid aggregate (fibril) structures based on IR spectra was confusing. According to Sarroukh et al. [26], FTIR spectroscopy has been used to study fibril structures since the early 70ties. However, as was pointed out in the mentioned review, in the earliest articles describing the infrared spectra of fibrils, the major components of the amide I band were misinterpreted. The first paper which helped with the proper assignment of the amide I region peaks on the FTIR spectra of fibrils was released in 2000 by Bouchard et al. [39]. Tracking the insulin aggregation process over time (at pH 2.3, 70 °C), it was shown that the high frequency component at 1690 cm^{-1} appears in the spectrum only for short incubation times and vanishes after 18 h of incubation. It became clear that the ca. 1690 cm^{-1} amyloid IR spectrum component can be assigned to structures other than fibrils with the cross-β structure present in the sample. In general, for amyloid peptide aggregation, the spectroscopic marker of the aggregation process and mature fibril formation is the presence of a ca. 1630 cm^{-1} strong peak in the amide I region attributed to parallel β-sheets. At the same time, the approximately 5-fold weaker peak at ca. 1695–1685 cm^{-1} characteristic for antiparallel β-sheets is not present in the spectrum. The simultaneous occurrence of two components at ca. 1695–1685/1630–1610 cm^{-1} attributed to antiparallel β-sheets is visible in the IR spectrum when the aggregation process is in progress, at its initial stages, and oligomers are still present in the sample [40]. The IR marker spectroscopic bands of neurodegenerative peptide aggregation are presented in Table 2.

The aggregation of amyloid β (Aβ) peptides using IR spectroscopy has been extensively studied, especially in the past two decades. IR characteristic peaks occurring in the amide I region for the major component of amyloid β deposits, amyloid β$_{1-42}$ [40,41] and amyloid β$_{1-40}$ peptide [42–44] have already been described. To better understand the Aβ aggregation process, Juszczyk et al. [31] followed the aggregation of synthetic 11–28 fragment of the Aβ peptide with FTIR spectroscopy. This sequence is considered to be responsible for Aβ aggregation. The authors also investigated the aggregation of Aβ 11–28 fragments after the introduction of five different mutations (E22K, A21G, D23N, E22G and E22Q) in 21–23 position. The process was studied in HFIP–D$_2$O mixtures with increasing water content.

The cytotoxicity of neurodegenerative proteins is considered to be strongly correlated with the secondary structure of entities occurring at subsequent stages of the amyloid aggregation process [8,24,28,45,46]. To confirm stronger cytotoxicity of oligomers with antiparallel β-sheet conformation over fibrils, Sandberg et al. [32] incorporated the double-cysteine mutation into the amyloid β$_{1-42}$ and amyloid β$_{1-40}$ (called Aβ$_{42}$CC and Aβ$_{40}$CC, respectively), which prevents mature fibril formation by stabilizing the oligomer structure with an additional intramolecular disulfide bridge. Both mutants create stable oligomeric molecules but within a different aggregation pathway. Aβ$_{42}$CC oligomers/protofibrils turned out to be ca. 50 times more efficient in apoptosis induction than Aβ$_{1-42}$ monomers or mature fibrils.

The aggregation of α-synuclein, the 140-amino-acid protein most abundant in Lewy bodies occurring as fibrillar intraneuronal inclusions in Parkinson's Disease (PD), has also been studied with IR spectroscopy [28,30,47,48]. In general, infrared spectral markers of α-synuclein aggregation are similar to those found for amyloid β (see Table 2). An interesting work concerning α-synuclein aggregation was released lately by Ruggeri et al. [30]. Due to medical evidence of missense mutations in the *SCNA* gene encoding α-synuclein involvement in the PD pathogenesis, the aggregation of a wild type α-synuclein, as well as its variants with the one amino acid replacement in the protein sequence, has been studied. Results showed that amyloid fibrils formed by different variants of α-synuclein were varying in the percentage ratio of secondary structure content. This was due to alternative mechanisms of the aggregation pathway for studied protein variants. Another recent work concerning α-synuclein aggregation describes the influence of the ionic strength on the β-sheets orientation in fibrils studied with 1D- and 2D-IR spectroscopy [49]. 2D-IR spectroscopy provides information about the specific residues of interest and is sensitive to more ordered structures in general [49,50]. The β-sheet arrangement in fibrils turned out to be correlated with salt concentration during fibrilization. α-synuclein aggregation in low ionic strength conditions (NaCl concentration ≤ 25 mM) results in parallel β-sheet orientation in fibrils, while the fibrilization upon high salt concentration, including physiological conditions, contributes to the antiparallel β-sheet arrangement.

Natalello et al. [51] studied the aggregation pathway of prion peptide PrP$_{82-146}$ characteristic for another neurodegenerative amyloid disease, called Gerstmann–Sträussler–Scheinker syndrome (GSS). The major components occurring in brain amyloid plaques are the PrP peptide fragments consisting of 81–82 to 144–153 amino acids. The FTIR spectra of the 82–146 PrP fragment at the initial stages of aggregation revealed two components, typical for oligomers displaying antiparallel structures: a low-frequency band around 1623 cm^{-1} simultaneously with a high-frequency band around 1690 cm^{-1}.

At the final stages of aggregation, when the sample was rich in the cross-β structure fibrils, only the 1626 cm^{-1} peak was determined.

Transthyretin (TTR), a biologically relevant protein occurring in human plasma, serves as thyroxin carrier, or a retinol binding molecule. The misfolding or abnormal aggregation of TTR leads to amyloidosis such as senile systemic amyloidosis (SSA), familial amyloid polyneuropathy (FAP), and familial amyloid cardiomyopathy (FAC) [52]. TTR fibril formation was studied by Cordeiro et al. [52] and earlier by Zandomeneghi et al. [53]. The spectral maker for TTR fibril formation is the band at 1625 cm^{-1}.

Table 2. IR marker bands of abnormal protein aggregation.

Marker Band of the Aggregation [cm^{-1}]	Assignment	Peptide	Reference
amide I (1700-1600 cm^{-1})			
1695-1685/1633-1623	antiparallel β-sheet	amyloid β$_{1-42}$, amyloid β$_{1-40}$, α-synuclein, PrP$_{82-146}$ oligomers	[40,41] [42-44] [28,47,48] [51]
1693-1685/1623-1613	antiparallel β-sheet	amyloid β$_{11-28}$ fragment and its mutants in 21-23 position	[31]
1692/1620	antiparallel β-sheet	HET$_{218-289}$	[55]
1691/1630	antiparallel β-sheet	Aβ 42CC oligomers/protofibrils	[32]
1688/1620	antiparallel β-sheet	human lysozyme, oligomers	[29]
1686/1616	antiparallel β-sheet	transthyretin (TTR) soluble aggregates	[52]
1686/1614	antiparallel β-sheet	hen egg white lysozyme (HEWL)	[56]
1684/1616		β2-microglobulin (short curved structures)	[54]
1684/1612	antiparallel β-sheet	SH3 domain, amorphous aggregates, non-fibrilar	[57]
1683/1612 ↑→	antiparallel β-sheet	insulin oligomer	[39]
1678-1670	β-turns	amyloid β$_{11-28}$ fragment and its mutants in 21-23 position	[31]
1670	β-turns	amyloid β$_{1-42}$, α-synuclein oligomers and fibrils	[40] [28]
1669	β-turns	HET$_{218-289}$	[55]
1667-1661	3$_{10}$-helix	E22K and A21G mutants of Aβ(11-28) fragment	[31]
1664	β-turns	SH3 fibrils/pepsin digested	[57]
1660-1650	random coil and/or helical structures	amyloid β$_{1-42}$, α-synuclein oligomers and fibrils	[40] [28]
1659-1652	α-helix	amyloid β$_{11-28}$ fragment and its mutants in 21-23 position	[31]
1658	Turns	human lysozyme monomers, oligomers, fibrils	[29]
1655	random coil	HET$_{218-289}$	[55]
1649	unstructured	SH3 amorphous aggregates	[57]
1648-1639	random coil	amyloid β$_{11-28}$ fragment and its mutants in 21-23 position	[31]
1648	random coil	PrP$_{82-146}$	[51]
1644-1641	disordered/loops	human lysozyme oligomers, fibrils	[29]
1641	disordered structures	SH3 fibrils/pepsin digested	[57]
1635-1624	β-sheet	amyloid β$_{11-28}$ fragment and its mutants in 21-23 position	[31]
1633	parallel β-sheet	Sup35 crystals, prion-like	[58]
1630	parallel β-sheets	HET$_{218-289}$	[55]
1630-1623	parallel β-sheet	amyloid β$_{1-42}$, amyloid β$_{1-40}$ fibrils	[40,41] [42-44]
1630-1614	parallel β-sheet	human lysozyme fibrils	[29]
1628	parallel β-sheet	α-synuclein fibrils	[28]
1626 ↑	parallel β-sheet	PrP$_{82-146}$ fibrils	[51]
1626 ↑→	parallel β-sheet	insulin fibrils	[39]
1625	parallel β-sheet	transthyretin (TTR) fibrils	[52]
1620-1618		β2-microglobulin, fibrils	[54]
1620-1600	β-sheets	hen egg white lysozyme (HEWL)	[56]
1618 ←	parallel β-sheet	SH3 fibrils/pepsin digested	[57]

↑ increase in intensity, ← shift towards higher wavenumbers after aggregation, → shift towards lower wavenumbers after aggregation.

In addition to neurodegenerative peptide aggregation, the secondary structure of amyloid fibrils of other non-neurodegenerative peptides has also been extensively studied. The IR marker bands of abnormal protein aggregation have been determined for β2-microglobulin (β2m), whose deposits occurred in dialysis-related amyloidosis within the musculoskeletal system [54], a prion-forming 218–289 domain of HET protein [55], wild-type human lysozyme [29] and hen egg white lysozyme (HEWL) [56] (see Table 2). These proteins have been studied as great models of amyloid aggregation. Zurdo and co-workers [57] characterized the fibrils formed by the SH3 domain of the α-subunit of bovine phosphatidylinositol-3′-kinase. The FTIR spectra were collected for amorphous aggregates as well as for fibrils (created at low pH) exposed to pepsin to receive a sample containing only a fibrillar structure with characteristic low frequency band for parallel β-sheets (1618 cm^{-1}). Another amyloid-like structure formed from a prion-like protein, Sup35, derived from yeast, was studied by Balbirnie et al. [58]. FTIR spectra of aggregated protein crystals showed the 1633 cm^{-1} parallel β-sheet band. The aggregation kinetics of human islet amyloid polypeptide (hIAPP) was studied using 2D infrared spectroscopy combined with site-specific isotope labeling [59]. This methodology allowed to follow the intensity growth of the 1617 cm^{-1} peak related to the increase of β-sheet content upon aggregation and fibril formation. Ami et al. [60] incorporated the FTIR microscopy to study aggregates of amyloidogenic immunoglobulin light chains (LCs) occurring in the light chain (AL) amyloidosis pathology. The applied methodology involved the FTIR in situ studies of unfixed tissues (hear and subcutaneous abdominal fat) derived from AL amyloidosis affected patients as well as the research of in vitro aggregated peptide (derived from a patient). The infrared β-sheet signature characteristic for amyloid aggregation was possible to detect in situ in the spectra of tissues.

2.1. Infrared Spectroscopy at the Nano Scale in Studies of Abnormal Proteins/Peptide Aggregation

Due to the diffraction limit, the resolution of conventional IR spectroscopy does not make it possible to track changes concerning single molecules. The signal reaching the detector contains bulk information, averaged over many molecules of the studied sample. To overcome this limitation and follow the IR absorbance spectra at the single molecule level, the novel nano-FTIR and AFM-IR techniques have been implemented.

2.1.1. Nano-FTIR In Studies of Abnormal Proteins/Peptide Aggregation

The nano-FTIR technique makes it possible to achieve simultaneous infrared chemical and topographic characteristics of a sample at nanoscale resolution. It became possible due to the invention of scattering-type scanning near-field optical microscopy (s-SNOM), which is a unique combination of atomic force microscopy (AFM) with optical imaging and IR spectroscopy [61]. Nano-FTIR combines the nanometric spatial resolution of AFM with the chemical sensitivity of IR spectroscopy. The local probing of molecular vibrations involves the incident light scattering at an AFM tip apex (Figure 2). The electric charge accumulated on the probe apex generates a localized electric field. When the AFM probe (usually metallic) is in contact with the sample, it is possible to detect the near-field interaction of infrared light with the sample due to the induction of a local evanescent field [61]. Depending on the measurement purposes, the nano-FTIR technical setup provides a broadband mid-infrared laser (fiber laser plus difference frequency generator) or tunable single line laser (quantum cascade laser, QCL). Both of these light sources enable chemical imaging, the broadband laser makes it possible to obtain the hyperspectral maps and the QCL provides maps at the fixed wavelength. When the interferometer is operating as a Fourier transform spectrometer, it is possible to collect single FTIR spectra at points of interest on the sample, selected precisely based on the AFM image [61].

Nano-FTIR was employed to study amyloid aggregates by Amenabar et al. [61]. In this work, it was shown for the first time that nano-FTIR can be used to study the secondary structure of individual proteins. Amenabar and co-workers explored the conformation of single insulin fibrils. The nano-FTIR spectra of insulin fibrils (Figure 3) revealed the bands characteristic for β-sheets at 1639 cm^{-1}, α-helical structures at 1671 cm^{-1} and the band at 1697 cm^{-1} whose authors attributed to β-turns or antiparallel β-sheet. The weak band at 1609 cm^{-1} was assigned to amino acid side chains. The images at single wavelengths provided by QCL laser (Figure 3c) as well as single broadband spectra (Figure 3d) confirmed the presence of α-helices in 3-nm and 9-nm thick fibrils.

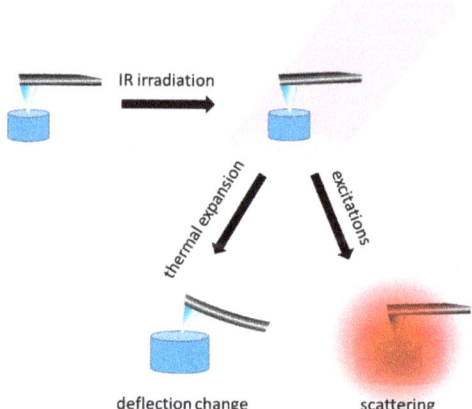

Figure 2. The schematic representation of infrared nanospectroscopy principles: thermal extension of a sample and a change of AFM cantilever deflection, as a consequence of the interaction with the incident IR light (AFM-IR) and scattering of the near infrared field from the sample at the metal/metalized probe apex (nano-FTIR).

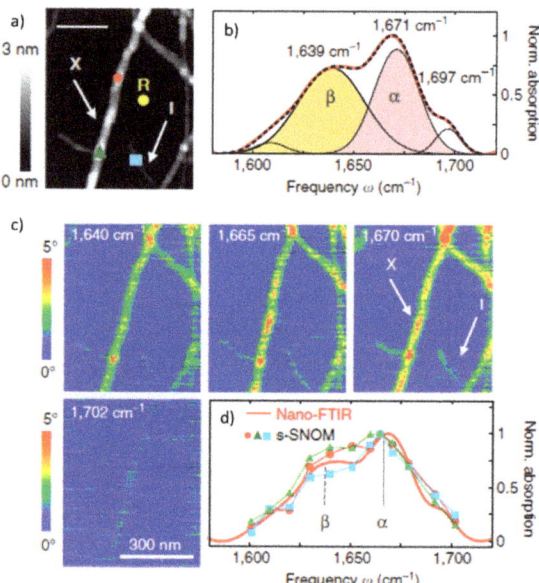

Figure 3. Infrared nanospectroscopy and nanoimaging of insulin fibrils. (**a**) AFM topography image of insulin fibrils, scale bar: 200 nm; (I) 3-nm thick type I fibril, (X) a 9-nm-thick fibril. (**b**) the nano-FTIR spectrum (red curve) based on five absorption bands. (**c**) s-SNOM phase images of the fibrils at fixed wavelengths, scale bar: 300 nm. (**d**) Single infrared absorption spectra at the positions marked in (a), topography image [61]. Images reproduced under CC BY license.

2.1.2. Infrared Spectroscopy Combined with Atomic Force Microscopy (AFM-IR) in Studies of Abnormal Proteins/Peptide Aggregation

Infrared nanospectroscopy, also called Photothermally Induced Resonance (PTIR) or infrared spectroscopy coupled with atomic force microscopy (AFM-IR), enables the measurement of a local infrared light absorption [62]. Thanks to the use of tunable infrared lasers it is possible to obtain single spectra or maps showing the spatial distribution of selected molecules and their functional groups. Conventional spectroscopic methods, such as IR or Raman spectroscopy, are limited by the diffraction criterion. The spatial resolution of these techniques depends on the wavelength and is in the range of 5–10 μm [63]. Therefore, they cannot be applied in imaging of micro- or nanometric biological objects. The PTIR method has been first demonstrated in 2005, by prof. A. Dazzi from Université Paris Sud [62–64]. This technique is based on the detection of infrared radiation absorption using atomic force microscopy (AFM). Transient and local sample extension are detected as a change of AFM cantilever deflection (Figure 2). Photothermally induced signal increases when the laser is tuned to frequencies which are absorbed by the sample. The limitation of AFM-IR spatial resolution is determined by scanning and sampling rates and is related to several factors such as the tip apex size and thermomechanical properties of the investigated sample [62–64].

An application of AFM-IR in the structural organization of individual fibril aggregates is of central importance for several scientific groups. Henry and co-workers studied the amyloid β ($A\beta_{1-42}$) peptide and its G37C mutant at the nanoscale [65]. They used AFM-IR to examine the interaction of lipids such as 1-palmitoyl-2-oleoylphosphatidylcholine, sphingomyelin, or cholesterol with amyloid aggregates. It was discovered that aggregation of amyloid β_{1-42} and its mutant essentially changes the structural organization of the fibrils in the presence of the investigated phospholipids. Ruggeri et al. applied AFM-IR to investigate single aggregates of a highly fibrillogenic domain of ataxin-3 called "Josephine" and followed its whole aggregation pathway [66]. They observed that already at the first aggregation, abnormal folding of Josephine was formed. It was possible to detect conformational transitions starting from native oligomers to misfolded oligomers and finally to mature amyloid fibrils. Ruggeri and co-workers also studied Exon1 aggregation which is the first exon of Hungtinton protein [67]. They determined the influence of the polyQ content and the Nt17 domain occurrence on the biophysical features of Exon1 fibrils. In the presence of the Nt17 domain, the IR spectra of fibrils revealed the peak characteristic for β-turn rich secondary structure at 1684 cm^{-1} while the lack of this domain changed the secondary structure of fibrils which displayed antiparallel β-sheet structure. Rizevsky et al. [68] studied insulin fibrils using AFM-IR. In the following work, the formation of two different types of fibers was described. This phenomenon is associated with the insulin aggregation pathway. The first fibril type had an β-sheet-rich secondary structure, whereas the second polymorph revealed mainly an unordered secondary structure. Galante and co-workers [69] studied the influence of the most toxic, pyroglutamylated isoform of amyloid β (AβpE3-42) on the biophysical features and biological activity of amyloid β_{1-42} [70,71]. It was confirmed that the mixture of ApE3-42/A1-42 with the 5% content of AβpE3-42 isoform was the most toxic. Ramer et al. [72] applied AFM-IR for the first time to investigate the chemical structure of amyloid aggregates at the nanoscale in water. They demonstrated that it was possible to collect PTIR spectra in liquid with a high signal-to-noise ratio. They studied aggregates of diphenylalanine (FF, Figure 4b), which is the core amyloid β peptide, and its tert-butoxycarbonyl derivative (Boc-FF, Figure 4a). It was confirmed that the detection of differences in the studied fibrils is possible for measurements conducted in liquids. Boc-FF appeared to have a more helical conformation in comparison to FF. These results are promising, and make it possible to state that AFM-IR can be widely used for very composed aggregation systems in liquids. AFM-IR marker bands related with the neurodegenerative peptide aggregation are presented in Table 3.

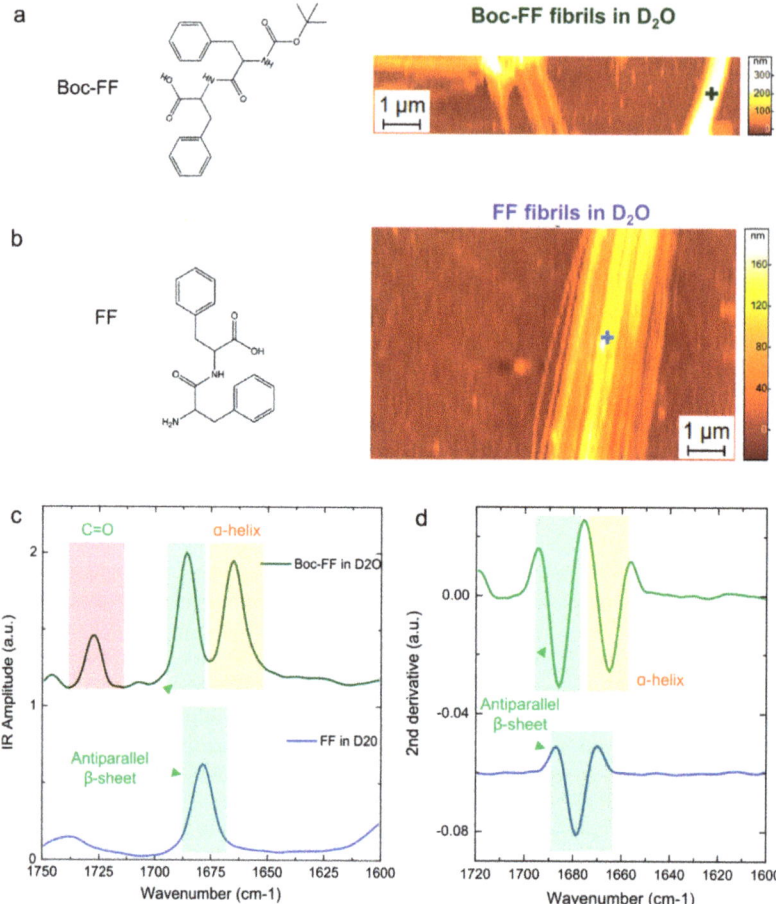

Figure 4. PTIR structural comparison of FF and Boc-FF fibrils in D_2O. Chemical structure and AFM morphology maps of (**a**) Boc-FF and (**b**) FF fibrils. (**c**) Comparison of the fibrils averaged PTIR spectra in the amide I (green and yellow) and C=O stretching vibration (red) spectra ranges. (**d**) Comparison of the second derivatives spectra in the amide band I region [72].

Table 3. AFM-IR marker bands of the neurodegenerative peptide aggregation.

Marker Band of the Aggregation [cm^{-1}]	Assignment	Peptide	Reference
amide I (1730–1600 cm^{-1})			
1730	C=O	Boc-FF, FF	[72,73]
1700–1690	anti-parallel β-sheet	amyloid β (AβpE3-42)	[69]
1700–1600	ν(C=O) bk	unexpanded Exon1 (22Q)	[26,34,35,67,74]
1696–1690	anti-parallel β-sheet, carbamate group	Boc-FF, FF	[72,73]
1695–1665	β-turn and antiparallel β-sheets	ataxin-3	[38,66,70,71,75]
1695,1684	β-turn, anti-parallel β-sheet	amyloid β (AβpE3-42)	[69]
1695	β-turn	insulin	[68]
1692	anti-parallel β-sheet	amyloid β with 5% pyroglutamylated peptide	[69]
1692	anti-parallel β-sheet	unexpanded Exon1 (22Q)	[26,34,35,67,74]
1689,1625	anti-parallel β-sheet	mutant oligomer G37C	[65,76]
1684	glutamine side chain vibrations and β-turn	expanded Exon1 (42Q)	[26,34,35,67,74]
1664,1655	α-helix, 3-helix	Boc-FF, FF	[72,73]
1662	β-turn	amyloid β	[65,76]
1660–1650	α- helix	ataxin-3	[38,66,70,71,75]
1660	α-helix/unordered protein secondary structures	Insulin	[68]
1658	poor α-helix	amyloid β (AβpE3-42)	[69]
1658	α-helix	expanded Exon1 (42Q)	[26,34,35,67,74]
1645–1630	random coil	ataxin-3	[38,66,70,71,75]
1645–1635	random coil	unexpanded Exon1 (22Q)	[26,34,35,67,74]
1640–1600	residual water absorption, NH$_3^+$ group	Boc-FF, FF	[72,73]
1638	random coil	amyloid β (AβpE3-42), amyloid β with 5% pyroglutamylated peptide	[69]
1635–1610	low density native/high density amyloid β-sheets	ataxin-3	[38,66,70,71,75]
1635	β-sheet	unexpanded Exon1 (22Q), expanded Exon1 (42Q)	[26,34,35,67,74]
1631	parallel β-sheet secondary structure	amyloid β	[65,76]
1623	high β-sheet	amyloid β (AβpE3-42)	[69]
1620	β-sheet	Insulin	[68]
amide II (1600–1500 cm^{-1})			
1580–1510	bk δ(N-H), ν(C-N)	ataxin-3	[38,66,70,71,75]
1555,1520	NH vibrations	Boc-FF, FF	[72,73]
C-C ring vibrations			
1605,1495,1452,1430	C-C ring vibrations	FF	[72,73]
amide III (1400–1200 cm^{-1})			
1350–1200	ν(C-N), δ(N-H),ν(C-C), δ(C=O)	ataxin-3	[38,66,70,71,75]

ν—stretching, δ—bending, bk—backbone.

3. Raman Spectroscopy in Studies of Abnormal Proteins/Peptide Aggregation

Raman spectroscopy is a non-destructive analytical technique that makes it possible to obtain information about the chemical structure and composition of the investigated sample. The basic phenomenon used in this method is an inelastic light scattering, which contains information about the energies of the vibrational states [77] (the light scattering mechanism is shown in Figure 5). The energy difference between incident radiation and scattered radiation is corresponding to the energy levels of vibrational modes in functional groups of the investigated molecules. This process causes the excitation of characteristic vibrations, which are observed in the spectrum as specific bands [77,78].

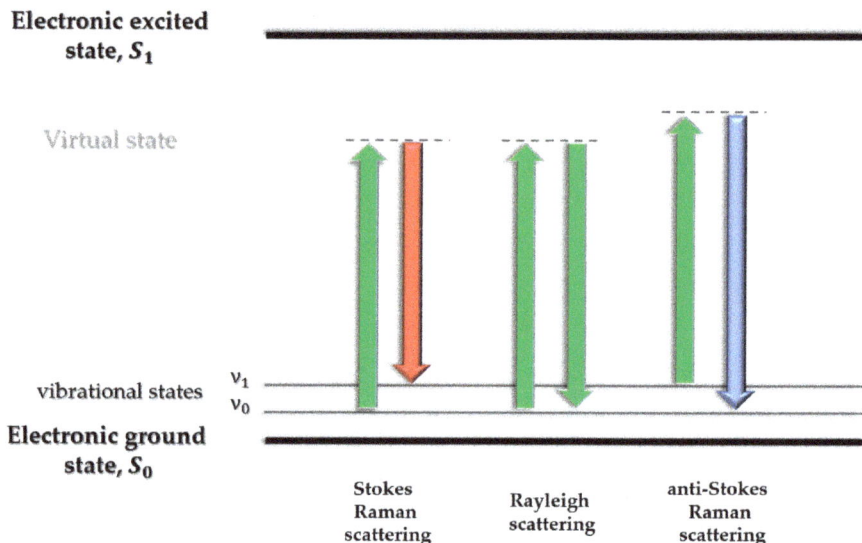

Figure 5. During non-resonant light scattering, three scenarios are possible: in the middle, elastic scattering, called Rayleigh scattering (without energy changes), on sides inelastic light scattering (causing energy changes), called Raman scattering including anti-Stokes (higher energy of scattered photon), and Stokes (lower energy) lines.

In respect to these relationships there is a selection rule, and according to this, the polarity of the molecule must change. Inelastic scattering, also called Raman scattering, is very rare. Approximately only one photon per 10^{10} is scattered inelastically [78]. For this reason, various methods to amplify the signal are proposed, such as surface and tip-enchanted Raman spectroscopy, which are described in the next paragraphs of this article. Raman spectroscopy is widely used in the study of various types of biological material [79], and its use in medical diagnostics is being considered [80,81].

Deep UV Resonance Raman (DUVRR) Spectroscopy is a variant of the Raman spectroscopy very commonly used in the research of protein aggregation. In contrast to Non-resonance or Normal Raman (NR) Spectroscopy, the radiation used in the experiment (in this case UV spectral range) transfers the electron to a higher energy state, which causes resonant amplification of the scattered light signal (Figure 6). Thus, it becomes possible to observe bands that are less visible in the NR method [82,83].

As a result of the aggregation process, structural transformations occur in proteins, which manifest as characteristic changes in the Raman spectra (Table 4). An increase in the intensity of the bands from amide I and II is observed, due to the formation of β-sheets. This process was observed in various forms, depending on the proteins tested [84–86]. Kurouski et al. observed an increase in the intensity of amide I and II bands in the formation of insulin fibrils [86], while in the case of hen egg white lysozyme (HEWL) studied by Rosario-Alomar et al. only the change of amide II was visible [85]. In addition, a shift of amide I and II bands towards higher energies was observed [85,86]. The second characteristic marker that almost always appears is an increase in the intensity of the C_α-H band at 1390 cm^{-1}, which is explained by the decay of α-helix into a disordered structure [85,87]. In addition, there are changes in the intensity of the complex amide III band. Xu et al. described an increase in intensity in the range of 1270–1320 cm^{-1} and a decrease in the range of 1230–1270 cm^{-1} [88].

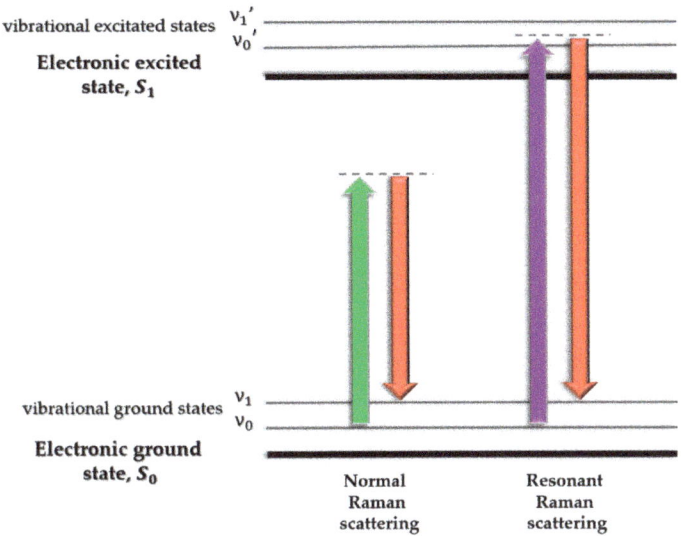

Figure 6. Comparison of NR and DUVRR spectroscopy. In NR, the electron is not transferred to a higher energy state, but only to a virtual state (lower energy than S_1). In DUVRR, the radiation scattered on the sample excites the electron to the S_1 state energy, causing the resonant signal amplification.

Table 4. Raman marker bands of peptide aggregation.

Marker Band of the Aggregation [1]	Assignment	Peptide	Reference
amide I (1700–1600 cm^{-1})			
1690–1600 ↑1672→	β-sheets formation	insulin, hen egg white lysozyme (HEWL)	[85,86,88]
amide II (1580–1480 cm^{-1})			
1580–1480 ↑ 1550→	β-sheets formation	HEWL 1-SS-carboxymethyl lactalbulin (1-SS-LA), HEWL	[84–86]
Cα H			
1390 ↑		insulin, HEWL	[85,86,88]
amide III (1283–1218 cm^{-1})			
1320–1270 ↑1270–1230 ↓		HEWL	[88]
disulfide (S-S) (550–450 cm^{-1})			
523, 507 ↓490 540, 510 ↓508		HEWL apo-α-lactabulin (LA)	[84,85]
phenylalanine (Phe)			
1000 ↓		HEWL	[87,88]

[1] ↑ an increase of peak intensity, ↓ a decrease of peak intensity, → a peak shift to higher energy

In the spectra of aggregated proteins, there is often a change in the band's intensity from individual amino acids. In particular, a characteristic peak for phenyloalanine (1004 cm^{-1}) is often described as one of the spectroscopic markers of fibrillation [88]. In some cases, there is a significant decrease of Phe band intensity [87,88], which is explained by the change in the chemical environment of this amino acid. However, this change is not always observed [86].

Interesting changes in Raman spectra were detected for the disulfide band in the range of 450–550 cm^{-1}. Kurouski and Lednev observed significant changes in the spectrum for apo-α-lactalbumin (LA) and 1-SS-carboxymethyl lactalbumin (1-SS-LA) [84]. In the spectra of native LA, two peaks were observed at 510 cm^{-1} and 530 cm^{-1} due to various secondary structures. In LA fibrils, the β-sheets structure is mainly predominant; therefore, in their spectra, one peak at 508 cm^{-1} was observed. In turn, 1-SS-LA showed a slightly more complex structure, but also in this case a significant change in the spectra related to secondary structure transition was visible. Similar studies were carried out by Rosario-Alomar et al. for HEWL. Along with ongoing fibrillation, two peaks at 507 cm^{-1} and 523 cm^{-1} merged into one at 490 cm^{-1} [85].

3.1. SERS In Studies of Abnormal Proteins/Peptide Aggregation

Surface-Enhanced Raman Spectroscopy (SERS) is based on the enhancement phenomenon in Raman scattering by the application of nanostructures consisting of noble metals, transition metals, or semiconductors. For all molecules adsorbed onto nanostructured metal surfaces inelastic light scattering is greatly enhanced (enhancement factor can be up to 10^{10}) in comparison to free molecules [89]. It is considered that the SERS effect is a combination of two mechanisms: an electromagnetic field enhancement (EM) and chemical surface interactions. EM field enhancement is caused by the excitation of surface plasmons on the surface due to the interaction between the electrons in metal nanostructure with the incident electromagnetic radiation. The locally enhanced electromagnetic field on the nanoparticle surface strongly increases the intensity of Raman scattering, because the Raman scattering cross-section is proportional to the electromagnetic field. On the other hand, the Raman scattering cross-section also strongly depends on the polarizability of the investigated molecule. Chemical enhancement is attributed to a significant increase in the polarization of the molecules due to its absorption onto the metallic surface, which results in the charge transfer affecting significantly the polarizability. New energy states are created that can be excited with the laser beam. This Raman resonance enables electron transfer from the Fermi level of the metal to the lowest unoccupied molecular orbital of the molecule (LUMO) and from the highest occupied molecular orbital (HOMO) to the Fermi level of the metal. The chemical enhancement is less effective and strongly depends on the type of adsorbed molecules, while the electromagnetic enhancement is universal for all molecules [90,91].

During the last decade, SERS has become quite popular as one of the most sensitive analytical techniques in chemistry, material science and biotechnology. It provides information about conformational and structural changes in molecules at very low concentrations. However, protein investigation still remains challenging. First of all, SERS measurements suffer from low spectral reproducibility. The SERS enhancement factor is strongly determined by the distribution of nanoparticles. Variations in spectral characteristics are therefore induced by inconsistent aggregation and collocation of the nanoparticles. Thus, the uncontrolled enhancement of signal from proteins adsorbed on the SERS-active surface generates SERS spectra with low reproducibility. Moreover, proteins, as the intrinsic molecules, often form complex SERS patterns, which makes identifying characteristic Raman fingerprints difficult [92–94]. To overcome this limitation, it is important to ensure that protein binding to a SERS-active substrate is consistent and well understood. Constant enhancement and reproducibility of SERS substrate preparation are therefore of central importance.

Due to the fact that SERS technology provides information about the secondary structure, it allows for an understanding of properties and functionality of the abnormal protein aggregates at very low concentrations. SERS marker bands associated with the abnormal amyloid aggregation are summarized in Table 5. One of the first SERS studies on protein aggregation concerned amyloid β [95]. Choi et al. used a nanofluidic device with SERS active gold nanoparticles to determine and characterize various stages of amyloid β aggregation. To understand conformational changes induced by protein aggregation, SERS spectra were collected as a function of time and protein concentration. The analysis of the amide III band allowed to investigate the formation of amyloid aggregates, the decrease in α-helics and the increase in β-sheet at micromole to femtomole concentrations [96]. A similar study was conducted by a group from Texas University [97]. Further research related to the conformational changes of amyloid aggregates was provided. Bhowmik et al. used Surface-Enhanced Raman Spectroscopy with silver nanoparticles to determine the Aβ oligomers structure that spontaneously bind to lipid bilayer. The secondary structure information of individual residues was examined based on Raman isotope shifts. The authors determined that the structure of Aβ$_{40}$ oligomers bound to the membrane, display β-turn in the 23-28 region, situated between antiparallel β-sheets [98]. Insulin aggregation has also been investigated. Kurouski et al. used surface-enhanced Raman scattering to study the kinetics of insulin aggregation followed via detection of the insulin prefibrillar oligomers secondary structure. Insulin at different stages of aggregation was mixed with 90 nm of Au nanoparticles. SERS analysis demonstrated a more than two-fold increase in the number of insulin oligomers after the first hour of incubation. Further protein incubation resulted in a slow decrease in oligomers amount. The observed decrease in the quantity of prefibrillar species was linked to their conversion into fibrils [99]. Li et al. discovered that bromophenol blue (BPB) potentially inhibits the insulin fibrillation process [100]. Karabelli et al. used the electrochemical SERS (EC-SERS) method to study interaction between intermediates occurring at various stages of insulin aggregation and biomimetic membrane. It was observed that protofibrils and oligomers evoked significant membrane perturbation, whereas the native protein seems to have a protective role. The authors presented one of the first applications of the EC-SERS technique to investigate protein aggregation intermediates–biomembrane interactions [101]. Yu et al. discovered a method that makes it possible to distinguish Aβ$_{40}$ and Aβ$_{42}$ peptides easily. SERS combined with principal component analysis (PCA) revealed changes in peptide conformation and self-association [102]. Undeniably, SERS has great potential in peptide/protein studies, continuous work on increasing spectrum repeatability will certainly provide a deeper understanding of protein aggregation.

Table 5. SERS marker bands of the neurodegenerative peptide aggregation.

Marker Band of the Aggregation	Assignment	Peptide	Reference
amide I (1700–1600 cm^{-1})			
1678–1664↑ 1664–1640↓	β-sheets α-helics, unordered	insulin	[99]
amide III (1283–1218 cm^{-1})			
1244 ↑ 1266 ↓	β-sheets α-helics	amyloid β$_{1-42}$	[95,96]
CCN stretching			
1144 ↓		amyloid β$_{1-42}$	[96]
C-C stretching			
960 ↑		amyloid β$_{1-42}$	[95]

↓ a decrease of peak intensity, ↑ an increase of peak intensity

3.2. TERS In Studies of Abnormal Proteins/Peptide Aggregation

Tip-enhanced Raman Spectroscopy (TERS) takes the advantage of the nanometric spatial resolution from Scanning Probe Microscopy (SPM) including Atomic Force Microscopy (AFM) or Scanning Tunnelling Microscopy (STM), and the chemical selectivity of Raman spectroscopy. Therefore, it provides information about the molecular structure and composition with nanometric spatial resolution [103]. The excitation of surface plasmons in a metal nanostructure, which can be deposited on the apex of an AFM tip (AFM-TERS) or in STM tip apex itself (STM-TERS), modifies the electromagnetic field of incident laser light [103,104]. The Raman scattering cross-section is proportional to this electromagnetic field, and due to the enhancement, it can be dramatically improved. What is more, the generated electromagnetic field is highly localized; therefore, spectra are acquired from the small amount of sample located directly under the tip [103,104]. TERS tip can be understood as a nanoantenna.

It converts the electric field of the Raman laser into highly spatially confined energy [105], breaking the Rayleigh's diffraction limit, and improving the spatial resolution to a few nanometers. On the other hand, TERS probe converts the near field of the sample to a far field, accessible to a microscopic objective, increasing the sensitivity to the single molecule level [106–108]. In recent years, TERS has mainly been applied for samples with a high natural Raman scattering cross-section including carbon nanotubes and dyes [106–108]. The main limitation of TERS is actually related to its high sensitivity. The Raman scattering cross-section of carbonaceous substances and contaminations is exceptionally high [109]. Even thin traces of carbon contaminations can be easily detected as rapidly fluctuating and sharp peaks, which averaged over many spectra result in the broad D- (1360 cm^{-1}) and G-bands (1580 cm^{-1}) [109]. A very strong electromagnetic field, however, could damage delicate biological samples and cause their photochemical/thermal decomposition.

TERS delivers highly resolved topographic information and provides a chemical composition of the investigated sample. Therefore, this analytical technique is potentially ideal for studying protein secondary structure and its modification upon the fibrillation process. However, TER marker bands required for secondary structure identification such as amides I and III or C-H/N-H motions, allowing for conformation detection based on spectral positions, are mostly not well resolved. In particular, the most informative amide I band is absent in TER data. This problem has been broadly discussed [110–112], and finally it was proved that peptide backbone bonds may dissociate in the laser hot spot, which causes the lack of the amide I band. An application of the mid measurement condition (reduced laser exposition time and laser power) prevents the decomposition of the peptide backbone improving the spectra quality.

Due to the TERS limitations mentioned in the previous paragraphs, measurements of biological samples including proteins are still very limited. However, the results obtained so far are undoubtedly very valuable and should be presented in this review. For clarity, TER marker bands related to the abnormal amyloid aggregation are presented in Table 6. One of the first neurodegenerative peptides characterized with TERS was β2-microglobulin [113]. Several polypeptides were intensively studied by the group of prof. Deckert from Jena University. Deckert and co-workers applied TERS to analyze the surface molecular structure of human islet amyloid polypeptide fibrils, proving that the highly heterogeneous fibril surface mainly contains α-helical or unordered structures, in contrast to the fibril core, which is built of β-sheets [114]. Insulin fibrils were also investigated. Kurouski et al. probed the chemical composition and the secondary structure of insulin fibrils [115], demonstrating the high capability of TERS in characterization of heterogeneous fibrils. Insulin was also applied to test the efficiency of several aggregation inhibitors including: benzonitrile, dimethylsulfoxide, quercetin, and β-carotene [116]. Analysis of the shape of the amide I band in the TERS spectra makes it possible to detect the influence of these inhibitors on fibrils secondary structure [116]. It was proven that all insulin aggregates showed similar morphology but various inhibition results in their different β-sheet structure content, suggesting serious modification in the aggregation pathway. Specifically, the authors suggested that prevention of the β-sheet stacking of the peptide chains played a major role in the aggregation inhibition [116]. Additionally, the group from Jena identified the secondary structure of insulin aggregates via the amide III band and what is more, performed successful detection of hydrophobic and hydrophilic domains on the surface [117]. Bonhommeau et al. characterized Aβ$_{1-42}$ and its two synthetic, less toxic mutants at the nanoscale. Based on the examination of amides I and III bands, detection of fibers organized in parallel β-sheets and in anti-parallel β-sheets was possible.

Table 6. TER marker bands of the neurodegenerative peptide aggregation.

Marker Band of the Aggregation [cm^{-1}]	Assignment	Peptide	Reference
amide I (1700–1600 cm^{-1})			
1680–1660 ↑	β-sheets	amyloid β$_{1-42}$, insulin, β2-microglobulin	[111,113,115–118]
1674 ↑	β-sheets	hIAPP (Amylin)	[112]
1655–1630 ↓	random coil	amyloid β$_{1-42}$	[117]
1640 ↓, 1664–1640 ↓	α-helics, unordered	hIAPP (amylin), insulin	[112,113,115,118]
Cα H/N-H (1370–1360 cm^{-1})			
1364 ↑	β-sheets	amyloid β$_{1-42}$	[116]
amide III (1283–1218 cm^{-1})			
1228–1218 ↑	parallel β-sheets	amyloid β$_{1-42}$	[117]
1242–1233, 1250 ↑	antiparallel β-sheets	insulin	[116,117]
1235–1230 ↑	β-sheets	amyloid β$_{1-42}$, β2-microglobulin	[111,115]
1261–1248, 1258 ↓	random coil		[116,117]

↓ a decrease of peak intensity with the ongoing fibrillation, ↑ an increase of peak intensity with the ongoing fibrillation.

In previous studies, single-point TER spectra were mainly acquired at selected locations. However, recently, hyperspectral TER mapping, which consists of a full TERS spectrum in each pixel of the image, was performed. Paulite et al. successfully applied STM-TERS to map Aβ$_{1-40}$ through integration of the phenylalanine ring breathing mode [119]. The highly reproducible and intense peak from phenylalanine at 1004 cm^{-1} dominates each STM-TER spectrum; however, amides were not well resolved. Recently, AFM-TERS was applied for nano-spectroscopic hyperspectral mapping of amyloid β$_{1-42}$. An application of AFM-TERS allowed for localization of β-sheet and unstructured coil conformation distribution in oligomers and along fibrils and protofibrils [118]. These studies confirmed the high capability of TERS in the mapping of amyloids, and what is more, they allowed for direct monitoring of aggregation pathways by following the distribution of the secondary structure in individual aggregates: oligomers, protofilaments, and fibrils. Figure 7 demonstrates the AFM image of A β$_{1-42}$ deposited on a gold surface. The zoomed-in area is partially overlapped by the TERS map in the central part. Two types of TER spectra, presented below the maps, were acquired in the area of the individual fibril. The distribution of each of those spectra is superimposed on the zoomed AFM topography. The main difference between the two types of spectra acquired from the fibril is a significant shift of the amide III band from 1260 cm^{-1} (green spectrum) to 1250 cm^{-1} (blue spectrum). The position of the amide III band indicates the particular peptide secondary structure: turns/random coil and β-sheet conformation, respectively. The distribution of these conformations along a single fibril is presented in the zoomed-in area. Another characteristic marker of β–sheet structure, which can be observed in the blue spectrum, is a high intensity of the Cα- H/N-H bending mode at 1364 cm^{-1} (Figure 7). Despite the absence/low intensity of the amide I band in the studies described above, it was confirmed that TERS is an efficient tool for investigation of the amyloids secondary structure and for direct verification of the aggregation pathway [118].

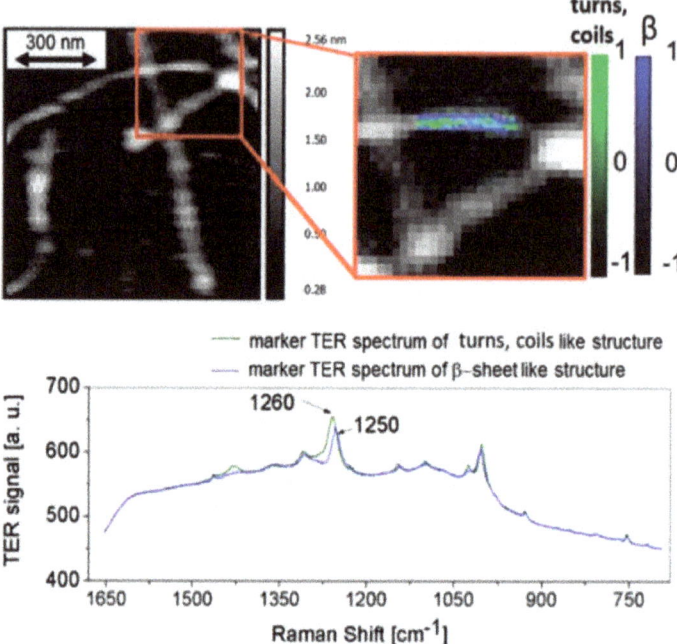

Figure 7. AFM-TER mapping of Aβ$_{1-42}$ fixed on gold: AFM topography with overlapped TERS map in zoomed area showing the distribution of β-sheets (blue) and turns/unstructured coils (green) with corresponding averaged TERS spectra, characteristic for both conformations; adapted with permission from [118].

4. Multivariate Data Analysis in Studies of Abnormal Proteins/Peptide Aggregation

The main purpose of performing a multidimensional statistical analysis (MSA) on spectroscopic data is to prove that measured spectral differences exist according to different populations and this information is statistically significant. The subject of considerations in multidimensional statistical analysis is a large set of data describing the studied phenomenon by means of vectors. MSA methods make it possible to organize this data, establish relationships between variables, as well as to estimate parameters and verify hypotheses of interest. Classical multivariate statistical analysis was developed assuming that the observable random vectors have normal distributions. However, in recent decades, there has been a rapid development of nonparametric multidimensional statistical analysis. Principal Component Analysis (PCA) and Hierarchical Cluster Analysis (HCA) are the most popular. However, in contrast to the PCA method, the use of HCA is not limited to all-numeric data. In the case of the spectroscopic methods with the single-molecule sensitivity as SERS, TERS, AFM-IR or nano-FTIR, the details obtained by performing the MSA analysis cannot be immediately apparent from the separate analysis of peak intensities. The different approach is commonly used in Raman or IR techniques due to the greater signal-to-noise ratio in those methods.

Many techniques have been developed to simplify data analysis, especially in the case of difficulties with interpretation of large datasets. Principal component analysis (PCA) is one of the oldest and most widely used techniques in this area. The main idea of PCA is to drastically reduce dataset dimensionality in an interpretable way without a significant loss of statistical information. The basic requirement in PCA is to successively maximize variance and finally to find new variables that are linear functions of those in the original dataset. Therefore, it finally leads to an eigenvalue/eigenvector problem solution. PCA was firstly incorporated for data analysis by Pearson [120] and Hotelling [121].

Although PCA needs no distributional assumptions, a multivariate normal distribution of the dataset establishment is commonly used. This method is an exploratory adaptive and can be used on various numerical data types. The PCA method makes it possible to reduce the noise and to detect the sub-groupings within the spectra. The most important results of PCA are the score plots and the loading plots. The score plots represent the degree of variability within the totality of spectra and the loading plots show which variables in the data set are responsible for the greatest degree of separation [122,123].

The overall goal of hierarchical cluster analysis (HCA) is similar to that of PCA, but the mathematical approach is different. The HCA is typically applied to determine how entities can be grouped into clusters, that exhibit high similarity within-group and low similarity to other groups. The result of HCA is usually presented as a dendrogram, which is a plot that shows the relationships between the samples in a tree form. In HCA, two main approaches (agglomerative and divisive) are used to resolve the grouping problem. In the agglomerative approach, each sample is initially considered as a cluster, and subsequently, pairs of clusters are merged. The divisive approach is contrary, the algorithm starts with one cluster including all of the samples, and recursive splits are performed. Clustering is achieved by the use of an appropriate metric of sample distance and linkage criterion among groups. The most commonly used, in the case of the datasets obtained from mentioned techniques, are Euclidean or Manhattan distance in combination with Ward's method linkage. Clustering can be used for the identification of interesting patterns and distributions in the data structure. Therefore, it is a useful technique for discovering and extracting information that may previously have been unnoticed [124]. Cluster analysis was proposed in 1930; however, it has found applications across a wide variety of disciplines since the beginning of the early 1960s. One of the primary difficulties with the application of HCA for the extraction of significant information is the choice of the optimal number of clusters. This problem is still considered to be important despite several activities concerning this field. Although several criteria are available to assist in the proper selection of the appropriate number of clusters, the results depend on the researcher.

PCA was used to analyze spectroscopic data concerning amyloid aggregation for the first time by Ruggeri et al. [66]. The authors performed the PCA by means of the nonlinear iterative partial least squares algorithm with cross-validation and mean-centered data analysis of underivatized spectra and second derivatives calculated after an application of the Savitzky–Golay algorithm for smoothing. This analysis confirmed that shifts of the amide I (from 1655–1620 cm^{-1} to 1710–1680 cm^{-1}) and the amide III bands (from 1380 to 1295 cm^{-1}) are spectral markers of the Josephine domain of ataxin-3 protein aggregation. The secondary structure of the studied domain changed from the random coil/α-helical structure in native protein to β-turn/antiparallel β-sheet conformations in aggregated forms [66].

The HCA analysis was performed to obtain the averaged TERS spectra used as the reference for assigning the two conformations, turns/random coils, and β-sheet in the analysis of TER maps [99]. The correlations between each single spectrum from the acquired maps and the marker spectra determined by HCA were estimated with Pearson's correlation coefficient. Calculations were performed using not the raw spectra, but rather their second derivatives to avoid any influence from the baseline. The same maps were also analyzed using principal component analysis for the statistical comparison. The multidimensional statistical analysis allowed to demonstrate the distribution of the turns/random coils and β-sheets secondary structures in individual amyloid $β_{1-42}$ aggregates, therefore, Lipiec et al. verified spectroscopically several hypothetically proposed aggregation pathways [118].

The PCA performed on the SERS data allowed to distinguish amyloid β isoforms, $Aβ_{1-40}$ associated with classical vascular and $Aβ_{1-42}$, related to parenchymal vascular plaques in the Alzheimer's Disease. Five different wavelengths were analyzed using the PCA method to maximize the between-group (data from each point) variation. During the mentioned analysis, the observation of a minimum of nine differentiable clusters within PCA spaces was implied, which corresponds to at least nine different assembly states in the fibril formation pathway. This approach finally proved to be extremely useful for determining structure-activity relationships [102].

5. Conclusions

In conclusion, the work presented in this review shows the enormous progress in the state of knowledge concerning amyloid aggregation. Spectroscopic studies of abnormal protein aggregation revealed characteristic marker bands related to specific secondary structure content occurring at successive steps of the aggregation process. The spectroscopic research carried out so far has made it possible to link the ratio fluctuations of individual secondary structures content in fibrils with the occurrence of alternative aggregation pathways. In particular, molecular nanospectroscopy has made it possible to study the distribution of particular conformations along individual amyloid fibrils that appeared to have a highly heterogeneous structure at the nanoscale. This tremendous progress allows one to go one step further and conclude about the specific aggregation mechanism of the studied system. Now, the greatest limitation of nanospectroscopy seems to be related to technical difficulties with transferring measurements into the liquid conditions. The drying process of biological components may influence their native structure. The efforts of nanospectroscopy community are focused on overcoming this limitation and the progress in this direction had already been presented in the work of Ramer et al. [72].

The research concerning abnormal protein aggregation has been focused so far on studying the aggregation of isolated/synthesized peptides and fibrils formed in vitro. That gave one the solid benchmark to study more complicated systems such as isolated amyloid plaques from neurodegenerative diseases or amyloidosis affected patients. Such works have been limited so far due to the complexity of such systems, but now it seems to be one of the future directions of spectroscopic studies concerning abnormal protein aggregation.

Author Contributions: N.W., M.C., S.S., K.S.-N., M.S., E.L. and K.S. discussed the conception of the review and contributed to the final manuscript preparation. All authors have read and agreed to the published version of the manuscript.

Funding: This research was funded by the National Science Centre grant number No. UMO-2018/31/B/ST4/02292 and The APC was funded by the National Science Centre grant number No. UMO-2018/31/B/ST4/02292.

Acknowledgments: This work is supported by the National Science Centre, Poland under the "OPUS 16" project (Reg. No. UMO-2018/31/B/ST4/02292).

Conflicts of Interest: The authors declare no conflict of interest.

References

1. Knowles, T.P.J.; Vendruscolo, M.; Dobson, C.M. The amyloid state and its association with protein misfolding diseases. *Nat. Rev. Mol. Cell Biol.* **2014**, *15*, 384–396. [CrossRef] [PubMed]
2. Jucker, M.; Walker, L.C. Self-propagation of pathogenic protein aggregates in neurodegenerative diseases. *Nature* **2013**, *501*, 45–51. [CrossRef] [PubMed]
3. Riek, R.; Eisenberg, D.S. The activities of amyloids from a structural perspective. *Nature* **2016**, *539*, 227–235. [CrossRef]
4. Alzheimer's Disease International Dementia statistics | Alzheimer's Disease International. *Alzheimer's Dis. Int.* **2016**, *1*. [CrossRef]
5. Stefani, M. Protein misfolding and aggregation: New examples in medicine and biology of the dark side of the protein world. *Biochim. Biophys. Acta—Mol. Basis Dis.* **2004**, *1739*, 5–25. [CrossRef] [PubMed]
6. Wang, W. Protein aggregation and its inhibition in biopharmaceutics. *Int. J. Pharm.* **2005**, *289*, 1–30. [CrossRef] [PubMed]
7. Tutar, L.; Tutar, Y. Heat Shock Proteins; An Overview. *Curr. Pharm. Biotechnol.* **2010**, *11*, 216–222. [CrossRef]
8. Kumar, V.; Sami, N.; Kashav, T.; Islam, A.; Ahmad, F.; Hassan, M.I. Protein aggregation and neurodegenerative diseases: From theory to therapy. *Eur. J. Med. Chem.* **2016**, *124*, 1105–1120. [CrossRef]

9. Chung, C.G.; Lee, H.; Lee, S.B. Mechanisms of protein toxicity in neurodegenerative diseases. *Cell. Mol. Life Sci.* **2018**, *75*, 3159–3180. [CrossRef]
10. Lee, C.; Yu, M.H. Protein folding and diseases. *J. Biochem. Mol. Biol.* **2005**, *38*, 275–280. [CrossRef]
11. Kolarova, M.; García-Sierra, F.; Bartos, A.; Ricny, J.; Ripova, D. Structure and pathology of tau protein in Alzheimer disease. *Int. J. Alzheimers Dis.* **2012**, *2012*. [CrossRef]
12. Buée, L.; Bussière, T.; Buée-Scherrer, V.; Delacourte, A.; Hof, P.R. Tau protein isoforms, phosphorylation and role in neurodegenerative disorders. *Brain Res. Rev.* **2000**, *33*, 95–130. [CrossRef]
13. Martin, L.; Latypova, X.; Terro, F. Post-translational modifications of tau protein: Implications for Alzheimer's disease. *Neurochem. Int.* **2011**, *58*, 458–471. [CrossRef]
14. Murphy, M.P.; Levine, H. Alzheimer's disease and the amyloid-β peptide. *J. Alzheimer's Dis.* **2010**, *19*, 311–323. [CrossRef]
15. Atwood, C.S.; Obrenovich, M.E.; Liu, T.; Chan, H.; Perry, G.; Smith, M.A.; Martins, R.N. Amyloid-β: A chameleon walking in two worlds: A review of the trophic and toxic properties of amyloid-β. *Brain Res. Rev.* **2003**, *43*, 1–16. [CrossRef]
16. Meng, F.; Bellaiche, M.M.J.; Kim, J.Y.; Zerze, G.H.; Best, R.B.; Chung, H.S. Highly Disordered Amyloid-β Monomer Probed by Single-Molecule FRET and MD Simulation. *Biophys. J.* **2018**, *114*, 870–884. [CrossRef]
17. Vivekanandan, S.; Brender, J.R.; Lee, S.Y.; Ramamoorthy, A. A partially folded structure of amyloid-beta(1-40) in an aqueous environment. *Biochem. Biophys. Res. Commun.* **2011**, *411*, 312–316. [CrossRef] [PubMed]
18. Li, M.; Chen, L.; Lee, D.H.S.; Yu, L.C.; Zhang, Y. The role of intracellular amyloid β in Alzheimer's disease. *Prog. Neurobiol.* **2007**, *83*, 131–139. [CrossRef]
19. Ma, Q.L.; Chan, P.; Yoshii, M.; Uéda, K. α-synuclein aggregation and neurodegenerative diseases. *J. Alzheimer's Dis.* **2003**, *5*, 139–148. [CrossRef]
20. Etezadi, D.; Warner, J.B.; Ruggeri, F.S.; Dietler, G.; Lashuel, H.A.; Altug, H. Nanoplasmonic mid-infrared biosensor for in vitro protein secondary structure detection. *Light Sci. Appl.* **2017**, *6*, 17029. [CrossRef]
21. Shastry, B.S. Neurodegenerative disorders of protein aggregation. *Neurochem. Int.* **2003**, *43*, 1–7. [CrossRef]
22. Trześniewska, K.; Brzyska, M.; Elbaum, D. Neurodegenerative aspects of protein aggregation. *Acta Neurobiol. Exp. (Wars)* **2004**, *64*, 41–52.
23. Moreno-Gonzalez, I.; Soto, C. Misfolded protein aggregates: Mechanisms, structures and potential for disease transmission. *Semin. Cell Dev. Biol.* **2011**, *22*, 482–487. [CrossRef]
24. Chiti, F.; Dobson, C.M. Protein Misfolding, Amyloid Formation, and Human Disease: A Summary of Progress Over the Last Decade. *Annu. Rev. Biochem.* **2017**, *86*, 27–68. [CrossRef] [PubMed]
25. Bemporad, F.; Chiti, F. Protein misfolded oligomers: Experimental approaches, mechanism of formation, and structure-toxicity relationships. *Chem. Biol.* **2012**, *19*, 315–327. [CrossRef] [PubMed]
26. Sarroukh, R.; Goormaghtigh, E.; Ruysschaert, J.M.; Raussens, V. ATR-FTIR: A "rejuvenated" tool to investigate amyloid proteins. *Biochim. Biophys. Acta Biomembr.* **2013**, *1828*, 2328–2338. [CrossRef] [PubMed]
27. Powers, E.T.; Ferrone, F.A. Kinetic Models for Protein Misfolding and Association. In *Protein Misfolding Diseases: Current and Emerging Principles and Therapies*; John Wiley & Sons: Hoboken, NJ, USA, 2010; pp. 73–92.
28. Celej, M.S.; Sarroukh, R.; Goormaghtigh, E.; Fidelio, G.D.; Ruysschaert, J.M.; Raussens, V. Toxic prefibrillar α-synuclein amyloid oligomers adopt a distinctive antiparallel β-sheet structure. *Biochem. J.* **2012**, *443*, 719–726. [CrossRef]
29. Frare, E.; Mossuto, M.F.; de Laureto, P.P.; Tolin, S.; Menzer, L.; Dumoulin, M.; Dobson, C.M.; Fontana, A. Characterization of Oligomeric Species on the Aggregation Pathway of Human Lysozyme. *J. Mol. Biol.* **2009**, *387*, 17–27. [CrossRef]
30. Ruggeri, F.S.; Flagmeier, P.; Kumita, J.R.; Meisl, G.; Chirgadze, D.Y.; Bongiovanni, M.N.; Knowles, T.P.J.; Dobson, C.M. The Influence of Pathogenic Mutations in α-Synuclein on Biophysical and Structural Characteristics of Amyloid Fibrils. *ACS Nano* **2020**. [CrossRef]
31. Juszczyk, P.; Kolodziejczyk, A.S.; Grzonka, Z. FTIR spectroscopic studies on aggregation process of the β-amyloid 11-28 fragment and its variants. *J. Pept. Sci.* **2009**, *15*, 23–29. [CrossRef]
32. Sandberg, A.; Luheshi, L.M.; Söllvander, S.; De Barros, T.P.; Macao, B.; Knowles, T.P.J.; Biverstål, H.; Lendel, C.; Ekholm-Petterson, F.; Dubnovitsky, A.; et al. Stabilization of neurotoxic Alzheimer amyloid-β oligomers by protein engineering. *Proc. Natl. Acad. Sci. USA* **2010**, *107*, 15595–15600. [CrossRef]
33. Rygula, A.; Majzner, K.; Marzec, K.M.; Kaczor, A.; Pilarczyk, M.; Baranska, M. *Raman Spectroscopy of Proteins: A Review*; John Wiley & Sons, Ltd.: Hoboken, NJ, USA, 2013; Volume 44, pp. 1061–1076.

34. Stuart, B.H. *Infrared Spectroscopy: Fundamentals and Applications*; John Wiley & Sons, Ltd.: Hoboken, NJ, USA, 2005.
35. Barth, A. Infrared spectroscopy of proteins. *Biochim. Biophys. Acta Bioenerg.* **2007**, *1767*, 1073–1101. [CrossRef] [PubMed]
36. Sofińska, K.; Wilkosz, N.; Szymoński, M.; Lipiec, E. Molecular spectroscopic markers of DNA damage. *Molecules* **2020**, *25*, 561. [CrossRef] [PubMed]
37. Haris, P.I.; Chapman, D. Does Fourier-transform infrared spectroscopy provide useful information on protein structures? *Trends Biochem. Sci.* **1992**, *17*, 328–333. [CrossRef]
38. Shivu, B.; Seshadri, S.; Li, J.; Oberg, K.A.; Uversky, V.N.; Fink, A.L. Distinct β-sheet structure in protein aggregates determined by ATR-FTIR spectroscopy. *Biochemistry* **2013**, *52*, 5176–5183. [CrossRef]
39. Bouchard, M.; Zurdo, J.; Nettleton, E.J.; Dobson, C.M.; Robinson, C.V. Formation of insulin amyloid fibrils followed by FTIR simultaneously with CD and electron microscopy. *Protein Sci.* **2000**, *9*, 1960–1967. [CrossRef]
40. Cerf, E.; Sarroukh, R.; Tamamizu-Kato, S.; Breydo, L.; Derclayes, S.; Dufrênes, Y.F.; Narayanaswami, V.; Goormaghtigh, E.; Ruysschaert, J.M.; Raussens, V. Antiparallel β-sheet: A signature structure of the oligomeric amyloid β-peptide. *Biochem. J.* **2009**, *421*, 415–423. [CrossRef]
41. Eckert, A.; Hauptmann, S.; Scherping, I.; Meinhardt, J.; Rhein, V.; Dröse, S.; Brandt, U.; Fändrich, M.; Müller, W.E.; Götz, J. Oligomeric and fibrillar species of β-amyloid (Aβ42) both impair mitochondrial function in P301L tau transgenic mice. *J. Mol. Med.* **2008**, *86*, 1255–1267. [CrossRef]
42. Sarroukh, R.; Cerf, E.; Derclaye, S.; Dufrêne, Y.F.; Goormaghtigh, E.; Ruysschaert, J.M.; Raussens, V. Transformation of amyloid β(1-40) oligomers into fibrils is characterized by a major change in secondary structure. *Cell. Mol. Life Sci.* **2011**, *68*, 1429–1438. [CrossRef]
43. Itkin, A.; Dupres, V.; Dufrêne, Y.F.; Bechinger, B.; Ruysschaert, J.M.; Raussens, V. Calcium ions promote formation of amyloid β-peptide (1-40) oligomers causally implicated in neuronal toxicity of Alzheimer's disease. *PLoS ONE* **2011**, *6*. [CrossRef]
44. Habicht, G.; Haupt, C.; Friedrich, R.P.; Hortschansky, P.; Sachse, C.; Meinhardt, J.; Wieligmann, K.; Gellermann, G.P.; Brodhun, M.; Götz, J.; et al. Directed selection of a conformational antibody domain that prevents mature amyloid fibril formation by stabilizing Aβ protofibrils. *Proc. Natl. Acad. Sci. USA* **2007**, *104*, 19232–19237. [CrossRef]
45. Lashuel, H.A.; Hartley, D.; Petre, B.M.; Walz, T.; Lansbury, P.T. Neurodegenerative disease: Amyloid pores from pathogenic mutations. *Nature* **2002**, *418*. [CrossRef]
46. Habchi, J.; Arosio, P.; Perni, M.; Costa, A.R.; Yagi-Utsumi, M.; Joshi, P.; Chia, S.; Cohen, S.I.A.; Müller, M.B.D.; Linse, S.; et al. Neuroscience: An anticancer drug suppresses the primary nucleation reaction that initiates the production of the toxic Ab42 aggregates linked with Alzheimer's disease. *Sci. Adv.* **2016**, *2*. [CrossRef]
47. Hong, D.P.; Fink, A.L.; Uversky, V.N. Structural Characteristics of α-Synuclein Oligomers Stabilized by the Flavonoid Baicalein. *J. Mol. Biol.* **2008**, *383*, 214–223. [CrossRef]
48. Uversky, V.N.; Li, J.; Fink, A.L. Evidence for a Partially Folded Intermediate in α-Synuclein Fibril Formation. *J. Biol. Chem.* **2001**, *276*, 10737–10744. [CrossRef]
49. Roeters, S.J.; Iyer, A.; Pletikapiä, G.; Kogan, V.; Subramaniam, V.; Woutersen, S. Evidence for Intramolecular Antiparallel Beta-Sheet Structure in Alpha-Synuclein Fibrils from a Combination of Two-Dimensional Infrared Spectroscopy and Atomic Force Microscopy. *Sci. Rep.* **2017**, *7*, 1–11. [CrossRef]
50. Middleton, C.T.; Marek, P.; Cao, P.; Chiu, C.C.; Singh, S.; Woys, A.M.; De Pablo, J.J.; Raleigh, D.P.; Zanni, M.T. Two-dimensional infrared spectroscopy reveals the complex behaviour of an amyloid fibril inhibitor. *Nat. Chem.* **2012**, *4*, 355–360. [CrossRef]
51. Natalello, A.; Prokorov, V.V.; Tagliavini, F.; Morbin, M.; Forloni, G.; Beeg, M.; Manzoni, C.; Colombo, L.; Gobbi, M.; Salmona, M.; et al. Conformational Plasticity of the Gerstmann-Sträussler-Scheinker Disease Peptide as Indicated by Its Multiple Aggregation Pathways. *J. Mol. Biol.* **2008**, *381*, 1349–1361. [CrossRef]
52. Cordeiro, Y.; Kraineva, J.; Suarez, M.C.; Tempesta, A.G.; Kelly, J.W.; Silva, J.L.; Winter, R.; Foguel, D. Fourier transform infrared spectroscopy provides a fingerprint for the tetramer and for the aggregates of transthyretin. *Biophys. J.* **2006**, *91*, 957–967. [CrossRef]
53. Zandomeneghi, G.; Krebs, M.R.H.; McCammon, M.G.; Fändrich, M. FTIR reveals structural differences between native β-sheet proteins and amyloid fibrils. *Protein Sci.* **2009**, *13*, 3314–3321. [CrossRef]

54. Fabian, H.; Gast, K.; Laue, M.; Misselwitz, R.; Uchanska-Ziegler, B.; Ziegler, A.; Naumann, D. Early stages of misfolding and association of β2- microglobulin: Insights from infrared spectroscopy and dynamic light scattering. *Biochemistry* **2008**, *47*, 6895–6906. [CrossRef] [PubMed]
55. Berthelot, K.; Ta, H.P.; Géan, J.; Lecomte, S.; Cullin, C. In vivo and in vitro analyses of toxic mutants of HET-s: FTIR antiparallel signature correlates with amyloid toxicity. *J. Mol. Biol.* **2011**, *412*, 137–152. [CrossRef] [PubMed]
56. Zou, Y.; Li, Y.; Hao, W.; Hu, X.; Ma, G. Parallel β-sheet fibril and antiparallel β-sheet oligomer: New insights into amyloid formation of hen egg white lysozyme under heat and acidic condition from FTIR spectroscopy. *J. Phys. Chem. B* **2013**, *117*, 4003–4013. [CrossRef] [PubMed]
57. Zurdo, J.; Guijarro, J.I.; Dobson, C.M. Preparation and characterization of purified amyloid fibrils. *J. Am. Chem. Soc.* **2001**, *123*, 8141–8142. [CrossRef] [PubMed]
58. Balbirnie, M.; Grothe, R.; Eisenberg, D.S. An amyloid-forming peptide from the yeast prion Sup35 reveals a dehydrated β-sheet structure for amyloid. *Proc. Natl. Acad. Sci. USA* **2001**, *98*, 2375–2380. [CrossRef]
59. Shim, S.H.; Gupta, R.; Ling, Y.L.; Strasfeld, D.B.; Raleigh, D.P.; Zanni, M.T. Two-dimensional IR spectroscopy and isotope labeling defines the pathway of amyloid formation with residue-specific resolution. *Proc. Natl. Acad. Sci. USA* **2009**, *106*, 6614–6619. [CrossRef]
60. Ami, D.; Lavatelli, F.; Rognoni, P.; Palladini, G.; Raimondi, S.; Giorgetti, S.; Monti, L.; Doglia, S.M.; Natalello, A.; Merlini, G. In situ characterization of protein aggregates in human tissues affected by light chain amyloidosis: A FTIR microspectroscopy study. *Sci. Rep.* **2016**, *6*, 1–12. [CrossRef]
61. Amenabar, I.; Poly, S.; Nuansing, W.; Hubrich, E.H.; Govyadinov, A.A.; Huth, F.; Krutokhvostov, R.; Zhang, L.; Knez, M.; Heberle, J.; et al. Structural analysis and mapping of individual protein complexes by infrared nanospectroscopy. *Nat. Commun.* **2013**, *4*. [CrossRef] [PubMed]
62. Dazzi, A.; Prazeres, R.; Glotin, F.; Ortega, J.M. Local infrared microspectroscopy with subwavelength spatial resolution with an atomic force microscope tip used as a photothermal sensor. *Opt. Lett.* **2005**, *30*. [CrossRef]
63. Unger, M.; Marcott, C. Recent Advances and Applications of Nanoscale Infrared Spectroscopy and Imaging. In *Encyclopedia of Analytical Chemistry*; John Wiley & Sons, Ltd.: Hoboken, NJ, USA, 2017; pp. 1–26.
64. Dazzi, A.; Prazeres, R.; Glotin, F.; Ortega, J.M. Subwavelength infrared spectromicroscopy using an AFM as a local absorption sensor. *Infrared Phys. Technol.* **2006**, *49*, 113–121. [CrossRef]
65. Henry, S.; Bercu, N.B.; Bobo, C.; Cullin, C.; Molinari, M.; Lecomte, S. Interaction of Aβ1-42 peptide or their variant with model membrane of different composition probed by infrared nanospectroscopy. *Nanoscale* **2018**, *10*, 936–940. [CrossRef] [PubMed]
66. Ruggeri, F.S.; Longo, G.; Faggiano, S.; Lipiec, E.; Pastore, A.; Dietler, G. Infrared nanospectroscopy characterization of oligomeric and fibrillar aggregates during amyloid formation. *Nat. Commun.* **2015**, *6*. [CrossRef] [PubMed]
67. Ruggeri, F.S.; Vieweg, S.; Cendrowska, U.; Longo, G.; Chiki, A.; Lashuel, H.A.; Dietler, G. Nanoscale studies link amyloid maturity with polyglutamine diseases onset. *Sci. Rep.* **2016**, *6*, 1–11. [CrossRef] [PubMed]
68. Rizevsky, S.; Kurouski, D. Nanoscale Structural Organization of Insulin Fibril Polymorphs Revealed by Atomic Force Microscopy–Infrared Spectroscopy (AFM-IR). *ChemBioChem* **2020**, *21*, 481–485. [CrossRef] [PubMed]
69. Galante, D.; Ruggeri, F.S.; Dietler, G.; Pellistri, F.; Gatta, E.; Corsaro, A.; Florio, T.; Perico, A.; D'Arrigo, C. A critical concentration of N-terminal pyroglutamylated amyloid beta drives the misfolding of Ab1-42 into more toxic aggregates. *Int. J. Biochem. Cell Biol.* **2016**, *79*, 261–270. [CrossRef] [PubMed]
70. Kumosinski, T.F.; Unruh, J.J. Quantitation of the global secondary structure of globular proteins by FTIR spectroscopy: Comparison with X-ray crystallographic structure. *Talanta* **1996**, *43*, 199–219. [CrossRef]
71. Kong, J.; Yu, S. Fourier transform infrared spectroscopic analysis of protein secondary structures. *Acta Biochim. Biophys. Sin. (Shanghai)* **2007**, *39*, 549–559. [CrossRef]
72. Ramer, G.; Ruggeri, F.S.; Levin, A.; Knowles, T.P.J.; Centrone, A. Determination of Polypeptide Conformation with Nanoscale Resolution in Water. *ACS Nano* **2018**, *12*, 6612–6619. [CrossRef]
73. Creasey, R.C.G.; Louzao, I.; Arnon, Z.A.; Marco, P.; Adler-Abramovich, L.; Roberts, C.J.; Gazit, E.; Tendler, S.J.B. Disruption of diphenylalanine assembly by a Boc-modified variant. *Soft Matter* **2016**, *12*, 9451–9457. [CrossRef]
74. Sivanandam, V.N.; Jayaraman, M.; Hoop, C.L.; Kodali, R.; Wetzel, R.; Van Der Wel, P.C.A. The aggregation-enhancing huntingtin N-terminus is helical in amyloid fibrils. *J. Am. Chem. Soc.* **2011**, *133*, 4558–4566. [CrossRef]

75. Haris, P.I.; Severcan, F. FTIR spectroscopic characterization of protein structure in aqueous and non-aqueous media. *Proc. J. Mol. Catal. B Enzym.* **1999**, *7*, 207–221. [CrossRef]
76. Vignaud, H.; Bobo, C.; Lascu, I.; Sörgjerd, K.M.; Zako, T.; Maeda, M.; Salin, B.; Lecomte, S.; Cullin, C. A structure-toxicity study of Aß42 reveals a new anti-parallel aggregation pathway. *PLoS ONE* **2013**, *8*. [CrossRef] [PubMed]
77. McCreery, R.L. Raman Spectroscopy for Chemical Analysis. *Meas. Sci. Technol.* **2001**, *12*, 653–654. [CrossRef]
78. Long, D.A. *The Raman Effect: A Unified Treatment of the Theory of Raman Scattering by Molecules*; John Wiley & Sons, Ltd.: Chichester, UK, 2002; Volume 8.
79. Movasaghi, Z.; Rehman, S.; Rehman, I.U. Raman spectroscopy of biological tissues. *Appl. Spectrosc. Rev.* **2007**, *42*, 493–541. [CrossRef]
80. Choo-Smith, L.P.; Edwards, H.G.M.; Endtz, H.P.; Kros, J.M.; Heule, F.; Barr, H.; Robinson, J.S.; Bruining, H.A.; Puppels, G.J. Medical applications of Raman spectroscopy: From proof of principle to clinical implementation. *Biopolym. Biospectrosc. Sect.* **2002**, *67*, 1–9. [CrossRef]
81. Nabiev, I.; Chourpa, I.; Manfait, M. Applications of Raman and surface-enhanced Raman scattering spectroscopy in medicine. *J. Raman Spectrosc.* **1994**, *25*, 13–23. [CrossRef]
82. Shashilov, V.A.; Sikirzhytski, V.; Popova, L.A.; Lednev, I.K. Quantitative methods for structural characterization of proteins based on deep UV resonance Raman spectroscopy. *Methods* **2010**, *52*, 23–37. [CrossRef]
83. Kurouski, D.; Van Duyne, R.P.; Lednev, I.K. Exploring the structure and formation mechanism of amyloid fibrils by Raman spectroscopy: A review. *Analyst* **2015**, *140*, 4967–4980. [CrossRef]
84. Kurouski, D.; Lednev, I.K. The impact of protein disulfide bonds on the amyloid fibril morphology. *Int. J. Biomed. Nanosci. Nanotechnol.* **2011**, *2*, 167–176. [CrossRef]
85. Rosario-Alomar, M.F.; Quiñones-Ruiz, T.; Kurouski, D.; Sereda, V.; Ferreira, E.B.; De Jesús-Kim, L.; Hernández-Rivera, S.; Zagorevski, D.V.; López-Garriga, J.; Lednev, I.K. Hydrogen sulfide inhibits amyloid formation. *J. Phys. Chem. B* **2015**, *119*, 1265–1274. [CrossRef]
86. Kurouski, D.; Washington, J.; Ozbil, M.; Prabhakar, R.; Shekhtman, A.; Lednev, I.K. Disulfide bridges remain intact while native insulin converts into amyloid fibrils. *PLoS ONE* **2012**, *7*. [CrossRef] [PubMed]
87. Xu, M.; Ermolenkov, V.V.; He, W.; Uversky, V.N.; Fredriksen, L.; Lednev, I.K. Lysozyme fibrillation: Deep UV Raman spectroscopic characterization of protein structural transformation. *Biopolymers* **2005**, *79*, 58–61. [CrossRef] [PubMed]
88. Xu, M.; Ermolenkov, V.V.; Uversky, V.N.; Lednev, I.K. Hen egg white lysozyme fibrillation: A deep-UV resonance Raman spectroscopic study. *J. Biophotonics* **2008**, *1*, 215–229. [CrossRef] [PubMed]
89. Stiles, P.L.; Dieringer, J.A.; Shah, N.C.; Van Duyne, R.P. Surface-Enhanced Raman Spectroscopy SERS: Surface-enhanced Raman spectroscopy Raman scattering: Inelastic scattering of a photon from a molecule in which the frequency change precisely matches the difference in vibrational energy levels. *Annu. Rev. Anal. Chem.* **2008**, *1*, 601–626. [CrossRef]
90. Willets, K.A.; Van Duyne, R.P. Localized Surface Plasmon Resonance Spectroscopy and Sensing. *Annu. Rev. Phys. Chem.* **2007**, *58*, 267–297. [CrossRef]
91. Kneipp, K.; Kneipp, H.; Itzkan, I.; Dasari, R.R.; Feld, M.S. Ultrasensitive Chemical Analysis by Raman Spectroscopy. *Chem. Rev.* **1999**, *99*, 2957–2975. [CrossRef]
92. Xie, W.; Schlücker, S. Medical applications of surface-enhanced Raman scattering. *Phys. Chem. Chem. Phys.* **2013**, *15*, 5329–5344. [CrossRef]
93. Cialla, D.; März, A.; Böhme, R.; Theil, F.; Weber, K.; Schmitt, M.; Popp, J. Surface-enhanced Raman spectroscopy (SERS): Progress and trends. *Anal. Bioanal. Chem.* **2012**, *403*, 27–54. [CrossRef]
94. Han, X.X.; Zhao, B.; Ozaki, Y. Surface-enhanced Raman scattering for protein detection. *Anal. Bioanal. Chem.* **2009**, *394*, 1719–1727. [CrossRef]
95. Beier, H.T.; Cowan, C.B.; Chou, I.H.; Pallikal, J.; Henry, J.E.; Benford, M.E.; Jackson, J.B.; Good, T.A.; Coté, G.L. Application of surface-enhanced raman spectroscopy for detection of beta amyloid using nanoshells. *Plasmonics* **2007**, *2*, 55–64. [CrossRef]
96. Choi, I.; Huh, Y.S.; Erickson, D. Ultra-sensitive, label-free probing of the conformational characteristics of amyloid beta aggregates with a SERS active nanofluidic device. *Microfluid. Nanofluid.* **2012**, *12*, 663–669. [CrossRef]

97. Chou, I.H.; Benford, M.; Beier, H.T.; Coté, G.L.; Wang, M.; Jing, N.; Kameoka, J.; Good, T.A. Nanofluidic biosensing for β-amyloid detection using surface enhanced raman spectroscopy. *Nano Lett.* **2008**, *8*, 1729–1735. [CrossRef] [PubMed]
98. Bhowmik, D.; Mote, K.R.; MacLaughlin, C.M.; Biswas, N.; Chandra, B.; Basu, J.K.; Walker, G.C.; Madhu, P.K.; Maiti, S. Cell-Membrane-Mimicking Lipid-Coated Nanoparticles Confer Raman Enhancement to Membrane Proteins and Reveal Membrane-Attached Amyloid-β Conformation. *ACS Nano* **2015**, *9*, 9070–9077. [CrossRef] [PubMed]
99. Kurouski, D.; Sorci, M.; Postiglione, T.; Belfort, G.; Lednev, I.K. Detection and structural characterization of insulin prefibrilar oligomers using surface enhanced Raman spectroscopy. *Biotechnol. Prog.* **2014**, *30*, 488–495. [CrossRef]
100. Li, Z.; Wang, H.; Chen, Z. Monitoring and modulation of insulin fibers by a protein isomerization targeting dye bromophenol blue. *Sens. Actuators B Chem.* **2019**, *287*, 496–502. [CrossRef]
101. Karaballi, R.A.; Merchant, S.; Power, S.R.; Brosseau, C.L. Electrochemical surface-enhanced Raman spectroscopy (EC-SERS) study of the interaction between protein aggregates and biomimetic membranes. *Phys. Chem. Chem. Phys.* **2018**, *20*, 4513–4526. [CrossRef]
102. Yu, X.; Hayden, E.Y.; Xia, M.; Liang, O.; Cheah, L.; Teplow, D.B.; Xie, Y.H. Surface enhanced Raman spectroscopy distinguishes amyloid B-protein isoforms and conformational states. *Protein Sci.* **2018**, *27*, 1427–1438. [CrossRef]
103. Stöckle, R.M.; Suh, Y.D.; Deckert, V.; Zenobi, R. Nanoscale chemical analysis by tip-enhanced Raman spectroscopy. *Chem. Phys. Lett.* **2000**, *318*, 131–136. [CrossRef]
104. Lipiec, E.W.; Wood, B.R. Tip-Enhanced Raman Scattering: Principles, Instrumentation, and the Application toe Biological Systems. In *Encyclopedia of Analytical Chemistry*; John Wiley & Sons, Ltd.: Hoboken, NJ, USA, 2017; pp. 1–26.
105. Novotny, L.; Van Hulst, N. Antennas for light. *Nat. Photonics* **2011**, *5*, 83–90. [CrossRef]
106. Chen, C.; Hayazawa, N.; Kawata, S. A 1.7 nm resolution chemical analysis of carbon nanotubes by tip-enhanced Raman imaging in the ambient. *Nat. Commun.* **2014**, *5*, 1–5. [CrossRef]
107. Zhang, R.; Zhang, Y.; Dong, Z.C.; Jiang, S.; Zhang, C.; Chen, L.G.; Zhang, L.; Liao, Y.; Aizpurua, J.; Luo, Y.; et al. Chemical mapping of a single molecule by plasmon-enhanced Raman scattering. *Nature* **2013**, *498*, 82–86. [CrossRef] [PubMed]
108. Liao, M.; Jiang, S.; Hu, C.; Zhang, R.; Kuang, Y.; Zhu, J.; Zhang, Y.; Dong, Z. Tip-enhanced raman spectroscopic imaging of individual carbon nanotubes with subnanometer resolution. *Nano Lett.* **2016**, *16*, 4040–4046. [CrossRef] [PubMed]
109. Domke, K.F.; Zhang, D.; Pettinger, B. Enhanced Raman spectroscopy: Single molecules or carbon? *J. Phys. Chem. C* **2007**, *111*, 8611–8616. [CrossRef]
110. Blum, C.; Schmid, T.; Opilik, L.; Metanis, N.; Weidmann, S.; Zenobi, R. Missing amide i mode in gap-mode tip-enhanced raman spectra of proteins. *J. Phys. Chem. C* **2012**, *116*, 23061–23066. [CrossRef]
111. Kurouski, D.; Postiglione, T.; Deckert-Gaudig, T.; Deckert, V.; Lednev, I.K. Amide i vibrational mode suppression in surface (SERS) and tip (TERS) enhanced Raman spectra of protein specimens. *Analyst* **2013**, *138*, 1665–1673. [CrossRef]
112. Szczerbiński, J.; Metternich, J.B.; Goubert, G.; Zenobi, R. How Peptides Dissociate in Plasmonic Hot Spots. *Small* **2020**, *16*. [CrossRef]
113. Lipiec, E.; Perez-Guaita, D.; Kaderli, J.; Wood, B.R.; Zenobi, R. Direct Nanospectroscopic Verification of the Amyloid Aggregation Pathway. *Angew. Chem.* **2018**, *130*, 8655–8660. [CrossRef]
114. Hermann, P.; Fabian, H.; Naumann, D.; Hermelink, A. Comparative study of far-field and near-field raman spectra from silicon-based samples and biological nanostructures. *J. Phys. Chem. C* **2011**, *115*, 24512–24520. [CrossRef]
115. Kurouski, D.; Deckert-Gaudig, T.; Deckert, V.; Lednev, I.K. Structure and composition of insulin fibril surfaces probed by TERS. *J. Am. Chem. Soc.* **2012**, *134*, 13323–13329. [CrossRef]
116. Deckert-Gaudig, T.; Deckert, V. High resolution spectroscopy reveals fibrillation inhibition pathways of insulin. *Sci. Rep.* **2016**, *6*. [CrossRef]
117. Deckert-Gaudig, T.; Kurouski, D.; Hedegaard, M.A.B.; Singh, P.; Lednev, I.K.; Deckert, V. Spatially resolved spectroscopic differentiation of hydrophilic and hydrophobic domains on individual insulin amyloid fibrils. *Sci. Rep.* **2016**, *6*, 1–9. [CrossRef] [PubMed]

118. vandenAkker, C.C.; Deckert-Gaudig, T.; Schleeger, M.; Velikov, K.P.; Deckert, V.; Bonn, M.; Koenderink, G.H. Nanoscale Heterogeneity of the Molecular Structure of Individual hIAPP Amyloid Fibrils Revealed with Tip-Enhanced Raman Spectroscopy. *Small* **2015**, *11*, 4131–4139. [CrossRef] [PubMed]
119. Paulite, M.; Blum, C.; Schmid, T.; Opilik, L.; Eyer, K.; Walker, G.C.; Zenobi, R. Full spectroscopic tip-enhanced raman imaging of single nanotapes formed from β-Amyloid(1-40) peptide fragments. *ACS Nano* **2013**, *7*, 911–920. [CrossRef] [PubMed]
120. Pearson, K. LIII. On lines and planes of closest fit to systems of points in space. *Lond. Edinb. Dublin Philos. Mag. J. Sci.* **1901**, *2*, 559–572. [CrossRef]
121. Hotelling, H. Analysis of a complex of statistical variables into principal components. *J. Educ. Psychol.* **1933**, *24*, 417–441. [CrossRef]
122. Lastovicka, J.L.; Jackson, J.E. A User's Guide to Principal Components. *J. Mark. Res.* **1992**, *29*. [CrossRef]
123. Jolliffe, I.T. Principal Component Analysis, Second Edition. *Encycl. Stat. Behav. Sci.* **2002**, *30*. [CrossRef]
124. Halkidi, M.; Batistakis, Y.; Vazirgiannis, M. On clustering validation techniques. *J. Intell. Inf. Syst.* **2001**, *17*, 107–145. [CrossRef]

© 2020 by the authors. Licensee MDPI, Basel, Switzerland. This article is an open access article distributed under the terms and conditions of the Creative Commons Attribution (CC BY) license (http://creativecommons.org/licenses/by/4.0/).

Review

Trends in the Hydrogen−Deuterium Exchange at the Carbon Centers. Preparation of Internal Standards for Quantitative Analysis by LC-MS

Paulina Grocholska and Remigiusz Bąchor *

Faculty of Chemistry, University of Wrocław, F. Joliot-Curie 14, 50-383 Wrocław, Poland; paulina.grocholska@chem.uni.wroc.pl
* Correspondence: remigiusz.bachor@chem.uni.wroc.pl; Tel.: +48-71-375-7218; Fax: +48-71-328-2348

Abstract: The application of internal standards in quantitative and qualitative bioanalysis is a commonly used procedure. They are usually isotopically labeled analogs of the analyte, used in quantitative LC-MS analysis. Usually, ^{2}H, ^{13}C, ^{15}N and ^{18}O isotopes are used. The synthesis of deuterated isotopologues is relatively inexpensive, however, due to the isotopic effect of deuterium and the lack of isotopologue co-elution, usually they are not considered as good internal standards for LC-MS quantification. On the other hand, the preparation of ^{13}C, ^{15}N and ^{18}O containing standards of drugs and their metabolites requires a complicated multistep de novo synthesis, starting from the isotopically labeled substrates, which are usually expensive. Therefore, there is a strong need for the development of low-cost methods for isotope-labeled standard preparations for quantitative analysis by LC-MS. The presented review concentrates on the preparation of deuterium-labeled standards by hydrogen−deuterium exchange reactions at the carbon centers. Recent advances in the development of the methods of isotopologues preparation and their application in quantitative analysis by LC-MS are evaluated.

Keywords: hydrogen−deuterium exchange; liquid chromatography-mass spectrometry; *N*-methylated amino acids; quantitative LC-MS analysis

1. Introduction

Hydrogen−deuterium exchange (HDX) is a process involving the substitution of a hydrogen atom by a deuterium atom in a molecule of a chemical compound, in the presence of a deuterating agent, e.g., deuterium oxide (D_2O), or another source of dissociating deuterons [1,2]. Labile hydrogens in the backbone and side-chain functional groups of chemical compounds undergo exchange with protons of the solvent within a few minutes [1]. Due to the basic mechanism of the H/D exchange reaction, including acid-base catalysis, the degree of the reaction depends strongly on the pH of the solution [2]. The decimal logarithm of the reaction rate constant versus pH curve takes a V-shape with a characteristic minimum between pH 2 and 3, where the average half-time of the exchange at $0°$ C is tens of minutes. The rate of H/D exchange increases significantly with increasing pH. The rate of the H/D exchange reaction depends also on the acidity of the hydrogens bounded with heteroatoms, which is a consequence of the electronegativity differences between these atoms. For this reason, hydrogens from carboxyl or hydroxyl groups undergo the H/D exchange reaction much more easily than amide ones. The role of the inductive effects of the individual functional groups present in the vicinity of the hydrogen undergoing H/D exchange, and of steric hindrance, hindering the access of the acid or basic catalyst to the exchangeable proton is also described [1,2].

The influence of the molecular structure of the compound on the isotope exchange has also been observed, especially in the case of peptides and proteins. It was found that amide hydrogens participating in intramolecular hydrogen bonds were much less susceptible

to isotopic exchange reactions. Therefore, the H/D exchange was found as a tool in the conformation analysis of biologically active compounds. Additionally, the isotope exchange reactions enabled the analysis of the mechanisms of chemical processes [3]. The compelling advantage of mass spectrometric analysis of the H/D exchange of peptides and proteins is the high sensitivity of the method, the low concentrations of the analyte which can be used and the ability to simultaneously analyze individual components of the complex mixtures [4].

In contrast to the hydrogens attached to heteroatoms, the hydrogen atoms bound to the carbon atoms are usually not exchangeable, however, the specific reaction conditions, including pH-dependent and metal-dependent catalysis, may promote the replacement of carbon-attached protons by deuterons (Figure 1) [5–8]. Such compounds may serve as internal standards in quantitative LC-MS analysis. Therefore, the exchange of hydrogen attached to the carbon by its heavier isotopes are of interest for mechanistic, product-orientated research and quantitative analysis [3]. The existing methods of isotope exchange on the α-carbon atoms of amino acids are expensive and time-consuming, because they require multistep de novo synthesis with the application of isotopically labeled substrates. Usually preparation of such deuterated derivatives by hydrogen−deuterium exchange is easier and more cost effective than by classical de novo synthesis [7]. Therefore, the development of new, 'gentle' methods of isotope exchange on the α-C carbon atoms of amino acid residues in peptides seems to be an important issue. At present, isotopically labeled compounds at the carbon atoms are of interest, especially due to their application as internal standards in quantitative mass spectrometry in the analysis of fragmentation mechanisms.

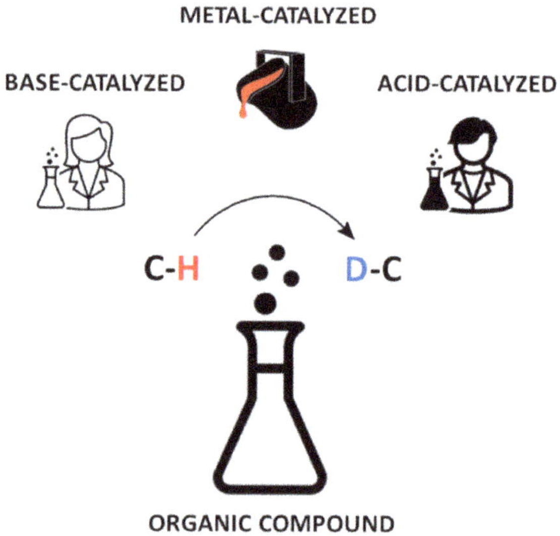

Figure 1. Hydrogen−deuterium exchange catalyzed by base, acid or metal catalyst.

Liquid chromatography-mass spectrometry (LC-MS) quantification frequently is performed in the presence of isotopically labeled standards which in most cases have to be synthesized de novo [9,10]. There are certain requirements for these standards: isotopologues should be characterized by identical chromatographic behavior, the mass difference between isotopologues should be at least 2 Daltons and the introduced isotopes cannot undergo back exchange during LC-MS separation conditions [11]. Quantification is performed by MS analysis, by comparing the extracted ion chromatograms (peak area) of the isotope-labeled and nonlabeled compounds. Although various stable isotope-labeled quantification reagents have been described containing ^{2}H, ^{13}C, ^{15}N and ^{18}O isotopes, due to their complicated and expensive chemical synthesis, there is still a strong need to develop a method of preparation of new isotopically-labeled standards. Usually, the preparation of deuterated standards is relatively inexpensive, however, the possibility of isotope effect affecting their co-elution during LC-MS would limit their possible application in LC-MS quantification [12,13]. It was found that the isotope effect on the chromatographic behavior of deuterated and nondeuterated isotopologues depends on the number and place of the introduced deuterons [14,15]. Additionally, deuterium-labeled compounds cannot be used for in vivo studies due to the possible loss of deuterium or different metabolism pathways. Therefore, there is a strong need for low-cost methods for isotope-labeled standard preparation for quantitative analysis by LC-MS. Although, the incorporation of deuterium into the target molecules may present some drawbacks, nevertheless the most important advantage of such a procedure is its low cost and simplicity of preparation.

In this review, we present the methods of deuterium incorporation into the molecules of compounds by exchange reactions and the possible application of deuterated standards in quantitative analysis by LC-MS.

2. Hydrogen−Deuterium Exchange at Carbon Centers

2.1. Acid- and Base-Catalyzed HDX

The pH-dependent hydrogen−deuterium exchange reactions are the first reported methods in the presented field [16]. The acid and base-catalyzed HDX process involves enolization which makes the H/D exchange at activated carbon centers possible in the presence of a source of deuterium, including deuterated Brønsted acids or bases. The back exchange of introduced deuterons is of course possible, therefore further deactivation of the analyzed compound is required.

In most cases, acid-catalyzed H/D-exchange reactions are used to incorporate the deuterium into the aromatic molecules. In these cases, strong deuterated Brønsted or Lewis acids, in the presence of a deuterium source are commonly used. The application of Lewis acids, including $AlBr_3$ or $MoCl_5$ is mostly restricted to the nonpolar arenes. In acid-catalyzed HDX, the incorporation of the ^2H isotope to the aromatic compounds exhibits limited regioselectivity. The effect of the aromatic ring substituents on the deuteration regioselectivity was analyzed in the acid-catalyzed HDX on ferrocenes. It was found that the electrophilic aromatic deuteration of the cyclopentadienyl rings was favored by alkyl groups whereas enolization of the carbonyl group in ketones led to the selective and complete H−D exchange of all three hydrogens of the acyl residue [17].

Base-catalyzed HDX is also a facile method for deuterium incorporation by means of keto–enol equilibria. Due to the higher acidity of carbon-bound hydrogen atoms in carbonyl compounds, including ketones [18], aldehydes [19], esters [20] and carboxylic acids [21], they undergo H/D exchange with high selectivity (>90% D) and yield. The γ hydrogens in α,β-unsaturated carbonyl compounds are also able to exchange through conjugation, as presented on the steroid framework of androstenedione, testosterone, and cortisone [22].

The deuteration of the methyl group in aryl methyl ketones and aryl methyl sulfones under basic conditions was presented by Berthelette and Scheigetz [23]. It was found that the reaction efficiency and rate depended on the base, the substrate and the solvent nature. Whereas 1,8-diazabicyclo[5.4.0]undec-7-ene (DBU) was found as a base allowing high deuteration efficiency, N,N,N-triethylamine (TEA) was less effective in the corresponding HDX processes (Figure 2). TEA was found as a base allowing methyl group deuteration in the base-sensitive ketones without decomposition.

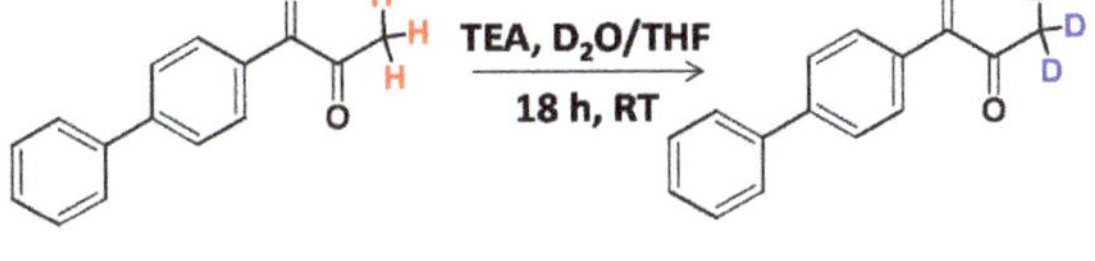

Figure 2. Deuteration of diketone presented by Berthelette and co-workers [23].

The base-catalyzed HDX reaction is a simple method for the acidic hydrogen exchange for deuterium by keto-enol equilibria [16]. The carbon-bound acidic hydrogens in carbonyl compounds, including N-substituted acetamides or diketopiperazines are usually highly exchanged [5]. The presented H/D exchange involved the application of acetone-d_6, N,N,N-triethylamine (TEA) or a stronger base in the form of diazabicycloundec-7-ene (DBU) and incubation at higher temperatures (35 °C to 50 °C). The acidity of α-carbon hydrogens of amino acids and peptides is of interest because the corresponding enolates play an important role in nonenzymatic racemization during peptide synthesis and enzyme-catalyzed racemization in different biochemical transformations [5,24]. Till now only a few quantitative investigations on their stability in water have been published. Their formation rate constants have been analyzed and determined in studies of α-hydrogen exchange or racemization reactions of amino acids and peptides at high temperatures [25,26] for the development of base-catalyzed methods for the preparation of α-C-deuterated amino acids is an important research task [8,27]. Generally, synthetic methods are based on the glycine derivatives application, which are subjected to a basic HDX and the stereoselective insertion of the desired side chain.

The acidity of α-C hydrogens of various amino acids and their derivatives have been extensively investigated. Ho et al. informed that α-C hydrogens in N-methylated analogs of cyclic dipeptides are more acidic than those in nonmethylated compounds [5,28]. It was also confirmed that the substitution of the amino group with electron-withdrawing substituents, such as the acetyl group, facilitated the HDX of α-C hydrogens. It was found by Rios and co-workers that the acidity of the α-C hydrogens depended on the ionization state of the amino acid, additionally the exhaustive methylation of α amino group also affected the pKa of the presented hydrogens [29,30]. Up to now, several different compounds containing N-methylated amino acids in their chemical structure have been described. These compounds belongs to the group of important drugs and natural tissue metabolites or substances commonly used in industry and households. In many cases there is a need to quantify them.

Moozeh and co-workers presented the stereoinversion of L-alanine to α-C deuterated D-alanine by base catalysis [31]. In the presence of salicylaldehyde and a chiral base, 87% deuterium incorporation at α-C of L-alanine and an inversion to D-alanine, was obtained. The enantiomeric excess (ee) was 67% (Figure 3). The developed method was also successfully applied for the deuteration of another 11 natural amino acids (threonine, tryptophan, phenylalanine, methionine, glutamic acid, glycine, glutamine, asparagine, serine, lysine and leucine). Stereoinversion for the presented examples was not reported.

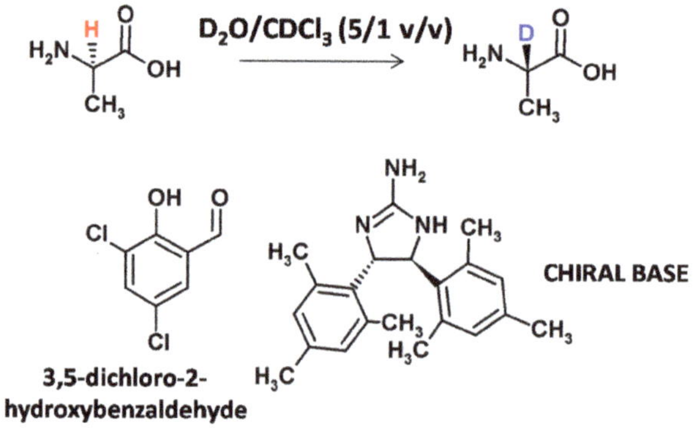

Figure 3. Catalytic stereoselective deuteration of L-alanine to deuterated D-alanine in the presence of 3,5-dichloro-2-hydroxybenzaldehyde and a chiral base [31].

Mitulovi and co-workers [32] presented the method of acid-catalyzed deuteration of α-amino acids (i.e., alanine, leucine, phenylalanine). In the presence of [D1]acetic acid (excess) and catalytic amounts of aldehyde, the reaction is characterized by good yield and the deuterium incorporation at the level of 95% via the corresponding Schiff's base formation (Figure 4). The obtained compound was converted into the tertbutoxycarbonyl (Boc) protected derivative and the resulting enantiomeric mixture was separated by preparative high-performance liquid chromatography on a chiral stationary phase.

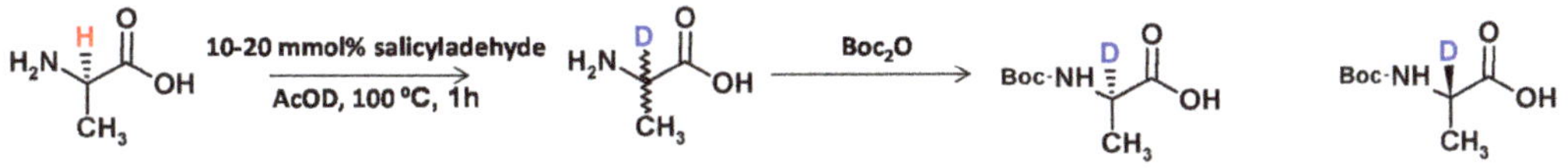

Figure 4. Deuteration of amino acids under acidic conditions [16].

The methods for the preparation of enantiomerically pure α-C-deuterated amino acids in the presence of a base are based on the application of glycine or its derivatives, which are subjected to a basic HDX. Finally, the side chain is inserted stereoselectively with the aid of chiral auxiliaries [6–8].

Lankiewicz and co-workers [7], described the method of the preparation of deuterated glycine derivative in the mixture of MeOD/D_2O and the presence of catalytic amounts of Na_2CO_3 as a base. After three reaction steps of reaction the obtained derivative was characterized by a deuterium content greater than 98%. Then the reaction with the Oppolzer sultam provide an intermediate for the subsequent stereoselective alkylation [33]. After removal of the auxiliary, the chiral Boc-protected amino acid (glycine, alanine, leucine, phenylalanine, O-benzyltyrosine) was isolated almost enantiomerically pure (>99% ee) in high yield (Figure 5).

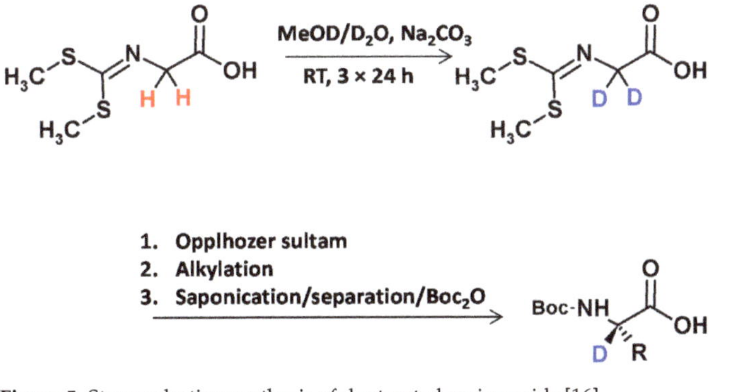

Figure 5. Stereoselective synthesis of deuterated amino acids [16].

Rose and et al. [8], inspired by the bislactim ether method developed by Schöllkopf and co-workers [34], described a base-catalyzed method of C6-position deuteration of the dihydropyrazine in boiling mixture of MeOD/D_2O (Figure 6). No hydrogen−deuterium exchange was observed at the C3-position due to the steric hindrance of the isopropyl group in the transition state. The obtained [6-D2]isotopologue was stereoselectively alkylated at the C6-position, thereby giving access to a series of α-C-deuterated amino acids (serine, phenylalanine, allylglycine, aspartic acid) in good yields, high degrees of deuteration and enantiomeric excesses (>95%) [8].

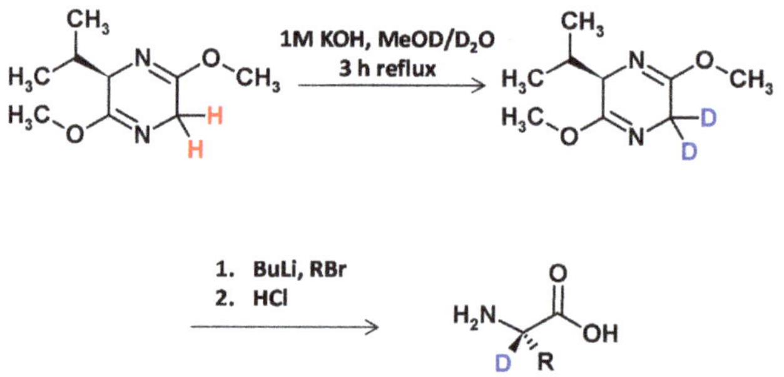

Figure 6. Schöllkopf bislactim ether application in the preparation of deuterated amino acids [16].

A further enantioselective synthesis of α-C-deuterated (Figure 7) proceeds by asymmetric alkylation of the activated glycine in the presence of the chiral phase-transfer catalyst. The HDX in the presence of KOD in D_2O and the introduction of the side chain were performed in a single reaction step. After imine hydrolysis the amino acid *tert*-butyl esters were isolated in good yields and with high deuterium incorporation of more than 90% [6].

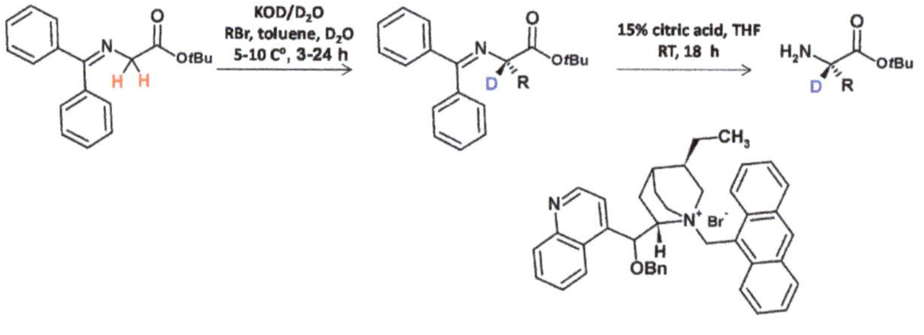

Figure 7. Enantioselective deuteration of amino acids in the presence of chiral phase transfer catalyst [16].

In addition to hydrogen−deuterium exchange based on the keto-enol tautomerism, the deuterolysis of an organometallic compound is also the chemical tool used for the synthesis of deuterated derivatives. In this reaction, the intermediate organometallic compound is formed by deprotonation in the presence of strong bases (i.e., Grignard reagent or alkyl–lithium compound), and subsequently deuterated with electrophiles in the form of D_2O, MeOD, or AcOD; which formally correspond to the H/D exchange [35]. Using this approach complete ortho-deuteration of aromatic amides and aromatic carbamates was achieved. Moreover, due to the large kinetic isotope effect (KIE), incorporated deuterium served as a protecting group for the carbon center, allowing the control of the regioselectivity of the subsequent lithiation.

Hydrogen−deuterium exchange reactions can also be performed without the addition of acids and bases. Such transformations are characteristic for the acidic CH centers which may be deuterated simply by the incubation of compound in deuterium oxide. The autoprotolysis equilibrium of D_2O, makes it possible to act as either an acid or a base. For example, the synthesis of [1,1,3,3-D4]2-indanone was achieved by the heating of the compound of interest in the D_2O [36]. Other reactions, depending on the compound, which were suspected to be HDX, required sometimes drastic reaction conditions which cannot be applied for most organic molecules.

A simple strategy was presented by Pacchioni et al. [37]. It was found that the formation of N,N,N-trinitroso derivative of the 1,4,7-triazacyclononane made the α-methylene hydrogen atoms more acidic allowing HDX in the presence of base.

A variant of acid-catalyzed HDX, which uses only D$_2$O during the deuteration process and is accelerated by microwave irradiation, was presented by Barthez et al. [38]. The developed strategy was also successfully applied to aminopyridine derivative preparation [39]. In order to avoid any proton sources, the labile hydrogen atoms bound to the nitrogen were exchanged to deuterons in the presence of D$_2$O. The applied strategy allowed complete deuteration within a few minutes and a high deuterium content at the *ortho* and *para* positions in the amino group.

Very recently, the application of microwaves in HDX processes significantly increased, especially due to the higher degree of deuteration, shorter reaction times as compared to the classical heating conditions. Based on this technique several MS standards of bleomycin A2 for quantitative MS analysis were successfully prepared in D$_2$O after a two-minute heating at 165 °C [40]. Additionally, some physicochemical reports have been published in which the kinetics of noncatalyzed HDX and energetic investigations were described [41].

It was previously reported that the base-catalyzed hydrogen–deuterium exchange at the carbon centers of aldehydes and ketones, thioesters and oxygen esters or amides occurred via a stepwise mechanism involving the enolate intermediate formation when the enolate was sufficiently stable to exist for the time of a bond vibration [30].

N-methylglycine, also called sarcosine, represents a natural, achiral compound with a methylated amino group which plays an important role in biological systems [42]. This amino acid residue is present in cyclosporine A, a cyclic nonribosomal peptide, commonly used as immunosuppressant [43]. Methods for the quantitative analysis of cyclosporine A and its metabolites have been developed [44] however, due to the necessity of preparation of its isotopically labeled standards for MS quantification, the costs of such analysis were very high. Joining the mainstream of the base-catalyzed HDX, previously, we developed several methods of preparation of deuterated standards of compounds containing *N*-substituted glycine derivative in their chemical structure, including denatonium benzoate, peptomers, cyclosporin A and creatinine. Additionally, the applicability of the obtained deuterated standards were tested in the quantitative analysis of these compounds by LC-MS.

In our previous work, the base-catalyzed HDX of α-C hydrogens in sarcosine residue and specific hydrogen scrambling in such peptides were investigated [45]. We found the unusual hydrogen–deuterium exchange at the α-carbon in *N*-methy- and *N*-benzylglycine residues in the presence of 1% solution of *N,N,N*-triethylamine in D$_2$O (Figure 8). We found that the observed HDX proceeded at a much slower rate as compared to the hydrogen–deuterium exchange of hydrogens present in amines or amides. Moreover, we observed the hydrogen scrambling during the collision-induced dissociation experiment which suggested the lability of such hydrogens. The presented work opened a wider possibility of application of the presented HDX reaction in peptide chemistry and mass spectrometry.

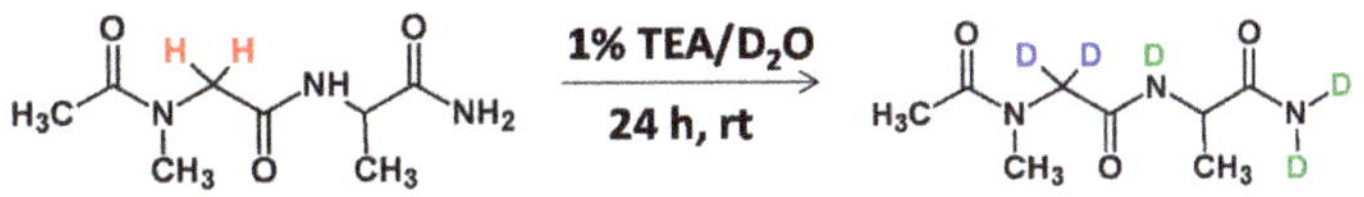

Figure 8. Schematic presentation of H/D exchange reaction at the α carbon in *N*-methylglycine moiety [45]. TEA-*N,N,N*-triethylamine.

Our investigation on compounds containing sarcosine residue revealed its presence in cyclosporin A molecule. The analysis of the possibility of the α-C deuteration of cyclosporine A (CsA) in *N*-methylated amino acid residues was performed [46]. The proposed reaction is based on the method previously reported by us [45], proceeds under basic conditions in the presence of TBD (1,5,7-triazabicyclo[4.4.0]dec-5-ene) or MTBD (7-methyl-1,5,7-triazabicyclo[4.4.0]dec-5-ene) at pH 13.4 (Figure 9). The obtained results revealed that there is a possibility of three deuteron incorporation, two at the α-C of *N*-methylglycine and one at the α-C of 2-*N*-methyl-(R)-((E)-2-butenyl)-4-methyl-L-threonine (MeBmt) residue. The prepared isotopologues were stable (did not undergo back exchange) under neutral and acidic conditions. Additionally, the deuterated and nondeuterated derivatives revealed co-elution, which make their application for the quantitative analysis by using isotope dilution strategy possible. The developed strategy of the CsA deuteration is rapid, cost-efficient and does not require special reaction conditions, other reagents or further purification [46].

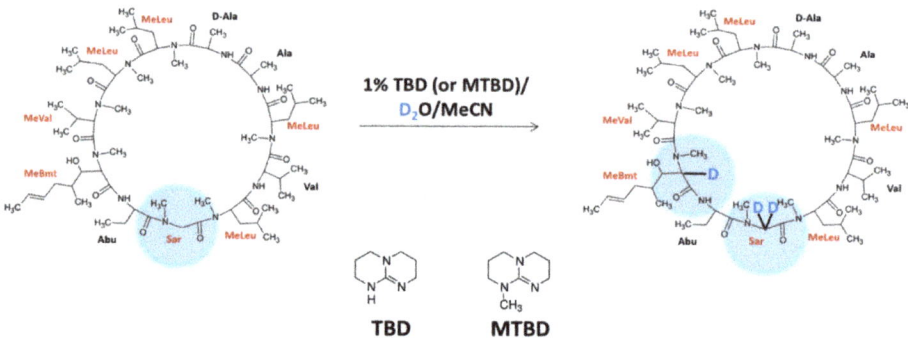

Figure 9. Preparation of deuterated analogs of cyclosporin A as described by Bąchor et al. [46].

The synthesis of the α-C deuterium-labeled N-substituted glycine residues in peptomers—oligomers composed of both α-amino acids and N-substituted glycine monomers—at basic conditions at room temperature was also analyzed by Bąchor et al. [47]. The developed method covered the deuterium labeling of peptomers at the α-C atom of N-substituted glycine residues by using simple HDX. The proposed labeling procedure is easy, inexpensive, and does not require any derivatization reagents or further purification. The introduced deuterons do not undergo a back-exchange under neutral and acidic conditions during LC-MS separation. The LC-MS analysis of an isotopologue mixture showed their co-elution. Therefore the developed strategy may be applied in the quantitative isotope dilution analysis of peptoids and other derivatives of N-substituted glycines.

Very recently we developed a method of deuterium-labeled standard preparation of creatinine, a breakdown product of creatine phosphate in muscle and a molecular biomarker of renal function [48]. The N-methylated glycine moiety was also presented within the creatinine molecule. The performed investigation allowed the doubly deuterated Cre analogue to be obtained, even after 60 min incubation in 1% TEA/D$_2$O solution at room temperature (Figure 10). We found that the introduced deuterons were stable under acidic and neutral conditions and any back exchange was observed. The obtained results suggest that the obtained deuterated Cre analogue may serve as a good internal standard for quantitative analysis by ESI-MS by using the isotope dilution method. The proposed methodology is a new, inexpensive and simple way for creatinine quantification. Additionally the performed quantification in the presence of the obtained deuterated Cre standard correlates with the Jaffe test method.

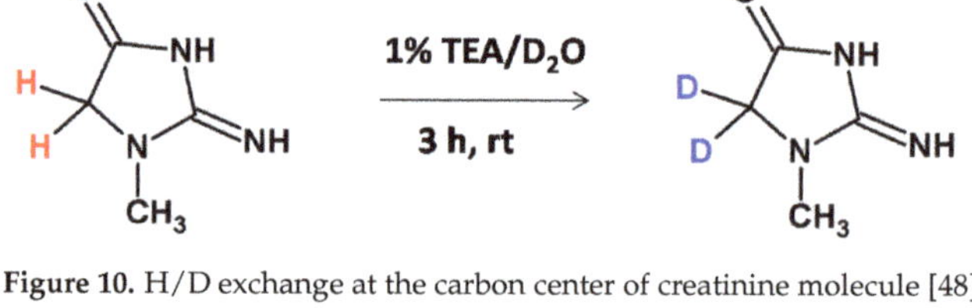

Figure 10. H/D exchange at the carbon center of creatinine molecule [48].

In 2015, the method of denatonium benzoate (Bitrex) deuteration via HDX of α-carbon hydrogens located in the CH$_2$ group, situated between carbonyl carbon and quaternary nitrogen atom (Figure 11) [49]. The reaction proceeded at room temperature and the deuteration was completed after 1 h of sample incubation in 1% TEA/D$_2$O mixture. We found that the introduced deuterons did not undergo back exchange under acidic and neutral conditions. We also found that the isotopologues—deuterated and nondeuterated—denatonium cation co-elute during the chromatographic separation. The applicability of the obtained deuterated denatonium cation as the internal standard for quantitative analysis of Bitrex was confirmed by the LC-MS analysis of various Bitrex-containing household products. The proposed strategy is a new and simple solution for sensitive Bitrex quantification by LC-MS method. We found that the presence of a quaternary nitrogen atom connected with the α-C atom facilitated the H/D exchange. Based on this observation, we focused our attention on the compounds containing quaternary ammonium groups in the form of N-substituted glycine derivatives. We reported the influence of the quaternary ammonium group on HDX at the α-C of sarcosine and N-methylalanine in peptides [50–52]. The significant acceleration of the HDX in sarcosine residue caused by the presence of a fixed charge tag was found. The effect depended on the distance between the sarcosine residue and the quaternary nitrogen atom. The deuterium atoms introduced at the α-C, did not undergo back-exchange under acidic aqueous solution. The tandem mass spectrometry analysis of the deuterated analogs of quaternary ammonium-tagged oligosarcosine peptides without mobile hydrogen showed the mobilization of the hydrogens localized at α-C atom of sarcosine residue.

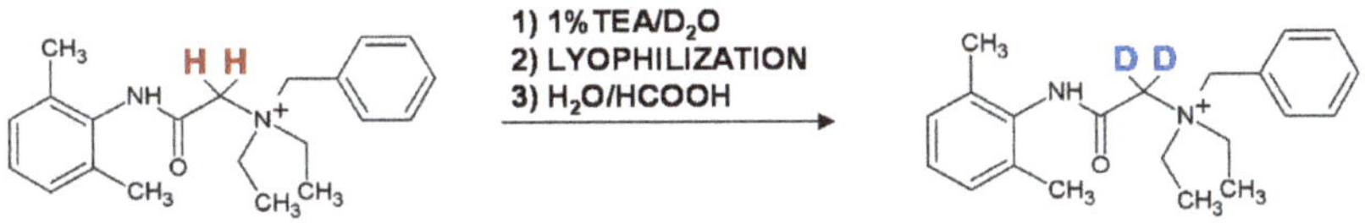

Figure 11. Preparation of deuterated Bitrex standard for quantitative analysis by mass spectrometry [49].

It was presented previously that the racemization and hydrogen−deuterium exchange at the α-amino carbon atoms in dipeptides may proceed via the reversible diketopiperazine intermediate formation [53,54]. It was also assumed by Rios and co-workers that the exchange of hydrogens into deuterons at the α carbon atom in amino acid residues occurs via a stepwise mechanism catalyzed by DO$^-$ anion where the enolate intermediate is formed [30]. The most important advantage of the acid or base catalyzed HDX is their simplicity of preparation, relatively low costs to perform, and mostly high efficiency. It should be also be pointed out that sometimes the hard reaction conditions, including high or low pH values, may lead to the compound decomposition.

2.2. Metal-Catalyzed HDX Adjacent to Oxygen Atom

The first example of H/D exchange adjacent to oxygen was discovered in 1974 by Regen. In this study, the deuteration of 1-butanol at the α carbon atom was performed in the presence of a catalyst in the form of (tris(triphenylphosphine) ruthenium dichloride [RuCl$_2$(PPh$_3$)$_3$] [55]. In the proposed method 1-butanol was incubated at 200 °C for one hour in the presence of a catalyst, which allowed the exchange of hydrogen at the α carbon position by deuterium bonded to the oxygen atom (Figure 12). The proposed reaction conditions were also applied to other deuterated alcohols at the α-C atom. The obtained results also revealed that the addition of D$_2$O to the reaction mixture significantly increased the deuteration efficiency and that the degree of deuteration depended on the D$_2$O/alcohol ratio.

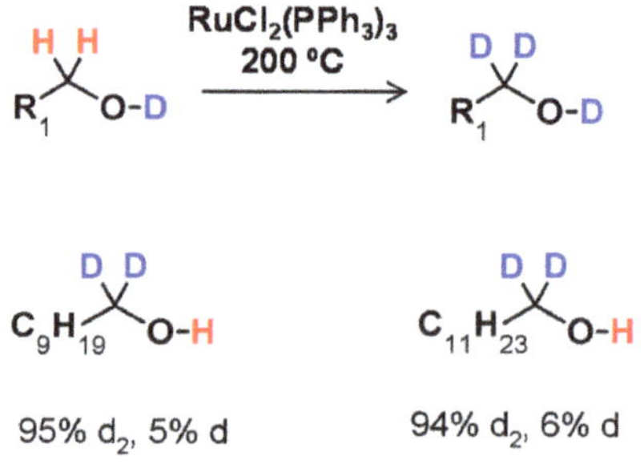

Figure 12. H/D exchange adjacent to oxygen atom, presented by Regen [55].

Ishibashi et al. presented the efficient deuteration of compounds containing electron donors in the form of double bonds, hydroxyl groups, in the presence of ruthenium catalyst. It was found that alkenols were efficiently deuterated in D_2O by the migration of the double bond and isomerization to ketones in the presence of ruthenium catalyst (Figure 13) [56]. It was found that primary alcohols were oxidized to aldehydes on the selective way in the presence of $RuCl_2(PPh_3)_3$ catalyst which was hampered in the presence of small amounts of water. The reaction temperature around 150 °C in microwave synthesizer, allowed an efficient HDX at the α carbon atom of the primary alcohols with small epimerization observed in the case of chiral compounds. Under lower temperatures around 100 °C the epimerization was sufficiently suppressed [57].

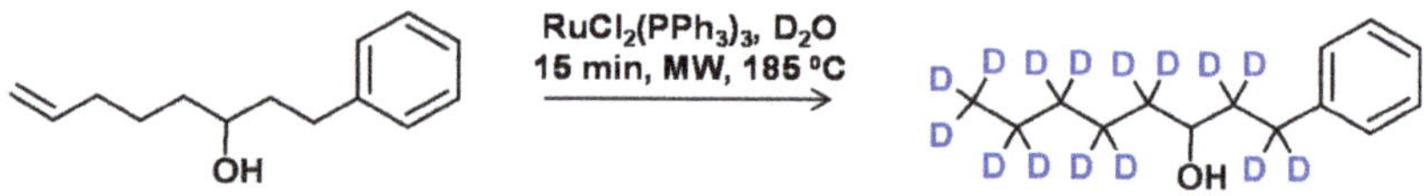

Figure 13. Ruthenium-mediated deuteration according to Ishibashi et al. [56].

In 2011, Bossi and co-authors presented a method for selective deuteration of alcohols at the α position in the presence of ruthenium and osmium pincer catalysts [58]. In the presence of isopropanol-d8 as a source of deuterium, Bossi was able to obtain high deuterium incorporation at the C1 carbon atom of primary and secondary alcohols, and within the case of secondary alcohols also the deuteration at the C2 position was observed. In 2013, Khaskin and Milstein proposed another ruthenium pincer catalyst allowing deuteration at the carbon centers in the presence of D_2O and with a lower catalyst loading [59]. Other substrates, including secondary alcohols also presented the possibility for β deuteration, which was frequent, with deuterium incorporation up to 97%.

In 2015, Bai et al. [60] proposed a selective α and β deuteration method of alcohols in the presence of D_2O as a deuterating agent. The reaction optimization process revealed that octahedral ruthenium complexes with the amine ligand presented higher activity in HDX promotion (Figure 14). It was also found that the (η6-cymene) ruthenium complex allowed deuteration only at the β position. The mechanism of this process involves oxidation of the alcohol to an aldehyde followed by base-mediated β deuteration. The formed aldehyde is then reduced to an alcohol by means of the deuterated ruthenium complex which results in the formation of α,β-deuterated derivative.

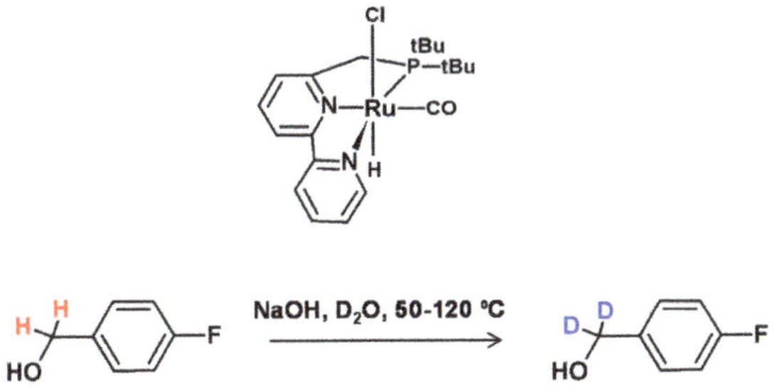

Figure 14. Khaskin's work on alcohol deuteration catalyzed by ruthenium pincer catalyst [59].

In the same year, Chatterje et al. presented the method of selective α and α,β deuteration of alcohols in the presence of a low-loading and commercially available ruthenium pincer catalyst (Ru-MACHO), the base in the form of potassium *tert*-butoxide and D_2O as the deuterating agent (Figure 15) [61]. The proposed reaction condition optimization on aromatic benzylic alcohols revealed that mild heating (60–100 °C) was sufficient to obtain 95% of deuterium incorporation. It was also presented, that in this method linear primary alcohols were also able to undergo β deuteration at 10–20%, while in the case of secondary alcohols, deuteration at the β carbon was higher. The mechanism of this reaction involves the oxidation and reduction of the alcohol. Based on this, the authors proposed that the intermediate, in the form of ketones, was a more long-living species than the aldehydes and therefore, β deuteration proceeded at a higher rate in the case of secondary alcohols. What is interesting, in the case of diols, only the deuteration at the α position was observed.

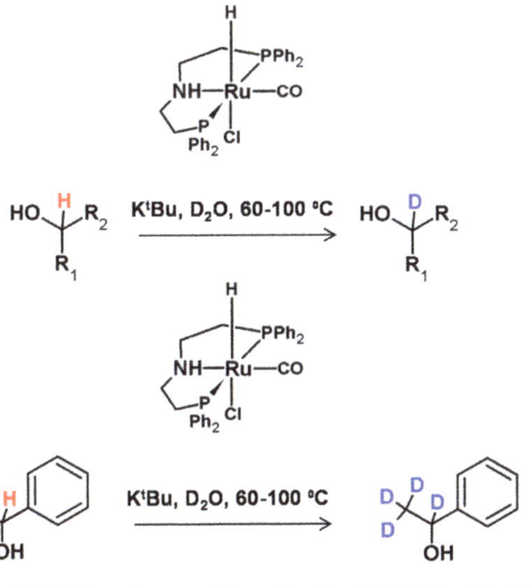

Figure 15. H/D exchange catalyzed by Ru-MACHO catalyst [61].

Similarly to Khaskin [59], in 2016 Gauvin and co-workers focused their attention on the application of ruthenium pincer complexes in the transformation of alcohols to carboxylic acids. They performed the reaction in a closed vessel to displace the reaction equilibrium toward the substrate (alcohol), thus allowing deuteration in the presence of D_2O [62]. They also decided to optimize the catalyst by switching from the Ru-MACHO to its analog containing cyclohexyl substituents, resulting in the higher activity and selectivity in α carbon deuteration.

Using a similar strategy to Regen's method, Koch and Stuart found that there was a possibility for primary and secondary alcohol deuteration at the α carbon by refluxing the alcohol in the presence of D_2O and Raney nickel as the catalyst [63]. The proposed strategy was successfully applied in the preparation of deuterated nonreducing carbohydrates. Additionally, it was proposed that the alcohol should undergo a redox process in the presence of Raney nickel as hydrogen-transfer catalyst. The observed retention of the configuration was explained by the polyhydric cyclic structure of the carbohydrate [64]. The isomerization of methyl α-D-mannopyranoside and methyl β-D-galactopyranoside to corresponding D-gluco isomers was also found after several days of reaction.

As described by Cioffi and co-workers, the activation of Raney nickel by sonication allowed a microwave-assisted deuteration of nonreducing carbohydrates without racemization [65]. The 1-O-methyl-β-d-galactopyranoside, used as a model compound, was heated in 15-s intervals up to 36 times, using a microwave oven. It was found that the deuterium was incorporated without epimerization or compound decomposition. Further optimization of the Stuart's method may be achieved by application of sonication which makes a higher level of deuterium incorporation possible (Figure 16) [66]. Microwave irradiation may also improve the deuteration however in this case the epimerization and substrate degradation was observed [65]. In the proposed solution deuteration was regioselective and occurred at the C2, C3 and C4 positions at the carbon atoms connected with hydroxyl group which was necessary for the HDX [67].

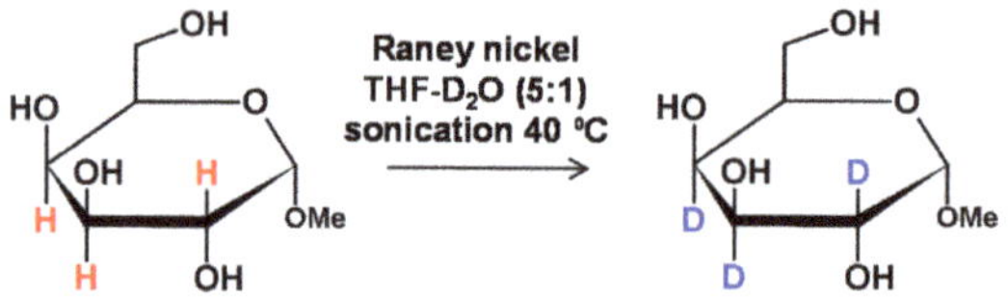

Figure 16. Methyl α-D-glucopyranoside deuteration catalyzed by Raney nickel under sonication as presented by Cioffi and co-workers [66].

Vert and co-workers applied the HSCIE (high-temperature solid-state catalytic isotope exchange) technique to obtain the selectively deuterated lactide [68] and glycolide derivatives [69], which were applied as substrates for the synthesis of isotopically labeled biocompatible absorbable poly-α-hydroxy acids. It was found that the optimal reaction temperature was close to the melting point of the substrate. The hydrogen−deuterium exchange of L-lactide at 120 °C in the presence of the $Pd/CaCO_3$ catalyst resulted in incomplete deuterium incorporation (Figure 17), but occurred without epimerization. Additionally, the reaction conditions were suitable for the tritium incorporation [70].

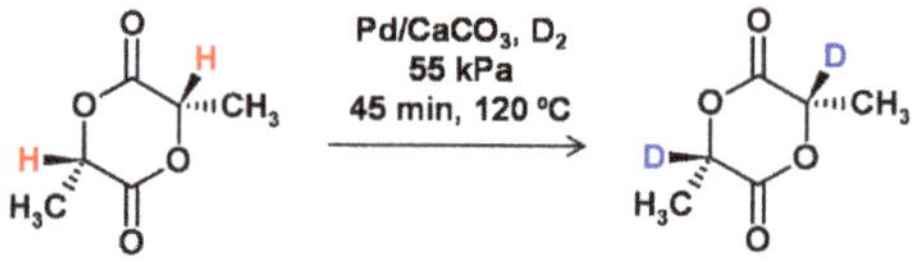

Figure 17. HSCIE method of selective H/D exchange at the carbon centers [68].

In 1990, Möbius and Schaaf [71] developed a method for the preparation of deuterated aliphatic hydrocarbons by metal-catalyzed HDX at higher temperatures (up to 290 °C). The reaction was performed in an autoclave, where a wire basket with the catalyst was placed above the substrate to be deuterated. The D_2/D_2O atmosphere was used to obtained deuterated derivatives under the pressure of around 25 Mpa. Under these conditions, water dissociates much more rapidly than at room temperature [72] and therefore Pd^0 is able to insert oxidatively into the H-OH bond with the formation of a Pd(II) derivative [73].

Based on the method developed by Matsubara and co-workers [74], complete deuteration of aromatic or aliphatic hydrocarbons was achieved by decarboxylation of carboxylic acids under hydrothermal reaction conditions. In this method, the model lactone molecule in D$_2$O allowed the formation of the phenol derivative with a high yield in the presence of 10% Pd/C (5 mol %) at 250 °C and a pressure of 4–5 Mpa (Figure 18).

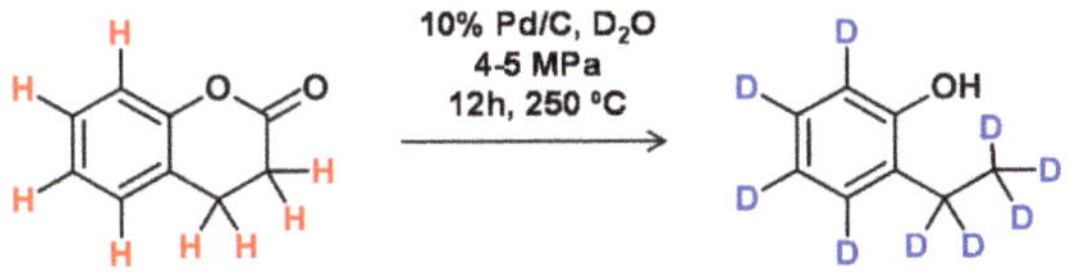

Figure 18. Lactone deuteration as presented by Matsubara et al. [74].

Previous studies by Sajiki and co-workers [75] on 5-phenylvaleric acid revealed the influence of reaction temperatures on the regioselectivity and deuteration. It was found that the benzylic hydrogen atoms were selectively exchanged to deuterons at room temperature, whereas at the higher temperature (160 °C), the deuteration in less reactive positions was also found, resulting in the formation of multideuterated derivatives. The proposed reaction conditions were compatible with compounds containing different functional groups in the form of carboxy, keto or hydroxyl groups, but the described reaction was characteristic for those with aryl-linked side chains. The proposed Pd/C–H$_2$/D$_2$O system may be also applied for the preparative formation of the phenylalanine selectively deuterated at the β carbon atom which takes place at 110 °C (6 h, 96% D) without racemization [76]. It was also found that at 160 °C the α position is also able for HDX, but these reaction conditions promote racemization (17% ee).

In 2005, Proszenyák et al. [77] developed a method allowing higher efficiency of HDX for the benzylic hydrogen atoms of the piperidine derivative in the presence of Pd/C–H$_2$–D$_2$O, deuterated alcohols and DCl. Earlier, in 1986, Stock and Ofosu-Asante [78] presented a method for selective benzylic deuteration of the tetrahydronaphthalene carboxylic acid in the presence of a Pd/C catalyst under D$_2$ atmosphere and deuterated acetic acid as the deuterating agent (Figure 19).

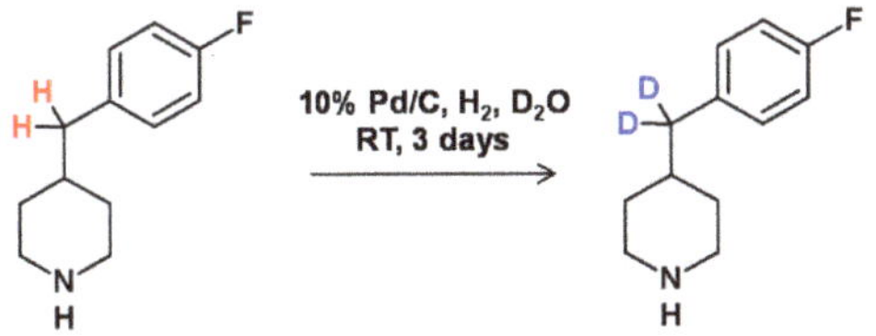

Figure 19. HDX reaction at the carbon center of piperidine derivative as presented by Proszenyák et al. [77].

In 2005, Sajiki and co-workers [79] revealed that platinum catalysts present a higher tendency towards the deuteration of aromatic positions, whereas palladium catalysts prefer mostly aliphatic ones (Figure 20). Using this method, the efficient deuteration of phenol was obtained in the presence of 5% Pt/C at room temperature, whereas the palladium-catalyzed reaction needed a higher temperature (180 °C) to obtain the same level of deuteration.

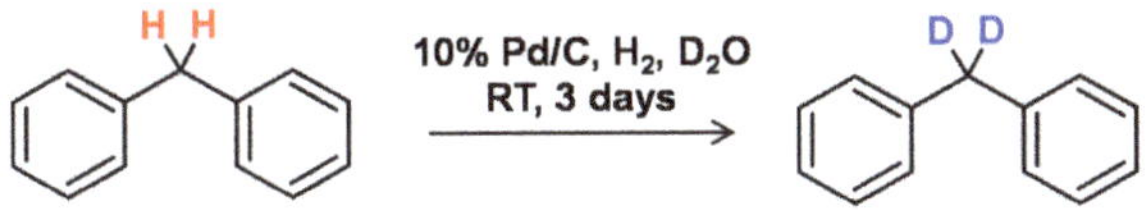

Figure 20. HDX at the carbon center presented by Sajiki et al. [79].

Palladium and platinum catalysts may also be applied in a mixed catalyst system for the preparation of deuteration derivatives on the sterically hindered aromatic positions. It was found that the deuterium incorporation at the *ortho* position in 5-phenylvaleric acid in the presence of palladium (10% Pd/C) was only 14%, with platinum (5% Pt/C) 19%. In a mixed catalyst, the same reaction was characterized by almost complete deuteration (97% D) of the *ortho* position. Additionally, the synergistic effect of palladium and platinum complexes in stepwise deuteration was postulated as a useful tool in the case of a low degree of deuterium incorporation. [80].

HDX at the carbon center of alcohol molecules may also be catalyzed by molybdocenes. In this case, the reaction mechanism involves C–H bond activation by the metal catalyst. Deuteration may occur in different ranges, depending on the chemical structure of the alcohol used and can reach up to 99% in the case of benzylic hydrogens [81].

Most of the presented metal-catalyzed deuteration methods involve the application of second-row transition metals. In 2018, Prakash and co-workers developed a method of the regioselective deuteration of primary and secondary alcohols in the presence of first-row transition-metal catalysts [82]. In 2007 Hamid et al. presented the 'borrowing hydrogen' method of deuteration [83] which allows deuteration at the α and β carbon atoms of alcohols and amines by using a hydrogen-transfer catalyst. It was reported that manganese and iron pincer catalysts increase the deuterium incorporation in primary and secondary alcohols especially in the presence of a base. The presented mechanism of HDX involves the amido-complex formation with the used base which is a key factor for catalytic cycle initiation. The formed aldehyde is subsequently reduced by the deuterated metal complex formed by HDX on the catalyst mediated by D_2O. This method was successfully applied in the deuteration on the α and β carbons of primary and secondary alcohols, as well as diols.

As presented by Bergman et al., the cationic iridium complexes may also activate C-H bonds [84], and therefore iridium-mediated HDX represents the largest number of published examples in the field of homogeneous metal catalysis. The most exploited area is the ortho-deuteration of aryl ketones and acetanilides. Starting with the investigations of the Heys [85] and Hesk [86] research groups, several different studies related to the effects of complex ligands [87], the deuterating agent [88], solvent [88–90], addition of bases [85] the amount of catalyst [36], the temperature, and the duration of the reaction [85] on the degree of deuteration and the substitution pattern in the substrate (Figure 21).

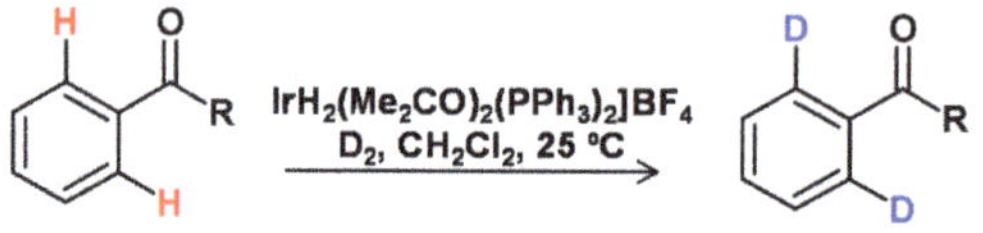

Figure 21. H/D exchange catalyzed by iridium catalyst as shown by Heys [85].

Further, Kröger et al. presented that the unsaturated carbonyl compounds were also suitable substrates for the above presented deuteration method reacting through a similar mechanism [88]. It was shown that β-hydrogen atoms underwent the H/D exchange with a good yield. It was also pointed out that the regioselectivity of the labeling depended on which deuterium source was used.

Buchanan et al., who developed catalysts for C-H bond activation [84] also presented the applicability of the soluble iridium complexes for the specific deuterium incorporation in aliphatic and nonfunctionalized aromatic substrates. A high degree of deuterium incorporation was obtained with hydrocarbons, alcohols, phenols, ethers, carboxylic acids, esters, and amides with D_2O, [D6]acetone, or [D6]benzene [91].

In 2006, Peris et al. demonstrated efficient HDX with diethyl ether, ethyl methyl ketone, isopropanol, and styrene with the *N*-heterocyclic iridium–carbene complexes in the presence of [D4]methanol (Figure 22) [92].

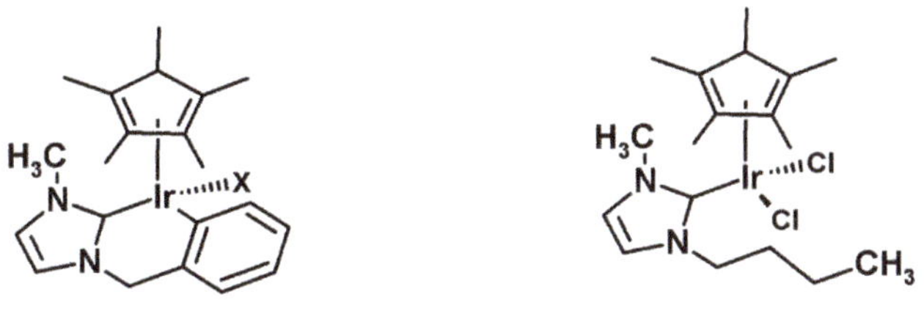

Figure 22. Iridium-carbene complexes presented by Peris et al. [92].

2.3. Metal-Catalyzed HDX Adjacent to Nitrogen Atom

A first example of deuterium incorporation into the carbon centers in amines was reported in 1977, by Maeda et al. [93]. The hydrogen−deuterium exchange on the carbon atoms of primary and tertiary amines was performed in the presence of a deuterated form of Adam's catalyst (platinum oxide treated under reductive conditions by D_2 in D_2O) (Figure 23). Such a solution allowed the selective deuteration at the β carbon of primary amines and at the α-C of the tertiary. The applied catalyst was activated by UV light or γ-radiation irradiation. PtO_2, after activation with D_2, was applied for the selective deuterium incorporation in nucleosides [94]. A strong dependency of the exchange selectivity upon the number and steric demand of the substituents on the nitrogen atom has been observed for the exchange of α-hydrogen atoms of aliphatic amines and amino acids with Adam's catalyst ($PtO_2 \cdot H_2O$). It was found that the nitrogen atom bound to the surface of the applied catalyst. The efficiency of H/D exchange decreased in the following series tertiary > secondary > primary amines [93]. A first example of chiral carbon atom deuteration in amines with retention of configuration was presented by Jere et al. in 2003 [95]. In this case, the HDX of alanine or alaninol with complete retention was performed in the presence of ruthenium on carbon under D_2 in D_2O.

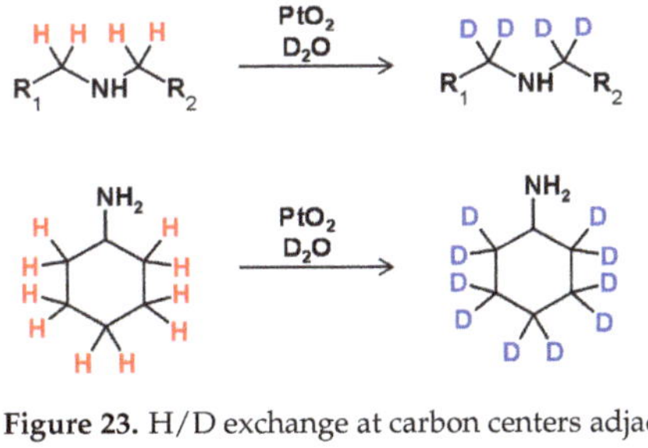

Figure 23. H/D exchange at carbon centers adjacent to nitrogen atom observed by Maeda et al. [93].

The possibility of secondary amine deuteration via H/D exchange in the presence of $RuCl_2(PPh_3)_3$ as the catalyst was investigated by Matsubara et al. [57]. It was found that the applied reaction conditions allowed selective deuteration at the α carbon atom with deuterium incorporations of up to 94% (Figure 24) The proposed reaction conditions were also applied in the deuteration of tertiary amines, resulting in 12% deuterium incorporation. As described, there was a possibility of selective amine deuteration at the α carbon atom under reaction conditions similar to those presented for alcohols. The configuration on the stereocenter in the β position depended on the temperature isotope exchange process and remained unaffected as long as the temperature did not exceed 100 °C [57].

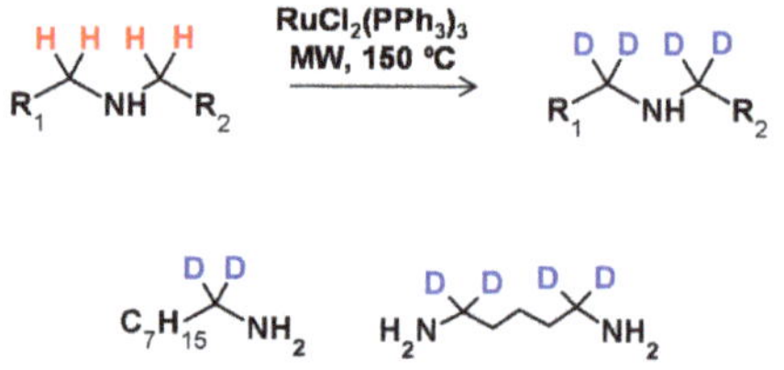

Figure 24. HDX at the carbon centers adjacent to nitrogen proposed by Matsubara and co-workers with selected examples [57].

In 2015, Taglang and co-workers, reported that hydrogen−deuterium exchange at the chiral carbon centers of amino acid derivatives turned out to be stereoretentive (Figure 25) [96]. The authors proposed a mechanism of reaction in which the key step is the coordination of the nitrogen atom of the amino group to the surface of the applied catalyst which makes a ruthenium site and enables the C–H activation at the α carbon atom. A molecule thus-activated undergoes HDX, resulting in a selective α carbon deuterated derivative.

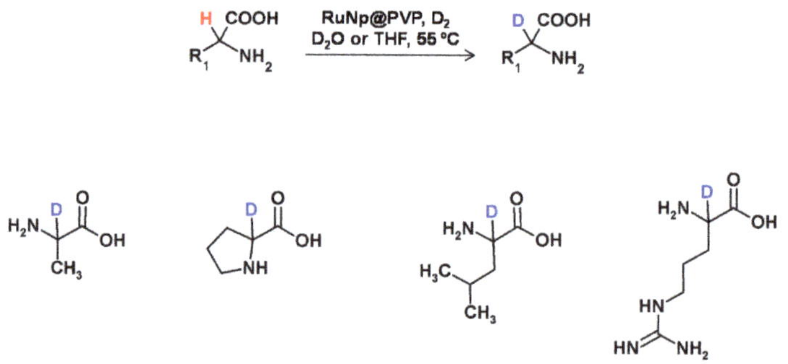

Figure 25. Schematic presentation of H/D exchange using ruthenium nanoparticles presented by Taglang et al. [96] with selected examples.

Alexakis et al. in 2005 analyzed different ruthenium(II) catalysts in the deuterium labeling reactions of piperidines, piperazines and several different dialkylamines in the presence of D_2O as the deuterium source [97]. They found that while $RuCl_2(PPh_3)_3$ was active for the deuteration of primary alcohols and amines, in the presence of $RuCl_4(CO)_6$ only secondary amine labeling was possible. The study on the deuteration of 4-benzylpiperidine in the presence of $[Ru_2Cl_4(CO)_6]$ as catalyst revealed the incorporation of an average of five deuterium atoms per molecule, however, the positions of the introduced deuterons were not precisely described [97].

The study of the increase of the deuterated yield led Roche's group [98] to demonstrate that similar deuteration as shown by Taglang and co-workers [96] could be achieved in the presence of a Ru/C catalyst instead of Ru nanoparticles, H_2 atmosphere in D_2O and a base under mild temperature (70 °C) [99]. The presented methodology allowed the stereoretentive preparation of large quantities (up to 0.2 mols) of hydrosoluble amines with chiral centers, showing the applicability of this process for industrial application.

After successful deuteration of α, β carbons of alcohols in the presence of Ru-pincer complexes, Gunanathan and co-workers [100] presented the α deuteration of amines and amino acids by using D_2O as deuterium source. The presented protocol is characterized by its selectivity for primary and secondary amines (no deuteration occurs at the α carbon of tertiary amines or alcohols). In the case of amino acids epimerization was observed.

In 2012, Neubert et al. applied 'borrowing hydrogen' catalysis to perform the deuteration of tertiary amines [101]. In this study the Shvo catalyst was used to optimize the reaction conditions (Figure 26) [102]. The mechanism involved the formation of an intermediate in the form of iminium ion (in equilibrium with the corresponding enamine) mediated by the monomer of the catalyst bearing an available position in the coordination sphere of the metal atom. The reduced form of the applied catalyst underwent HDX in the presence of deuterating agent (D$_2$O, deuterated alcohols) and subsequently transferred the deuterium atoms to the unsaturated bond of the enamine, generating a doubly labeled tertiary amine. It was found in the case of N,N,N-trihexylamine as the substrate that the deuterium incorporation was as high as 93% at the α and β positions in the presence of isopropanol-d8 as the solvent. This protocol was also successfully applied in the formation of deuterated pharmaceutical compounds, such as sunitinib (a kinase inhibitor).

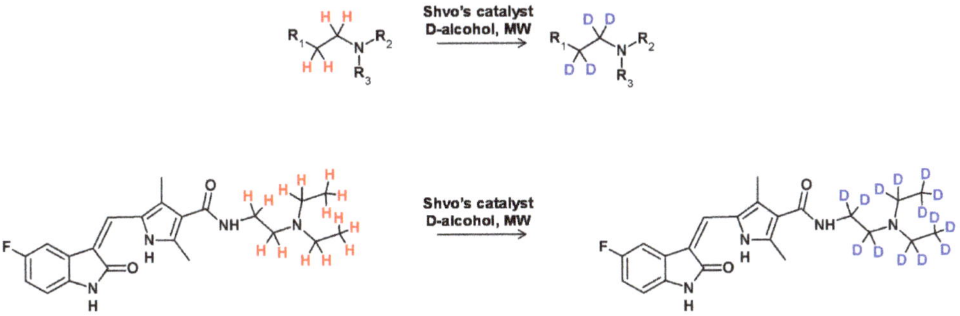

Figure 26. Schematic presentation of HDX in the presence of Shvo's catalyst [102].

In 2014, Pieters et al. developed a method of regioselective α deuteration of nitrogen-containing bioactive compounds in the presence of ruthenium nanoparticles supported on polyvinylpyrrolidone under the D$_2$ atmosphere [103] (Figure 27). Using this method, the authors obtained high deuterium incorporation with complete regioselectivity.

In 2016, Jackson et al. [105] developed a method of electrocatalytic deuteration of amines and alcohols which overcame the problem of the D$_2$ solubility in an aqueous environment, by in situ generation of D$_2$. Using this method, different deuterated derivatives were prepared with high level of deuterium incorporation, but with low yield.

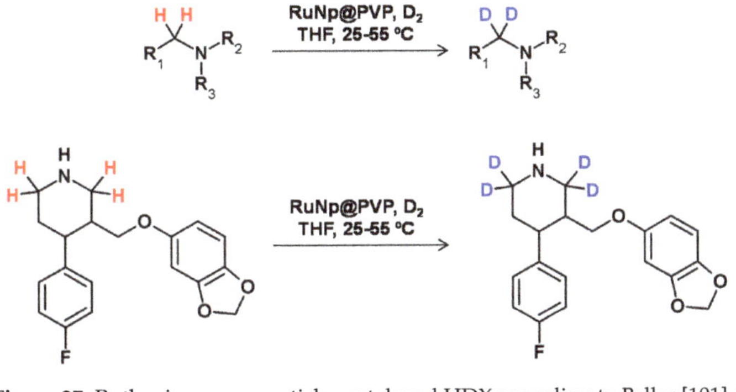

Figure 27. Ruthenium nanoparticles-catalyzed HDX according to Beller [101] and Pieters [103].

In 2016, Hale and Szymczak proposed a method of stereoretentive deuteration in the presence of Ru-bMepi complex that avoided the use of D$_2$. In this method the complete stereoretention in the case of a chiral amine was observed. However, in the case of chiral alcohol deuteration, complete racemization occurred (Figure 28) [104].

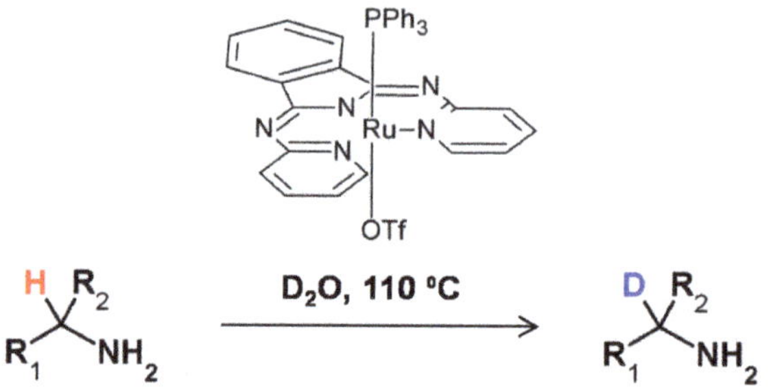

Figure 28. Model metal-catalyzed HDX reactions presented by Szymczak et al. [104].

A main challenge in this field was the development a stereoretentive hydrogen−deuterium exchange at the chiral carbon centers. Although the Rousseau reaction can be considered as stereospecific, an example of deuteration at the chiral carbon atom was presented in his paper. Maeda et al. also investigated the stereospecificity of the HDX in amines and found that at temperatures above 100 °C, L-alanine started to racemize [94]. During the analysis of selective HDX at the β carbon of phenylalanine in the presence of Pd/C catalyst under H_2, the additional deuteration at the α carbon was observed at 160°C, with partial racemization of the chiral carbon atom.

Pyrimidine bases, including uracil or cytosine, may also be deuterated in the 5- and 6-positions in the presence of a Pd/C–H_2/D_2O mixture at 110 °C [106]. The deuterium was incorporated in the 5-methyl group of thymine in addition to the 6-position, and no side products have been reported. Purine nucleosides, including adenosine or inosine, were successfully and chemoselectively deuterated in the 2- and 8-positions [107]. The lower HDX levels were noted for the pyrimidine bases and analyzed nucleoside in the presence of CD_3OD as a solvent and deuterating agent in the form of D_2O.

In 2001, Hardacre and co-workers reported the application of a catalyst activated by hydrogen reduction in the preparation of deuterated imidazoles and imidazole salts. The substrate solution in D_2O was treated by the reaction mixture, degassed by several cooling/thawing cycles [108].

Raney nickel catalyst in the presence of [D6]acetone, [D3]acetonitrile, [D1]chloroform, D_2O or [D8] 2-propanol allowed the selective deuteration of specific positions in tryptophan derivatives, based on their nucleophilicity [109]. The influence of the differing nucleophilic potential of these positions and the indole ring orientation on the catalyst surface influenced by the solvent were responsible for the selectivity of deuteration (Figure 29). In the case of deuteration catalyzed by Raney nickel, it was found that only the hydrogen atoms at the α-carbons underwent selective deuteration in quinuclidine at higher temperatures, a longer reaction time and in the presence of D_2O (100 °C, 40 h, 2 reaction cycles; ≥99.7% D). It was also reported that less than 1% of deuterium was incorporated into the β and γ carbon centers [110].

Previously Hickey et al. reported the extended spectrum of the substrate to aniline and benzylamine derivatives [111]. The application of [Ir(cod)(acac-F6)] (acac-F6= hexafluoroacetylacetonate) and D_2 allowed *ortho* HDX, relative to the position of the amino or methylamino group.

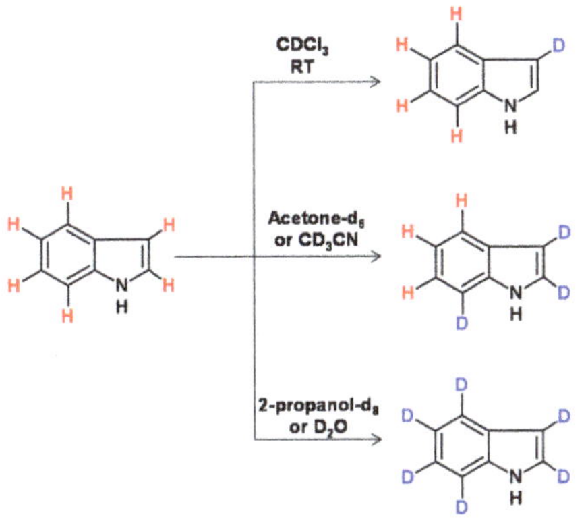

Figure 29. H/D exchange described by Gavrisch Yau and Gawrisch [109].

In 2017, Loh et al. [112] proposed a new deuteration strategy, based on the application of photoredox deuteration and tritiation. The method was successfully applied in the formation of isotopically labeled derivatives of pharmaceuticals containing alkyl amine moieties. The developed method involved the formation of an α-amino radical via a single-electron transfer occurred in the presence of an iridium(III) catalyst, previously excited by visible blue light. The application of a hydrogen atom transfer (HAT) catalyst allowed the abstraction of deuterium (or tritium) to form labeled derivatives. It was also found that application of thiols as HAT catalysts was crucial for the preparation of labeled compounds. This process was optimized and then applied in the preparation of gram scale products. All of the reported examples met all the requirement characteristic for internal standards which may be applied in quantitative analysis by mass spectrometry (more than 4 D and less than 0.1% of the unlabeled compound). Moreover, the proposed process may be performed in the presence of several functional groups, presenting a high selectivity towards exchange at the $C(sp^3)$–H centers adjacent to the nitrogen atom and retention of the stereochemistry, even when HDX proceeds at the chiral center. In 2015, Hu and co-workers described an example of deuteration based on the single-electron transfer mechanism [113]. In the presented process, deuteration took place only at the α carbon atom of the amine when the radical was stabilized at a benzylic position. It was also found that the presence of the nitrogen was not required for the deuterium incorporation.

2.4. pH Dependent and Metal-Catalyzed HDX Adjacent to Sulfur Atom

Sulfur-directed deuteration is not so common in scientific literature. Only a few papers report deuteration at the α carbon position of sulfur center. In 2018, the application of ruthenium on carbon under D_2 in the H/D exchange in thioesters was presented by Gao et al. [114] By using this method a number of deuterium-labeled drugs or amino acids and peptides have been successfully prepared with a good regioselectivity towards the α carbon to the sulfur center. Additionally, in the case of chiral molecules involved in this process, the retention of the configuration was found. It was also suggested that the mechanism of the described process was the same as that proposed for the ruthenium nanoparticle-mediated deuteration of amino acids [96].

A great review presenting the possibility of pH dependent and metal-catalyzed hydrogen−deuterium exchange adjacent to the sulfur atom was presented by Michelotti and co-workers [115]. Rauk et al. [116] demonstrated a deuteration process in the presence of NaOD in D_2O or just in D_2O for substances containing the methylsulfonyl or methylsulfinyl groups. They reported that sulfonyl (-SO_2-) and sulfinyl (-SO-) groups should present similar conjugative ability with adjacent carbanionic centers (Figure 30). It was also presented that -SO- group was much less effective than a -SO_2- group in promoting deuteration at the α carbon center. The previously reported HDX of dimethyl sulfoxide in the presence of deuteroxide may have resulted from the enhanced reactivity of applied bases in dimethyl sulfoxide solution. Additionally, an unexpected difference in the deuteration level at the two methylene hydrogens was found [101].

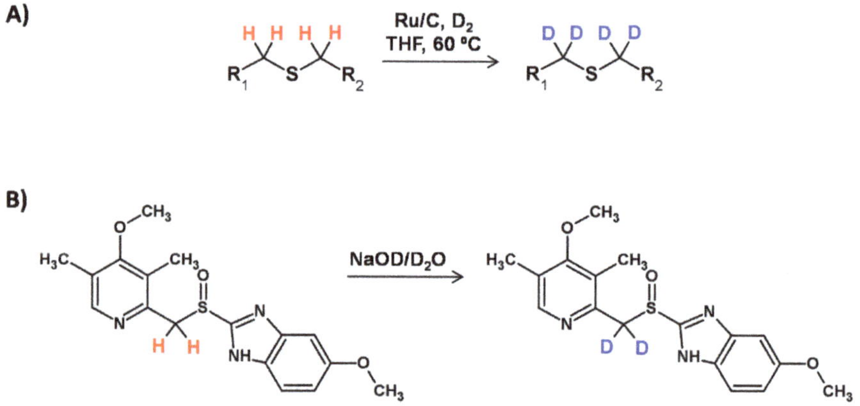

Figure 30. Schematic presentation of H/D exchange at the carbon center connected with the sulfur atom presented by Gao et al. [114] (**A**) and Rauk et al. [116] (**B**).

The possibility of complete deuteration of methylene hydrogens adjacent to the sulfinyl group was presented by Redondo and co-workers [117]. The reaction proceeded at room temperature when the sodium salts of compounds containing [(pyridylmethyl)sulfinyl] benzimidazole structural core were dissolved in a solvent serving as a deuterium source, including D_2O and CD_3OD. The presented process resulted from the weak acidity of the methylene hydrogen atoms, and was also observed in a nondeuterating solvent like DMSO-d_6 in the presence of aa catalytic amount of NaOH. The described HDX was monitored by ^1H NMR which also revealed the stereoselectivity of deuterium incorporation [116,118].

Transition metal-catalyzed H/D exchange can be classified into three categories. The first one involves the C−H bond activation which generates an organometallic intermediate, bearing a metal−carbon bond as a result of the carbon−hydrogen bond activation. The second one is denoted as C−H insertion catalysis as a key working mode, whereas the third one is related to the photoredox catalysis [119].

The advantages of metal catalyzed hydrogen−deuterium exchange processes are that they are usually easy to perform and most of the presented catalysts are commercially available. The limitation of the presented methods is the high price of the catalyst, necessity of compound purification, long reaction times or the necessity for the application of high pressure.

3. Conclusions

The presented manuscript provides an overview of the methods of H/D exchange reactions at the carbon centers. From the first reports of HDX at the carbon center in alcohols and amines, the state-of-the-art in this field has expanded enormously, mostly due to the application of metal catalysis. proving its validity and utility. Nowadays, the deuterated compounds play a crucial role as the internal standards in the quantitative analysis by the LC-MS technique. Although the possibilities of deuterated standards in quantitative investigation of compounds are beyond doubt, their application in medical diagnosis may be limited, due to the high costs of their preparation. Therefore, despite significant progress there is still a strong need for the development of new, simple, rapid and cost-efficient methods of carbon deuteration. It may be speculated that the new techniques of hydrogen−deuterium exchange at the carbon centers will be developed in the near future. It may be expected that they will be characterized by a low price of preparation, the stability of the introduced deuterons, long-term storage of the deuterated derivatives at room temperature, an appropriate mass difference for MS quantification, stability of introduced deuterons during chromatographic separation and co-elution with nondeuterated isotopologues. It may be expected that further work in the presented area will be forthcoming.

Author Contributions: Conceptualization, R.B.; writing—original draft preparation, P.G., R.B. The manuscript was written through contributions of all authors. Both authors commented on the manuscript. Both authors have read and agreed to the published version of the manuscript.

Funding: This research was funded by NATIONAL SCIENCE CENTRE POLAND, grant number UMO-2016/23/B/ST4/01036.

Institutional Review Board Statement: Not applicable.

Informed Consent Statement: Not applicable.

Data Availability Statement: Not applicable.

Acknowledgments: The authors would like to thank Andrzej Reszka (Shim-Pol, Poland) for providing access to LCMS-8050, LCMS-9030 and LCMS IT-TOF instruments.

Conflicts of Interest: The authors declare no conflict of interest.

References

1. Englander, S.W.; Sosnick, T.R.; Englander, J.J.; Mayne, L. Mechanisms and uses of hydrogen exchange. *Curr. Opin. Struct. Biol.* **1996**, *6*, 18–23. [CrossRef]
2. Bai, Y.; Milne, J.S.; Mayne, L.; Englander, S.W. Primary structure effects on peptide group hydrogen exchange. *Proteins* **1993**, *17*, 75–86. [CrossRef]
3. Junk, T.; Catallo, W.J. Hydrogen isotope exchange reactions involving C-H (D, T) bonds. *Chem. Soc. Rev.* **1997**, *26*, 401–406. [CrossRef]
4. Szewczuk, Z.; Konishi, Y.; Goto, Y. A two-process model describes the hydrogen exchange behavior of cytochrome c in the molten globule state with various extents of acetylation. *Biochemistry* **2001**, *40*, 9623–9630. [CrossRef]
5. Ho, J.; Coote, M.L.; Easton, C.J. Validation of the distal effect of electron-withdrawing groups on the stability of peptide enolates and its exploitation in the controlled stereochemical inversion of amino acid derivatives. *J. Org. Chem.* **2011**, *76*, 5907–5914. [CrossRef] [PubMed]
6. Lygo, B.; Humphreys, D.L. Enantioselective synthesis of α-carbon deuterium-labelled L- α-amino acids. *Tetrahedron Lett.* **2002**, *43*, 6677–6679. [CrossRef]
7. Lankiewicz, L.; Nyasse, B.; Fransson, B.; Grehn, L.; Ragnarsson, U. Synthesis of amino acid derivatives substituted in the backbone with stable isotopes for application in peptide synthesis. *J. Chem. Soc. Perkin Trans. 1* **1994**, *17*, 2503–2510. [CrossRef]
8. Rose, J.E.; Leeson, P.D.; Gani, D. Stereospecific synthesis of α-deuteriated α-amino acids: Regiospecific deuteriation of chiral 3-isopropyl-2,5-dimethoxy-3,6-dihydropyrazines. *J. Chem. Soc. Perkin Trans. 1* **1995**, *2*, 1563–1565. [CrossRef]
9. Jung, P.G.; Kim, B.; Park, S.-R.; So, H.-Y.; Shi, L.H.; Kim, Y. Determination of serum cortisol using isotope dilution-liquid chromatography-mass spectrometry as a candidate reference method. *Anal. Bioanal. Chem.* **2004**, *380*, 782–788. [CrossRef]
10. Gries, W.; Küpper, K.; Leng, G. Rapid and sensitive LC–MS–MS determination of 2-mercaptobenzothiazole, a rubber additive, in human urine. *Anal. Bioanal. Chem.* **2015**, *407*, 3417–3423. [CrossRef]
11. Brun, V.; Masselon, C.; Garin, J.; Dupuis, A. Isotope dilution strategies for absolute quantitative proteomics. *J. Proteom.* **2009**, *72*, 740–749. [CrossRef]
12. Guo, K.; Ji, C.; Li, L. Stable-isotope dimethylation labeling combined with LC-ESI MS for quantification of amine-containing metabolites in biological samples. *Anal. Chem.* **2007**, *79*, 8631–8638. [CrossRef]
13. Di Palma, S.; Raijmakers, R.; Heck, A.J.; Mohammed, S. Evaluation of the deuterium isotope effect in zwitterionic hydrophilic interaction liquid chromatography separations for implementation in a quantitative proteomic approach. *Anal. Chem.* **2011**, *83*, 8352–8356. [CrossRef] [PubMed]
14. Turowski, M.; Yamakawa, N.; Meller, J.; Kimata, K.; Ikegami, T.; Hosoya, K.; Tanaka, N.; Thornton, E.R. Deuterium isotope effects on hydrophobic interactions: The importance of dispersion interactions in the hydrophobic phase. *J. Am. Chem. Soc.* **2003**, *125*, 13836–13849. [CrossRef]
15. Iyer, S.S.; Zhang, Z.P.; Kellogg, G.E.; Karnes, H.T. Evaluation of deuterium isotope effects in normal-phase LC–MS–MS separations using a molecular modeling approach. *J. Chromatogr. Sci.* **2004**, *42*, 383–387. [CrossRef]
16. Atzrodt, J.; Derdau, V.; Fey, T.; Zimmermann, J. The renaissance of H/D exchange. *Angew. Chem. Int. Ed.* **2007**, *46*, 7744–7765. [CrossRef]

17. Evchenko, S.V.; Kamounah, F.S.; Schaumberg, K. Efficient synthesis of deuterium-labelled ferrocenes. *J. Label. Compd. Radiopharm.* **2005**, *48*, 209–218. [CrossRef]
18. Kusumoto, T.; Sato, K.; Kumaraswamy, G.; Hiyama, T.; Isozaki, T.; Suzuki, Y. Synthesis and properties of deuterated antiferroelectric liquid crystals. *Chem. Lett.* **1995**, *12*, 1147. [CrossRef]
19. Hill, R.K.; Abacherli, C.; Hagishita, S. Synthesis of (2S,4S)- and (2S,4R)-[5,5,5-^{2}H$_3$]leucine from (R)-pulegone. *Can. J. Chem.* **1994**, *72*, 110–113. [CrossRef]
20. Coumbarides, G.S.; Dingjan, M.; Eames, J.; Flinn, A.; Northen, J. An efficient laboratory synthesis of α-deuteriated profens. *J. Label. Compd. Radiopharm.* **2006**, *49*, 903–914. [CrossRef]
21. Castell, J.V.; Martinez, L.A.; Miranda, M.A.; Tarrega, P. A general procedure for isotopic (deuterium) labelling of non-steroidal antiinflammatory 2-arylpropionic acids. *J. Label. Compd. Radiopharm.* **1994**, *34*, 93–100. [CrossRef]
22. Shibasaki, H.; Furuta, T.; Kasuya, Y. Preparation of multiply deuterium-labeled cortisol. *Steroids* **1992**, *57*, 13–17. [CrossRef] [PubMed]
23. Scheigetz, J.; Berthelette, C.; Li, C.; Zamboni, R.J. Base-catalyzed deuterium and tritium labeling of aryl methyl sulfones. *J. Label. Compd. Radiopharm.* **2004**, *47*, 881–889. [CrossRef]
24. Rios, A.; Richard, J.P. Biological enolates: Generation and stability of the enolate of N-protonated glycine methyl ester in water. *J. Am. Chem. Soc.* **1997**, *119*, 8375–8376. [CrossRef]
25. Bohak, Z.; Katchalski, E. Synthesis, characterization, and racemization of poly-L-serine. *Biochemistry* **1963**, *2*, 228–237. [CrossRef]
26. Fridkin, M.; Wilchek, M. NMR studies of H-D exchange of α-CH group of amino acid residues in peptides. *Biochem. Biophys. Res. Commun.* **1970**, *38*, 458–464. [CrossRef]
27. Elemes, Y.; Ragnarsson, U. Synthesis of enantiopure α-deuteriated Boc-L-amino acids. *J. Chem. Soc. Perkin Trans. 1* **1996**, *6*, 537–540. [CrossRef]
28. Ho, J.; Easton, C.J.; Coote, M.L. The distal effect of electron-withdrawing groups and hydrogen bonding on the stability of peptide enolates. *J. Am. Chem. Soc.* **2010**, *132*, 5515–5521. [CrossRef]
29. Rios, A.; Amyes, T.L.; Richard, J.P. Formation and stability of organic zwitterions in aqueous solution: Enolates of the amino acid glycine and its derivatives. *J. Am. Chem. Soc.* **2000**, *122*, 9373–9385. [CrossRef]
30. Rios, A.; Amyes, T.L.; Richard, J.P. Formation and stability of peptide enolates in aqueous solution. *J. Am. Chem. Soc.* **2002**, *124*, 8251–8259. [CrossRef]
31. Moozeh, K.; So, S.M.; Chin, J. Catalytic stereoinversion of L-alanine to deuterated D-alanine. *Angew. Chem. Int. Ed.* **2015**, *54*, 9381. [CrossRef] [PubMed]
32. Mitulovi, G.; Lämmerhofer, M.; Maier, N.M.; Lindner, W. Simple and efficient preparation of (R)- and (S)-enantiomers of α-carbon deuterium-labelled α-amino acids. *J. Label. Compd. Radiopharm.* **2000**, *43*, 449–461. [CrossRef]
33. Oppolzer, W.; Pedrosa, R.; Moretti, R. Asymmetric syntheses of α-amino acids from α-halogenated 10-sulfonamido-isobornyl esters. *Tetrahedron Lett.* **1986**, *27*, 831–834. [CrossRef]
34. Schöllkopf, U.; Hartwig, W.; Groth, U. Enantioselektive synthese von α-methyl-α-amino-carbonsäuren durch alkylierung des lactimethers von cyclo-(L-Ala-L-Ala). *Angew. Chem.* **1979**, *91*, 922–923. [CrossRef]
35. Beak, P.; Brown, R.A. The tertiary amide as an effective director of ortho lithiation. *J. Org. Chem.* **1982**, *47*, 34–46. [CrossRef]
36. Edlund, U.; Bergson, G. Enamines from 1-Methyl-2-indanone. *Acta Chem. Scand.* **1971**, *25*, 3625–3631. [CrossRef]
37. Pacchioni, M.; Bega, A.; Faretti, A.C.; Rovai, D.; Cornia, A. Post-synthetic isotopic labeling of an azamacrocyclic ligand. *Tetrahedron Lett.* **2002**, *43*, 771–774. [CrossRef]
38. Barthez, J.M.; Filikov, A.V.; Frederiksen, L.B.; Huguet, M.-L.; Jones, J.R.; Lu, S.-Y. Microwave enhanced deuteriations in the solid state using alumina doped sodium borodeuteride. *Can. J. Chem.* **1998**, *76*, 726–728. [CrossRef]
39. Anto, S.; Getvoldsen, G.S.; Harding, J.R.; Jones, J.R.; Lu, S.-Y.; Russell, J.C. Microwave-enhanced hydrogen isotope exchange studies of heterocyclic compounds. *J. Chem. Soc. Perkin Trans.* **2000**, *2*, 2208–2211. [CrossRef]
40. de Keczer, S.A.; Lane, T.S.; Masjedizadeh, M.R. Uncatalyzed microwave deuterium exchange labeling of bleomycin A$_2$. *J. Label. Compd. Radiopharm.* **2004**, *47*, 733–740. [CrossRef]
41. Bai, S.; Palmer, B.J.; Yonker, C.R. Kinetics of deuterium exchange on resorcinol in D$_2$O at high pressure and high temperature. *J. Phys. Chem. A* **2000**, *104*, 53–58. [CrossRef]
42. Harris, R.J.; Meskys, R.; S0utcliffe, M.J.; Scrutton, N.S. Kinetic studies of the mechanism of carbon-hydrogen bond breakage by the heterotetrameric sarcosine oxidase of Arthrobacter sp. 1-IN. *Biochemistry* **2000**, *39*, 1189–1198. [CrossRef] [PubMed]
43. Cohen, D.J.; Loertscher, R.; Rubin, M.F.; Tilney, N.L.; Carpenter, C.B.; Strom, T.B. Cyclosporine—A new immunosuppressive agent for organ-transplantation. *Ann. Intern. Med.* **1984**, *101*, 667–682. [CrossRef] [PubMed]
44. Tszyrsznic, W.; Borowiec, A.; Pawlowska, E.; Jazwiec, R.; Zochowska, D.; Bartlomiejczyk, I.; Zegarska, J.; Paczek, L.; Dadlez, M. Two rapid ultra performance liquid chromatography/tandem mass spectrometry (UPLC/MS/MS) methods with common sample pretreatment for therapeutic drug monitoring of immunosuppressants compared to immunoassay. *J. Chromatogr. B* **2013**, *928*, 9–15. [CrossRef] [PubMed]
45. Bąchor, R.; Setner, B.; Kluczyk, A.; Stefanowicz, P.; Szewczuk, Z. The unusual hydrogen-deuterium exchange of α-carbon protons in N-substituted glycine-containing peptides. *J. Mass Spectrom.* **2014**, *49*, 43–49. [CrossRef]
46. Bąchor, R.; Kluczyk, A.; Stefanowicz, P.; Szewczuk, Z. Preparation of novel deuterated cyclosporin A standards for quantitative LC-MS analysis. *J. Mass Spectrom.* **2017**, *52*, 817–822. [CrossRef]

47. Bąchor, R.; Dębowski, D.; Łęgowska, A.; Stefanowicz, P.; Rolka, K.; Szewczuk, Z. Convenient preparation of deuterium labeled analogs of peptides containing N-substituted glycines for a stable isotope dilution LC-MS quantitative analysis. *J. Pept. Sci.* **2015**, *21*, 819–825. [CrossRef]
48. Bąchor, R.; Konieczny, A.; Szewczuk, Z. Preparation of isotopically labelled standarts of creatinine via H/D exchange and their application in quantitative analysis by LC-MS. *Molecules* **2020**, *25*, 1514. [CrossRef]
49. Bąchor, R.; Kluczyk, A.; Stefanowicz, P.; Szewczuk, Z. Facile synthesis of deuterium-labeled denatonium cation and its application in the quantitative analysis of Bitrex by liquid chromatography-mass spectrometry. *Anal. Bioanal. Chem.* **2015**, *407*, 6557–6561. [CrossRef]
50. Bąchor, R.; Rudowska, M.; Kluczyk, A.; Stefanowicz, P.; Szewczuk, Z. Hydrogen-deuterium exchange of α-carbon protons and fragmentation pathways in N-methylated glycine and alanine-containing peptides derivatized by quaternary ammonium salts. *J. Mass Spectrom.* **2014**, *49*, 529–536. [CrossRef]
51. Bąchor, R.; Kluczyk, A.; Stefanowicz, P.; Szewczuk, Z. Synthesis and mass spectrometry analysis of quaternary cryptando-peptidic conjugates. *J. Pept. Sci.* **2015**, *21*, 879–886. [CrossRef] [PubMed]
52. Setner, B.; Stefanowicz, P.; Szewczuk, Z. Quaternary ammonium isobaric tag for a relative and absolute quantification of peptides. *J. Mass Spectrom.* **2018**, *53*, 115–123. [CrossRef] [PubMed]
53. Smith, G.G.; Evans, R.C.; Baum, R. Neighboring residue effects: Evidence for intramolecular assistance to racemization or epimerization of dipeptide residues. *J. Am. Chem. Soc.* **1986**, *108*, 7327–7332. [CrossRef]
54. Smith, G.G.; Baum, R. First-order rate constants for the racemization of each component in a mixture of isomeric dipeptides and their diketopiperazine. *J. Org. Chem.* **1987**, *52*, 2248–2255. [CrossRef]
55. Regen, S.L. Ruthenium-catalyzed hydrogen-deuterium exchange in alcohols. Method for deuterium labeling of primary alcohols. *J. Org. Chem.* **1974**, *39*, 260–261. [CrossRef]
56. Ishibashi, K.; Takahashi, M.; Yokota, Y.; Oshima, K.; Matsubara, S. Ruthenium-catalyzed isomerization of alkenol into alkanone in water under irradiation of microwaves. *Chem. Lett.* **2005**, *34*, 664–665. [CrossRef]
57. Takahashi, M.; Oshima, K.; Matsubara, S. Ruthenium catalyzed deuterium labelling of alpha-carbon in primary alcohol and primary/secondary amine in D_2O. *Chem. Lett.* **2005**, *34*, 192–193. [CrossRef]
58. Bossi, G.; Putignano, E.; Rigo, P.; Baratta, W. Pincer Ru and Os complexes as efficient catalysts for racemization and deuteration of alcohols. *Dalton Trans.* **2011**, *40*, 8986–8995. [CrossRef]
59. Khaskin, E.; Milstein, D. Simple and efficient catalytic reaction for the selective deuteration of alcohols. *ACS Catal.* **2013**, *3*, 448–452. [CrossRef]
60. Bai, W.; Lee, K.-H.; Tse, S.K.S.; Chan, K.W.; Lin, Z.; Jia, G. Ruthenium-catalyzed deuteration of alcohols with deuterium oxide. *Organometallics* **2015**, *34*, 3686–3698. [CrossRef]
61. Chatterjee, B.; Gunanathan, C. Ruthenium catalyzed selective α- and α,β-deuteration of alcohols using D_2O. *Org. Lett.* **2015**, *17*, 4794–4797. [CrossRef]
62. Zhang, L.; Nguyen, D.H.; Raffa, G.; Desset, S.; Paul, S.; Dumeignil, F.; Gauvin, R.M. Efficient deuterium labelling of alcohols in deuterated water catalyzed by ruthenium pincer complexes. *Catal. Commun.* **2016**, *84*, 67–70. [CrossRef]
63. Koch, H.J.; Stuart, R.S. A novel method for specific labelling of carbohydrates with deuterium by catalytic exchange. *Carbohydr. Res.* **1977**, *59*, C1–C6. [CrossRef]
64. Harness, J.; Hughes, N.A. Epimerisations accompanying the reductive desulphurisation of some 5-S-alkyl-5-thiopentose dialkyl dithioacetals. *J. Chem. Soc. Perkin Trans. 1* **1972**, 38–41. [CrossRef]
65. Cioffi, E.A.; Bell, R.H.; Le, B. Microwave-assisted C–H bond activation using a commercial microwave oven for rapid deuterium exchange labeling (C–H→C–D) in carbohydrates. *Tetrahedron Asymmetry* **2005**, *16*, 471–475. [CrossRef]
66. Cioffi, E.A.; Willis, W.S.; Suib, S.L. Ultrasonically induced enhancement of isotope-exchange catalysts: Surface analysis of Raney nickel alloys. *Langmuir* **1988**, *4*, 697–702. [CrossRef]
67. Bokatzian-Johnson, S.S.; Maier, M.L.; Bell, R.H.; Alston, K.E.; Le, B.Y.; Cioffi, E.A. Facile C–H bond activation for deuterium and tritium labeling of glycoconjugates conducted in ultrasonic and microwave fields: A review. *J. Label. Compd. Radiopharm.* **2007**, *50*, 380–383. [CrossRef]
68. Dos Santos, I.; Morgat, J.-L.; Vert, M. Hydrogen isotope exchange as a means of labeling lactides. *J. Label. Compd. Radiopharm.* **1998**, *41*, 1005–1015. [CrossRef]
69. Dos Santos, I.; Morgat, J.-L.; Vert, M. Glycolide deuteriation by hydrogen isotope exchange using the HSCIE method. *J. Label. Compd. Radiopharm.* **1999**, *42*, 1093–1101. [CrossRef]
70. Shevchenko, V.P.; Nagaev, Y.; Myasoedov, N.F.; Susan, A.B.; Switek, K.-H.; Braunger, H. The efficiency of solvent-free catalyst systems in the synthesis of tritium-labelled biologically active compounds. *J. Label. Compd. Radiopharm.* **2006**, *49*, 421–427. [CrossRef]
71. Möbius, G.; Schaaf, G. Verfahren zur Darstellung Deuterierter Organischer Verbindungen. DD279376A3, 6 June 1990.
72. Siegbahn, P.E.M.; Blomberg, M.R.A.; Svenson, M. Theoretical study of the activation of the oxygen-hydrogen bond in water by second-row transition-metal atoms. *J. Phys. Chem.* **1993**, *97*, 2564–2570. [CrossRef]
73. Reardon, P.; Metts, S.; Crittendon, C.; Daugherity, P.; Parsons, E.J. Palladium-catalyzed coupling reactions in superheated water. *Organometallics* **1995**, *14*, 3810–3816. [CrossRef]

74. Matsubara, S.; Yokota, Y.; Oshima, K. Palladium-catalyzed decarboxylation and decarbonylation under hydrothermal conditions: decarboxylative deuteration. *Org. Lett.* **2004**, *6*, 2071–2073. [CrossRef] [PubMed]
75. Sajiki, H.; Aoki, F.; Esaki, H.; Maegawa, T.; Hirota, K. Efficient C−H/C−D exchange reaction on the alkyl side chain of aromatic compounds using heterogeneous Pd/C in D_2O. *Org. Lett.* **2004**, *6*, 1485–1487. [CrossRef] [PubMed]
76. Maegawa, T.; Akashi, A.; Esaki, H.; Aoki, F.; Sajiki, H.; Hirota, K. Efficient and selective deuteration of phenylalanine derivatives catalyzed by Pd/C. *Synlett* **2005**, *5*, 845–847. [CrossRef]
77. Proszenyák, A.; Bela, A.; Tarkanyi, G.; Vida, L.; Faigl, F. Convenient methods for the synthesis of d_4, d_2 and d_6 isotopomers of 4-(4-fluorobenzyl)piperidine. *J. Label. Compd. Radiopharm.* **2005**, *48*, 421–427. [CrossRef]
78. Ofosu-Asante, K.; Stock, L.M. A selective method for deuterium exchange in hydroaromatic compounds. *J. Org. Chem.* **1986**, *51*, 5452–5454. [CrossRef]
79. Sajiki, H.; Ito, N.; Esaki, H.; Maegawa, T.; Hirota, K. Aromatic ring favorable and efficient H–D exchange reaction catalyzed by Pt/C. *Tetrahedron Lett.* **2005**, *46*, 6995–6998. [CrossRef]
80. Ito, N.; Watahaki, T.; Maesawa, T.; Maegawa, T.; Sajiki, H. Synergistic effect of a palladium-on-carbon/platinum-on-carbon mixed catalyst in hydrogen/deuterium exchange reactions of alkyl-substituted aromatic compounds. *Adv. Synth. Catal.* **2006**, *348*, 1025–1028. [CrossRef]
81. Balzarek, C.; Weakley, T.J.R.; Tyler, D.R. C−H bond activation in aqueous solution: kinetics and mechanism of H/D exchange in alcohols catalyzed by molybdocenes. *J. Am. Chem. Soc.* **2000**, *122*, 9427–9434. [CrossRef]
82. Kar, S.; Goeppert, A.; Sen, R.; Kothandaraman, J.; Prakash, G.K.S. Regioselective deuteration of alcohols in D_2O catalysed by homogeneous manganese and iron pincer complexes. *Green Chem.* **2018**, *20*, 2706–2710. [CrossRef]
83. Hamid, M.H.S.A.; Slatford, P.A.; Williams, J.M. Borrowing hydrogen in the activation of alcohols. *J. Adv. Synth. Catal.* **2007**, *349*, 1555–1575. [CrossRef]
84. Buchanan, J.M.; Stryker, J.M.; Bergman, R.G. A structural, kinetic and thermodynamic study of the reversible thermal carbon-hydrogen bond activation/reductive elimination of alkanes at iridium. *J. Am. Chem. Soc.* **1986**, *108*, 1537–1550. [CrossRef]
85. Heys, J.R. Investigation of $[IrH_2(Me_2CO)_2(PPh_3)_2]BF_4$ as a catalyst of hydrogen isotope exchange of substrates in solution. *J. Chem. Soc. Chem. Commun.* **1992**, *9*, 680–681. [CrossRef]
86. Hesk, D.; Das, P.R.; Evans, B. Deuteration of acetanilides and other substituted aromatics using $[Ir(COD)(Cy_3P)(Py)]PF_6$ as catalyst. *J. Label. Compd. Radiopharm.* **1995**, *36*, 497–502. [CrossRef]
87. Kingston, L.P.; Lockley, W.J.S.; Mather, A.N.; Spink, E.; Thompson, S.P.; Wilkinson, D.J. Parallel chemistry investigations of ortho-directed hydrogen isotope exchange between substituted aromatics and isotopic water: Novel catalysis by cyclooctadienyliridium(I)pentan-1,3-dionates. *Tetrahedron Lett.* **2000**, *41*, 2705–2708. [CrossRef]
88. Kröger, J.; Manmontri, B.; Fels, G. Iridium-catalyzed H/D exchange. *Eur. J. Org. Chem.* **2005**, *7*, 1402–1408. [CrossRef]
89. Ellames, G.J.; Gibson, J.S.; Herbert, J.M.; McNeill, A.H. The scope and limitations of deuteration mediated by Crabtree's catalyst. *Tetrahedron* **2001**, *57*, 9487–9497. [CrossRef]
90. Salter, R.; Bosser, I. Application of 1-butyl-3-methylimidazolium hexafluorophosphate to Ir(I)-catalyzed hydrogen isotope exchange labelling of substrates poorly soluble in dichloromethane. *J. Label. Compd. Radiopharm.* **2003**, *46*, 489–498. [CrossRef]
91. Golden, J.T.; Andersen, R.A.; Bergman, R.G. Exceptionally low-temperature carbon−hydrogen/carbon−deuterium exchange reactions of organic and organometallic compounds catalyzed by the $Cp*(PMe_3)IrH(ClCH_2Cl)^+$ cation. *J. Am. Chem. Soc.* **2001**, *123*, 5837–5838. [CrossRef]
92. Corberán, R.; Sanaú, M.; Peris, E. Highly stable Cp*-Ir(III) complexes with N-heterocyclic carbene ligands as C-H activation catalysts for the deuteration of organic molecules. *J. Am. Chem. Soc.* **2006**, *128*, 3974–3979. [CrossRef] [PubMed]
93. Maeda, M.; Ogawa, O.; Kawazoe, Y. Studies on hydrogen exchange. XIV. Selective hydrogen-deuterium exchange in aliphatic amines and amino acids catalyzed by platinum. *Chem. Pharm. Bull.* **1977**, *25*, 3329–3333. [CrossRef]
94. Maeda, M.; Kawazoe, Y. Chemical alteration of nucleic acids and their components. XI. Hydrogen-deuterium exchange of nucleosides and nucleotides catalyzed by platinum. *Tetrahedron Lett.* **1975**, *16*, 1643–1646. [CrossRef]
95. Jere, F.T.; Miller, D.J.; Jackson, J.E. Stereoretentive C−H bond activation in the aqueous phase catalytic hydrogenation of amino acids to amino alcohols. *Org. Lett.* **2003**, *5*, 527–530. [CrossRef] [PubMed]
96. Taglang, C.; Martínez-Prieto, L.M.; del Rosal, I.; Maron, L.; Poteau, R.; Philippot, K.; Chaudret, B.; Perato, S.; Lone, A.S.; Puente, C.; et al. Enantiospecific C–H Activation Using Ruthenium Nanocatalysts. *Angew. Chem. Int. Ed.* **2015**, *54*, 10474–10477. [CrossRef]
97. Alexakis, E.; Hickey, M.J.; Jones, J.R.; Kingston, L.P.; Lockley, W.J.S.; Mather, A.N.; Smith, T.; Wilkinson, D.J. One-step exchange-labelling of piperidines, piperazines and dialkylamines with deuterium oxide: Catalysis by various ruthenium complexes. *Tetrahedron Lett.* **2005**, *46*, 4291–4293. [CrossRef]
98. Taglang, C.; Korenchan, D.E.; von Morze, C.; Yu, J.; Najac, C.; Wang, S.; Blecha, J.E.; Subramaniam, S.; Bok, R.; VanBrocklin, H.F.; et al. Late-stage deuteration of 13C-enriched substrates for T1 prolongation in hyperpolarized ^{13}C MRI. *Chem. Commun.* **2018**, *54*, 5233–5236. [CrossRef]
99. Michelotti, A.; Rodrigues, F.; Roche, M. Development and scale-up of stereoretentive α-deuteration of amines. *Org. Process Res. Dev.* **2017**, *21*, 1741–1744. [CrossRef]
100. Chatterjee, B.; Krishnakumar, V.; Gunanathan, C. Selective α-deuteration of amines and amino acids using D_2O. *Org. Lett.* **2016**, *18*, 5892–5895. [CrossRef]

101. Neubert, L.; Michalik, D.; Bähn, S.; Imm, S.; Neumann, H.; Atzrodt, J.; Derdau, V.; Holla, W.; Beller, M. Ruthenium-catalyzed selective α,β-deuteration of bioactive amines. *J. Am. Chem. Soc.* **2012**, *134*, 12239–12244. [CrossRef]
102. Conley, B.L.; Pennington-Boggio, M.K.; Boz, E.; Williams, T. Discovery, applications, and catalytic mechanisms of Shvo's catalyst. *J. Chem. Rev.* **2010**, *110*, 2294–2312. [CrossRef]
103. Pieters, G.; Taglang, C.; Bonnefille, E.; Gutmann, T.; Puente, C.; Berthet, J.-C.; Dugave, C.; Chaudret, B.; Rousseau, B. Regioselective and stereospecific deuteration of bioactive aza compounds by the use of ruthenium nanoparticles. *Angew. Chem. Int. Ed.* **2014**, *53*, 230–234. [CrossRef]
104. Hale, L.V.A.; Szymczak, N.K. Stereoretentive deuteration of α-chiral amines with D_2O. *J. Am. Chem. Soc.* **2016**, *138*, 13489–13492. [CrossRef] [PubMed]
105. Bhatia, S.; Spahlinger, G.; Boukhumseen, N.; Boll, Q.; Li, Z.; Jackson, J.E. Stereoretentive H/D exchange via an electroactivated heterogeneous catalyst at sp^3 C–H sites bearing amines or alcohols. *Eur. J. Org. Chem.* **2016**, *24*, 4230–4235. [CrossRef]
106. Sajiki, H.; Esaki, H.; Aoki, F.; Maegawa, T.; Hirota, K. Palladium-catalyzed base-selective H-D exchange reaction of nucleosides in deuterium oxide. *Synlett* **2005**, *9*, 1385–1388. [CrossRef]
107. Esaki, H.; Aoki, F.; Maegawa, T.; Hirota, K.; Sajiki, H. Synthesis of base-selectively deuterium-labeled nucleosides by the Pd/C-catalyzed H-D exchange reaction in deuterium oxide. *Heterocycles* **2005**, *66*, 361–369. [CrossRef]
108. Hardacre, C.; Holbrey, J.D.; McMath, S.E. A highly efficient synthetic procedure for deuterating imidazoles and imidazolium salts. *J. Chem. Commun.* **2001**, *4*, 367–368. [CrossRef]
109. Yau, W.-M.; Gawrisch, K. Deuteration of indole and N-methylindole by Raney nickel catalysis. *J. Label. Compd. Radiopharm.* **1999**, *42*, 709–713. [CrossRef]
110. Dinnocenzo, J.P.; Banach, T.E. Quinuclidine dimer cation radical. *J. Am. Chem. Soc.* **1988**, *110*, 971–973. [CrossRef]
111. Hickey, M.J.; Jones, J.R.; Kingston, L.P.; Lockley, W.J.S.; Mather, A.N.; McAuley, B.M.; Wilkinson, D.J. Iridium-catalysed labelling of anilines, benzylamines and nitrogen heterocycles using deuterium gas and cycloocta-1,5-dienyliridium(I) 1,1,1,5,5,5-hexafluoropentane-2,4-dionate. *Tetrahedron Lett.* **2003**, *44*, 3959–3961. [CrossRef]
112. Loh, Y.Y.; Nagao, K.; Hoover, A.J.; Hesk, D.; Rivera, N.R.; Colletti, S.L.; Davies, I.W.; MacMillan, D.W.C. Photoredox-catalyzed deuteration and tritiation of pharmaceutical compounds. *Science* **2017**, *358*, 1182–1187. [CrossRef]
113. Hu, Y.; Liang, L.; Wei, W.; Sun, X.; Zhang, X.; Yan, M. A convenient synthesis of deuterium labeled amines and nitrogen heterocycles with KOt-Bu/DMSO-d_6. *Tetrahedron* **2015**, *71*, 1425–1430. [CrossRef]
114. Gao, L.; Perato, S.; Garcia-Argote, S.; Taglang, C.; Martínez-Prieto, L.M.; Chollet, C.; Buisson, D.-A.; Dauvois, V.; Lesot, P.; Chaudret, B.; et al. Ruthenium-catalyzed hydrogen isotope exchange of C(sp^3)–H bonds directed by a sulfur atom. *Chem. Commun.* **2018**, *54*, 2986–2989. [CrossRef]
115. Michelotti, A.; Roche, M. 40 Years of Hydrogen–Deuterium Exchange Adjacent to Heteroatoms: A Survey. *Synthesis* **2019**, *51*, 1319–1328. [CrossRef]
116. Rauk, A.; Buncel, E.; Moir, R.Y.; Wolfe, S. Hydrogen exchange in benzyl methyl sulfoxide. Kinetic and spectroscopic nonequivalence of methylene protons. *JACS* **1965**, *87*, 5498–5500. [CrossRef]
117. Rodondo, A.; Jaime, C.; Margués, A. Selective deuteration of [(pyridylmethyl)sulfinyl]benzimidazole antisecretory drugs. A NMR study where DMSO-d_6 acts as deuteration agent. *J. Pharma. Biomed. Anal.* **2016**, *131*, 454–463. [CrossRef] [PubMed]
118. Sakaguchi, U.; Morito, K.; Yoneda, H. Solution pH-induced reversal of stereoselectivity. Deuteration of malonate methylenes in some bis(malonato)cobalt(III) complexes. *JACS* **1979**, *101*, 2767–2768. [CrossRef]
119. Park, Y.; Kim, Y.; Chang, S. Transition metal-catalyzed C−H amination: Scope, mechanism, and applications. *Chem. Rev.* **2001**, *3*, 5052–5058. [CrossRef] [PubMed]

Review

Fluorescent Probes for Live Cell Thiol Detection

Shenggang Wang [†], Yue Huang [†] and Xiangming Guan *

Department of Pharmaceutical Sciences, College of Pharmacy and Allied Health Professions, South Dakota State University, Box 2202C, Brookings, SD 57007, USA; Shenggang.wang@fda.hhs.gov (S.W.); Yue.huang@sdstate.edu (Y.H.)
* Correspondence: Xiangming.Guan@sdstate.edu; Tel.: +1-605-688-5314; Fax: +1-605-688-5993
† Shenggang Wang and Yue Huang contributed equally to this article.

Abstract: Thiols play vital and irreplaceable roles in the biological system. Abnormality of thiol levels has been linked with various diseases and biological disorders. Thiols are known to distribute unevenly and change dynamically in the biological system. Methods that can determine thiols' concentration and distribution in live cells are in high demand. In the last two decades, fluorescent probes have emerged as a powerful tool for achieving that goal for the simplicity, high sensitivity, and capability of visualizing the analytes in live cells in a non-invasive way. They also enable the determination of intracellular distribution and dynamic movement of thiols in the intact native environments. This review focuses on some of the major strategies/mechanisms being used for detecting GSH, Cys/Hcy, and other thiols in live cells via fluorescent probes, and how they are applied at the cellular and subcellular levels. The sensing mechanisms (for GSH and Cys/Hcy) and bio-applications of the probes are illustrated followed by a summary of probes for selectively detecting cellular and subcellular thiols.

Keywords: live cell thiol fluorescence imaging; cellular thiols; subcellular thiols; non-protein thiols; protein thiols; mitochondrial thiols; Lysosomal thiols; glutathione; cysteine; homocysteine; thiol specific agents; thiol-sulfide exchange reaction; thiol-disulfide exchange reaction; Michael addition reaction

1. Introduction

Thiols, compounds that contain a sulfhydryl group (-SH), in the biological system are referred as biological thiols or biothiols [1,2]. Biothiols can be categorized into two classes: protein thiols (PSH) and non-protein thiols (NPSH). NPSH are mainly small molecule thiols including cysteine (Cys), homocysteine (Hcy), and glutathione (GSH), while PSH are primarily the cysteine residues in proteins [3]. Thiols play vital and irreplaceable roles in various cellular functions in the biological system [4] thanks to their unique chemical properties of strong nucleophilicity, reductivity, and chelating ability for metals. These roles include detoxification to remove reactive electrophiles, as an antioxidant to remove reactive oxygen species (ROS), and the ability of metal chelation. For example, GSH is the key compound in glutathione conjugation reaction—a phase II drug metabolism in the body to remove reactive electrophiles which are toxic. GSH is also the most abundant intracellular NPSH and serves as the most important antioxidant to prevent damages caused by ROS via termination of ROS. In addition, thiols are also involved in many other aspects of cellular functions, such as being part of enzyme active sites, being involved in signal transduction, being involved in cell division, etc. Abnormality of thiol levels has been linked with many diseases and biological disorders, such as cardiovascular diseases, cancer, Alzheimer's disease and etc [5].

Thiols are distributed heterogeneously and their levels change dynamically in the biological system. Subcellular organelles such as nucleus, mitochondria, lysosomes, endoplasmic reticulum (ER), Golgi apparatus, cell surfaces, etc., are important structures that play critical roles in the normal function of cells. Disruption of the homeostasis of these structures can interrupt the normal function of cells resulting in many disorders and diseases [6,7]. Thiols distribute unevenly in these subcellular organelles [8]. Their levels in different subcellular organelles can serve as an indicator to reflect the status of the subcellular organelles. Thus, tools or methods that can map thiol distribution and monitor their status changes will provide valuable information to understand the relationship between thiols and cells' functions/dysfunctions, and help guide the correlation/treatment of these dysfunctions and their related diseases [9].

Many analytical methods have been developed for detecting thiols in the biological system. These methods include high performance liquid chromatography (HPLC/UV) [10–12], mass spectrometry (MS), HPLC-MS [13], colorimetric assay [14], enzyme assays, capillary electrophoresis, and gel electrophoresis [15–18]. Most of these methods need a complicated and lengthy sample preparation step and involve a homogenization process to break cells or tissues before an analytical method can be applied. Due to chemical instability of thiols, the sample process step quite often can introduce potential artifacts and cause a potential loss of the analyte.

In the past two decades, one of the most popular approaches being developed to address analyte loss during cell/tissue homogenization is to analyze the analyte in live cells using the fluorescence imagining technique via the use of a fluorescent probe [19]. Fluorescence imaging outstands itself by its simplicity, high sensitivity, and capability of visualizing the analytes in live cells in a non-invasive way [1]. The non-invasive fluorescence imaging provides the advantage of visualizing an analyte without breaking the cell [20–27], enabling the determination of intracellular distribution and dynamitic movement of an analyte in its intact native environments, and reveal information that cannot be revealed after cells/tissues are homogenized.

Significant efforts have been made for the detection of thiols in the biological system via fluorescence imaging by developing a large number of fluorescent probes for thiols. These probes include BODIPY [28,29], rhodamine [30,31], monochlorobimane [32], fluorescein [33,34], tetraphenylethylene [35], coumarin [36,37], curcumin [38], mercury orange [32], rosamine-based [39], naphthalimide-based [40], polymethine [41], benzofurazan [42], and pyrene [43] (Figure 1). The optical spectra of these fluorescence probes cover visible fluorescence, near infra-red (NIR) fluorescence, single-photon or two-photon excitation fluorescence, or aggregation-induced emission (AIE) fluorescence. Some of these probes are developed as an "off-on" agents in a single channel which themselves are non-fluorescent due to a variety of quenching mechanisms (e.g., FRET, PET, and ICT) but can be turned on to a fluorescent product by thiols. The others are "ratiometric" probes utilizing dual-channel emission which themselves exhibit fluorescence but can be switched to a totally different fluorescence upon reaction with thiols and the thiol's level can be reflected by calculating the ratio of the two fluorescence intensities. Compared with "off-on" probes which measure the absolute intensity at only one wavelength, ratiometric fluorescent probes are believed to be more accurate since the use of two different emission wavelength can provide a built-in correction [44].

Among the reported probes, majority of them were designed to detect thiols from the whole cell despite the fact that thiols are distributed unevenly inside cells. Only a few probes were for the detection of mitochondrial thiols while fewer for lysosomes and endoplasmic reticulum (ER). Finding a method to detect thiols in subcellular organelles remains to be a challenge.

For a fluorescence imaging agent to image thiols in live cells, the agent needs to be cell membrane permeable, capable of turning thiols selectively, ideally specifically, into fluorescence strong enough for detection, and noncytotoxic or at least noncytotoxic during the imaging experiment. For a subcellular thiol imaging agent, the agent needs to be cell membrane permeable, capable of turning thiols selectively, ideally specifically, in the targeted subcellular organelle into fluorescence strong enough for detection, and noncytotoxic or at least noncytotoxic during the imaging experiment. For a cellular or subcellular thiol quantifying fluorescence imaging agent, it requires the reagent to quantitatively turn thiols into fluorescence strong enough for detection within a reasonable time in addition to the requirements for a cellular or subcellular thiol fluorescence imagining agent.

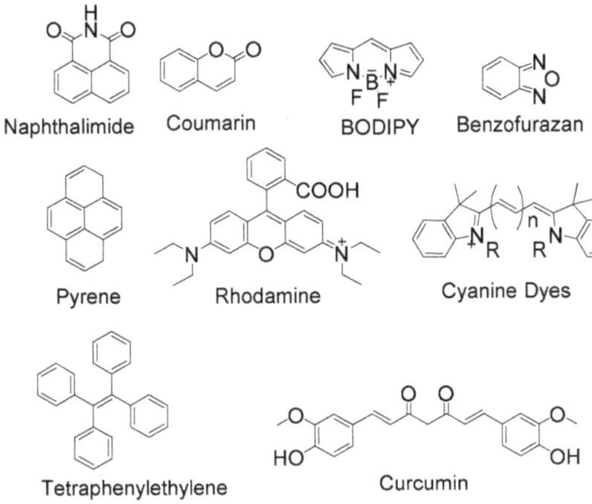

Figure 1. Different fluorophores used for construction of thiol detection probes.

This review will provide an overview of the fluorescent probes that are developed for detecting thiols in live cells. These probes will be mainly classified into two categories: i. probes used for detection of GSH or total thiols; ii. probes used to detect Cys/Hcy. Intracellularly, GSH is much more abundant (1–10 mM) than Cys (30–200 µM) and Hcy (5–12 µM) [45]. The detection of intracellular GSH is less likely to be interfered by Cys/Hcy. As a result, GSH is often used to reflect the total NPSH in cells. Therefore, the probes claimed for GSH or total NPSH are summarized as one category in this review although some of the probes were claimed to be selective for GSH. In contrast, the detection of Cys/Hcy can be interfered by GSH. For that reason, the probes designed for distinguishing Cys/Hcy over GSH are summarized into a different category.

The purpose of this review is not trying to be inclusive in terms of thiol fluorescence probes since they have been covered by other comprehensive reviews [2,46–49]. Rather, this review focuses on summarizing some of the major strategies or mechanisms that are used for detecting GSH, Cys/Hcy, and how they are applied for the detection of cellular or subcellular thiols. The fluorophores being used, sensing mechanisms (for GSH or Cys/Hcy) and bio-applications of the probes will be illustrated followed by a summary of probes for selectively detecting cellular and subcellular thiols.

2. Detection of Thiols in Cells

2.1. NPSH, GSH, Cys, and Hcy

NPSH are thiols other than PSH. They include GSH, Cys, Hcy, and other small molecule thiols. Among them, GSH is the most abundant thiol with a concentration of millimolar in the biological system. GSH serves as the most important antioxidant to prevent damages caused by ROS [50,51] via terminating ROS with itself being converted to its oxidized form glutathione disulfide (GSSG). GSSG is then converted back to GSH by glutathione reductase. Cells maintain a normal ratio (>100) of GSH/GSSG through the GSH/GSSG cycle [52,53]. The abnormal level of GSH has been associated with various diseases such as cancer, AIDS, growth delay, neurodegenerative disease, and cardiovascular disease [54–57]. Thus, determination of GSH level in live cells is of particular interest.

Numerous fluorescent probes have been developed for the detection of intracellular GSH or thiols in live cells based on different mechanisms. Most of these methods were designed based on the high nucleophilicity of thiols [58,59].

Distinguishing GSH from Cys and Hcy remains a challenge due to the structure and reactivity similarities GSH shared with Cys/Hcy [60–62]. The majority of the thiol probes being developed are not able to distinguish GSH from other thiols including Cys/Hcy, since most of them undergo similar reactions and similar emission changes upon reaction with a thiol. However, as mentioned earlier Cys or Hcy is less likely to interfere with the detection of GSH since intracellular GSH level (1–10 mM) is much higher than Cys (30–200 µM), and Hcy (5–15 µM). Although some of the probes have been reported to selectively detect GSH and some reported for detecting NPSH, all these methods are included in the category of GSH and total thiols. This part will focus on the main strategies/mechanisms that have been developed for the detection of GSH and total thiols. The strategies/mechanisms included in these methods are: A. Michael addition reaction with α, β -unsaturated carbonyl derivatives [63,64]; B. Nucleophilic substitution with substrates bearing a leaving group (e.g., halogen, ether, or thioether) [29,65,66]; C. Cleavage of sulfonamide or sulfonate ester [67]; D. Disulfide–thiol exchange reaction [68]; E. others [30,46] (Figure 2). It is noted that although reaction with a thiol is preferred in all these reactions, other nucleophiles such as -OH, or -NH can also be involved in these reactions. The involvement of other nucleophiles in the reaction will affect thiol imaging especially quantitative thiol imaging. Nevertheless, thiol–disulfide exchange reaction (D) and thiol–sulfide exchange reaction (B) have been demonstrated to be thiol specific. Thiol specific fluorescence imaging agents based on the thiol-sulfide reaction have been developed and used successfully to image and quantify cellular thiols and subcellular thiols in live cells (refer to Sections 2.1.2, 3.1.2 and 3.3.2)

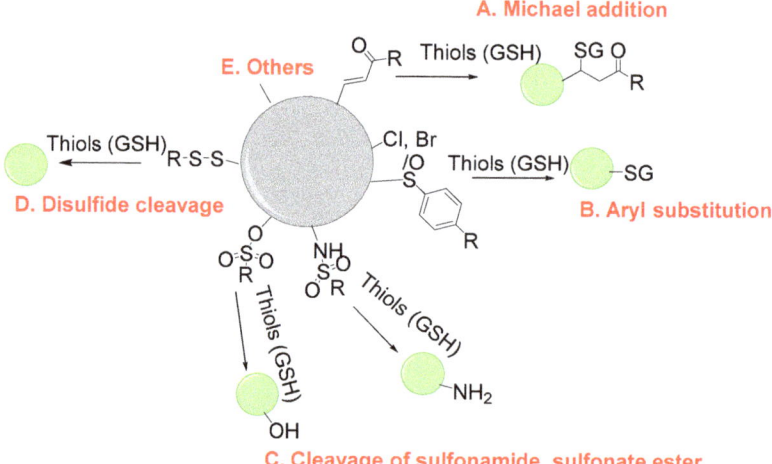

Figure 2. Representative sensing mechanisms used for thiol detection.

2.1.1. Detection of NPSH via a Michael Addition Reaction

Michael addition is one of the most commonly used strategies for detection of thiols. Michael addition reaction has a high selectivity for thiols though it can also occur with other nucleophiles such as -OH, NH, or -COOH. Xiaoqiang Chen et al. [34] published their work on the development of a thiol imaging tool bearing fluorescein as the fluorophore (**Probe 1**). As shown in Figure 3, the process of turning on the fluorescence starts from a 1,4-addition of thiols to the α, β-unsaturated ketone in the probe. Subsequent spiro ring opening of fluorescein lights up the fluorescence (Ex. 485 nm, Em. 520 nm) which provides fluorescence enhancement by 61-fold compared with probe itself. The probe exhibits a similar fluorescence enhancement towards Cys, Hcy, and GSH. With fluorescein as a standard, the quantum yields of the thiols adducts with probe were determined to be 0.65, 0.91, and 0.47 for GSH, Cys, and Hcy, respectively. The probe is featured with a rapid response and high sensitivity with a detection limit of 10^{-7}–10^{-8} M. Further, the probe was proved to be capable of detecting the thiol status change in live P19 cells. More importantly, the probe is also the first reagent being used to monitor the thiols status in zebrafish. A significant fluorescence intensity decrease was observed when the cells or zebrafish were treated with a thiol trapping reagent, N-methylmaleimide (NMM) (40–50 µM).

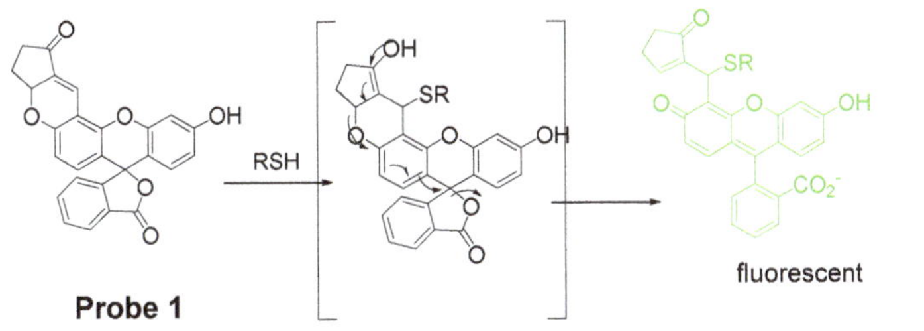

Figure 3. Structure of **Probe 1** and its reaction mechanism with RSH.

In 2011, Gun-Joong Kim et al. [36] reported a Michael addition reaction based fluorescent probe (**Probe 2**) for ratiometric imaging of cellular GSH. As illustrated in Figure 4, the probe has three parts, a coumarin moiety serving as the signaling fluorophore, a thiol reactive enone linker, and an O-hydroxyl group as an activation unit through a resonance-assisted hydrogen bond [69,70]. Once a reaction occurs between thiol (2-mercaptoethanol as a representative agent) and the Michael acceptor structure of the probe, the fluorescence emission band will shift from 553 to 466 nm (Ex. 420 nm). The ratio of the fluorescence intensity under these wavelength ($F_{466\ nm}/F_{553\ nm}$) changes ratiometrically upon reaction with thiol such as Hcy, Cys, and GSH. The probe was successfully applied for monitoring cellular thiol status in live cells. The fluorescence ratio ($F_{blue\ 410\ nm–460\ nm}/F_{green\ 490–540\ nm}$) increased significantly when the cells were pretreated with R-lipoic acid, an enhancer of GSH. In contrast, a remarkable ratio decrease was observed when cells were pretreated with NEM, a scavenger of GSH.

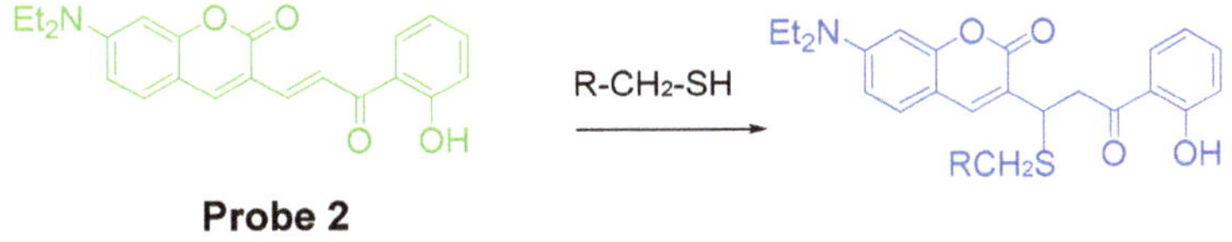

Figure 4. Structure of **Probe 2** and its reaction with RSH.

It is known that maleimide can react selectively toward thiols than other nucleophiles via a Michael addition reaction. In the meantime, maleimide can also serve as a fluorescence quencher in their conjugated form. Thus, maleimide has been widely utilized by many groups to build the thiols selective probes. For instance, in 2017, Hai Shu et al. [71] reported a rhodamine B-based probe (L1, **Probe 3**) with maleimide serving as a thiol recognition unit. Upon reaction with GSH, the probe can be turned on from non-fluorescent to a strong fluorescent thiol-rhodamine adduct (586 nm) due to a Michael addition reaction followed by a spirolactam ring opening (Figure 5). A more than 200-fold increase of fluorescence intensity was observed when excessive GSH was added. In addition, the fluorescence intensity was found to be linearly increased with an increase in GSH (2–26 µM) with a detection limit of 0.219 µM. The probe shows a high selectivity toward GSH than other thiols. Further, cellular experiments demonstrated that L1 can be used to monitor intracellular GSH in HepG-2 and HUVEC cells. Another similar approach was reported by Tao Liu et al. [72] in 2016 for selective detection of thiols in live cells with a high sensitivity (**Probe 4**). The detection limits were determined to be 0.085, 0.12, and 0.13 µM for GSH, Hcy, and Cys, respectively. Similarly, the probe was also successfully applied in the determination of intracellular GSH in live HepG2 cells.

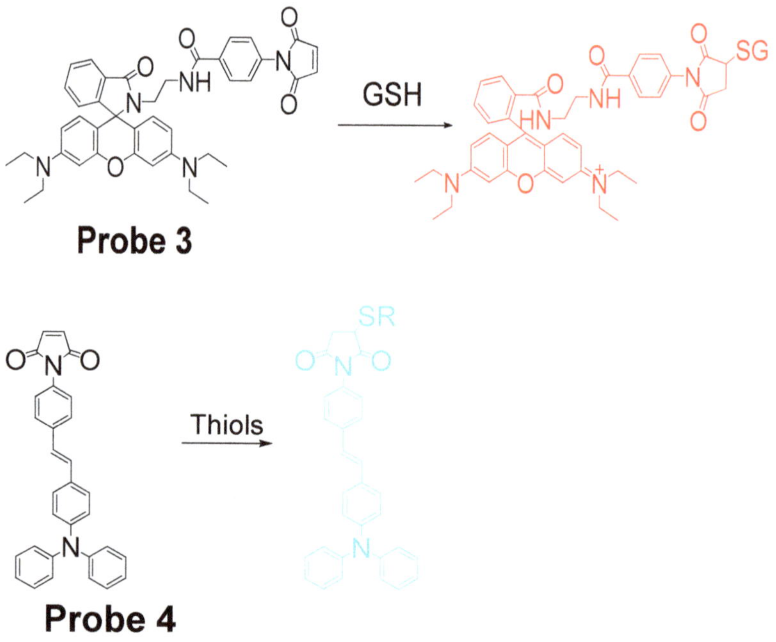

Figure 5. Structures of **Probes 3** and **4** and their reactions with GSH or thiols, respectively.

In contrast to traditional fluorescent methods, reagents which can detect a thiol in a reversible manner was developed. In 2015, Xiqian Jiang et al. [73] reported a reversible and ratiometric fluorescent probe-ThiolQuant Green (TQ Green, **Probe 5**) for quantitative imaging of GSH. The probe is composed of a 7-amino coumarin group as the fluorophore and an aromatic structure to extend the absorption wavelength of coumarin. These two parts are linked together by a thiol reactive group, Michael acceptor. The probe's reaction with GSH is reversible when GSH is depleted. TQ Green (Ex. 488 nm, Em. 590 nm) is turned into to TQ Green-GSH (Ex. 405 nm, Em. 463 nm) upon reaction with GSH (Figure 6). The ratio of signal intensities (absorbance or fluorescence) between TQ Green-GSH and TQ Green is proportional to the GSH concentration enabling the ratiometric detection of GSH with a detection limit of 20 nM. No reaction was observed when the probe was exposed to an excessive ratio of bovine serum albumin (BSA) indicating the probe does not react with PSH. Cellular experiments with multiple cell lines, including 3T3-L1, HepG2, PANC-1, and PANC-28 cells, proved that the GSH concentration obtained from TQ Green live imaging were well correlated with the values achieved from bulk lysate measurements. The probe can well reflect the GSH level changes in live cells when the GSH level was reduced by diethylmaleate (DEM). One drawback of the method is that the reverse reaction between GSH and TQ Green is slow leading to the inability of the probe for a quick detection of a decrease of GSH in live cells.

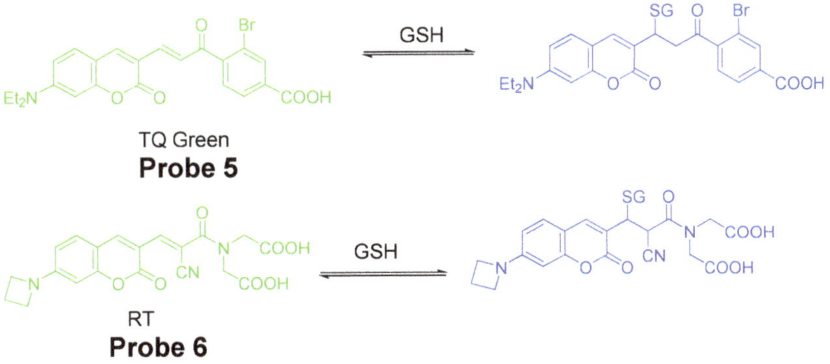

Figure 6. Structures of **Probes 5, 6** and their reversible reactions with GSH.

Continuing with their work, Xiqian Jiang et al. [74] published another reversible fluorescent probe named as RealThiol (RT, **Probe 6**) in 2017 for quantitatively real-time monitor of intracellular GSH in live cells with a much higher reaction rate (50-fold faster than TQ Green). RT was developed based on a series of optimization of TQ Green to enable a quantitative real-time imaging of GSH in living cells with a much-enhanced reaction rate for both forward and reverse directions. Similar to TQ Green, RT shows ratiometric fluorescence response with a superb linear relationship to various concentrations of GSH in a range of 1–10 mM. In contrast to TQ Green which exhibits a second-order forward reaction rate of 0.15 $M^{-1} s^{-1}$, the second-order reaction rate for RT is 7.5 $M^{-1} s^{-1}$ which is 50 times faster than TQ Green. No significant fluorescence change was observed when RT was mixed with cell lysates suggesting the inability of RT to react with PSH. Impressively, the probe could well reflect the GSH level change when HeLa cells were treated with different concentrations of H_2O_2. Later, the probe was demonstrated to quantitatively monitor GSH levels when cells were consecutively treated with H_2O_2 (500 μM) and GSH ester (100 μM). Further, with the aid of RT, an enhanced antioxidant capability of activated neurons and dynamic GSH changes during ferroptosis were observed. RT was successfully applied for high-throughput quantification of GSH in single cells via flow cytometry.

It should be noted that a number of other reversible fluorescent probes have been developed for quantitatively monitoring GSH in real time in live cells with various fluorophores and sensing mechanisms. These probes include probe QG-1 developed by Zhixue Liu et al. [75], probe RP-1 and RP-2 by Ming Tian et al. [76,77], and a number of single-molecule localization microscopy (SMLM) applicable probes developed from Urano groups [78].

2.1.2. Detection of NPSH via a Nucleophilic Aromatic Substitution (SNAr) Reaction Using Halogen, Ether, Thioether as a Leaving Group

SNAr reaction is another widely used mechanism for imaging thiols in live cells since the thiol group can easily displace a leaving group that is attached to an aromatic ring in a fluorophore by a substitution reaction to form a thioether. These leaving groups include halogen, ether, thioether, etc. To be noted here, SNAr substitution reactions have also been commonly used to discriminatively detect Cys/Hcy due to a rearrangement reaction occurring to the thioether formed in the substitution reaction. The rearrangement reaction only occurs to the thioether derived from Cys and Hcy (Figure 7).

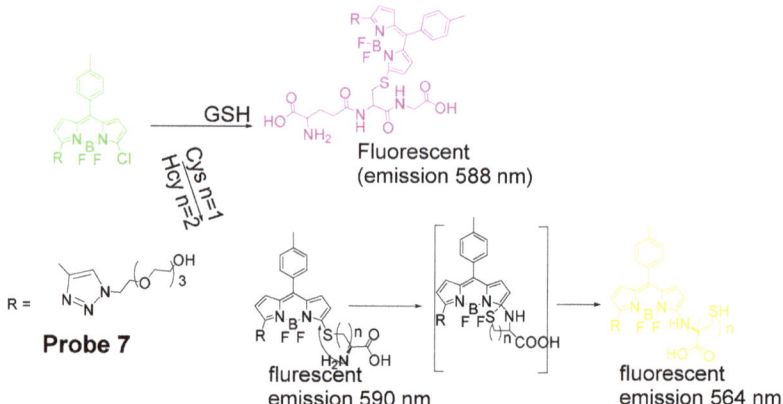

Figure 7. Probe 7 and its reaction with GSH or Cys (*n* = 1)/Hcy (*n* = 2).

Detection of NPSH via a SNAr Reaction Using Halogen as a Leaving Group

In 2012, Liya Niu and co-workers [29] reported their efforts on developing a fluorescent probe (**Probe 7**) for highly selective detection of GSH over Cys/Hcy based on monochlorinated BODIPY. As shown in Figure 7, a SNAr reaction occurred between the sulfur group (GSH, Cys, or Hcy) and the monochlorinated BODIPY resulting in a thioether. A fluorescence emission wavelength red shift (556 to ~588 nm) occurred after the reaction, and the ratio of fluorescence intensity (I_{588}/I_{556}) linearly increased as an increase in the concentration of GSH (0−60 µM) with the detection limit of 8.6×10^{-8} M, enabling the quantification of intracellular GSH in live cells. The probe was successfully applied for detecting the GSH levels' change when the cells were pretreated with diamide [79] in HeLa cells. The probe was claimed to be the first ratiometric fluorescent sensor developed for the selective detection of GSH over Cys/Hcy.

A quick intramolecular displacement occurred with the thioether generated from Cys and Hcy to replace the thiolate with the amino group to form amino-substituted BODIPY, while the thioether generated from GSH was not able to undergo the intramolecular displacement due to the unavailability of an amino group. Since the photophysics of BODIPY derivatives is sensitive to substituents, the amino-substituted BODIPY exhibit a relative weaker and blue shifted fluorescence which enable the selective detection of GSH over Cys and Hcy, while the rapid intramolecular displacement of S to N will shift the emission maxima back to around 564 nm for Cys and Hcy, with GSH remaining at 588 nm.

Detection of NPSH via a SNAr Reaction Using ether as a Leaving Group

In addition to halogen, other leaving groups, such as ether or thioether are also widely used as a thiol recognition unit for building a thiol selective probe. For example, Xinghui Gao et al. [80] developed a probe (**Probe 8**) in 2015 for detection of GSH and Cys in different emission channels by connecting 7-nitrobenzofurazan with resorufin through the ether bond in which the ether bond also served as a thiols recognition unit. Reaction with thiols (GSH, or Cys) yields a free strong fluorescent resorufin (Em. 585 nm) and a thiol conjugated nitrobenzofurazan. However, the thiol conjugated notrobenzofurazn derived from Cys will undergo an intramolecular rearrangement to generate a new fluorescent product with λem = 540 nm while the thiol conjugated notrobenzofurazn derived from GSH remained unchanged (Figure 8). Measuring the fluorescence intensity from resorufin (Em. 585 nm) enabled the determination of the total thiols (GSH, and Cys/Hcy) while measuring the fluorescence intensity of both resorufin and the thiol conjugated nitrobenzofurazan provided a method to distinguish GSH from Cys. The detection limits for GSH and Cys were reported to be 0.07 µM and 0.13 µM respectively with a linear relationship between the fluorescence intensity and the concentration of GSH (1–18 μM) and Cys (3.1–90 μM, 1–40 μM), respectively. The probe was employed to detect Cys and GSH in human plasma.

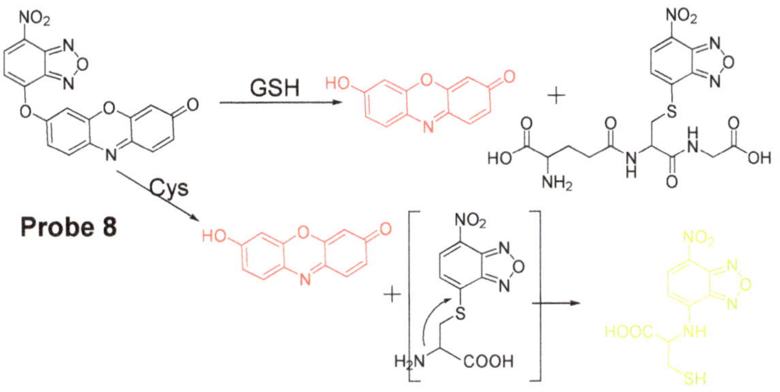

Figure 8. Structure of **Probe 8** and its reaction with GSH and Cys.

Xilei Xie et al. [81] reported a probe (Res-Biot, **Probe 9**) in 2017 for selective detection of GSH based on the ether bond. In this approach, pyrimidine moiety was for the first time investigated and optimized as a unique recognition unit for thiols. The pyrimidine moiety was connected with fluorophore resorufin via an ether linkage to generate probe Res-Biot. Upon reaction with a thiol, a high fluorescent resorufin is released with a strong fluorescence at 585 nm (Figure 9). Moreover, the fluorescence intensity at 585 nm was increased proportionally to the concentrations of GSH in a range of 0−20 μM with a detection limit of 0.29 μM. The probe was successfully used to detect a decrease in intracellular GSH concentration down-regulated by an oxidative stress inducer, 12-myristate 13-acetate (PMA), or by L-buthionine sulfoximine (BSO)—an inhibitor of γ-glutamylcysteine synthetase.

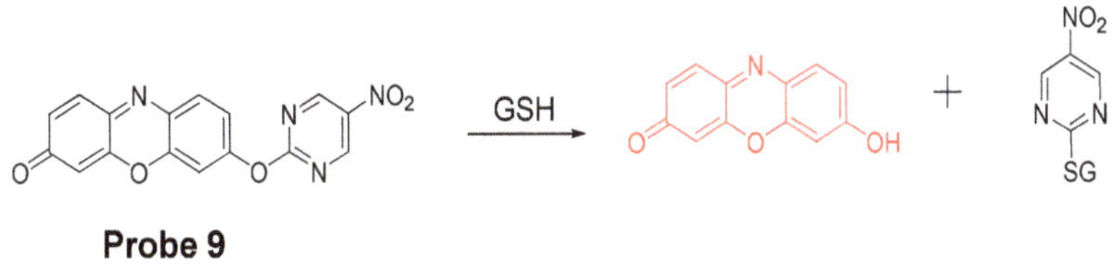

Figure 9. Probe 9 and its reaction with GSH.

Detection of NPSH via a SNAr Reaction Using Thioether as a Leaving Group

In 2012, Yinghong Li et al. from our group reported a thiol specific probe [42] (GUALY's reagent, **Probe 10**) based on a symmetric benzofurazan sulfide for imaging and quantifying total thiols in live cells. The symmetric benzofurazan sulfide was constructed by combining two benzofurazan fluorophores with a thioether group (Figure 10). The probe itself is not fluorescent due to the self-quenching of the symmetric structure but turns into a fluorescent product (Ex. 430 nm, Em. 520 nm) rapidly upon reaction with thiols through a thiol specific thiol–sulfide exchange reaction. GUALY's reagent has been successfully applied for imaging and quantifying total thiols (PSH+NPSH) in live cells [42,82].

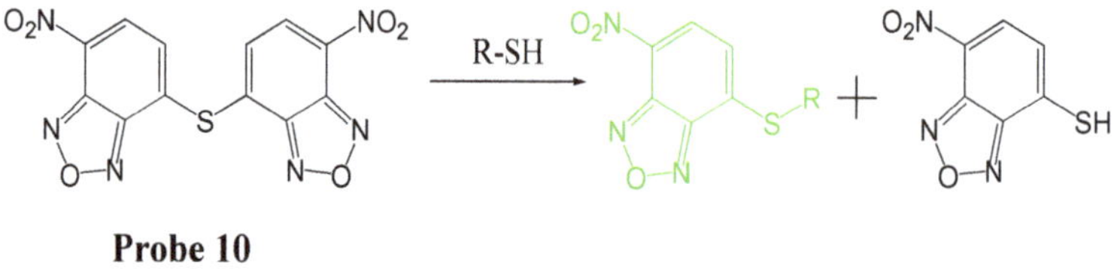

Probe 10

Figure 10. Probe 10 and its reaction with thiols.

Jing Liu et al. [83] reported a GSH discriminating fluorescent probe (**Probe 11**, Figure 11) in 2014 based on pyronin B as the fluorophore and thioether as the thiol recognition group. In the approach, A methoxythiophenol group was connected to the fluorescent pyronin moiety to serve as a fluorescence quencher via a photoinduced electron transfer (PET) process. Reaction with thiols (GSH, Cys, Hcy) through SNAr reaction generates a highly fluorescent thiol conjugated pyronin with λ_{em} for GSH-pyronin and Cys/Hcy-pyronin at 622 nm and 546 nm respectively due to the additional intramolecular rearrangements of Cys/Hcy. The probe has been successfully applied for simultaneously detection of Cys/Hcy and GSH in B16 cells.

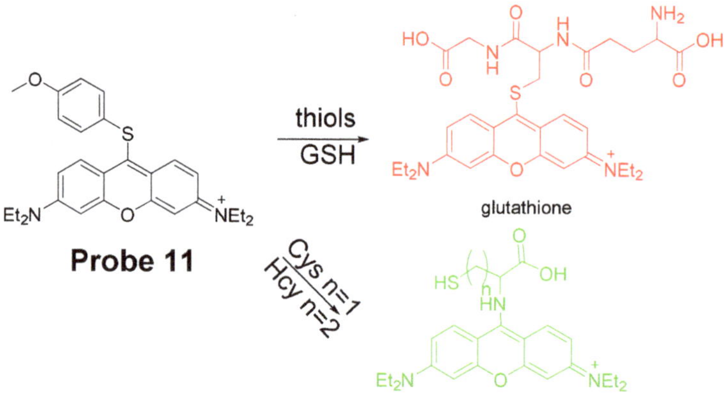

Figure 11. Probe 11 and its reaction with GSH or Cys/Hcy.

In 2015, Lun Song et al. [84] reported two water-soluble colorimetric and turn-on fluorescent probes (STP1, 2, **Probes 12, 13**) for selective detection of GSH using thioether as a thiol recognition group and naphthalimide as the fluorophore reporter. The probes are non-fluorescent due to a PET process from the naphthalimide electron donor to 4-nitrobenzene acceptor via the thioether bond (Figure 12). Upon reaction with GSH, a remarkable enhancement of fluorescence (~90-fold) at 487 nm was observed. A linear relationship was found between the fluorescence intensity (490 nm) and GSH concentration in the range of 0–100 μM with the detection limit of 84 nM. It should be noted that a linear relation was also found between the fluorescence intensity and thiol-containing proteins bovine serum albumin (BSA) and ovalbumin (OVA) suggesting that the probes can also imaging PSH. These probes have been demonstrated to be capable for fluorescence imaging of GSH in HeLa cells.

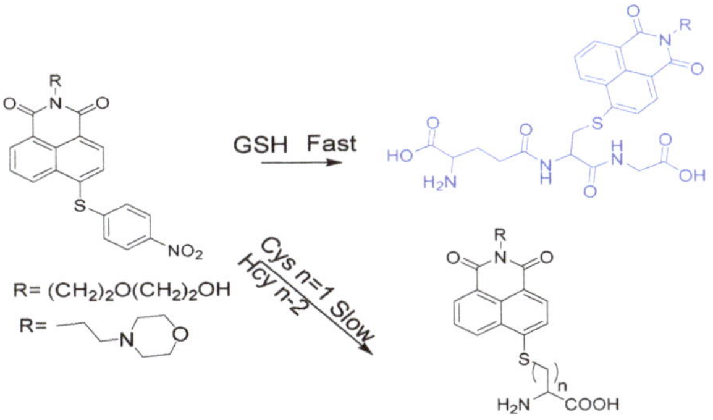

Figure 12. Structures of **Probes 12, 13** and their reaction with GSH or Cys/Hcy.

2.1.3. Detection of NPSH via Cleavage of Sulfonamide or Sulfonate Ester

Detection of NPSH via Cleavage of Sulfonamide

In 2014, Masafumi Yoshida et al. [85] reported a novel cell-membrane-permeable fluorescent probe (**Probe 14**) for detection of GSH based on the cleavage of sulfonamide as the thiol responsive mechanism. As illustrated in Figure 13, a fluorophore hydroxymethylrhodamine green (HMRG) and a 2,4-dinitrobenzenesulfonyl (DNBS) moiety were combined by a sulfonamide group in the approach. The fluorescence of the probe was quenched by two fluorescence quenching mechanisms, intramolecular spirocyclization (close-ring form of HMRG) and intramolecular PET between HMRG and DNBS. Reaction with a thiol breaks the quenching by cleaving the sulfonamide bond to release a free open-ring high fluorescent HMRG with an up to 7000-fold increase in fluorescence intensity (520 nm) and the increase was proportional to the concentration of GSH around the physiological concentration. The probe was successfully applied for the detection of GSH levels' change in live cells. Further, the probe was used to detect tumor nodules in tumor bearing mice of SHIN-3 ovarian cancer by taking advantage of the fact that GSH level in tumor tissues is higher than normal tissues.

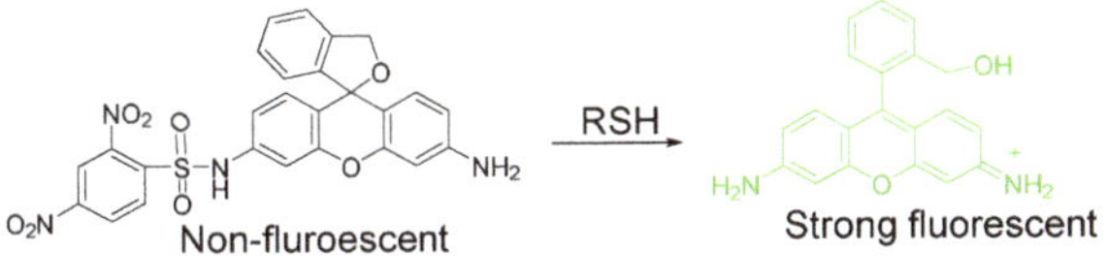

Figure 13. Structure of **Probe 14** and its reaction with thiols.

Fluorophores with emission in red or near NIR region have the advantages of low fluorescence background, deeper penetration, and less damages to live cells or tissues [86,87]. In 2014, a cyanine-based fluorescent probe (**Probe 15**, Figure 14) for highly selective detection of GSH in cells and live mouse tissues have been reported by Jun Yin et al. [88]. The probe utilizes a 5-(dimethylamino)naphthalenesulfonamide group for highly selective detection of GSH over Cys and Hcy. In vitro cellular experiments with HeLa cells proved that the probe was capable of monitoring GSH in live cells. When the cells were pretreated with N-methylmaleimide (NMM, a thiol-blocking agent) followed by treatment with the probe, it showed no red fluorescence while strong fluorescence was observed when no NMM was used confirming that the probe was detecting thiols. In addition, only a minor change of fluorescence intensity was observed when Cys (100 μM) or Hcy (100 μM) was used to the NMM-pretreated HeLa cells. In contrast, significant red fluorescence emission was observed when GSH (100 µM) was added suggesting that the probe can selectively detect GSH in the presence of Cys or Hcy. The probe was successfully applied for monitoring the GSH levels in mouse tissues such as liver, kidney, lung, and spleen.

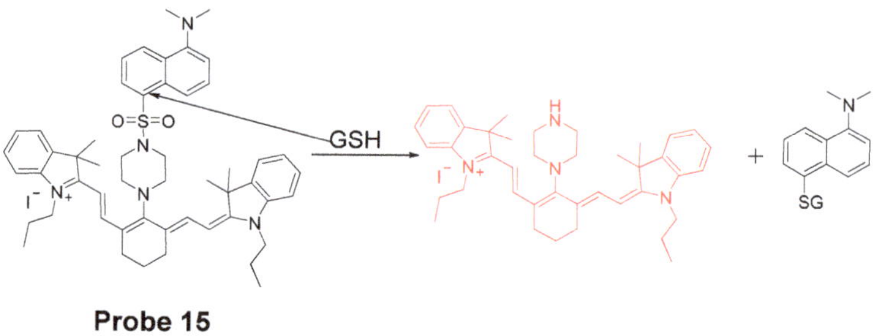

Probe 15

Figure 14. Structure of **Probe 15** and its reaction with GSH.

In 2019, Xie Han et al. [89] reported an aggregation-induced emission (AIE) probe (TPE-Np, **Probe 16**, Figure 15) for selective detection of GSH. The probe is designed by modifying the widely used AIE fluorophore tetraphenylethene with a sulfonyl-based naphthalimide. Cleavage of sulfonamide by GSH induces a remarkable increase of fluorescence intensity at 496 nm which shows a good linear relationship with GSH concentrations in the range of 0–50 µM with a detection limit of 1.9 µM. The probe shows a high selectivity toward GSH than Cys and Hcy in the presence of cetyltrimethylammonium bromide. In addition, poly(ethyleneglycol)–polyethylenimine (PEG-PEI) nanogel was used as a carrier to cross-link TPE-Np to improve its solubility and biocompatibility. The probe was successfully applied for imaging and monitoring intracellular GSH level in MCF-7 cells.

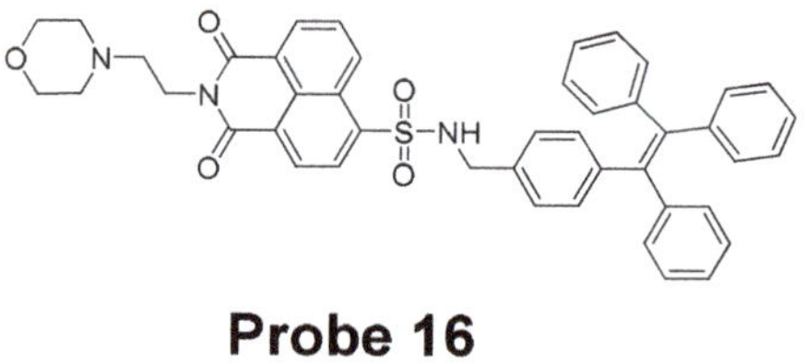

Probe 16

Figure 15. Structure of **Probe 16**.

Detection of NPSH via Cleavage of Sulfonate Ester

Dicyanomethylene-4H-pyran, one of the fluorophores with emission in red or NIR region, is featured with its better stability than other NIR fluorophores such as cyanine and squaraine. In 2014, Meng Li et al. [61] reported a colorimetric and NIR fluorescence turn-on thiol probe (**Probe 17**, Figure 16) based on dicyanomethylene-4H-pyran as the fluorophore and 2,4-dinitrobenzenesulfonyl (DNBS) as the thiol recognition group and also as the fluorescence quencher. A benzene unit was introduced to the dicyanopyran to extend its conjugated system to make its emission wavelength into the desired NIR region [90]. The fluorescence of the compound switched from off to on when the DNBS was cleaved by GSH (Ex. 560 nm, Em. 690 nm), and the fluorescence intensity increases linearly as the concentration of GSH increases (1 to 10 µM) with a detection limit of 1.8×10^{-8} M. A similar response was observed for other sulfhydryl-containing compounds such as Cys, Hcy, and dithiothreitol. The probe was applied for imaging GSH in HeLa cells.

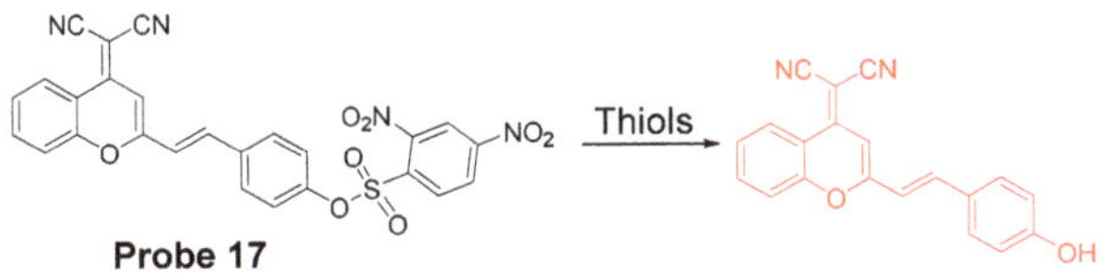

Figure 16. Structure of **Probe 17** and its reaction with thiols.

In 2016, Ling Huang et al. [91] reported a soluble, distyryl boron dipyrromethene (BODIPY)-based nanomicelles probe (**Probe 18**, Figure 17) with NIR properties. DNBS was attached to distyryl boron dipyrromethene (BODIPY) to serve as a fluorescence quencher through a PET mechanism and a thiols responsive unit. The probe will be lighted up by a thiol (Cys, GSH) with an NIR emission at 660 nm (Ex. 600 nm). A PLA-PEG unit is incorporated in the structure enabling micelle formation of the compound with an excellent water solubility. The probe can quantitatively detect the thiols' level changes in HeLa cells.

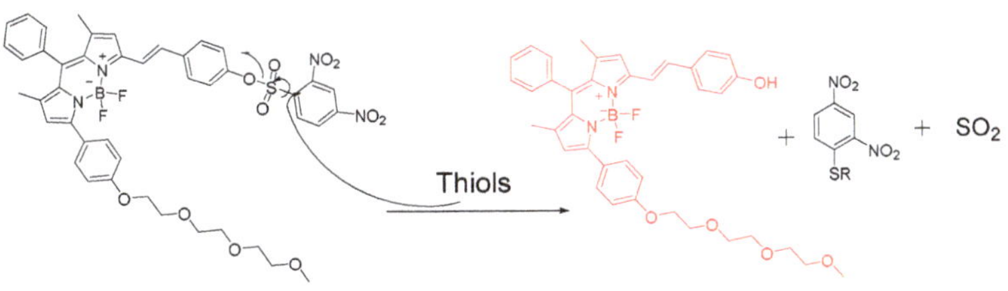

Figure 17. Structure of **Probe 18** and its reaction with thiols.

In 2018, Xiang Xia et al. [67] reported a BODIPY disulfonate probe (BODIPY-diONs, **Probe 19**, Figure 18) with a two-photon fluorescent turn-on effect for discriminately detecting GSH over Cys and Hcy. In the approach, the BODIPY dye was modified by extending the conjugated π-system to achieve a near-infrared emission. The DNBS moiety was utilized as the fluorescence quencher and also the thiol response group. A SNAr process will be initialized when the probe is exposed to GSH, resulting in the cleavage of DNBS and a significant increase of fluorescence (Em. 675 nm). Because of the double DNBS-functionalization, a different reaction rate between the probe and Cys/Hcy vs. GSH was observed with GSH being the fastest one. Later, the probe BODIPY-diONs was demonstrated to be capable of detecting GSH in live cells. In comparison with other reported fluorescent probes, BODIPY-diONs has the advantages of longer emission wavelength and low detection limit of 0.17 μM.

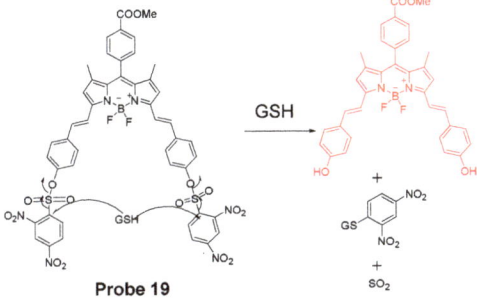

Figure 18. Probe 19 and its reaction with thiols.

2.1.4. Detection of NPSH via Cleavage of a Disulfide Bond

Disulfide bond (-S-S-) is another widely used thiol recognition unit for the designing of a thiol detection probe due to its minimal perturbation to the intracellular redox homeostasis. Many thiol probes were designed based on the reaction of cleavage of a disulfide bond through a thiol-disulfide exchange reaction which has been demonstrated to be thiol specific. For example, a naphthalimide derivative probe (**Probe 20**, Figure 19) made by connecting two naphthalimide units through a disulfide bond for selective and ratiometric detection of thiols were reported by Baocun Zhu et al. in 2010 [92]. The probe was switched from colorless to green color after reacting with GSH through a cleavage of a disulfide bond followed by a cyclization to release 4-aminonaphthalimide as a strong green fluorescent compound with a remarkable shift of emissions from 485 to 533 nm. The fluorescence intensity ratio of (F_{533}/F_{485}) (Ex. 400 nm) increased linearly with the GSH concentration in a range of 0.5 to 10 mM. The detection limit of the probe for GSH was 28 µM, enabling the ratiometric detection of GSH. The probe can successfully detect the intracellular GSH changes in live HeLa cells.

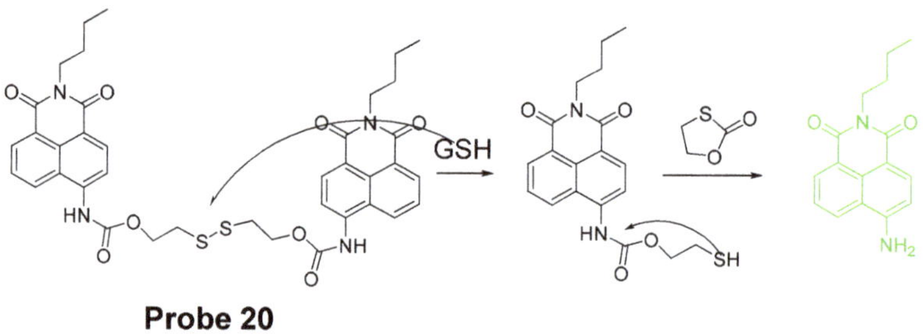

Figure 19. Structure of **Probe 20** and its reaction with GSH.

In 2016, Mingzhou Ye et al. [93] reported a probe (**Probe 21**, Figure 20) named as Cy-S-CPT (also a prodrug) in which a NIR cyanine dye and comptothecin (CPT) are connected through a disulfide bond. In the approach, the cleavage of the disulfide bond in Cy-S-CPT by thiols will induce a dramatic NIR fluorescence shift from 825 to 650 nm. In the meantime, the cleavage activates the anti-cancer drug CPT enabling a real time tracking of the prodrug distribution at 825 nm and the activated drug at 650 nm. The whole process involved the cleavage of disulfide and a cyclization to release both the active CPT and a highly fluorescent compound CyA-K. More impressively, a PEG-PLA nanoparticle loaded with the probe was able to inhibit the tumor growth rate as high as 94.0% which is much higher than the clinical approved CPT-11 (55.8%).

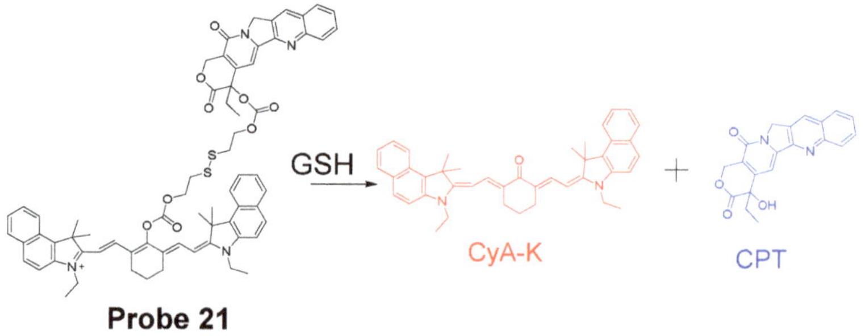

Figure 20. Structure of **Probe 21** and its reaction with GSH.

In 2018, Chao Yin et al. [94] developed a ratiometric photoacoustic imaging (PAI) fluorescence probe (**Probe 22**, Figure 21) for imaging GSH in vivo with a high resolution and deep penetration. The probe, IR806-pyridine dithioethylamine (PDA), employs a disulfide bond as the thiol recognition group and NIR cyanine as the fluorescence report unit. Upon reaction with GSH, the disulfide is cleaved followed by a subsequent intramolecular reaction to form a sulfhydryl group leading to a ratiometrically signal changes of NIR-absorption peak from 658 to 820 nm. The detection limit for GSH is as low as 0.86×10^{-6} M. The probe was successfully applied for in vivo ratiometric PAI of upregulated GSH in the tumor of mice.

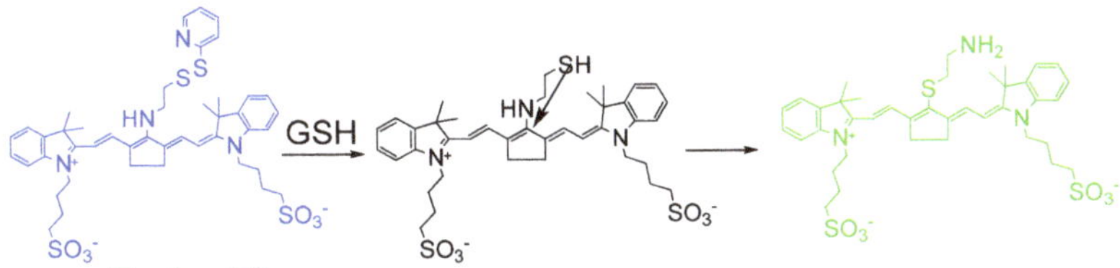

Probe 22

Figure 21. Structure of **Probe 22** and its reaction with GSH.

2.1.5. Detection of NPSH via Other Strategies

In addition to the thiol detection probes presented above, numerous other thiol detection probes have been developed based on other mechanisms or strategies [30,62,95]. Although not covered in this review, there are also a number of thiol detection probes which were developed based on nanoparticles and nanocomposites, and metal ion displacement and coordination [46]. These different strategies also emerged as promising strategies for detection of thiols.

2.2. Selective Detection of Cys and Hcy

Cysteine (Cys) is involved in many biological processes, such as protein synthesis, detoxification, metabolism, and post-translational modification [61]. Abnormal Cys level has been linked with many diseases such as slow growth, Alzheimer's disease and cardiovascular diseases et al. Hcy is associated with activation of multiple signal pathways and is found to be associated with cardiovascular diseases [61]. Accurate and effective detection of Cys and Hcy are essential for further exploring the roles these thiols in the biological system, A tool that can measure Cys/Hcy is in need. Although many fluorescent probes have been developed for thiols, special probes capable of discriminating Cys, Hcy, and GSH are still limited. The high GSH concentration (1–10 mM) makes it even more challenge to detect Cys and Hcy since the concentrations of Cys (30−200 μM) and Hcy (5–15 μM) are in μM.

Nevertheless, many Cys selective detection probes have been reported based on various mechanisms. These mechanisms include native chemical ligation (NCL) [96,97], aromatic substitution–rearrangement [98], and cyclization with aldehydes or acrylates [43,99–101]. The fluorophores employed in these probes include curcumin [38], coumarin [102,103], BODIPY [98,104,105], naphthalimide [106], rhodamine [107], fluorescein [108], and pyrene [43]. In the following, some of the mechanisms used to distinguish Cys/Hcy are presented. Most of the probes take two steps to achieve the selective detection of Cys vs. Hcy. Step one usually is the same for all the thiols (GSH, Cys, or Hcy) that is a reaction between the highly nucleophilic thiolate and a probe to form a thiol adduct. Step two commonly involved an intramolecular rearrangement reaction or cyclization reaction which occurs for Cys/Hcy adducts or Cys adduct only. Many selective probes for Cys or Hcy were developed based on the extended version of these two steps although some other probes achieve the discrimination of Cys and Hcy by the kinetic difference of the reaction between the different thiols and probes.

2.2.1. Selective Detection of Cys and Hcy through Cyclization of Cys/Hcy with Acrylates or Aldehydes

Cyclization of Cys/Hcy with aldehydes or acrylates is one of the most extensively used mechanism for selectively detection of Cys/Hcy over GSH. As mentioned above, a two-step reaction takes place for the selective detection of Cys/Hcy. The method starts with a Michael addition reaction from the thiolate of a thiol (GSH, Cys, or Hcy) to form a thioether. The thioether derived from Cys or Hcy will go through a cyclization to produce a seven members-ring cyclic amide (Cys) or eight-member ring cyclic amide (Hcy) at a different reaction rate while the thioether derived from GSH is not able to undergo the cyclization reaction. Many probes have been designed based on this mechanism to selectively detect Cys/Hcy over GSH (Figure 22). For example, Xiaofeng Yang et al. [99] reported in 2011 a benzothiazole derivative which utilized an α, β-unsaturated carbonyl recognition as the thiol responsive unit to selectively detect Cys.

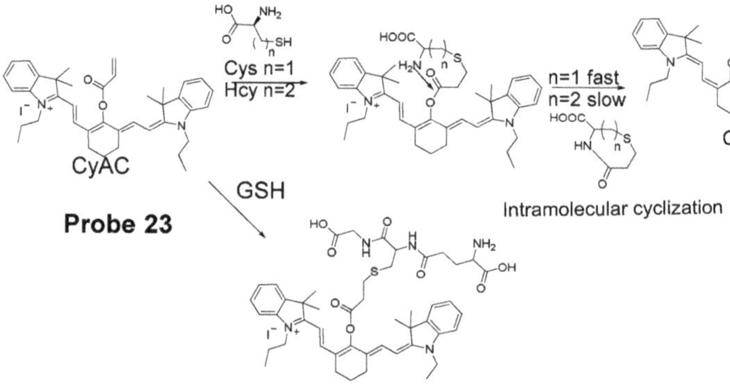

Figure 22. Probe 23 and its reaction mechanism for distinguishing Cys over GSH and Hcy.

Zhiqian Guo et al. [102] developed a ratiometric near-infrared fluorescent probe (CyAC, **Probe 23**) for selective detection of Cys over Hcy and GSH. As depicted in Figure 22, CyAC is composed of a NIR cyanine fluorophore and an acrylate group to serve as the thiol response unit. A significant optical property change (from emission 780 nm to emission 570 nm) was observed upon reaction with Cys, while GSH and Hcy were not able to induce the change confirming the selectivity of the probe for Cys. Further, a linear relationship was found between the fluorescence intensity ratio $I_{560\,nm}/I_{740\,nm}$ and Cys concentration in the range of 0–25 µM. The different kinetic of the cyclization reaction was the reason for inducing the difference between Cys and Hcy, since the seven-membered ring (Cys) should be formed easier and faster than eight-membered ring (Hcy). Under the same condition, GSH was not able to induce the cyclization reaction at all.

Colorimetric probes are another widely used reagents which enable the recognition of analytes by naked-eye and do not need an instrument [47,109]. Lanfang Pang et al. [38] published their work for development of a new curcumin-based fluorescent and colorimetric probe (CAC, **Probe 24**) for specific detection of Cys. Curcumin is used because of its excellent optical properties and great biocompatibility [110,111]. As depicted in Figure 23, CAC was made by incorporating two acrylate groups with curcumin to serve as thiol recognition sites. The CAC is easier to synthesize. The detection starts from a Michael addition reaction between the thiolate of Cys and the α, β-unsaturated ketone of CAC followed by a cyclization to produce a seven membered-ring cyclic amide and a free curcumin molecule [44,112,113]. Unlike other fluorescence probes, CAC itself carries a strong fluorescence ($\Phi F = 0.20$) which will be quenched after being activated by Cys ($\Phi F = 0.02$), whereas the fluorescence quenching effect was not observed with other thiols like Hcy and GSH. The probe itself is colorless but can be turned into yellow color by Cys which could be easily visualized by naked eyes. In vitro cell experiments in PC12 cells showed that the probe was able to detect the change in Cys levels in live cell. The probe was further successfully applied for detecting Cys status on zebrafish.

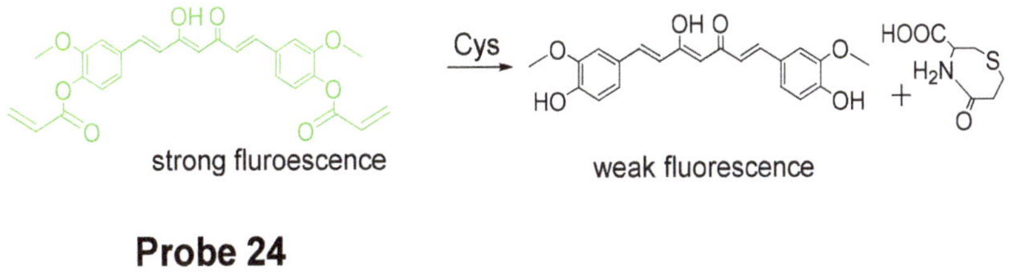

Probe 24

Figure 23. Structure of **Probe 24** and its detection mechanism with Cys.

Yao Liu et al. reported [44] a ratiometric fluorescent probe (**Probe 25**) for specific detection of Cys based on a visible-light excitable excited-state intramolecular proton transfer (ESIPT) dye (4′-dimethylami- no-3-hydroxyflavone) as the fluorophore. As illustrated in Figure 24, an acrylate group was anchored to the fluorophore to block the ESIPT and consequently quench the fluorescence. Recognition and cleavage of acrylate moiety by Cys will restore the ESIPT process and induce a remarkable fluorescence enhancement and emission wavelength shift (471 to 550 nm). The "off to on" fluorescence switch starts from a Michael addition reaction between the thiolate of Cys and the α, β-unsaturated ketone of the acrylate group followed by an intramolecular cyclization to produce a seven membered-ring amide. A kinetic difference of the intramolecular cyclization resulting in a selective detection of Cys over Hcy and GSH. A steric effect could also possibly elevate the kinetic difference of the reaction from Cys over Hcy. Upon completion of the reaction, a nearly 40-fold increase of ratiometric value of the emission intensities ($I_{550}/_{471}$) was induced. Additionally, the probe was able to detect Cys level changes in HeLa cells. Additional advantages of the probe include a large Stokes shift (>130 nm), fast response (within minutes) and high sensitivity (~0.2 μM) for Cys.

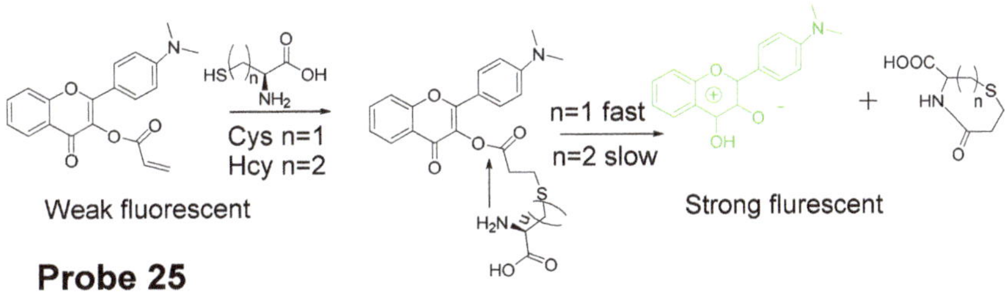

Probe 25

Figure 24. Structure of **Probe 25** and its reaction with Cys and Hcy.

Similar to the structure of acrylate, a bromoacetyl group was also used to develop probes for detection of Cys/Hcy employing the same sensing mechanism. In 2013, fluorescein chemodosimeter was developed by Keum-Hee Hong et al. [114] as a fluorescent probe (**Probe 26**, Figure 25) for detection of Cys over Hcy and GSH. The reagent was made by introducing a bromoacetyl group to fluorescein. The selectivity for Cys over Hcy and GSH in aqueous solution is due to the thermodynamically stable and kinetically rapid formation of six-member lactam ring in aqueous solution with Cys [115]. The reagent was capable of detecting Cys changes in HeLa cells.

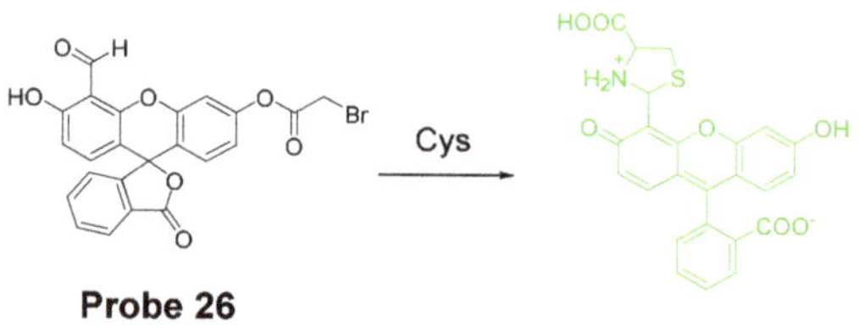

Figure 25. Structure of **Probe 26** and its reaction with Cys.

Hye Yeon Lee et al. [43] reported two new pyrene-based fluorescent probes (**Probes 27, 28**, Figure 26) for selective detection of Hcy over Cys and other thiols. The two probes were converted rapidly (10 min) to a high fluorescence (350 nm and 450 nm) form when mixed with Hcy, while no fluorescence response was observed when mixed with Cys and GSH even over a long reaction time. The detection limit of Hcy was determined to be 1.44×10^{-7} M with a linear range of 600–1000 nM in HEPES containing 10% DMSO (0.01 M and pH 7.4). The authors provided an explanation for the fluorescence response difference for Hcy over Cys and GSH. These probes have been applied for the detection of Hcy changes in mammalian and showed no response to the change of GSH.

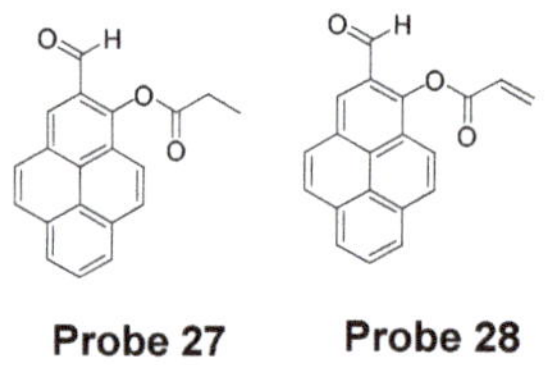

Figure 26. Structures of **Probes 27** and **28**.

2.2.2. Selective Detection of Cys and Hcy through Cys-Induced SNAr Substitution−Rearrangement Reaction

The Cys-induced SNAr substitution−rearrangement reaction used for selective discrimination of thiols has been widely explored. Many probes have been developed based on various fluorophores that possess the ability to undergo a S to N displacement rearrangement reaction presented early. The fluorophores employed include BODIPY, NBD, naphthalimide, and cyanine (Figure 27). The requirements for fluorophores to be applied for the purpose are that they should be able to form a thioether (GSH, or Cys/Hcy) first. Then the amino group of Cys/Hcy conjugated thioether will rearrange by displacing the sulfur of the thioether with the amino group to form amino-substituted product. As described early, the rearrangement does not occur with GSH thioether. An example of the mechanism is illustrated in Figure 28. The remarkable optical property differences between sulfur and amino substituted BODIPY enable the discrimination of GSH over Cys/Hcy.

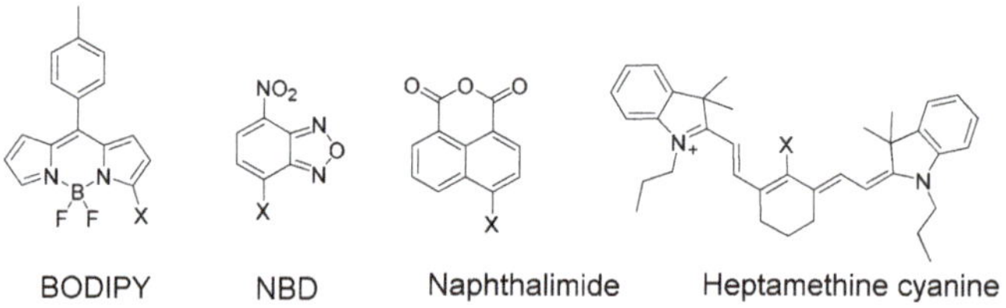

Figure 27. Representative fluorophores used for selective discrimination of thiols via Cys-induced SNAr substitution−rearrangement reaction.

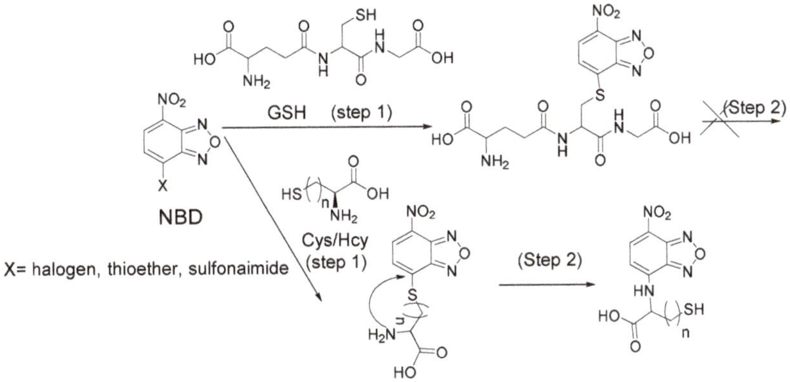

Figure 28. Illustration of the Cys-induced SNAr substitution−rearrangement reaction.

BODIPY

BODIPY fluorescent dyes featured with a high molar absorption coefficient, quantum yield, and photo stability which enable them to be widely applied in many fields [116].

In 2012, Liya Niu et al. [29] developed monochlorinated boron dipyrromethene (BODIPY) derivatives to selectively detect GSH over Cys/Hcy. However, the emission band of amino-substituted BODIPY (Cys and Hcy) is close to the emission band of the original monochlorinated boron dipyrromethene (BODIPY) probe, Thus, the probe failed to detect Cys and Hcy. In 2013, a structure modification [117] by adding a nitrothiophenyl or nitrophenyl group to the same BODIPY fluorophore was employed to address the problem (**Probe 29**). As depicted in Figure 29, the probe itself displays no fluorescence (PET) but can be turned on to fluorescent products upon reaction with thiols followed by the occurrence of the rapid intramolecular displacement of sulfur with the amino group of Cys and Hcy. The reaction of Cys is much faster than Hcy while no such intramolecular displacement occurred for GSH achieving the ability of discrimination of Cys from Hcy and GSH. The probe itself shows barely any fluorescence but can be turned on to strong fluorescence by thiols with the product derived from Cys exhibiting emission wavelength at 564 nm while the products derived from GSH and Hcy emitting at 588 nm. The fluorescence intensity at 565 nm followed a nicely linear relationship with the Cys concentration (0–100 µM) with a detection limit of 2.12×10^{-7} M. The probe was successfully shown to be capable of detecting Cys status change in live cells.

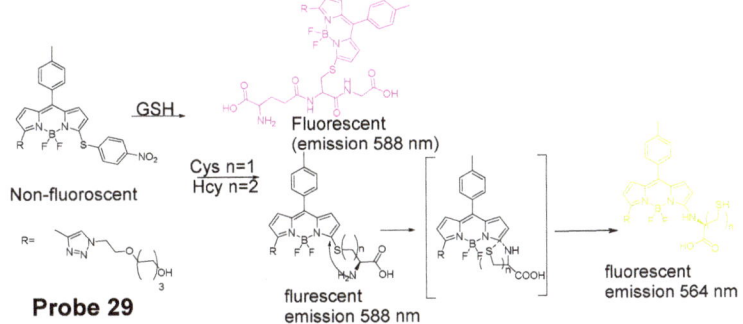

Figure 29. Structure of **Probe 29** and its sensing mechanisms for distinguish Cys/Hcy and GSH.

7-Nitrobenzofurazan (NBD)

7-Nitrobenzofurazan (NBD) is another widely used and extensively studied fluorophore that is able to go through the Cys-induced SNAr substitution−rearrangement reaction for selective detection of thiols (Figure 28).

In 2014, Liya Niu et al. [118] explored NBD as another fluorophore that can go through the intramolecular displacement mechanism to discriminatively detect different thiols. As shown in Figure 30, Cys/Hcy and GSH form different release fluorophores NBD, N-substitute products (Cys/Hcy) or S-substitute product (GSH). The N-substitute product (Cys/Hcy) exhibit significantly higher fluorescence than the S-substitute product (GSH). A simple probe NBD-Cl (**Probe 30**) was developed and successfully applied for detecting Cys in live cells. Pinaki Talukdar et al. [119] found the same phenomenon which further push the development of Cys/Hcy selective probes based on NBD.

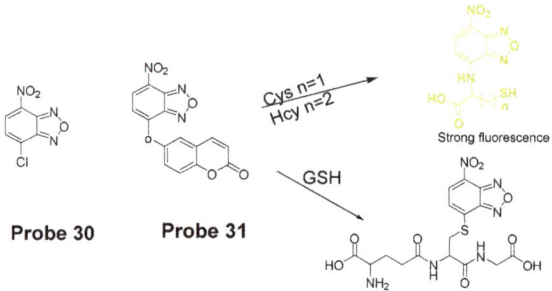

Figure 30. Structures of **Probes 30, 31** and their reaction toward GSH and Cys/Hcy.

In 2019, Lihui Zhai et al. [120] developed a dual emission turn on fluorescent probe (CA-NBD, **Probe 31**) for discriminative detection of Cys/Hcy and GSH simultaneously based on NBD. The probe was designed by combing another fluorophore coumarin together with 7-nitrobenzofurazan (NBD). Two different NBD derivatives formed, one is N-substitute products (Cys/Hcy) and the other is S-substitute product (GSH). At the excitation wavelength of 460 nm, GSH-NBD (S-substitute) derivative was non-emissive. In contrast, the Cys/Hcy-NBD derivatives exhibited strong yellow fluorescence enabling the discrimination of Cys/Hcy over GSH. The fluorescence intensity was linearly increased when the thiol concentration with a detection limit of 2.00×10^{-8} M for Cys and 1.02×10^{-8} M for Hcy, respectively. The probe can successfully detect the Cys status change in live cells.

A dual channel responsive NIR fluorescent probe (**Probe 32**, Figure 31) for selective detection of Cys in live cells was reported by Zhuo Ye et al. [121] in 2017. The probe named as BDY-NBD is composed of two fluorophores, a NIR BODIPY fluorophore, BDY-OH, and a NBD fluorophore. In the structure, NBD is employed as the thiol responsive unit and the fluorescence of BDY-OH is quenched by NBD owing to PET. Once the probe reacts with a thiol, the nucleophilic thiolate will cleave the NBD from BDY-OH and attach to NBD to form a NBD thiol adduct. The NIR fluorescence of BDY-OH (λex = 650 nm, λem = 735 nm) will be recovered because of the termination of PET. Different from GSH, the thiol adduct (Cys-NBD) produced in the reaction between Cys and probe will subsequently undergo an S to N acyl shift reaction to form a much stronger fluorescent product with a much different emission wavelength (Ex. 466 nm, Em. 540 nm) than BDY-OH while the shift reaction does not occur for GSH. Therefore, a dual-emission mode can be utilized for the detection of Cys. The probe has a much low detection limitation (22 nM, 540 nm emission) that is lower than most of the fluorescent probes developed for the detection of Cys previously [100,122,123]. In vitro cell experiments with HeLa cells demonstrated that the probe can be applied for multicolor imaging of intracellular Cys, and a significant differences of fluorescence imaging were observed for GSH and Cys in live cells when their levels are manipulated.

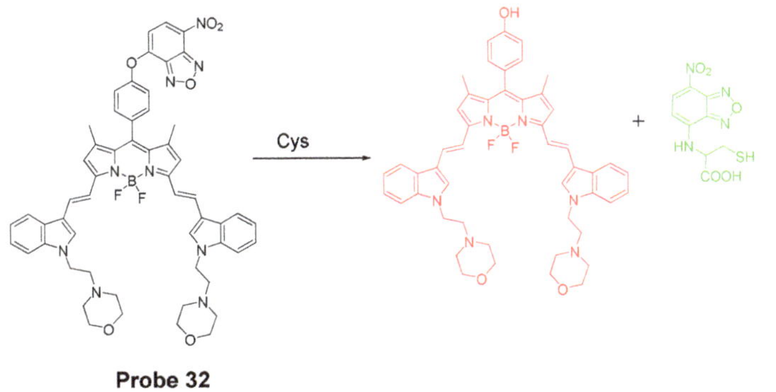

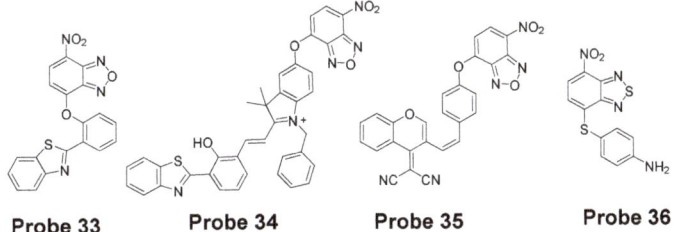

Probe 32

Figure 31. Structure of **Probe 32** and its reaction with Cys.

Many other probes have been designed and applied based on different thiol recognition groups or other modifications. Shuangshuang Ding et al. [124] developed the probe (**Probe 33**, Figure 32) by combing a benzothiazole molecule and 7-nitrobenzofurazan (NBD) molecule through an ether bond. Longwei He et al. [125] designed and synthesized a probe (**Probe 34**, Figure 32) by attaching NBD to another fluorophore hydroxyphenyl benzothiazole merocyanine (HBTMC) through an ether bond. The probe was able to detect Cys/Hcy, GSH, and H_2S at different emission wavelengths. Peng Wang et al. [126] developed a NIR probe (**Probe 35**, Figure 32) named as DCM-NBD which was made by combining a NIR fluorophore icyanomethylene-4H-pyran (DCM) derivatives and a NBD fluorophore via ether linkage. The probe was successfully being applied not only in vitro but also in animals.

Probe 33 **Probe 34** **Probe 35** **Probe 36**

Figure 32. Structures of **Probe 33, 34, 35,** and **36**.

Similar with the structure of NBD, a nitrobenzothiadiazole probe (**Probe 36**, Figure 32) was also developed by Dayoung Lee et al. [127] to achieve the goal of selective detection of Cys and Hcy over GSH based on the same mechanism. The probe was found to increase fluorescence at pH 7.4 in the presence of Cys or Hcy, while at weakly acidic conditions (pH 6.0), only Cys can induce the fluorescence enhancement.

Other Fluorophores

In addition to NBD, BODIPY, probes based on naphthalimide, and cyanine [118] have also been extensively developed, such as 4-nitro-1,8-naphthalic anhydride (NNA, **Probe 37**, Figure 33) which was reported by Limin Ma et al. [128] in 2012 based on nitro-naphthalimides as the fluorophore and aromatic nitro part as the thiol recognition group. The probe can selectively detect Cys with a detection limit of 0.3 µM by following the same mechanism (Figure 28) exhibited for NBD and BODIPY.

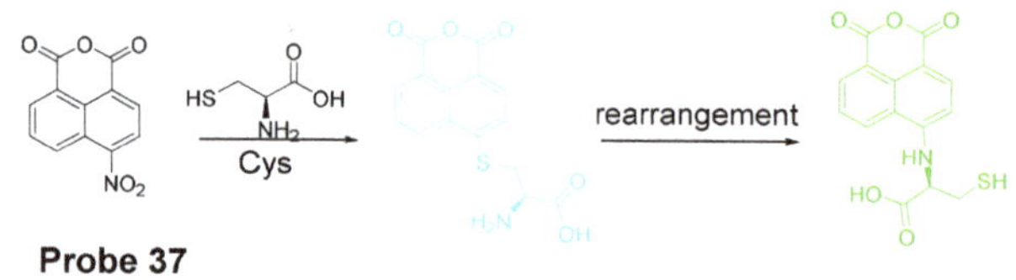

Figure 33. Structure of **Probe 37** and its reaction with Cys.

2.2.3. Selective Detection of Cys and Hcy via Other Mechanisms

Many other probes have been developed for discrimination of Cys/Hcy over GSH based on different sensing mechanism. For instance, Hyo Sung Jung et al. [129] developed a probe (**Probe 38**, Figure 34) in 2011 that can selectively detect intracellular Cys by the Michael addition reaction in combination with a steric hindrance factor. A bulky substituent was added to achieve the preferential response for Cys relative to GSH since the molecule of GSH or Hcy is larger than Cys. In addition, the lower pKa of Cys than Hcy and GSH is another factor leading to a higher nucleophilicity of Cys. The fluorescence intensity was found to be linearly increased with an increase in Cys concentration in a range of 0–0.9 mM. The detection limit was determined to be 10^{-7} M for Cys. Later, the probe was successfully applied for monitoring intracellular Cys in HepG2 cells. The fluorescence derived from the probe in live cells decreased after an increase in NEM concentration. Xin Zhou et al. [130] developed a Cys selective detection probe (**Probe 39**, Figure 34) based on a Michael addition reaction assisted by an electrostatic attraction.

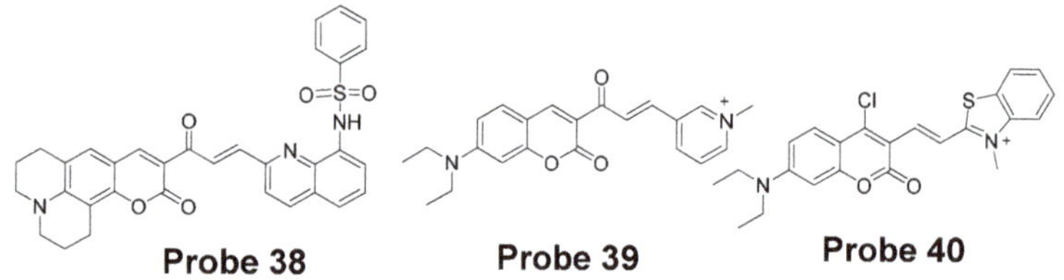

Figure 34. Structures of **Probes 38, 39**, and **40**.

While most of the probes being developed are capable of sensing only one of the thiols at a time, a probe which can selectively detect two or three thiols simultaneously with different emissions is highly desirable. In order to address the need, Jing Liu et al. [102] developed a chlorinated coumarin-hemicyanine probe (**Probe 40**, Figure 34) which can simultaneously detects Cys and GSH separately from two different emission and excitation channels. Similar with the two steps strategy for discrimination of thiols, this new coumarin-hemicyanine dye utilize the same strategy starting from a SNAr nucleophilic substitution first to produce a thio-coumarin-hemicyanine, followed by a different intramolecular rearrangement reaction (GSH vs. Cys/Hcy) or different kinetic of the same intramolecular reaction (Cys vs. Hcy) to selectively detect GSH or Cys. The probe is non-emissive and can be turned on by Cys and GSH with emission maximum at 420 and 512 nm, respectively, indicting the high selectivity of the probe for Cys or GSH. For both Cys and GSH, a linear relationship was found between the fluorescence intensity at their emission wavelength and their concentrations (GSH, 0 to 0.9 equiv; Cys, 0.4 to 0.8 equiv). The probe was demonstrated to be able to simultaneously monitor Cys and GSH in COS-7 cells in multicolor imaging.

3. Detection of Thiols in Subcellular Organelles

Thiols are known to distribute unevenly in subcellular organelles [8] such as mitochondria, lysosomes, endoplasmic reticulum (ER), golgi apparatus, nucleus, and cell surface.

Their levels in different organelles could be an indicator or biomarker to reflect the status of the organelles. Thus, tools or methods that can map thiol distribution and monitoring their status changes not only for the whole cell but also for subcellular organelles will provide valuable information to understand the relationship between thiols and cells' functions and dysfunctions, and reveal their correlation as well as find treatment for various thiol-associated diseases [9]. While extensive work has been made and plenty of probes have been developed for the detection of thiols for the whole cell, efforts have been made to develop tools for detecting thiols in subcellular organelles. A subcellular organelle targeting structure is usually needed for a subcellular thiol imaging probe to direct the probe into the targeted organelle. A few subcellular organelle targeting thiol probes have been developed based on the difference identified for subcellular organelles. Most of these probes are for mitochondria with a few for lysosome and ER. In the next part, the diverse targeting approaches and sensing mechanisms of subcellular organelle-targeted thiols probes will be discussed and summarized.

3.1. Mitochondrial Total Thiols and GSH

As one of the most important subcellular organelles in cells, mitochondria are double-membrane constructed organelles that serve as the energy generator and also involved in many other biological process including calcium circulation, protein synthesis, apoptosis pathways, etc. A large amount of reactive oxygen species (ROS) are generated in mitochondria during the energy generation process [131–133]. With thiols (especially GSH) being the most important antioxidant to terminate ROS in cells, it is of importance to have a tool that can monitor mitochondrial thiols in live cells. Inspired by the knowledge that mitochondrial membrane spans carry a negative charge [134], a lipophilic cationic structure are widely used for achieving the mitochondria targeting. These lipophilic cations include triphenylphosphonium (TPP), positive charged cyanine, and rhodamine (Figure 35). In addition to lipophilic cations, functional groups, such as peptides [135] have also been developed for mitochondria targeting [136]. In this section, probes developed for detection of mitochondria thiols will be illustrated based on their sensing mechanisms.

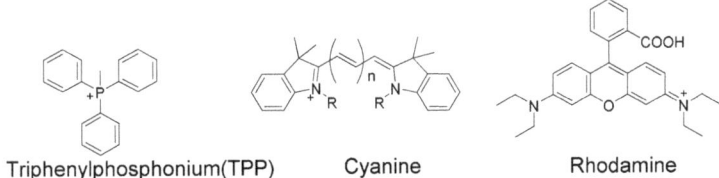

Figure 35. Representative mitochondrion-targeting.

3.1.1. Detection of Mitochondrial Thiols via a Michael Addition Reaction

In 2020, Yutao Yang et al. [137] developed a NIR fluorescent probe (NIR-HMPC, **Probe 41**, Figure 36) based on a Michael addition reaction followed by self-immolative reaction to selectively detect mitochondrial thiols. This NIR probe made by conjugating hemicyanine dye with a benzopylium was connected with a 7-hydroxymethyl-2,3-dihydro-1H-cyclopent-a[b]chromene-1-one with a carbonyl ester. The positive charged indoles iodized salt in the molecule attributes to the mitochondria selectivity. Upon exposure to thiols, nucleophilic addition of the sulfhydryl group will lead to the chromene ring-open to result in a released phenol anion. Subsequently, phenol anions will trigger a self-immolative reaction to generate carbon dioxide and release a strong fluorescent NIR product (λ_{em} 731 nm). The fluorescence intensity (731 nm) was found to be linearly increased as the concentration of GSH, Cys, and Hcy increased with the detection limit of 0.59 μM, 0.39 μM, and 0.54 μM, respectively. Colocalization studies with DAPI and Mito-Tracker Green in MCF-7 cells, HepG2 cells, and HeLa cells clearly confirmed that NIR-HMPC can effectively imaging mitochondrial thiols. When PMA (Phorbol 12-myristate

13-acetate, a cellular oxidative stress inducer) and BSO (a sulfoximine that can reduce GSH levels) were used to pretreat cells to induce an intracellular thiol decrease, a dramatic fluorescence decrease was observed revealing that the probe was able to visualize and reflect mitochondrial thiol change in live cells. More impressively, the probe can not only reflect thiol change in vitro but also reflect the change in vivo in a mouse model. The probe has also been successfully applied on sensing thiol change in a mouse mode of cerebral ischemia, warranting the application of the probe for monitoring the physiological and pathological processes at the cellular and animal levels.

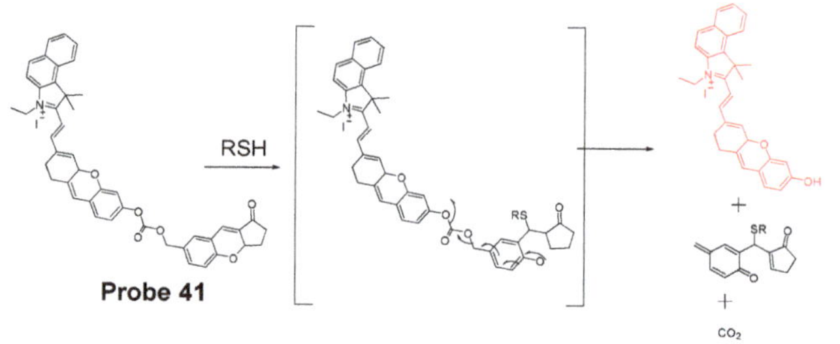

Figure 36. Structure of **Probe 41** and its reaction with thiols.

In 2017, Jianwei Chen et al. [138] developed a new mitochondrion-targeted probe (MitoRT, **Probe 42**, Figure 37) which could detected the mitochondrial GSH status reversibly. In the molecule, TPP, the positive charged mitochondria targeting group, was linked to a fluorophore via an optimized 4-carbon linker. The reagent is capable of reacting with GSH rapidly in both forward and reverse directions that enable the real time monitoring of mitochondrial GSH dynamically in live cells. In addition, MitoRT showed ratiometric fluorescence responses with a wide dynamic range when reacting with GSH. The probe has been demonstrated to be capable of monitoring mitochondrial GSH in live HeLa cells reversibly through fluorescence microscope. The probe can also be applied to monitor mitochondrial GSH levels in a high throughput manner (flow cytometry).

Figure 37. Structure of **Probe 42** and its reversible reaction with GSH.

In 2017, Keitaro Umezawa et al. [139] reported their work on developing reversible fluorescent probe (**Probe 43**) for live cell imaging and ratiometric quantification of fast GSH dynamics in mitochondria based on a cationic rhodamine fluorophore as an electrophile for Michael addition to reversibly react with thiols. The calculated rate constant $k = 560$ M^{-1} s^{-1} ($t_{1/2}$ = 620 ms at GSH = 1 mM) which is a 3900-fold increase than TQ green [73]. Similar with GSH, other thiol-containing molecules such as Cys, Hcy, and H_2S can also induce the fluorescence change if their concentration is in millimolar range. However, since their physiological concentration is much lower than GSH, their interference with GSH should be a minimum. As shown in Figure 38, a TMR (typical rhodamine structure) was introduced to the probe as a donor of the FRET. Based on that, two probes, QG0.6 and QG3.0 were designed and synthesized. Both probes exhibit ratiometric absorption/fluorescence changes in response to various concentrations of GSH. Co-staining experiments with MitoTracker Deep Red FM indicated that the compound could accumulate inside mitochondria. QG3.0 can quantitatively monitor GSH status change in real time when cells are consecutively treated with H₂O₂ or GSH ester. The probe is also capable of monitoring the real time GSH level change after the cells being glucose deprived.

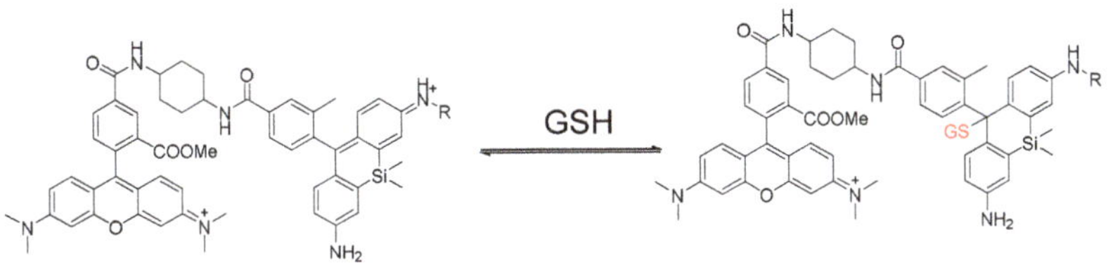

Probe 43

Figure 38. Structure of **Probe 43** and its reversible reaction with GSH.

3.1.2. Detection of Mitochondrial NPSH via a SNAr Reaction Using Halogen, Ether, or Thioether as a Leaving Group

Detection of Mitochondrial NPSH via a SNAr Reaction Using Halogen as a Leaving Group

In 2017, Xueliang Liu et al. [140] reported a mitochondria-targeting fluorescent probe (BODIPY-PPh₃, **Probe 44**) for selective and ratiometric detection of GSH over Cys and Hcy in line with the group's previous work [29]. As depicted in Figure 39, chlorinated BODIPY was used as the fluorophore and the chlorinated site serves as thiol (GSH, Cys, or Hcy) recognition site through a SNAr reaction to afford the thioether. As discussed early, the thioether formed with Cys and Hcy underwent an intramolecular rearrangement to yield the amino derivatives resulting in a remarkable change of photophysical properties, while the thioether formed with GSH could not undergo such an intramolecular rearrangement. The difference in the intramolecular rearrangement reaction for GSH vs. Cys and Hcy attribute to the selective detection of GSH over Cys and Hcy. To achieve mitochondria targeting, a TPP structure was attached to the probe. The probe itself shows an intrinsic emission at 557 nm which decreases dramatically upon reaction with GSH and a new emission at 588 nm (Ex. 550 nm) increased significantly. Their ratio (I_{588}/I_{557}) was found to increase linearly with GSH concentration (0–80 μM) with a detection limit of 1.1 μM. The linearity enables the quantitative detection of GSH. The fluorescence behavior of the probe reacting with Cys and Hcy was totally different with a much lower emission at 588 nm and much smaller (I_{588}/I_{557}) ratio change. The colocalization study with rhodamine 123 found that BODIPY-PPh₃ could accurately locate in the mitochondria. Pretreatment of the cells with NEM to deplete GSH resulted in a lower red/green fluorescence ratio (~2). In contrast, pretreatment of cells with GSH before incubation with BODIPY-PPh₃ led to a higher red/green channels fluorescence ratio (~5). The results reveals that the BODIPY-PPh₃ can effectively and ratiometrically monitor the status of mitochondrial thiols.

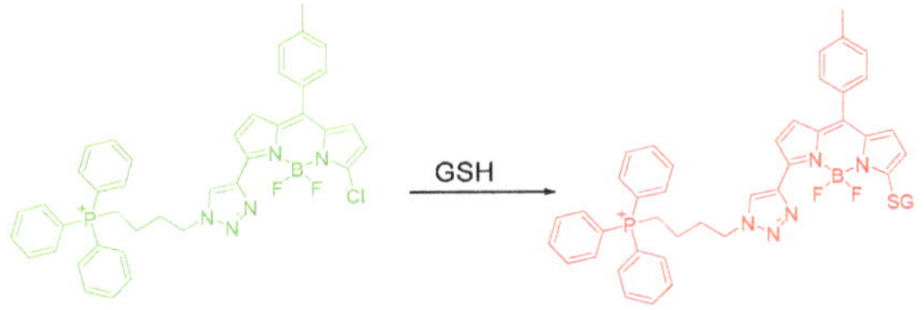

Probe 44

Figure 39. **Probe 44** and its reaction with GSH.

In 2018, Sujie Qi et al. [141] presented a water-soluble ratiometric fluorescent probe (**Probe 45**) for selective detection of GSH in mitochondria of live cells. The probe was built based on colorimetric hemicyanine dye and was capable of discriminating GSH over Cys/Hcy in a quantitative manner in a range of 1.0–15.0 mM with a low detection limit in aqueous solution. In detail, the hemicyanine dye was modified by an aldehyde group and a chloro-substitution which equip the probe the ability to selectively react with GSH since GSH reacts with the probe via a SNAr and an intramolecular aldimine condensation to form a ring (Figure 40). The reaction was much faster than the reaction with Cys and Hcy since an intramolecular aldimine condensation to form a ring could not occur with Cys and Hcy [65,142]. The quaternary ammonium cation serves as the mitochondria targeting moiety and a hydrophilic sulfonate was added to enhance the water-solubility of the probe. Upon reaction with GSH, significant optical property (absorption and fluorescence) changes were induced which significantly dropped fluorescence intensity at 607 nm along with appearance of a new fluorescence peak at 648 nm. The ratio of emission intensities (I_{648}/I_{607}) displayed a linear relationship with GSH concentration (1.0 to 15.0 mM) with a detection limit of 24.16 µM, enabling the quantitative detection of GSH. The probe displays a great mitochondria targeting ability and can well reflect the status changes of mitochondrial GSH in a ratiometrical manner.

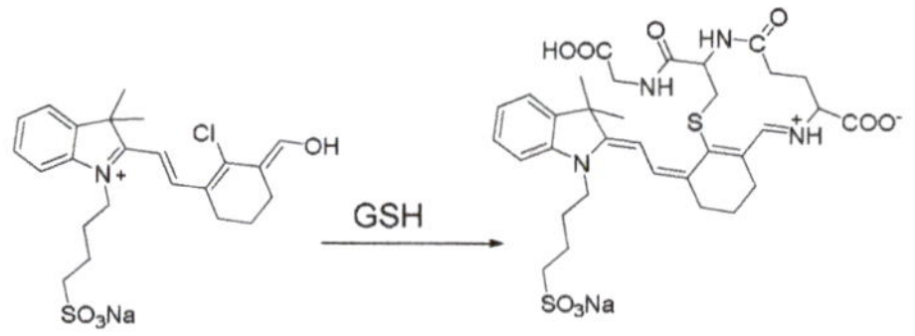

Figure 40. Structure of **Probe 45** and its reaction with GSH.

Detection of Mitochondrial NPSH via a SNAr Reaction Using Thioether as a Leaving Group

In 2018, we developed a mitochondrial-targeting rhodamine based probe [22] (TBROS, **Probe 46**) for selective imaging and quantification of mitochondrial thiols in live cells. The probe was designed based on a thiol specific thiol–sulfide exchange reaction reported by us in 2012 [42]. Similar to the GUALY's reagent [42], two benzofurazan moieties were connected together by a thioether to maintain a symmetric structure (Figure 41). Two cationic rhodamine B units were introduced symmetrically to the benzofurazan structure to provide the mitochondria selectivity and also serve as a strong fluorophore to enhance the fluorescence intensity of benzofurazan. The probe itself is not fluorescent due to the self-quenching of the symmetric structure but turns into strong fluorescent products (Ex. 550 nm, Em. 580 nm) upon reaction with thiols through the thiol specific thiol–sulfide exchange reaction. TBROS has been confirmed nonreactive toward PSH and successfully used to image and quantify NPSH in mitochondria in live cells.

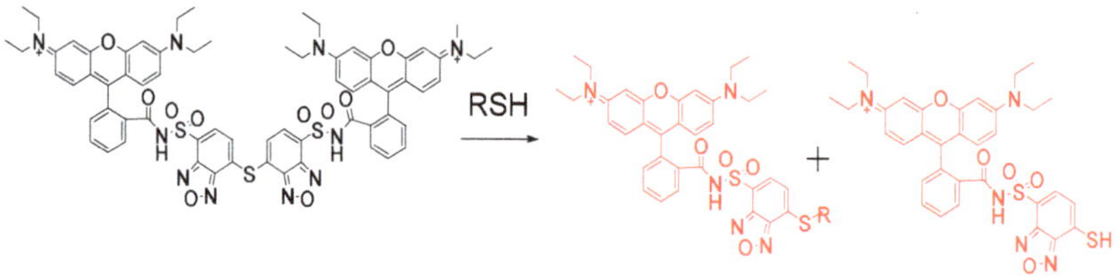

Figure 41. Structure of **Probe 46** and its reaction with thiols.

In 2019, Zhiqiang Xu et al. [143] developed a NIR probe (Cy-S-Np, **Probe 47**, Figure 42) based on thioether as the thiol response unit to visualize mitochondrial and lysosomal GSH in live cells and also in a mouse model. The probe was constructed by connecting a cationic cyanine IR-780 dye (mitochondria targeting) and a morpholine-coating naphthalimide (lysosome targeting) through a thioether (thiol response moiety). Cy-S-Np can effectively distinguish GSH over Cys and Hcy in mitochondria and lysosome, warranting its promising use for studying the interaction of cellular function between mitochondria and lysosome. After reaction with GSH, a remarkable fluorescence intensity enhancement at 812 nm (Ex. 710 nm) was observed. Impressively, the probe showed a detection limit of 11 nM. Cellular experiments showed that the probe was able to detect GSH changes both in mitochondria and in lysosome of live HeLa cells. Further, visualization of subcellular GSH level at organism level in mice was achieved as well.

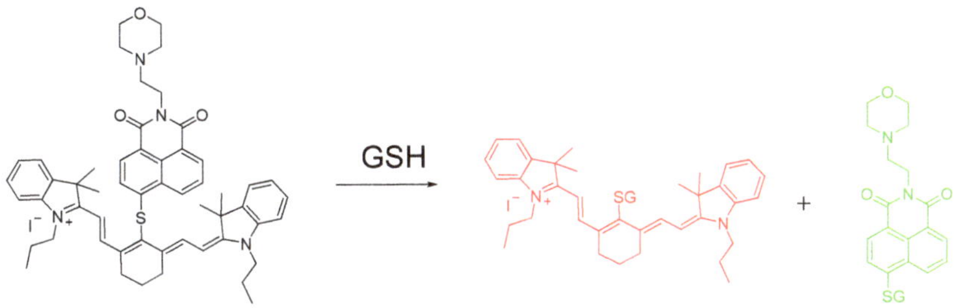

Figure 42. Structure of **Probe 47** and its reaction with GSH.

Detection of Mitochondrial NPSH via a SNAr Reaction Using Ether as a Leaving Group

Yuan Gu et al. [35] developed a ratiometric reagent which can be used in two-photon fluorescence microscopy (TPFM). The two photon excitation fluorescent imaging has the advantage of less interference from autofluorescence background, deeper penetration to tissues, and low phototoxicity [144–147]. The regent (**Probe 48**, Figure 43) named as TPE-PBP showed a high sensitivity and selectivity toward thiols, including GSH, Cys, Hcy, etc., with a high selectivity for mitochondria. Cell experiments shows that TPE-PBP was able to measure the mitochondrial thiol status change in a ratiometric manner. The two-photon-absorption cross section enables the regent to apply in live cells, in living skeletal muscle tissues, and in two day old fish larva. The success of this reagent provides a new strategy for the construction of ratiometric two-photon active reagens on the application of in vivo biosensing and bioimaging application [35]. It needs to be noted that the application of traditional fluorophore probes was hampered by the aggregation-caused quenching (ACQ) effect that is that the probes' fluorescence will be quenched or weakened if the probes are concentrated or aggregated [148–150]. The ACQ effect makes the quantification of analytes in live cells more challenging. Tang and co-workers observed an unusual phenomenon on a class of molecules with propeller shape (e.g., tetraphenylethylene (TPE), siloles) which can turn on their fluorescence in the aggregate or solid states while keep non-emissive in diluted solutions. This phenomenon was termed as "aggregation-induced emission (AIE)" and the mechanism can be explained by a restriction of intramolecular motions (RIMs), including restriction of intramolecular vibrations (RIVs) and restriction of intramolecular rotations (RIR) [151–159]. Herein, AIE probes can overcome the drawbacks of the traditional fluorophore probes.

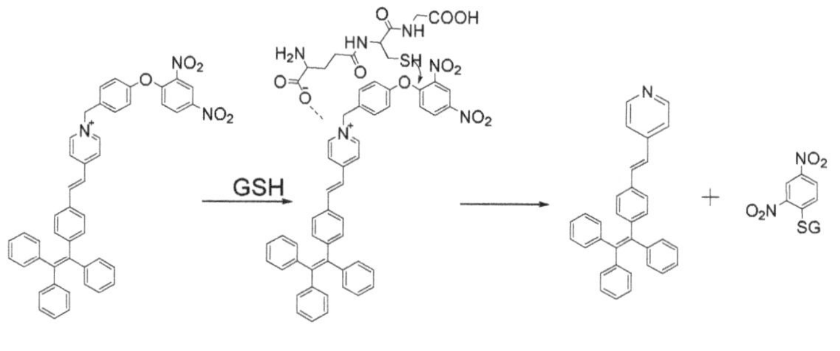

Probe 48

Figure 43. Structure of **Probe 48** and its reaction with GSH.

In 2016, Jian Zhang et al. [160] developed a probe (**Probe 49**, Figure 44) for detecting mitochondrial GSH with BODIPY serving as the fluorophore. In this approach, a self-immolative dinitrophenoxy benzyl pyridinium was connected with BODIPY to serve as the targeting ligand and also the GSH recognition group. Once being exposed to GSH, the dinitrophenyl moiety will be cleaved through a SNAr reaction, followed by a self-immolation reaction to release fluorophore BODIPY with strong fluorescence (599 nm). A linear relationship was observed between the fluorescence intensity and GSH concentration in a range of 1 to 15 µM with a detection limit of 109 nM. The selectivity of GSH over Cys and Hcy was possible due to an electrostatic interaction between the cationic pyridinium moiety and GSH: once the electrostatic interaction formed, GSH was believed to react much faster than Cys and Hcy. The probe has been successfully used to monitor the status of mitochondrial GSH in live HeLa cells.

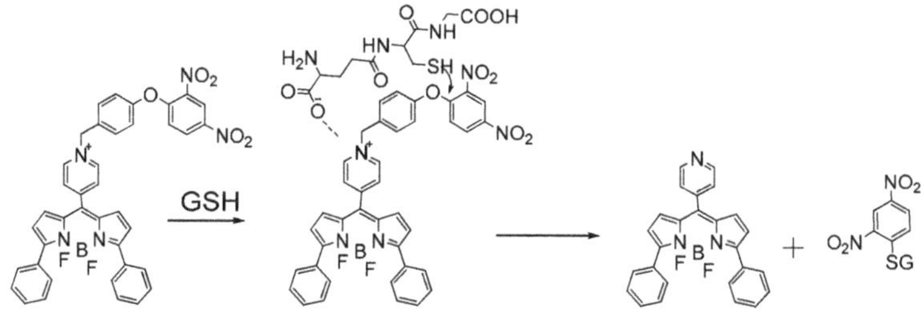

Probe 49

Figure 44. Structure of **Probe 49** and its reaction with GSH.

In 2019, Yue Xu et al. [1] reported a visible and near-infrared, dual emission fluorescent probe (Cy-DC, **Probe 50**) for monitoring mitochondrial thiol in vitro. The probe also exhibits impressive ability to detect solid tumor by both naked eye and near-infrared fluorescence. As depicted in Figure 45, two fluorophores, dicyanomethylene-4H-pyran and cyanine, with distinct wavelength emission bands are linked together with a thiol responsive unit ether linker aryl ether group. The positive charged cyanine is the driving force for mitochondria targeting. The fluorescence activation by thiols was achieved with a nucleophilic aromatic substitution-rearrangement reaction between the thiolate of a thiol with the aryl ether to form a thioether which the thiol molecule is attached to the cyanine with strong NIR fluorescence (Em. 810 nm). In the meantime, the other fluorophore, dicyanomethylene-4H-pyran, was released and restored its visible fluorescence (520 nm). In contrast to GSH, the thioether formed from Cys and Hcy will undergo an intramolecular rearrangement vis a 5-(Cys) or 6-(Hcy) membered transition state resulting in the replacement of S with N leading to a significant fluorescence elimination (810 nm). This provides the probe the ability for discrimination of GSH over Cys and Hcy. Briefly, GSH activated a remarkable increase of fluorescent intensity of Cy-DC both at 520 nm and 810 nm while Cys and Hcy only attribute to the fluorescent intensity of Cy-DC at 520 nm. The detection limits for GSH in two different channels were determined to be 24 nM (visible) and 32 nM (NIR). Additionally, an increase in the thiol blocking agent (NEM) led to a gradually decrease of fluorescence intensity in both channels. In line with the hypothesized mechanism, pretreatment of cells with NEM totally inhibited the fluorescence while addition of GSH to the cells induced a remarkable increase in fluorescence intensity in both channels (green visible channel and NIR red channel). In contrast, addition of Cys and Hcy only led to the fluorescence increase in green visible channel while NIR red fluorescence intensity was barely changed.

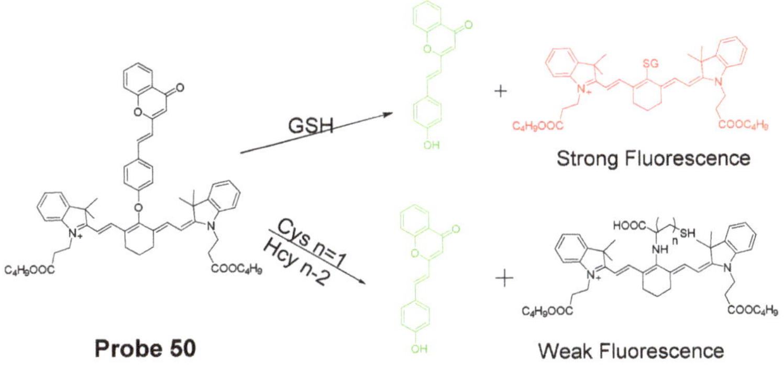

Figure 45. Structure of **Probe 50** and its reaction with GSH and Cys/Hcy.

In 2019, Mingming Cui et al. [161] reported a turn on fluorescent probe (**Probe 51**, Figure 46) for detection of mitochondrial GSH with twisted intramolecular charge transfer (TICT) and aggregation-enhanced emission (AEE) characteristics. Cleavage of the dinitrophenyl ether from QUPY-S by GSH followed by a self-immolation reaction results in a low water solute compound QUPY which aggregates to turn on the fluorescence at 516 nm. The detection limit was determined to 434 nM. The probe is featured with a large Stokes shift (131 nm) and capable of detecting GSH in HeLa cells.

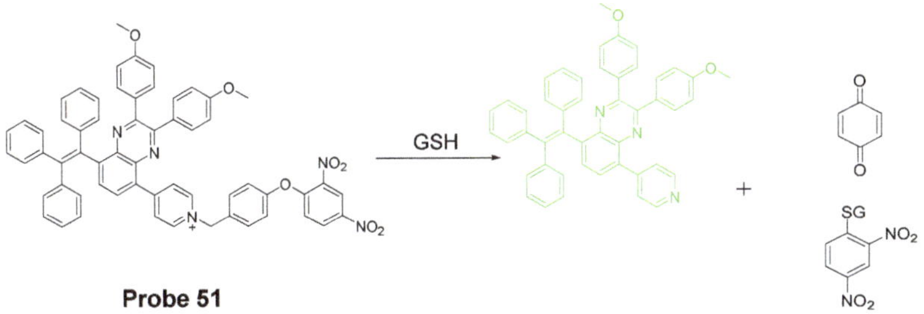

Figure 46. Structure of **Probe 51** and its reaction with GSH.

3.1.3. Detection of Mitochondrial NPSH via Cleavage of Sulfonamide or Sulfonate Ester
Detection of Mitochondrial NPSH via Cleavage of Sulfonamide

In 2018, Zhiqiang Xu et al. [162] developed a visible and near-infrared, dual-channel fluorescence-on robe (CyP-SNp, **Probe 52**, Figure 47) for monitoring mitochondrial GSH in a spatiotemporal and synchronous manner. In the approach, two widely used strong fluorophores, naphthalimide (a visible fluorophore) and cyanine (a NIR fluorophore), are bridged together by a thiol-reactive sulfonamide moiety. These two fluorophores were chosen since they are emitting in two distinctly different wavelength regions. The probe which displays a weak fluorescence will be turned on to two completely different fluorescence upon reaction with GSH, green fluorescence (Em. 495 nm, Ex. 370 nm) in the visible channel and red fluorescence (Em. 795 nm, Ex. 700 nm) in the near-infrared channel. The cationic cyanine dye accumulates in mitochondria. The detection mechanism involves a nucleophilic substitution reaction between the thiolate of GSH with sulfonamide to form an intermediate and subsequent cleavage of the sulfur–nitrogen bond of the intermediate will result in a stable NIR fluorescent product CyP and another reactive intermediate Np-GSH-SO$_2$ which will switch to a visible-light-emitting product Np-GSH after releasing SO$_2$. Cys and Hcy follow the similar mechanistic pathway in a much slower manner. The kinetic differences enable the probe to discriminatively detect GSH over Cys and Hcy. The detection limits in the two different channels are 1.53×10^{-7} M (visible channel) and 1.71×10^{-7} M (near-infrared channel) respectively. Cellular experiments showed that the probe display good mitochondria-targeting capacity.

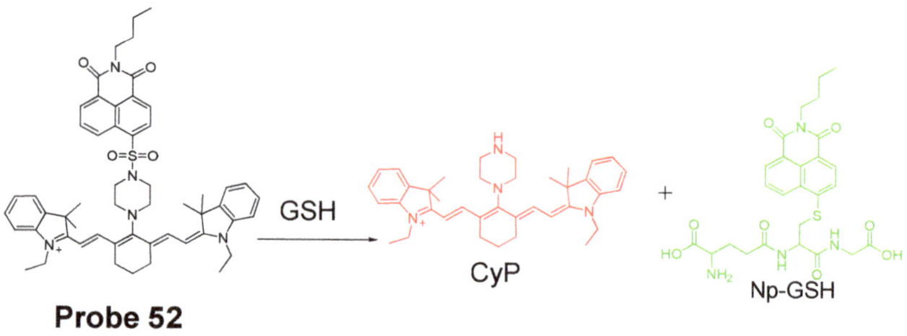

Figure 47. Structure of **Probe 52** and its reactions with thiols.

The probe was further demonstrated to be capable of monitoring GSH status in live cells with a minimum interference from Cys and Hcy. Results showed that addition of Cys and Hcy did not change the fluorescence intensity significantly while addition of GSH induced a remarkable intense fluorescence increase. Later the probe showed to be capable of tracking GSH levels in living tissues with imaging depths of up to 120 μm in the near-infrared channel [163].

In 2020, Zhengkun Liu et al. [164] reported a two-photon probe MT-1 (**Probe 53**, Figure 48) based on a two-photon and fluorescence resonance energy transfer (FRET) strategy. In the molecule, naphthalimide was used as the two-photon receptor as well as the FRET donor while a rhodamine B group was used to serve as the FRET acceptor as well as the mitochondria-targeting ligand. DNBS was attached to naphthalimide moiety via piperazine and utilized as thiol recognition group. In addition, DNBS is a high electron withdrawing group (EWG) which could withdraw the electron to induce intramolecular charge transfer (ICT) effect resulting in the quenching of the fluorescence [165,166]. More importantly, this is the first probe developed for mitochondrial thiols based on two-photon and Förster resonance energy transfer (TP-FRET) strategy, and the probe was successfully applied for detection of mitochondrial thiols in live cells and in mouse liver tissue slices.

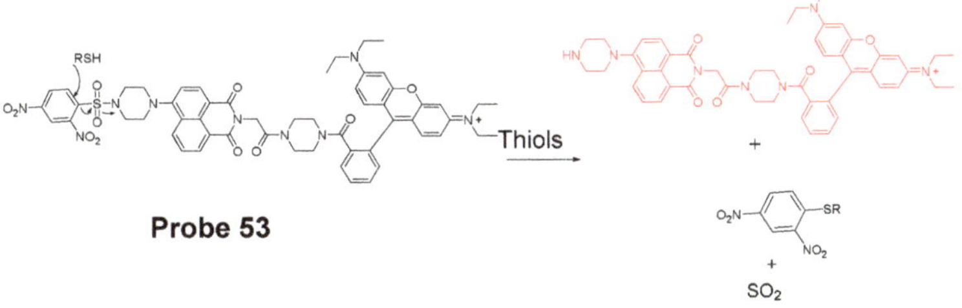

Probe 53

Figure 48. Structure of **Probe 53** and its reaction with thiols.

Detection of Mitochondrial NPSH via Cleavage of Sulfonate Ester

In 2018, Fangfang Wang et al. [167] presented a BODIPY-based fluorescent probe (**Probe 54**) for fast detection of intracellular mitochondrial thiols with high sensitivity and selectivity by introduction of a dual reactive group. Although BODIPY exhibits a number of advantages as a fluorophore, such as high fluorescence quantum yield, excellent photostability, and intense absorption of visible light [160,168–170], utilization of BODIPY for building a BODIPY-based turn-on fluorescent probes is hindered by the hardship of completely quenching the original fluorescence of BODIPY leading to a high background and low sensitivity. To overcome the challenge, two quenching groups, a DNBS moiety and a nitroolefin moiety (-CH=CH-NO$_2$) were introduced to the BODIPY fluorophore. These two groups also serve as the thiol recognition groups. In the molecule, DNBS serves as the acceptor of the PET and nitroolefin moiety (-CH=CH-NO$_2$) serves as a Michael acceptor for the PET as well as a receptor for the intramolecular charge transfer (ICT) process. Thus, the fluorescence of the probe was heavily quenched by these dual quenching systems. The dual reaction of GSH with the probe (Figure 49) breaks the PET and ICT process and restores the strong fluorescence of BODIPY (Em. 517 nm). A linearly increase of fluorescence was observed with an increase in GSH concentration (0–350 µM), Hcy (0–300 µM), and Cys (0–100 µM), enabling the quantitatively detection of thiols. The detection limits of the probe for Hcy, Cys, and GSH were determined to be 87 nM, 147 nM, and 129 nM, respectively. Later, the probe was demonstrated to selectively accumulated inside mitochondria. The ability of the probe to reflect the mitochondrial thiols change was confirmed with live HeLa cells, a significant decrease of fluorescence was observed when the cells were pretreated with NEM. Medium wash was not needed in cell imaging since the probe exhibited a low fluorescence background. The probe was able to image mitochondrial thiols in ca. 10 min.

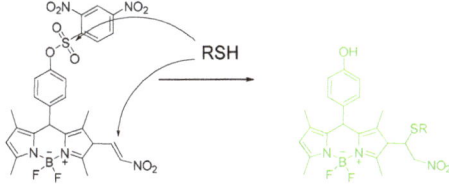

Probe 54

Figure 49. Structure of **Probe 54** and its reaction with thiols.

In 2017, three two-photon fluorescent mitochondrial thiol probes (**Probes 55–57**, Figure 50) were reported by Yi Li et al. [171] based on coumarin as the fluorophore. The probes employed an imidazolium cation to target mitochondria and a strong electron-withdrawing DNBS was introduced to serve as the thiol recognition site and to quench the fluorescence of the coumarin fluorophore via an intramolecular charge transfer (ICT) reaction. A thiol molecule will react with the probes and remove the DNBS structure to turn on the fluorescence (Em. 482 nm). The probes showed dramatic optical responses to various thiols, such as GSH, Cys, and Hcy. Moreover, the probes showed response to BSA confirming they can also react with PSH. In contrast, negligible response was observed for non-thiol amino acids. The authors demonstrated that the probes were located inside mitochondria via a colocalization study with mito-tracker red while probes without the imidazolium cation distributed everywhere inside the cells. These probes were capable of monitoring thiol level change in A549 cells. The ability of detecting thiol level changes was further confirmed in the zebrafish. The penetration depth of these two-photon probes was determined to be 160 µm in the vasculature of an anaesthetized mouse using two-photon microscopy. Overall, these probes showed high mitochondrial selectivity, large two-photon absorption cross-section, and good biocompatibility. They are promising reagents for application for thiol imaging in vitro and in vivo.

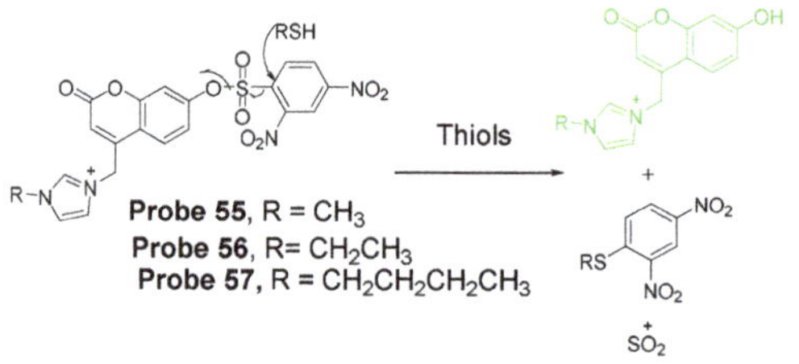

Figure 50. Structures of **Probes 55–57** and their reactions wih thiols.

In 2020, Xin Li et al. [172] developed a novel composite NIR dye (Hcyc-NO, **Probe 58**) for effective detection of thiols and oxidative stress inside mitochondria in live cells. In the approach (Figure 51), a novel NIR hemicyanine dye (Hcyc) composed by cyanine and coumarin were designed and synthesized. The probe itself exhibits weak fluorescence but can be turned on to a high fluorescence product (Ex. 723 nm, Em. 751 nm) upon reaction with thiols, such as GSH, Cys, Hcy H_2S, and dithiothreitol (DTT). Removal of DNBS group from the probe induced the fluorescence enhancement. The probe displays an instantaneous response (<5 s) to thiols with high selectivity and sensitivity. A linear relationship was observed when the probe reacted with various concentrations of GSH (0–5.0 µM), Cys (0–2.0 µM), and Hcy (0–1.0 µM) with the detection limits of 0.11, 0.08, and 0.20 µM, respectively. A cellular colocalization study with Mito-Tracker demonstrated that the probe was localized inside mitochondria. The probe demonstrated the ability to detect a change in thiol levels in live cells. Furthermore, Hcyc-NO was confirmed to be capable of quickly detecting the content of thiols or oxidative stress in vivo in a mouse model.

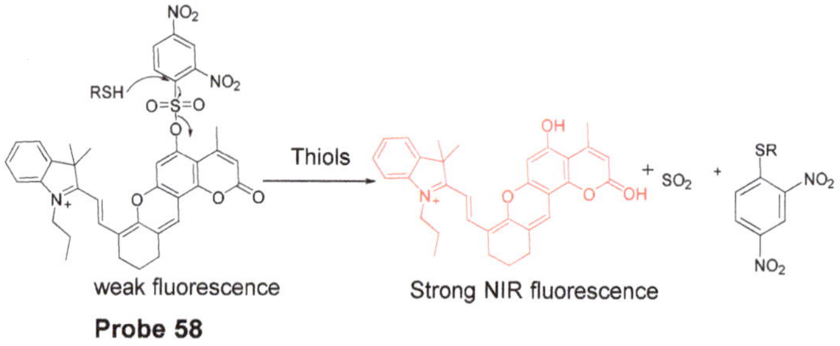

Figure 51. Structure of **Probe 58** and its reaction with thiols.

3.1.4. Detection of Mitochondria NPSH via Cleavage of Disulfide Bond

In 2011, Chang Su Lim et al. [173] developed a two-photon probe (SSH-Mito, **Probe 59**) for ratiometrically detecting of mitochondrial thiols. In the approach, 6-(benzo[d]thiazol-20-yl)-2-(N,N-dimethylamino)naphthalene (BTDAN) was used as the fluorophore, TPP as the mitochondrial-targeting site, and a disulfide group as the thiol recognition group. The reaction was illustrated in Figure 52. Thiol will first react with the disulfide bond followed by a cleavage of the C–N bond to afford the final product. The process will shift the fluorescence from blue color (462 nm (Φ = 0.82)) to yellow color (545 nm (Φ = 0.12)) and the ratio of the fluorescence intensity of F_{yellow} (525–575 nm) to F_{blue} (425–475 nm) elevated 42 to 77-fold in the presence of thiols (GSH, Cys, dithiothreitol (DTT), 2-mercaptoethanol (2-ME), and 2-AET). In contrast, no significant change was observed in the presence of non-thiol amino acids. In vitro cellular experiments demonstrated that SSH-Mito predominantly accumulated inside mitochondria. The probe was proved to be able to respond to thiols' changes in live cells. Moreover, the probe was successfully used for visualizing thiol levels in living tissue depths of 90–190 µm.

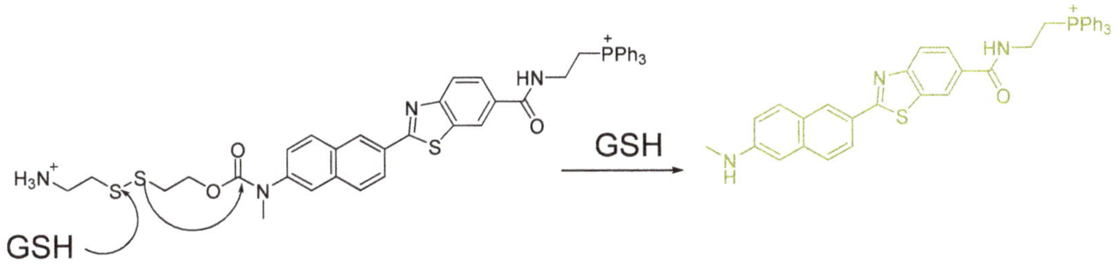

Probe 59

Figure 52. Structure of **Probe 59** and its reaction with GSH.

In 2020, Lu Wang et al. [174] reported a ratiometric fluorescent probe (**Probe 60**, Figure 53) for detection of mitochondria thiols. The probe was built based on the FRET strategy. BODIPY and rhodamine were used in this approach to achieve the FRET-based fluorescent probes since there is a clear overlap between the emission of BODIPY and absorption of rhodamine [175,176]. Thus, BODIPY and a modified near-infrared rhodamine [26,177] were connected together with a disulfide to build an effective FRET structure with BODIPY serving as the donor and rhodamine as the acceptor. In the meantime, the disulfide is the thiol recognition group and also the switch for turning on-off the FRET. When the probe was excited at 488 nm (BODIPY's excitation wavelength), a strong emission with a peak at 656 nm of rhodamine acceptor was observed and a weak emission peak at 512 nm (BODIPY's emission) due to the high efficiency FRET process. Once the probe reacted with a thiol molecule (GSH, Cys, or Hcy), the disulfide was cleaved by thiols and the FRET was turned off resulting in an increase of the emission intensity at 512 nm (corresponding BODIPY fluorescence) while the fluorescence intensity at 656 nm decreased. The fluorescence ratio ($F_{512\ nm}/F_{656\ nm}$) increased linearly as GSH concentration increased from 10 to 100 µM with a detection limit of 0.26 µM suggesting that probe can be used as a ratiometric fluorescent probe for quantitative detection of thiols. The probe was demonstrated to be well localized inside mitochondria and able to detect the mitochondrial thiols' change in live HeLa cells.

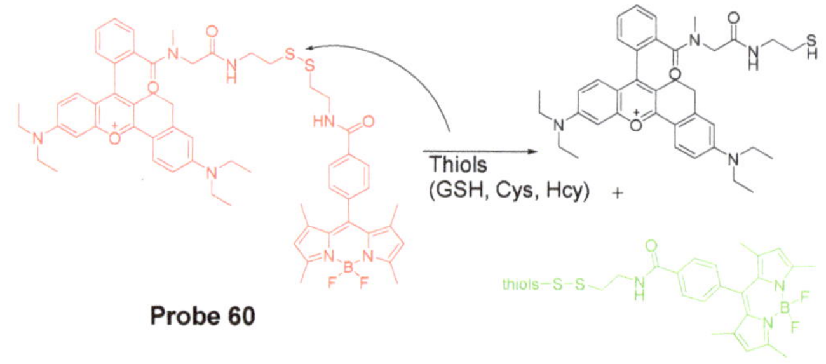

Probe 60

Figure 53. Structure of **Probe 60** and its reaction with thiols.

In 2017, Qingbin Zeng et al. [178] developed a mitochondria targeting and intracellular thiol triggered hyperpolarized [128] Xe magnetofluorescent biosensor (**Probe 61**, Figure 54) based on a disulfide as the thiol recognition group. The probe composed of a TPP group as mitochondrial targeting ligand, a Xe encapsulated in a cryptophane-A cage as a [128] Xe NMR reporter, and a naphthalimide moiety as a fluorescent reporter. After the disulfide bond being cleavage by thiols, a cyclization reaction will occur to afford an amine product, resulting in a remarkable increase of fluorescence intensity at 560 nm and a significant change in the [128] Xe chemical shift. The probe exhibits an extremely low detection limit (10^{-10} M) by using Hyper-CEST (chemical exchange saturation transfer) NMR. The probe was successfully applied to effectively detect mitochondrial thiol changes in live cells.

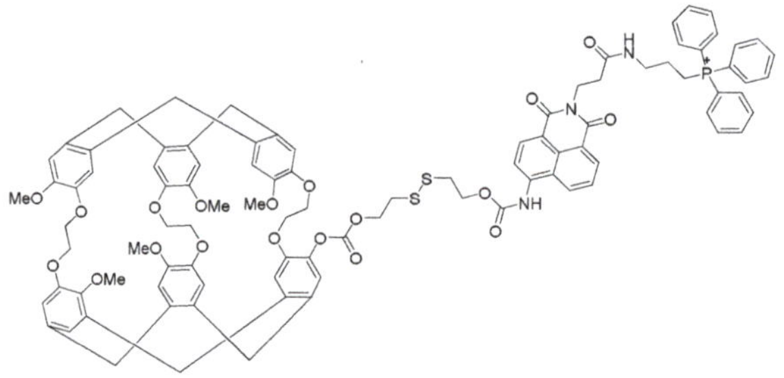

Probe 61

Figure 54. Structure of **Probe 61**.

3.1.5. Detection of Mitochondrial NPSH through Other Methods

In 2014, a mitochondrion-targeting NIR probe (**Probe 62**, Figure 55) named MitoGP was developed by Soo-Yeon Lim et al. [179] for imaging and detecting mitochondrial thiols. Within the molecule, a heptamethine group was utilized as the mitochondrial targeting moiety and also served as the fluorophore, while a nitroazo group was used as the GSH recognition group as well as the fluorescence quencher. The probe showed a great selectivity for GSH over other amino acids including Cys and Hcy. The probe can be turned on to a strong fluorescence in the presences of GSH at 810 nm. The selectivity for GSH over Cys and Hcy was achieved by a specific 1, 6-conjugate addition and subsequent elimination reaction which triggered the release of the fluorophore [136]. In vitro cell experiments with HeLa cells demonstrated that the probe could effectively reflect the GSH level change in mitochondria. The authors believe that the MitoGP is much more superior to the commercially available thiol probe (mCB) or GSH-silent (rhodamine 123) MitoTracker and anticipated the possibility that the probe can be further used as a therapeutic reagent for mitochondrial GSH-related pathology. Moreover, the NIR emissive fluorescence property provides the advantage of avoiding the interference from cellular autofluorescence when used as noninvasive in vivo imaging tool.

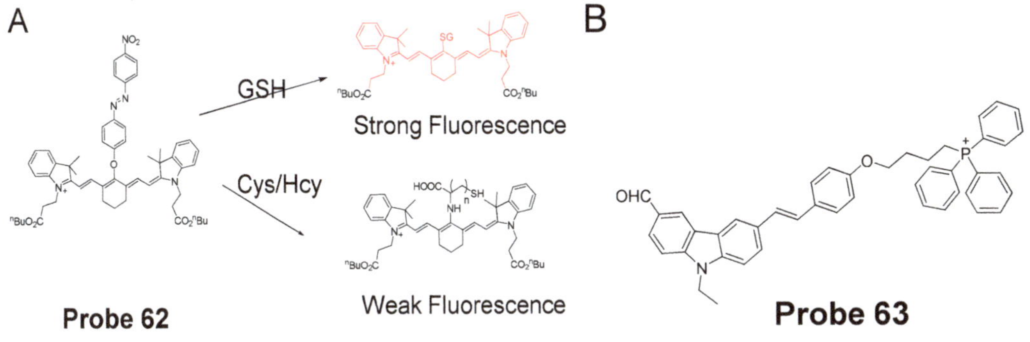

Figure 55. (**A**). Structure of **Probe 62** and its reaction with GSH and Cys/Hcy; (**B**). Structure of **Probe 63**.

In 2016, Fangfang Meng et al. [180] presented a two-photon fluorescent probe (**Probe 63**, Figure 55) for sensing mitochondrial thiols in live cells. Its two-photon properties enable the probe to be applied in mouse liver tissues. The probe named CA-TPP was developed by a combination of three parts, 9-ethyl-3-styryl-9*H*-carbazole (CA) as the TP fluorescent platform, aldehyde as the thiol recognition units, and a TPP as the mitochondrial targeting moiety. In vitro cellular experiments in A549 cells shows that pretreating the cells with NEM to deplete the mitochondrial thiol could lead to a decrease in fluorescence intensity by ~64%, indicating CA-TPP is able to reflect the mitochondrial thiols in live cells. CA-TPP was successfully applied for monitoring mitochondrial thiols in living tissues as well.

Hua Zhang et al. [181] developed an ultrasensitive ratiometric fluorescent probe (IQDC-M, **Probe 64**, Figure 56) for detection of ultratrace change of mitochondrial GSH in cancer cells with a detection limit to be as low as 2.02 nM. IQDC-M was designed by utilizing a modified sulfonamide as a thiol response moiety. Two naphthalene derivative fluorophores served as the fluorescence reporting groups are linked together by sulfonamide to maintain a fluorescence resonance energy transfer (FRET) process. In the presence of FRET, IQDC-M displays a maximum emission at 592 nm (Ex. 450 nm). Interestingly, IQDC-M show no fluorescence response when GSH changed at mM level due to the formation of N=N" group [182], while ultratrace change of GSH at nM could forbad the FRET and led to a significant fluorescence decrease at 592 nm and appearance of a new peak mainly at 520 nm from the product IQ-M which was a result of a thiol attack at the sulfonamide bond (addition reaction) followed by an elimination reaction to release IQ-M. The ratio of $I_{520\,nm}/I_{592\,nm}$ increased linearly as the GSH concentration increases in nM scale. The probe showed a high selectivity for GSH than other thiols, such as Cys, Hcy, or DTT. The probe was proved to be well located inside mitochondria via a colocalization study with MitoTrackers Green FM. Interestingly, the probe was found to selectively enter cancer cells. The compound has been successfully applied to monitor the ultrachange of mitochondrial GSH in cancer cells during the apoptosis for the first time.

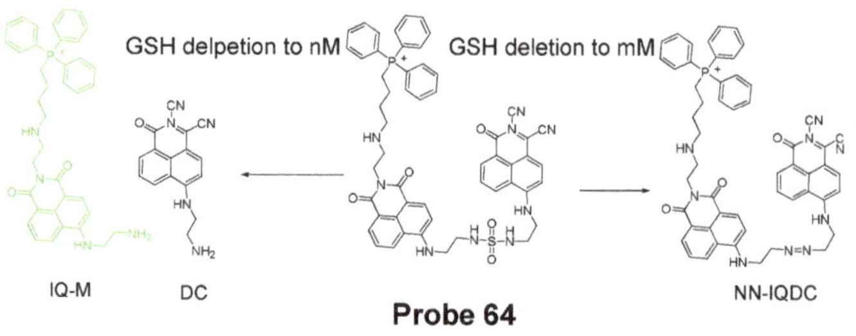

Figure 56. Probe **64** and its reaction with different levels of GSH.

In 2020, Pingru Su et al. [135] reported a TAT peptide-based ratiometric two-photon fluorescent probe (TAT-probe, **Probe 65**, Figure 57) which can detect thiols and simultaneously distinguish GSH in mitochondria. Cell penetrating peptide TAT (RRQRRKKRG) was widely used for delivering macromolecular substances into cells and mitochondria [183–185]. Thus, TAT was incorporated in the probe for achieving the mitochondria selectivity. In the meantime, a naphthalimide and a rhodamine B fluorophore were connected through a thioester bond to construct a Forster resonance energy transfer (FRET) system. A reaction with thiols breaks the thioester and releases a free green fluorescent naphthalimide derivative (Ex. 404/820 nm, Em. 520 nm) and a thiol-conjugated rhodamine. The GSH conjugated rhodamine displays a red fluorescence (Ex. 545 nm, Em. 585 nm) and Hcy/Cys conjugated rhodamine exhibit no fluorescence. TAT-probe was demonstrated to be able to detect GSH, Cys and Hcy with the detect limits of 5.15 μM, 0.865 μM, and 6.51 μM, respectively. Cellular experiments with HeLa cells showed that TAT-probe was able to detect thiols as well as discriminatively detect GSH over Cys/Hcy in mitochondria in live cells in different excitation channels.

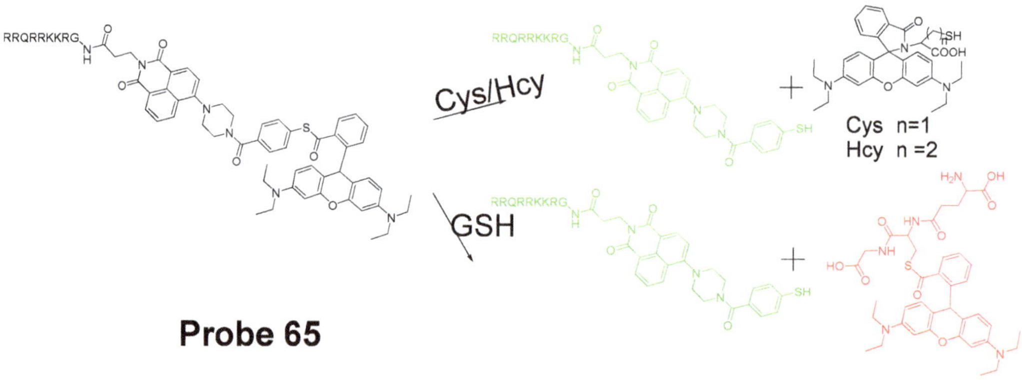

Figure 57. Probe **65** and its reaction with GSH and Cys/Hcy.

In 2017, Chunlong Sun et al. [186] reported a mitochondria-targeted two-photon fluorescent probe (TP-Se, **Probe 66**, Figure 58) to monitor changes of ONOO$^-$/GSH levels in cells based on a organoselenium moiety serving as the rection site for ONOO$^-$/GSH and a methyl pyridinium moiety as a mitochondria-targeting functional group. The emission intensity at 565 nm (Ex. 430 nm) linearly increased upon reaction with ONOO$^-$, while the fluorescence intensity increase was reversed when GSH was added. Cellular experiments revealed that TP-Se exhibited a good sensitivity and selectivity in monitoring ONOO$^-$ oxidation and GSH reduction events under physiological conditions in live cells.

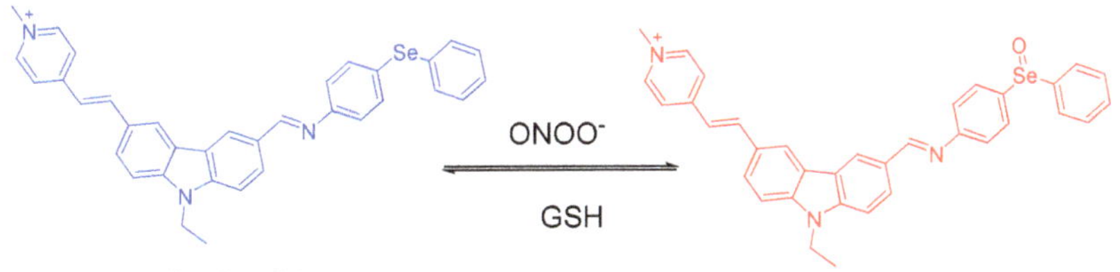

Probe 66

Figure 58. Structure of **Probe 66** and its reaction with ONOO⁻/GSH.

In 2004, George T. Hanson et al. [187] reported redox-sensitive green fluorescence protein indicators (roGFPs, **Probe 67**) for monitoring mitochondrial redox potential. RoGFPs were constructed by introducing of redox-reactive groups (disulfides) to green fluorescence protein through gene modification. The leader sequence of the E1_ subunit of pyruvate dehydrogenase was employed to direct the protein to mitochondria. The probe exhibits two maximum wavelength bands of around 400 and 490 nm while monitoring emission at 508 nm. The ratio of fluorescence intensity of two wavelengths responds to the changes in ambient redox potential rapidly, reversibly, and ratiometrically. RoGFPs were proved to be able to monitor the mitochondrial redox potential changes caused by cell treatment with a reductants (DTT) or an oxidants (H_2O_2) in these cells through fluorescence microscopy or in cell suspension using a fluorometer.

3.2. Selective Detection of Mitochondrial Cys/Hcy

3.2.1. Selective Detection of Mitochondrial Cys and Hcy through Cyclization of Cys/Hcy with Acrylates or Aldehydes

In 2015, Chunmiao Han et al. [188] presented a near-infrared mitochondria-targeting fluorescent probe (NFL_1, **Probe 68**, Figure 59) for selective detection of Cys over Hcy and GSH with a low detection limit of 14.5 nM. NFL_1 is composed of a semiheptamethine derivate group to serve as the NIR fluorophore and an acryloyl group to serve as a thiol responsive moiety. The fluorescence of the semiheptamethine derivate group was quenched by acryloyl group due to a PET process. The semiheptamethine derivate group also provides the probe the ability to target mitochondria due to a positive charge. The off-on mechanism triggered by Cys can be explained in two steps. First, a Michael addition of the thiolate from Cys to the acryloyl group to yield a thioether. Second, the thioether will go through a rapid intramolecular cyclization to form a seven-membered ring and release a free semiheptamethine derivate. This destroys the PET process and restores the strong NIR fluorescence of the semiheptamethine derivate with the maximum emission at 735 nm. The high selectivity of NFL_1 toward Cys over other thiols is due to the fact that the kinetic rate of the intramolecular cyclization reaction with Cys is higher (formation of a seven-membered ring) than Hcy (formation of an eight-membered ring). The intramolecular cyclization reaction does not occur for GSH products. The fluorescence intensity at 735 nm is linearly increased as the concentrations of Cys ascended from 5 to 15 µM with a detection limit of 14.5 nM, enabling the quantitative detection of Cys. The probe has been proved to successfully target mitochondria and monitor mitochondrial Cys change. The reagent is also capable of assessing mitochondrial oxidative stress that induced by LPS in cells, Moreover, NFL1 was successfully applied in vivo for the detection of mitochondrial Cys.

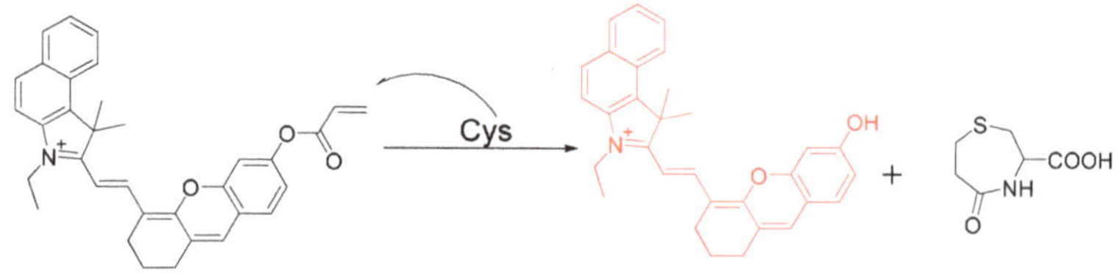

Probe 68

Figure 59. Structure of **Probe 68** and its reaction with Cys.

In 2016, Weifen Niu et al. [112] developed a two-photon fluorescent probe (**Probe 69**) for ratiometrically imaging and detection of Cys in mitochondria in live cells. As illustrated in Figure 60, the probe named ASMI is made by attaching an acrylate moiety to a highly two-photon active and biocompatible merocyanine fluorophore, in which the acrylate moiety is employed as a thiol responsive site and merocyanine as the mitochondria targeting group since it carries a positive charge. Same as other acrylate probes, the reaction between Cys and the probe started with a Michael addition of Cys to the acrylate followed by an intramolecular cyclization to release the merocyanine fluorophore and a seven membered-ring cyclic amide. The probe ASMI itself showed blue fluorescence with a maximum emission wavelength at 452 nm while the released free merocyanine exhibited green fluorescence with a maximum emission wavelength at 452 nm. In addition, both ASMI and merocyanine showed large two-photon action cross-section ($\Phi\sigma_{max}$) of 65.2 GM (Ex. 740 nm) and 72.6 GM (Ex. = 760 nm), respectively. In addition, the ratio of fluorescence intensity of emission at 518 nm and 452 nm (F_{518}/F_{452}) is linearly proportional to Cys concentrations in the range of 0.5–40 μM, suggesting that the probe can detect Cys' level ratiometrically. All of these features raise the high potential to use the probe for high contrast and brightness ratiometric two-photon fluorescence imaging of live samples. Compared to other thiols such as Hcy or GSH, ASMI reacted much faster with Cys. Moreover, the probe was able to ratiometrically detect the mitochondrial Cys change in live cells when its level is elevated by addition of Cys or decreased by NEM. ASMI was also demonstrated to selectively detect mitochondrial Cys and monitor Cys status change in intact living tissues at the depth of 150 μm by the two-photon fluorescence microscopy in a mouse model.

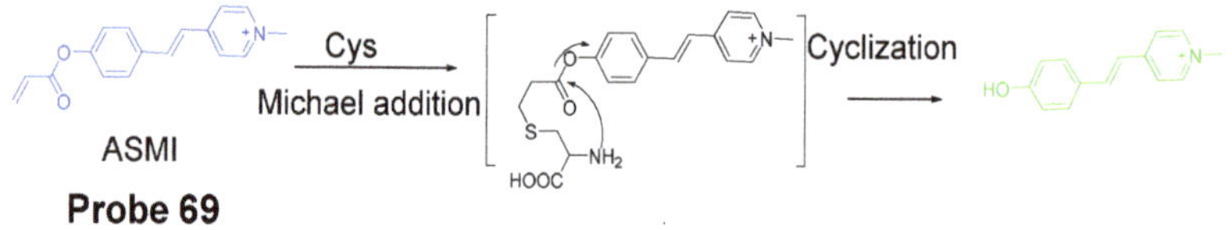

Probe 69

Figure 60. Structure of **Probe 69** and its reaction with Cys.

In addition, a number of other probes have been developed by using the same sensing mechanism. For instance, Chae Yeong Kim et al. [189] developed a mitochondrial Cys imaging probe by caged oxazolidinoindocyanine, Peng Zhang et al. [190] developed a NIR probe based on a difluoroboron curcuminoid scaffold, and LijunTang et al. [191] reported a far-red emissive fluorescence probe based on benzothiazole, etc.

3.2.2. Selective Detection of Mitochondrial Cys/Hcy through Cys-induced SNAr Substitution–Rearrangement Reaction

Mingwang Yang et al. [192] reported a colorimetric and ratiometric fluorescent chemosensor (**Probe 70**, Figure 61) for the detection of Cys/Hcy in mitochondria with a high selectivity and sensitivity (Cys, 22 nM; Hcy, 23 nM) in 2019. The probe is made by connecting a 4-methylthiophenol to a positive charged pyronin fluorophore with a thioether bond which also serves as the thiol recognition unit. The selectivity of Cys/Hcy over GSH is achieved by a SNAr substitution reaction followed by an intramolecular rearrangement reaction while GSH is not able to undergo the intramolecular rearrangement reaction. Upon reaction with Cys/Hcy, fluorescence intensity ratio of F_{540}/F_{620} increases linearly with the concentration of Cys/Hcy, enabling the raitometric detection of Cys/Hcy. The probe was not only able to reflect mitochondrial thiol change in MCF-7 cells induced by H_2O_2, but also able to ratiometrically image endogenous and exogenous thiols in living organisms.

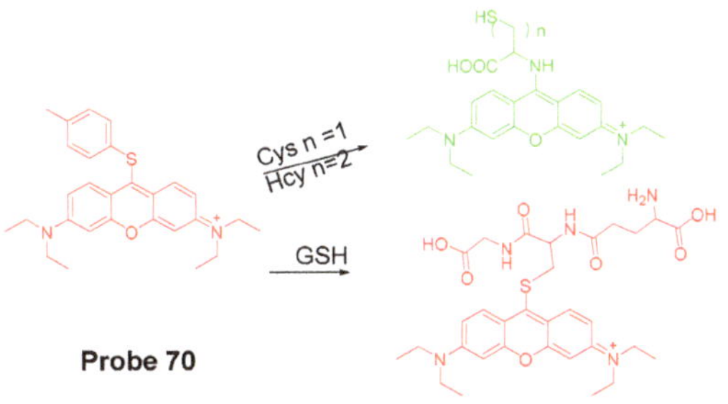

Figure 61. Structure of **Probe 70** and its reaction with GSH and Cys/Hcy.

In 2021, Xin Ji et al. [193] reported a pyridinium substituted BODIPY probe ((BDP-S-o-Py+, **Probe 71**, Figure 62) for selective and rapid detection of mitochondrial Cys in vitro and in vivo based on the SNAr substitution−rearrangement reaction. The pyridinium group was connected to a BODIPY fluorophore via a thioether bond to serve as a thiol recognition group and a mitochondrion targeting moiety. As a good leaving group, the pyridinium will be replaced by the thiolate of Cys upon reaction. Then an intramolecular rearrangement reaction occurred to produce an amine-substituted BODIPY with significantly different fluorescence properties. The detection limit of the compound was determined to be 72 nM with a linear relationship between the fluorescence intensity and Cys concentration in a range of 0 to 50 μM. The probe was successfully used to detect mitochondrial Cys change in HeLa cells with minimum interference from GSH and Hcy. The probe was capable of monitoring endogenous Cys in mice as well.

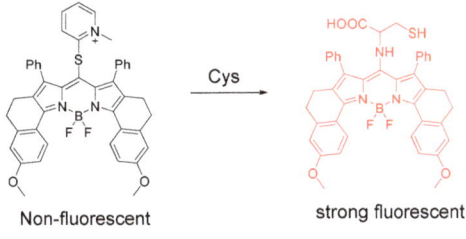

Figure 62. Structure of **Probe 71** and its reaction with Cys.

3.2.3. Selective Detection of Mitochondrial Cys/Hcy through Other Mechanisms

In 2019, Li Fan et al. [194] reported the first mitochondria-targeted ratiometric two-photon (DNEPI) probe (**Probe 72**, Figure 63) based on the DNBS group. The probe is prepared by combining the two-photon fluorophore, merocyanine, and a thiol reaction unit DNBS, while a positive charge of the structure was accounted for mitochondria targeting. A cleavage of the sulfonic acid ester bond by Cys leads to a significant red shift of both fluorescence (485 to 583 nm) and absorption properties (352 to 392 nm). The ratio of fluorescence intensity (F_{583} nm/F_{485} nm) displays a linear relationship with Cys concentration (2–10 μM), and the detection limit was determined to be 0.29 μM. A much higher reactivity of Cys than GSH and Hcy leads to the selectivity of the probe for Cys. The probe was capable of monitoring the level of intracellular Cys.

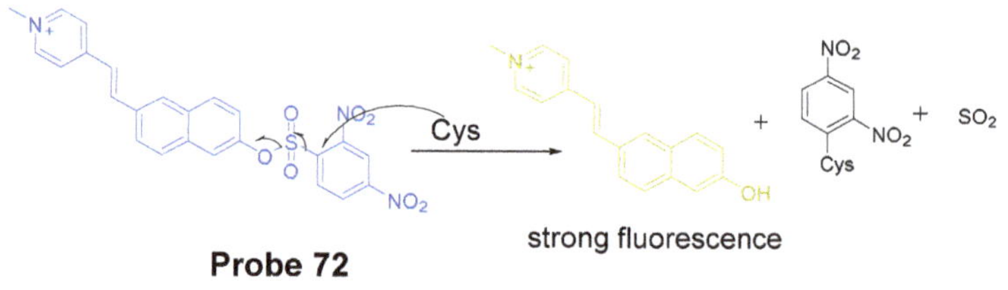

Figure 63. Structure of **Probe 72** and its reaction with Cys.

Kun Yin et al. [195] reported in 2015 a near-infrared ratiometric fluorescence probe Cy-NB (**Probe 73**) for detection of mitochondrial Cys. The probe is composed of heptamethine cyanine as fluorophore and *p*-nitrobenzoyl as the Cys recognition site. Heptamethine cyanine also serves as the mitochondria targeting moiety (Figure 64). The fluorescence properties of heptamethine cyanine are changed dramatically when *p*-nitrobenzoyl is attached to modulate the intramolecular polymethine π-electron system. Cleavage of *p*-nitrobenzoyl from the probe by Cys will restore the polymethine π-electron system and induce a remarkable fluorescence emission shift simultaneously. The probe itself shows a maximum excitation wavelength and maximum emission wavelength of 720 nm and 785 nm, respectively. Upon reaction with thiols, the maximum excitation and emission wavelengths change to 580 nm and 640 nm due to the released heptamethine cyanine. The emission peak at 785 nm (Ex. 720 nm) of probe Cy-NB decreases gradually while the emission peak at 640 nm (Ex. 560 nm) increases gradually in response to an increase in Cys concentration. The ratio of two peaks (F_{640} nm/F_{785} nm) followed a linearly change when the concentration of cysteine is in the range of 0–35 μM and 35–100 μM, and the detection limit is 0.2 μM in 5 min. The lower pKa of Cys (8.53) than Hcy (10.00) and GSH (9.20) is believed to be one of the reasons attributing to the Cys selectivity because more nucleophilic ionized form will be formed for Cys. Moreover, the steric hindrance of tripeptide GSH is believed to be another reason for the selectivity. The pseudo-first-order reaction constant *k* for Cys is 330-fold higher than GSH and 78-fold higher than Hcy. The probe was applied for detection of mitochondrial Cys' status in HepG2 cells successfully. The probe can reflect the Cys level change in mitochondria. The probe and is capable of detecting endogenous Cys level in mice.

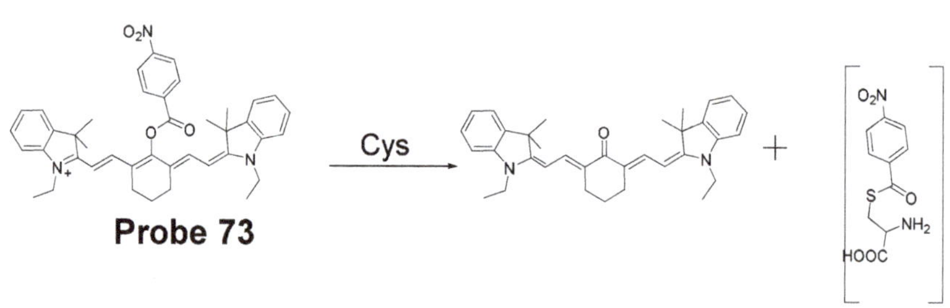

Figure 64. Structure of **Probe 73** and its reaction with Cys.

In addition to the probes developed above, many other probes have been developed as well for selective detection of mitochondrial Cys or Hcy., Xiaopeng Yang et al. [196]

developed a multi-signal fluorescent probe for selective detection of Cys and SO_2 based on a coumarin fluorophore. Jing Liu et al. [197] developed a highly sensitive mitochondrial Cys detection probe based on a coumarin-hemicyanine fluorophore.

3.3. Lysosomal Total Thiols and GSH

As another important subcellular organelle, lysosomes serve as the main digestive compartments with more than 50 hydrolases for digesting various exogenous and endogenous biomolecules. Thiols help with the processing of macromolecule degradation by the reduction of disulfide bonds [198–201]. Additionally, GSH has been reported to stabilize lysosome membranes while Cys is known for activating albumin degradation in liver lysosomes in mice [202]. Therefore, the importance of assessment of thiol levels in lysosomes has been well recognized. However, not many probes have been reported for detecting thiols in lysosomes in live cells.

The design of most of lysosome targeting thiol probes involves the incorporation of a weak basic group which can be trapped in lysosomes which are acidic (pH ~5.0). 4-(2-Aminoethyl)morpholine and *N,N*-dimethylethylenediamine are the most widely used base molecules for lysosomes targeting [26,136] (Figure 65).

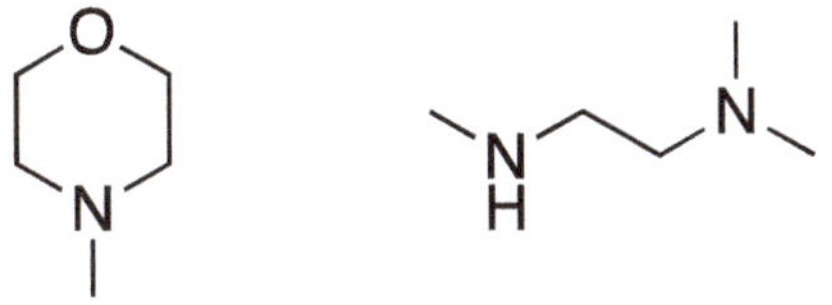

Figure 65. Representative lysosome-targeting.

3.3.1. Detection of Lysosomal NPSH via a Michael Addition Reaction

In 2016, Rong Huang et al. [203] reported a lysosome targeting rhodamine B-based fluorescent probe (**Probe 74**, Figure 66) for rapid and sensitive detection of GSH utilizing a Michael acceptor as the thiol response group. The probe itself is non-fluorescent due to PET but can be turned on to fluorescent (Ex. 520 nm, Em. 582 nm) by GSH within in 10 s. The "turn on" can be explained by a Michael addition reaction followed by the formation of an H-bond between the rhodamine's carbonyl group and the *N*-H group of GSH to open the ring of rhodamine resulting in a remarkable fluorescence increase. The weak basicity of the probe attributes to the lysosome selectivity. The probe is featured with a high sensitivity with a detection limit of 190 nM. Cellular experiments and colocalization studies showed that the probe effectively accumulated inside lysosomes and could detect the GSH level change in lysosomes in HeLa cells.

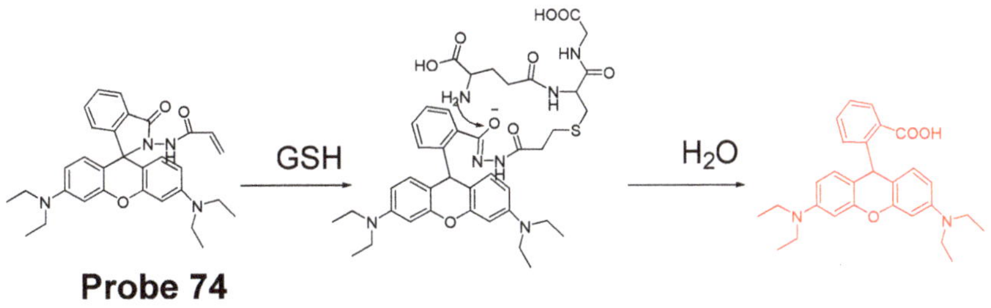

Figure 66. Structure of **Probe 74** and its reaction with GSH.

3.3.2. Detection of Lysosomal NPSH via SNAr Reactions Using Ether, Thioether as a Leaving Group

Detection of Lysosomal NPSH via SNAr Reactions Using Ether as a Leaving Group

In 2018, Hui Zhang et al. [204] reported a lysosome targeting, coumarin and resorufin based fluorescent probe (**Probe 75**, Figure 67) for simultaneous detection of Cys/Hcy and GSH. A morpholine moiety was introduced to serve as the lysosome targeting group, and the coumarin derivative and resorufin were connected together by an ether bond which was the thiol response site. Reaction with different thiols by a nucleophilic substitution reaction will lead to the same free red fluorescent resorufin but different thio-coumarin. The thio-coumarin of Cys/Hcy undergoes an intramolecular rearrangement to switch from thio-coumarin to amino-coumarin (blue) with a significantly fluorescence change while the thio-coumarin (green) derived from GSH will remain unchanged. Thus, the probe can be used to detect the total thiols by measuring the red fluorescence of red fluorescent resorufin and detect Cys/Hcy, and GSH simultaneously from their corresponding coumarin adducts. A nearly linear relationship was found between the fluorescence intensity at 480 nm and Cys/Hcy concentration with a detection limit determining of 27 nM (or 33 nM for Hcy). A linear relationship and a detection limit of 16 nM were found for GSH with its corresponding fluorescence intensity at 542 nm. Cellular experiments with HeLa cells demonstrated that the probe was cable of discriminating intracellular Cys/Hcy, GSH based on their different signal patterns.

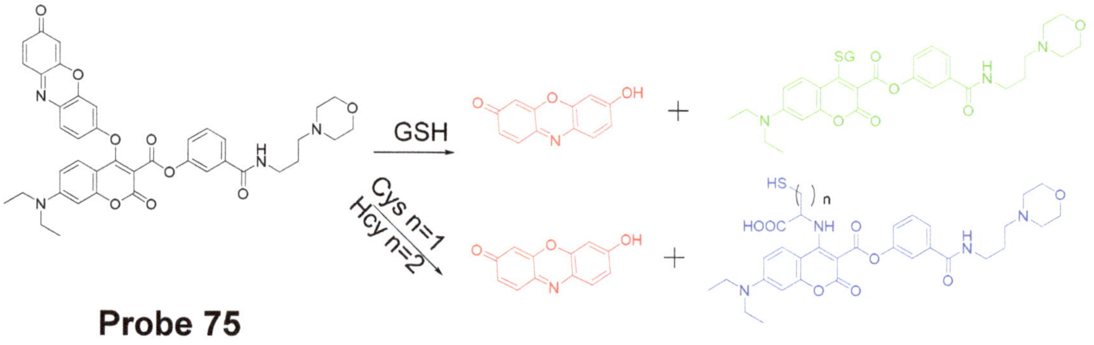

Figure 67. Structure of **Probe 75** and its reaction with GSH and Cys/Hcy.

Xufen Song et al. [205] published their work in 2020 for the development of a lysosome-targeting fluorescent probe (**Probe 76**, Figure 68) for simultaneous detection and discrimination of Cys/Hcy and GSH by dual channels based on a similar mechanism. The probe named Lyso-O-NBD was constructed by combining a coumarin fluorophore and a NBD fluorophore via an ether bond, and a lysosome targeting moiety morpholine was attached to the Coumarin fluorophore. Similar with the probe discussed above, an intramolecular rearrangement after the nucleophilic substitution reaction attributed to the fluorescence difference of the GSH adduct and the Cys/Hcy adducts. A linear relationship was also found between the thiol (GSH, Cys, or Hcy) concentration and fluorescence intensity at 490 nm with a detection limit of 3.3×10^{-8} M, 5.2×10^{-8} M, and 3.9×10^{-8} M for Cys, Hcy, and GSH, respectively. Later, Lyso-O-NBD was successfully applied to detect and discriminate Cys/Hcy and GSH in lysosomes in HeLa cells by confocal laser scanning microscopy.

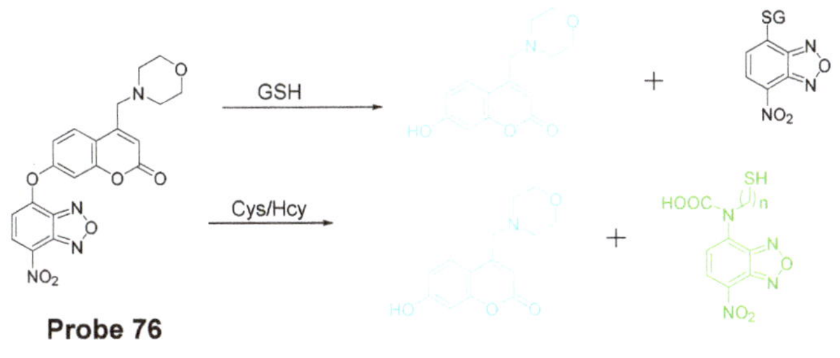

Figure 68. Structure of **Probe 76** and its reactions with GSH or Cys/Hcy.

Detection of Lysosomal NPSH via SNAr Reactions Using Thioether as a Leaving Group

Based on the thiol specific thiol–sulfide exchange reaction reported from our group [42], we developed a thiol specific and lysosome-selective fluorogenic agent (BISMORX, **Probe 77**, Figure 69) [206] in 2019 by introducing a lysosome targeting structure morpholine symmetrically to the symmetric benzofurazan sulfide. BSMORX uses a thioether moiety to serve as the thiol recognition units. Similar to the early thiol specific fluorescence imaging probes developed from our lab, BISMORX itself is non-fluorescent due to the self-quenching of the symmetric structure but reacts readily with a thiol to form a strong fluorescence thiol adducts (Ex. 380 nm, Em. 540 nm). Cellular experiments showed that BISMORX was able to image, quantify, and detect the change of NPSH in lysosomes in live cells. We also developed another lysosomal NPSH imaging agent TBONES (**Probe 78**) [207]. Interestingly, the lysosome selectivity of this probe was likely due to its selective entry of lysosomes through an endocytosis pathway which is different than other lysosome-targeting probes.

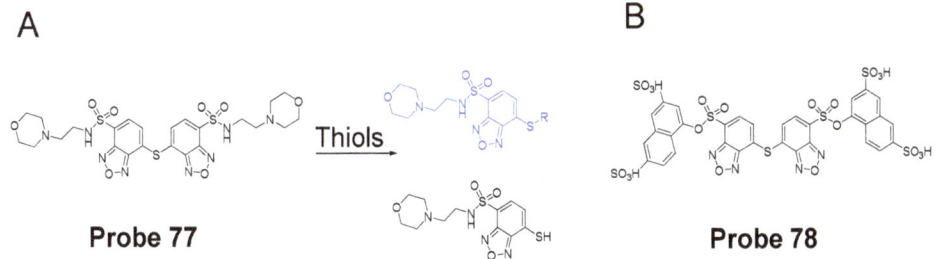

Figure 69. A. Structure of **Probe 77** and its reaction with thiols; B. Structure of **Probe 78**.

3.3.3. Detection of Lysosome NPSH via Cleavage of Sulfonamide and Sulfonate

Detection of Lysosomal NPSH via Cleavage of Sulfonamide

In 2016, Meijiao Cao et al. [208] reported a lysosome targeting naphthalimide-based fluorescence probe (**Probe 79**, Figure 70) for selective detection of GSH in live cells based on cleavage of sulfonamide as the thiol recognition mechanism. A morpholine was attached to the probe to provide the lysosome selectivity. Cleavage of sulfonamide by GSH will produce a strong fluorescent GSH conjugated naphthalimide derivate (Ex. 370 nm, Em. 495 nm). Colocalization experiments with LysoTracker Red on HepG2 cells demonstrated that the probe effectively localized inside lysosomes. Further cellular experiments showed that the probe effectively reflected lysosome GSH changes but not affected by a change in the levels of Cys and Hcy in HepG2 cells.

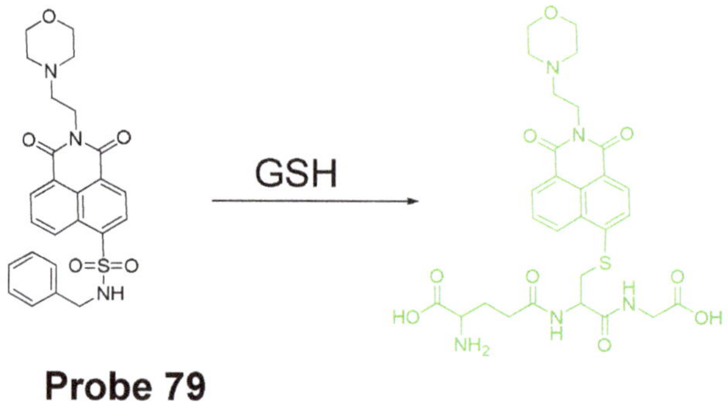

Probe 79

Figure 70. Structure of **Probe 79** and its reaction with GSH.

Jiangli Fan et al. [165] reported a similar work (**Probe 80**, Figure 71) in 2016 in which naphthalimide was used as the fluorophore. A DNBS group was attached to serve as a thiol recognition group and a morpholine was incorporated as the lysosome targeting ligand. PET between DNBS and naphthalimide quenches the fluorescence of the probe. Upon reaction with thiols, DNBS will be detached and terminate the PET process, leading to the restoration of the fluorescence of naphthalimide (Ex. 400 nm, Em. 540 nm). A linear relationship was found between the fluorescence intensity at 540nm and the concentration of thiols with a detection limit of 2.6×10^{-7} M, 2.41×10^{-6} M and 4.87×10^{-6} M for Cys, GSH, and Hcy, respectively. The probe was successfully used to detect lysosomal thiol change in live HeLa cells and MCF-7 cells as well as in tissues by a two-photon microscopy.

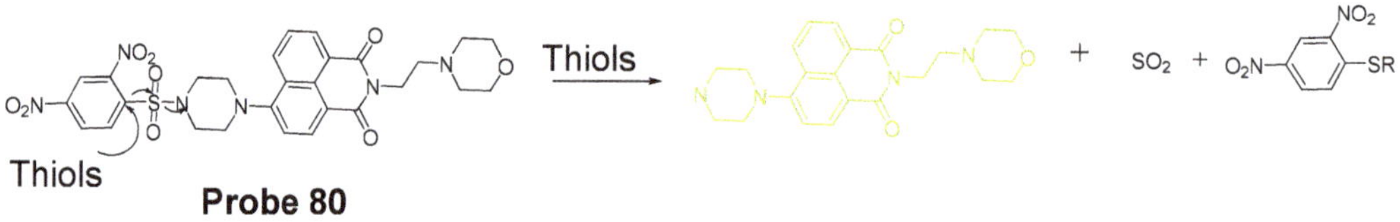

Probe 80

Figure 71. Structure of **Probe 80** and its reaction with thiols.

Detection of Lysosomal NPSH via Cleavage of Sulfonate

A lysosome targeting ruthenium(II) derivative probe (**Probe 81**, Figure 72) was developed by Quankun Gao et al. [209] in 2017 for selective detection of thiols based on the cleavage of sulfonate mechanism. A morpholine moiety was incorporated into the complex to achieve the lysosome selectivity, DNBS was attached to serve as an efficient PET quencher. Upon reaction with thiols, a phosphorescence enhancement was observed at 620 nm (Ex. 459 nm) due to the cleavage of sulfonate moiety. A linear relationship was observed between the phosphorescence intensity and the concentration of GSH in a range of 4.0 to 25 µM. The detection limit was determined to be 62 nM, 146 nM, and 115 nM for GSH, Cys, and Hcy, respectively. The probe was successfully applied for visualizations of thiols in lysosomes in live cells and in *Daphnia magna*.

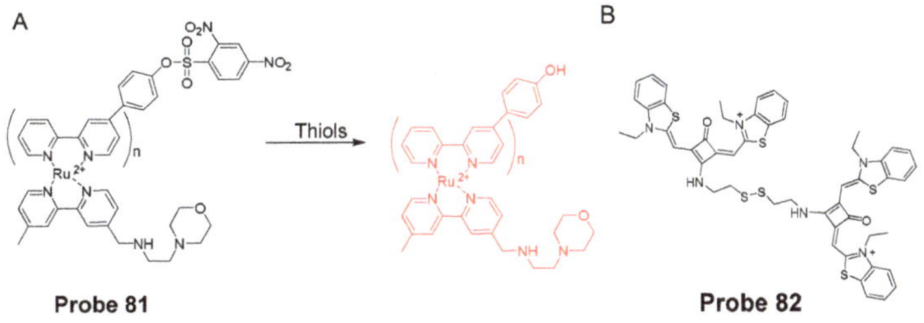

Figure 72. (**A**). Structure of **Probe 81** and its reaction with thiols. (**B**). Structure of **Probe 82**.

3.3.4. Detection of Lysosomal NPSH via Cleavage of Disulfide

In 2019, Ziming Zheng et al. [210] reported a disulfide bond based probe (SQSS, **Probe 82**) for selective detection of lysosome thiols in cells with a high signal-to-noise ratio. Two squaraines (NiR fluorophore) were linked together via a disulfide bond and a high fluorescence resonance energy transfer (FRET) quenching effect was generated between the two squaraines. The two weakly alkaline groups in the structure might attribute to the lysosome selectivity. Upon reaction with GSH, the double bond will be broken leading to the restoration of the fluorescence (Em. 665 nm, Ex. 610 nm). A linear relationship was observed between the concentration of GSH and fluorescence intensity (665 nm) with a detection limit of 0.15 µM, and the probe showed a higher selectivity toward GSH than Cys and Hcy by a possible electrostatic interaction between the two carbonate anions of GSH and the thiazole cations of SQSS. Cell imaging experiments revealed that SQSS can monitor endogenous and exogenous GSH in tumor or normal cells.

3.3.5. Detection of Lysosome NPSH via Other Methods

In 2020, Hong Wang et al. [211] built a dual targeting (cancer-specific and lysosome-targeted) fluorescence nanoprobe (DTFN, **Probe 83**) for GSH imaging in live cells. The nanoprobe was constructed by combining folic acid-modified photostable aggregation-induced emission dots with manganese dioxide (MnO_2) nanosheets (GSH responsive site) through electrostatic interactions. Folic acid was introduced to facilitate the probe to be taken by folate receptor (FR) over-expressed cancer cells, and the positively charged amino moiety helps the nanoparticles enter lysosomes. Fluorescence quenching is achieved vis a fluorescence resonance energy transfer (FRET) effect from AIE dots to the MnO_2 nanosheets. Reaction with GSH will reduce MnO_2 nanosheets to Mn^{2+} to break the FRET process resulting in the restoration of the fluorescence at 523 nm. The fluorescence intensity increased linearly as the concentration of GSH increased with a detection limit of 1.03 µM. Cellular experiments with live HeLa cells showed that the probe could successfully detect lysosomal GSH level in cancer cells.

3.4. Selective Detection of Lysosomal Cys and Hcy

Development of probes for detection of lysosomal Cys and Hcy are still in the early stage. In 2019, Jinhua Gao et al. [212] reported two lysosome targeting BODIPY fluorescence probes (**Probes 84–85**) for selective detection of Cys over GSH and Hcy based on a Cys induced SNAr substitution reaction followed by an intramolecular rearrangement. As illustrated in Figure 73, morpholinoethoxy group was introduced to achieve the lysosome selectivity, and a *p*-methoxyphenylmercapto moiety was attached to the meta position to serve as a Cys recognition unit. After reaction with Cys, the *p*-methoxyphenylmercapto moiety was replaced by Cys by a Cys-induced SNAr substitution followed by an intramolecular rearrangement to generate meso-amino-BODIPYs with strong fluorescence at 566 nm. The two probes displayed high selectivity for Cys over GSH and Hcy. A linear relationship was found between the fluorescence intensity (566 nm) and Cys concentration and a low detection limit was achieved for Cys (46 nM for Lyso-S and 76 nM for Lyso-D). Cellular experiments in HeLa cells showed that the probes effective localized in lysosomes and imaged lysosome thiols.

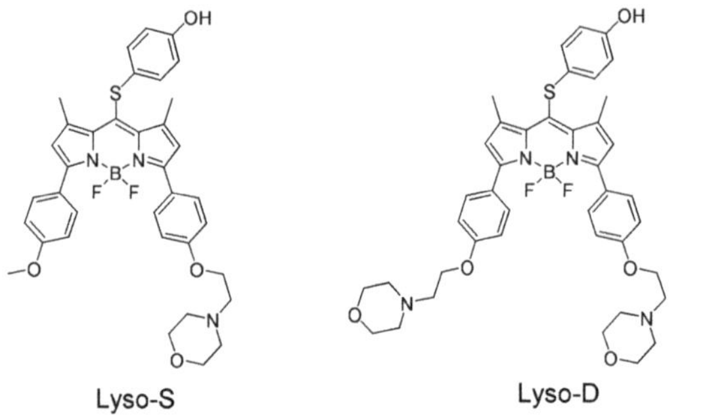

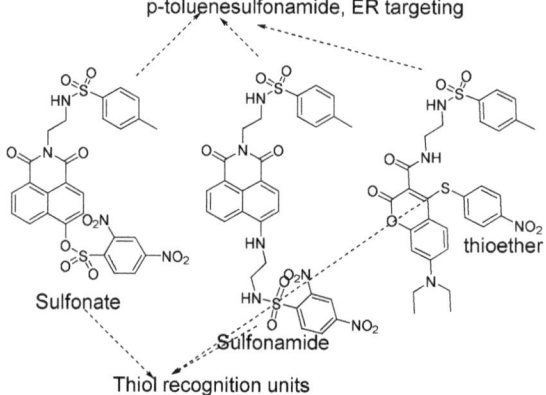

Probe 84 Lyso-S

Probe 85 Lyso-D

Figure 73. Structures of **Probes 84**, and **85**.

3.5. Detection of NPSH in Endoplasmic Reticulum

Endoplasmic reticulum plays important roles [136] in protein synthesis, lipid and carbohydrate metabolism, calcium signaling, etc. Fluorescent probes for imaging thiols inside endoplasmic reticulum have been rarely reported. Among a very few reported probes, methyl benzene sulfonamide is the most commonly used as an ER-targeting moiety (Figure 74) [115,213,214].

Figure 74. ER-targeting thiols imaging probes.

In 2019, Chengshi Jiang et al. [215] developed an naphthalimide-based fluorescent probe (**Probe 86**, Figure 75) for imaging cellular GSH in the ER in live cells. The probe named ER-G was constructed by introducing a *p*-toluenesulfonamide unit to the top of naphthalimide as an ER-targeting unit and a DNBS was attached to the tail as the thiol recognition unit. The non-fluorescent ER-G can be turned on to a highly fluorescent naphthalimide derivate (Em. 558 nm, Ex. 470 nm) after reaction with GSH in which the DNBS was cleaved. The fluorescence intensity was found to be linearly associated with the concentration of GSH (75–300 µM). Cellular experiments with live HepG2 cells revealed that the probe detected the thiol level change by addition of NEM or addition of GSH. The probe was capable of monitoring H_2O_2-induced cellular GSH concentration change in the ER.

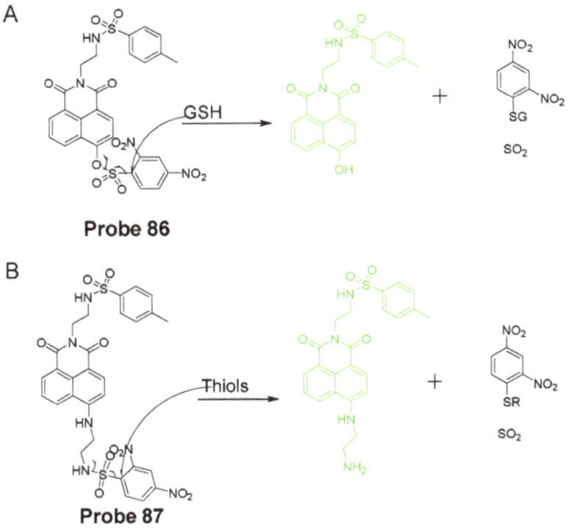

Figure 75. (**A**). Structure of **Probe 86** and its reaction with GSH; (**B**). Structure of **Probe 87** and its reactions with thiols.

A ER-targeted TP fluorescent probe named as ER-SH (**Probe 87**, Figure 75) based the mechanism of cleavage of sulfonamide has been developed for imaging of thiols in live cells and in vivo by Ping Li et al. in 2019 [214]. The probe was constructed similarly with the probe mentioned above in which a naphthalimide was used as a two-photon fluorophore, a methyl benzene sulfonamide was used as an ER targeting moiety and DNBS was used as a thiol recognition group. ER-SH exhibits minimum fluorescence since the fluorescence of naphthalimide was quenched by DNBS due to a PET mechanism. Reaction with thiol will cleave the DNBS and restore the fluorescence (535 nm). Fluorescence intensity at 535 nm increased linearly with an increase in the concentrations of Cys, GSH, and Hcy with the detection limit of 1.67×10^{-7} M, 4.70×10^{-6} M, and 9.62×10^{-7} M, respectively. In addition to the ability to reflect thiol levels' change in live cells, the probe was demonstrated to measure thiols levels in vivo by revealing that the thiol levels were reduced in brains of mice with depression phenotypes.

Xiuxiu Yue et al. [216] developed an ER selective fluorescent probe (ER-CP, **Probe 88**, Figure 76) to discriminatively detect Cys, Hcy, and GSH in live cells for the first time based on a SNAr substitution reaction followed by an intramolecular rearrangement reaction. ER-CP composed of three parts, a coumarin fluorophore, a methyl sulfonamide as an ER-targeting group, and 4-nitrobenzenethiol attached to the coumarin through a thioether bond as the thiol recognition group. PET presented between 4-nitrobenzenethiol and coumarin leads to a fluorescence quenching of ER-CP. A reaction with thiols will replace the nitrobenzenethiol moiety to form a green fluorescent thiol adducts, the same as other probes we described before. The thiol adducts with coumarin derived from Cys and Hcy will further undergo an intramolecular rearrangement reaction to afford a blue fluorescent amnio-substitution coumarin adducts (Ex. 370 nm, Em. 473 nm) although the reaction for Cys was much fast than Hcy while GSH adducts will remain unchanged as the green fluorescent thiol-substitution adducts (Ex. 427 nm, Em. 541 nm). A linear relationship was found between the concentration of thiols (Cys, Hcy, and GSH) and its corresponding fluorescence intensity with a detection limit of 14 nM, 16 nM, and 23 nM, respectively. The probe demonstrates an excellent ER-targeting property and is able to simultaneously detect Cys, Hcy, and GSH in HeLa cells.

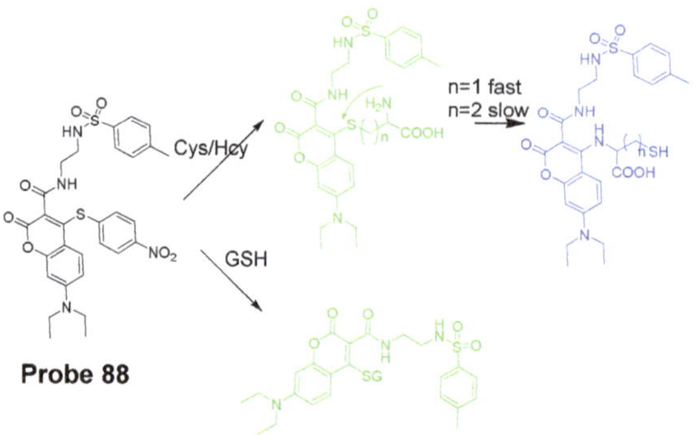

Figure 76. Structure of **Probe 88** and its different reactions mechanisms toward GSH or Cys/Hcy.

4. Conclusions

In the past two decades, extensive efforts have been made in thiol detection in live cells with different approaches and mechanisms. Numerous fluorescent probes have been developed for detection of thiols in the whole cell as well as in subcellular organelles. The probes for imaging thiols in the whole cells are usually made up of two parts: a structure that can react with a thiol selectively, ideally specifically, and a structure that can be fluorescent for detection. The probes for imaging thiol in subcellular organelles will need to add a subcellular organelle targeting part on top of the probe for imaging thiols in the whole cells. The selectivity and reactivity of a probe for thiol is critical in ensuring only thiols are imaged. This is especially true for quantification. As discussed early, most of the employed reactions for thiol detection are based on a nucleophilic reaction. Although thiol is preferred in these reactions based on the fact that thiol is superior in terms of its nucleophilicity when compared with other common nucleophilic groups present in the biological system such as -OH, -NH, and COOH, these reactions are not thiol specific. A few reactions for thiol were reported to be thiol specific. These reactions include thiol–disulfide exchange reaction and thiol–sulfide exchange reaction (Sections 2.1.4, 3.1.4 and 3.3.4 and Sections 2.1.2.3, 3.1.2.2 and 3.3.2.2). It is expected that significant progress will be made if more thiol specific reactions are identified.

This review summarizes various thiol imaging probes used for live cell thiol imaging. To help readers for the probes presented in this review, Table 1 lists the mechanisms of thiol detection, usage, detection limits, and references for most of the probes presented in this review.

Table 1. Thiol imaging probes presented in the review.

Probe Number	Mechanism	Usage	Detection Limit	Reference Number
Probe 1	Michael addition	Whole cells	GSH (53 nM); Cys (<50 nM); Hcy (~100 nM)	34
Probe 2	Michael addition	Whole cells	NA	36
Probe 3	Michael addition	Whole cells	0.219 µM	71
Probe 4	Michael addition	Whole cells	GSH (0.085 µM); Cys (0.13 µM); Hcy (0.12 µM)	72
Probe 5	Michael addition	Whole cells	NA	73
Probe 6	Michael addition	Whole cells	NA	74
Probe 7	SNAr reaction	Whole cells	GSH (8.6×10^{-8} M)	29
Probe 8	SNAr reaction		GSH (0.07 µM); Cys (0.13 µM)	80

Table 1. Cont.

Probe Number	Mechanism	Usage	Detection Limit	Reference Number
Probe 9	SNAr reaction	Whole cells	GSH (0.29 µM)	81
Probe 10	Thiol–sulfide exchange reaction	Whole cells	NA	42
Probe 11	SNAr reaction	Whole cells	NA	83
Probes 12, 13	SNAr reaction	Whole cells	GSH (84 nM)	84
Probe 14	cleavage of sulfonamide	Whole cells	NA	85
Probe 15	cleavage of sulfonamide	Whole cells	NA	88
Probe 16	cleavage of sulfonamide	Whole cells	GSH (1.9 µM)	89
Probe 17	cleavage of sulfonate ester	Whole cells	GSH (1.8×10^{-8} M)	61
Probe 18	cleavage of sulfonate ester	Whole cells	NA	91
Probe 19	cleavage of sulfonate ester	Whole cells	GSH (0.17 µM)	67
Probe 20	cleavage of a disulfide bond	Whole cells	GSH (28 µM)	92
Probe 21	cleavage of a disulfide bond	Whole cells	NA	93
Probe 22	cleavage of a disulfide bond	Whole cells	GSH (0.86×10^{-6} M)	94
Probe 23	cyclization of Cys/Hcy with acrylates or aldehydes	Whole cells (Cys selective)	NA	101
Probe 24	cyclization of Cys/Hcy with acrylates or aldehydes	Whole cells (Cys selective)	Cys (0.19 µM)	38
Probe 25	cyclization of Cys/Hcy with acrylates or aldehydes	Whole cells (Cys selective)	Cys (~0.2 µM)	44
Probe 26	cyclization of Cys/Hcy with bromoacetylfluorescein monoaldehyde	Whole cells (Cys selective)	Cys (0.51 µM)	114
Probes 27, 28	cyclization	Whole cells (Hcy selective)	Hcy (P-Hcy-1, 1.94×10^{-6} M; P-Hcy-2, 1.44×10^{-7} M)	43
Probe 29	Cys-induced SNAr substitution−rearrangement reaction	Whole cells (Cys selective)	Cys (2.12×10^{-7} M)	117
Probe 30	Cys-induced SNAr substitution−rearrangement reaction	Whole cells (Cys selective)	Cys (5.52×10^{-7} M)	118
Probe 31	Cys-induced SNAr substitution−rearrangement reaction	Whole cells (Cys/Hcy selective)	Cys (2.00×10^{-8} M) Hcy (1.02×10^{-8} M)	120
Probe 32	Cys-induced SNAr substitution−rearrangement reaction	Whole cells (Cys selective)	Cys (22 nM)	121
Probe 33	Cys-induced SNAr substitution−rearrangement reaction	Whole cells, discrimination of Cys and GSH	Cys (0.08 µM) GSH (0.06 µM)	124
Probe 34	Cys-induced SNAr substitution−rearrangement reaction	Whole cells, discrimination of Cys/Hcy, GSH, and H$_2$S	GSH (4.30 µM) Cys (4.25 µM) Hcy (5.11 µM) H2S (6.74 µM)	125
Probe 35	Cys-induced SNAr substitution−rearrangement reaction	Whole cells, discriminate Cys/Hcy from GSH	Cys (2.1×10^{-8} M) Hcy 1.7×10^{-8} M; GSH (2.6×10^{-8} M)	126

Table 1. Cont.

Probe Number	Mechanism	Usage	Detection Limit	Reference Number
Probe 36	Cys-induced SNAr substitution–rearrangement reaction	Whole cells, (Cys/Hcy selective)	Cys and Hcy (0.1 µM)	127
Probe 37	Cys-induced SNAr substitution–rearrangement reaction	Whole cells, (Cys selectvie)	Cys (0.3 µM)	128
Probe 38	Michael addition reaction in combination with a steric hinderance factor.	Whole cells (Cys selective)	Cys (10^{-7} M)	129
Probe 39	Michael addition assisted by an electrostatic attraction.	Whole cells (Cys selective)	Cys (25 nM)	130
Probe 40	SNAr substitution–rearrangement reaction	Whole cells simultaneously detects Cys and GSH	Cys (>0.4 µM) GSH (0.05 µM)	102
Probe 41	Michael addition followed by self-immolative reaction	Mitochondrial thiols	GSH (0.59 µM) Cys (0.39 µM) Hcy (0.54 µM)	137
Probe 42	Michael addition	Mitochondrial GSH	NA	138
Probe 43	Michael addition	Mitochondrial GSH	NA	139
Probe 44	SNAr reaction	Mitochondrial GSH	GSH (1.1 µM)	140
Probe 45	SNAr reaction	Mitochondrial GSH	GSH (24.16 µM)	141
Probe 46	Thiol–sulfide exchange reaction	Mitochondrial thiols	NA	22
Probe 47	SNAr reaction	Mitochondrial and lysosomal GSH	GSH (1 nM)	143
Probe 48	SNAr reaction	Mitochondrial thiols	GSH (0.61 µM).	35
Probe 49	SNAr reaction	Mitochondrial GSH	GSH (109 nM)	160
Probe 50	SNAr reaction	Mitochondrial GSH	GSH (24 nM, visible) and (32 nM, NIR)	1
Probe 51	Cleavage of the dinitrophenyl ether	Mitochondrial GSH	GSH (434 nM)	161
Probe 52	cleavage of sulfonamide	Mitochondrial GSH	GSH (1.53×10^{-7} M, visible channel) and (1.71×10^{-7} M, NIR channel)	162
Probe 53	cleavage of sulfonamide	Mitochondrial thiols	GSH (0.89 µM); Cys (0.47 µM), Hcy (2.4 µM)	164
Probe 54	cleavage of sulfonate ester	Mitochondrial thiols	Hcy (87 nM) Cys (147 nM) GSH (129 nM)	167
Probes 55, 56, 57	cleavage of sulfonate ester	Mitochondrial thiols	GSH (**Probe 55**, 31.4 nM; **Probe 56**, 29.2 nM; **Probe 57**, 29.6 nM)	171
Probe 58	cleavage of sulfonate ester	Mitochondrial thiols	GSH (0.11 µM) Cys (0.08 µM) Hcy (0.20 µM)	172

Table 1. Cont.

Probe Number	Mechanism	Usage	Detection Limit	Reference Number
Probe 59	cleavage of disulfide bond	Mitochondrial thiols	NA	173
Probe 60	cleavage of disulfide bond	Mitochondrial thiols	GSH (0.26 μM)	174
Probe 61	cleavage of disulfide bond	Mitochondrial thiols	GSH (10^{-10} M, using Hyper-CEST NMR)	178
Probe 62	1, 6-conjugate addition and subsequent elimination reaction	Mitochondrial GSH	GSH (26 nM)	179
Probe 63	Others	Mitochondrial thiols	Cys (0.2 μM)	180
Probe 64	cleavage of sulfonamide for ultratrace change of GSH	Mitochondrial GSH	GSH (2.02 nM)	181
Probe 65	Others	Mitochondrial GSH	GSH (5.15 μM) Cys (0.865 μM) Hcy (6.51 μM)	135
Probe 66	Others	Mitochondrial ONOO$^-$/GSH levels	NA	186
Probe 67	Others	Mitochondrial redox potential	NA	187
Probe 68	cyclization of Cys with acrylates or aldehydes	Mitochondrial Cys	Cys (14.5 nM)	188
Probe 69	cyclization of Cys with acrylates or aldehydes	Mitochondrial Cys	NA	112
Probe 70	Cys-induced SNAr substitution−rearrangement reaction	Mitochondrial Cys/Hcy	Cys (22 nM) Hcy (23 nM)	192
Probe 71	SNAr substitution−rearrangement reaction	Mitochondrial Cys	Cys (72 nM)	193
Probe 72	cleavage of sulfonamide	Mitochondrial Cys	Cys (0.29 μM)	194
Probe 73	Others	Mitochondrial Cys	Cys (0.2 μM)	195
Probe 74	Michael addition	Lysosomal GSH	GSH (190 nM)	203
Probe 75	SNAr reactions	Lysosomal thiols	GSH (16 nM) Cys (27 nM) Hcy (33 nM)	204
Probe 76	SNAr reactions	Lysosomal thiols	GSH (3.9×10^{-8} M); Cys (3.3×10^{-8} M); Hcy (5.2×10^{-8} M)	205
Probe 77	Thiol–sulfide exchange reaction	Lysosomal thiols	NA	206
Probe 78	Thiol–sulfide exchange reaction	Lysosomal thiols	NA	207
Probe 79	cleavage of sulfonamide	Lysosomal GSH	NA	208
Probe 80	cleavage of sulfonamide	Lysosomal thiols	GSH (2.41×10^{-6} M); Cys (2.6×10^{-7} M); Hcy (4.87×10^{-6} M)	165
Probe 81	cleavage of sulfonate	Lysosomal thiols	GSH (62 nM); Cys (146 nM); Hcy (115 nM)	209
Probe 82	cleavage of disulfide	Lysosomal GSH	GSH (0.15 μM)	210

Table 1. Cont.

Probe Number	Mechanism	Usage	Detection Limit	Reference Number
Probe 83	others	Lysosomal GSH	GSH (1.03 µM)	211
Probes 84, 85	Cys-induced SNAr substitution followed by a intramolecular rearrangement	Lysosomal Cys	Cys (46 nM for Lyso-S and 76 nM for Lyso-D)	212
Probe 86	cleavage of sulfonate	Endoplasmic reticulum GSH	NA	215
Probe 87	cleavage of sulfonamide	Endoplasmic reticulum thiols	GSH (4.70×10^{-6} M); Cys (1.67×10^{-7} M); Hcy (9.62×10^{-7} M)	214
Probe 88	SNAr substitution	Endoplasmic reticulum thiols	GSH (23 nM), Cys (14 nM), Hcy (16 nM)	216

Developing probes for imaging thiols in subcellular organelles obviously faces more challenges since it requires the availability of the subcellular targeting part. Thiol probes for many subcellular organelles including nucleus, ER, Golgi apparatus, and cell surface are still lacking due to a lack of an ideal targeting ligand for the subcellular organelle. Although a number of probes have been developed for imaging thiols in mitochondria and lysosomes, challenges still exist. For example, the majority of mitochondria targeting probes achieved their mitochondria targeting by attaching a lipophilic cationic structure which has the potential risk of decreasing negative mitochondrial transmembrane potential (MMP) [136]. The lysosome targeting group morpholine can induce certain toxicities by elevating the pH levels of lysosomes if a high amount of the probes are applied [136].

In addition, a low water solubility is another shortcoming for most of the probes. Organic solvents like DMSO are commonly used to improve water solubility to enable the probe to be applied in cells or tissues. Thus, a probe with adequate water solubility and sufficient permeability is desirable [136]. A membrane transporter can also be a problem. Although most of the probes can diffuse inside cells due to a good permeability, active transporters (e.g., P-gp, or MRP) on cell membrane can potentially pump the hydrophobic probes out of the cells rapidly, which can significantly affect the accuracy of thiol measurement in live cells. The potential interferences caused by active transporters should be studied.

Real-time thiol imaging in live cells is another challenge. Most of the probes developed are based on an irreversible reaction with thiols, which makes it difficult to monitor the dynamic change of thiols in real time. A few reagents that can detect the analysts of interest in a reversible manner have been developed in the past ~5 years to address this challenge [74–77,139,140,217]. These probes can monitor the dynamic status change of the analysts in real time. Some of the reversible probes have been well developed and characterized for quantifying and monitoring GSH changes at the cellular level and even subcellular levels with a number of probes showing the capability of quantifying and monitoring mitochondrial GSH changes [139,140].

More agents for different subcellular organelles are expected to be developed, and better fluorophores and sensing mechanisms are also anticipated to be discovered and applied in the future. Although majority of the developed probes can only be used for cells, quite a number of probes with NIR or two-photon fluorophores have been successfully applied at tissue levels or in in vivo models. More impressively, some of the probes were demonstrated to be capable of distinguishing tumor tissues from other tissues, warranting the possibility of employing these probes as clinical diagnostic tools in the future though much work is still needed.

In conclusion, we summarized the design strategies and sensing mechanisms of some of the most representative thiol imaging probes for live cell thiol imaging. We hope this review can provide the readers an overview and better understanding of the current status and challenges in the field.

Author Contributions: X.G. provided guidance and the outline of the review to co-authors and was responsible for the final editing and approval of the review. S.W. and Y.H. were responsible for the preparation of the manuscript draft and contributed equally to this article. All authors have read and agreed to the published version of the manuscript.

Funding: This research received no external funding.

Institutional Review Board Statement: Not applicable.

Informed Consent Statement: Not applicable.

Data Availability Statement: Not applicable.

Conflicts of Interest: The authors declare no conflict of interest.

References

1. Xu, Y.; Li, R.; Zhou, X.; Li, W.; Ernest, U.; Wan, H.; Li, L.; Chen, H.; Yuan, Z. A visible and near-infrared, dual emission fluorescent probe based on thiol reactivity for selectively tracking mitochondrial glutathione in vitro. *Talanta* **2019**, *205*, 120–125. [CrossRef] [PubMed]
2. Dai, J.; Ma, C.; Zhang, P.; Fu, Y.; Shen, B. Recent progress in the development of fluorescent probes for detection of biothiols. *Dye. Pigment.* **2020**, *177*, 108321. [CrossRef]
3. Hansen, R.E.; Roth, D.; Winther, J.R. Quantifying the global cellular thiol–disulfide status. *Proc. Natl. Acad. Sci. USA* **2009**, *106*, 422–427. [CrossRef] [PubMed]
4. Chen, H.; Tang, Y.; Lin, W. Recent progress in the fluorescent probes for the specific imaging of small molecular weight thiols in living cells. *Trac Trends Anal. Chem.* **2016**, *76*, 166–181. [CrossRef]
5. Liu, Y.; Teng, L.; Chen, L.; Ma, H.; Liu, H.-W.; Zhang, X.-B. Engineering of a near-infrared fluorescent probe for real-time simultaneous visualization of intracellular hypoxia and induced mitophagy. *Chem. Sci.* **2018**, *9*, 5347–5353. [CrossRef]
6. Kumar, N.; Bhalla, V.; Kumar, M. Development and sensing applications of fluorescent motifs within the mitochondrial environment. *Chem. Commun.* **2015**, *51*, 15614–15628.
7. Xu, W.; Zeng, Z.; Jiang, J.H.; Chang, Y.T.; Yuan, L. Discerning the chemistry in individual organelles with small-molecule fluorescent probes. *Angew. Chem. Int. Ed.* **2016**, *55*, 13658–13699. [CrossRef] [PubMed]
8. Hansen, J.M.; Go, Y.-M.; Jones, D.P. Nuclear and mitochondrial compartmentation of oxidative stress and redox signaling. *Annu. Rev. Pharmacol. Toxicol.* **2006**, *46*, 215–234. [CrossRef]
9. Moriarty-Craige, S.E.; Jones, D.P. Extracellular thiols and thiol/disulfide redox in metabolism. *Annu. Rev. Nutr.* **2004**, *24*, 481–509. [CrossRef]
10. Reed, D.; Babson, J.; Beatty, P.; Brodie, A.; Ellis, W.; Potter, D. HPLC analysis of nanogram levels of glutathione, glutathione disulphide and related thiols and disulphides. *Anal Biochem.* **1980**, *106*, 55–62. [CrossRef]
11. Newton, G.L.; Dorian, R.; Fahey, R.C. Analysis of biological thiols: Derivatization with monobromobimane and separation by reverse-phase high-performance liquid chromatography. *Anal. Biochem.* **1981**, *114*, 383–387. [CrossRef]
12. Chen, W.; Zhao, Y.; Seefeldt, T.; Guan, X. Determination of thiols and disulfides via HPLC quantification of 5-thio-2-nitrobenzoic acid. *J. Pharm. Biomed. Anal.* **2008**, *48*, 1375–1380. [CrossRef]
13. Guan, X.; Hoffman, B.; Dwivedi, C.; Matthees, D.P. A simultaneous liquid chromatography/mass spectrometric assay of glutathione, cysteine, homocysteine and their disulfides in biological samples. *J. Pharm. Biomed. Anal.* **2003**, *31*, 251–261. [CrossRef]
14. Wei, W.; Liang, X.; Hu, G.; Guo, Y.; Shao, S. A highly selective colorimetric probe based on 2,2′,2″-trisindolylmethene for cysteine/homocysteine. *Tetrahedron Lett.* **2011**, *52*, 1422–1425. [CrossRef]
15. Ren, M.; Zhou, K.; He, L.; Lin, W. Mitochondria and lysosome-targetable fluorescent probes for HOCl: Recent advances and perspectives. *J. Mater. Chem. B* **2018**, *6*, 1716–1733. [CrossRef]
16. Ren, T.-B.; Zhang, Q.-L.; Su, D.; Zhang, X.-X.; Yuan, L.; Zhang, X.-B. Detection of analytes in mitochondria without interference from other sites based on an innovative ratiometric fluorophore. *Chem. Sci.* **2018**, *9*, 5461–5466. [CrossRef] [PubMed]
17. Maeda, H.; Matsuno, H.; Ushida, M.; Katayama, K.; Saeki, K.; Itoh, N. 2, 4-Dinitrobenzenesulfonyl fluoresceins as fluorescent alternatives to Ellman's reagent in thiol-quantification enzyme assays. *Angew. Chem. Int. Ed.* **2005**, *44*, 2922–2925. [CrossRef] [PubMed]
18. Hurd, T.R.; Prime, T.A.; Harbour, M.E.; Lilley, K.S.; Murphy, M.P. Detection of reactive oxygen species-sensitive thiol proteins by redox difference gel electrophoresis implications for mitochondrial redox signaling. *J. Biol. Chem.* **2007**, *282*, 22040–22051. [CrossRef]
19. Mao, Z.; Jiang, H.; Li, Z.; Zhong, C.; Zhang, W.; Liu, Z. An N-nitrosation reactivity-based two-photon fluorescent probe for the specific in situ detection of nitric oxide. *Chem. Sci.* **2017**, *8*, 4533–4538. [CrossRef] [PubMed]

20. Rao, J.; Dragulescu-Andrasi, A.; Yao, H. Fluorescence imaging in vivo: Recent advances. *Curr. Opin. Biotechnol.* **2007**, *18*, 17–25. [CrossRef]
21. Giustarini, D.; Colombo, G.; Garavaglia, M.L.; Astori, E.; Portinaro, N.M.; Reggiani, F.; Badalamenti, S.; Aloisi, A.M.; Santucci, A.; Rossi, R. Assessment of glutathione/glutathione disulphide ratio and S-glutathionylated proteins in human blood, solid tissues, and cultured cells. *Free Radic. Biol. Med.* **2017**, *112*, 360–375. [CrossRef]
22. Wang, S.; Yin, H.; Huang, Y.; Guan, X. Thiol specific and mitochondria selective fluorogenic benzofurazan sulfide for live cell nonprotein thiol imaging and quantification in mitochondria. *Anal. Chem.* **2018**, *90*, 8170–8177. [CrossRef]
23. Ding, D.; Li, K.; Liu, B.; Tang, B.Z. Bioprobes based on AIE fluorogens. *Acc. Chem. Res.* **2013**, *46*, 2441–2453. [CrossRef]
24. Song, Z.; Mao, D.; Sung, S.H.; Kwok, R.T.; Lam, J.W.; Kong, D.; Ding, D.; Tang, B.Z. Activatable fluorescent nanoprobe with aggregation-induced emission characteristics for selective in vivo imaging of elevated peroxynitrite generation. *Adv. Mater.* **2016**, *28*, 7249–7256. [CrossRef]
25. Shi, L.; Li, K.; Li, L.-L.; Chen, S.-Y.; Li, M.-Y.; Zhou, Q.; Wang, N.; Yu, X.-Q. Novel easily available purine-based AIEgens with colour tunability and applications in lipid droplet imaging. *Chem. Sci.* **2018**, *9*, 8969–8974. [CrossRef]
26. Niu, G.; Zhang, P.; Liu, W.; Wang, M.; Zhang, H.; Wu, J.; Zhang, L.; Wang, P. Near-infrared probe based on rhodamine derivative for highly sensitive and selective lysosomal pH tracking. *Anal. Chem.* **2017**, *89*, 1922–1929. [CrossRef] [PubMed]
27. Caskey, C.T. Disease diagnosis by recombinant DNA methods. *Science* **1987**, *236*, 1223–1229. [CrossRef] [PubMed]
28. Li, X.; Qian, S.; He, Q.; Yang, B.; Li, J.; Hu, Y. Design and synthesis of a highly selective fluorescent turn-on probe for thiol bioimaging in living cells. *Org. Biomol. Chem.* **2010**, *8*, 3627–3630. [CrossRef] [PubMed]
29. Niu, L.-Y.; Guan, Y.-S.; Chen, Y.-Z.; Wu, L.-Z.; Tung, C.-H.; Yang, Q.-Z. BODIPY-based ratiometric fluorescent sensor for highly selective detection of glutathione over cysteine and homocysteine. *J. Am. Chem. Soc.* **2012**, *134*, 18928–18931. [CrossRef]
30. Tang, B.; Xing, Y.; Li, P.; Zhang, N.; Yu, F.; Yang, G. A rhodamine-based fluorescent probe containing a Se-N bond for detecting thiols and its application in living cells. *J. Am. Chem. Soc.* **2007**, *129*, 11666–11667. [CrossRef]
31. Shibata, A.; Furukawa, K.; Abe, H.; Tsuneda, S.; Ito, Y. Rhodamine-based fluorogenic probe for imaging biological thiol. *Bioorganic Med. Chem. Lett.* **2008**, *18*, 2246–2249. [CrossRef]
32. Hedley DW, C.S. Evaluation of methods for measuring cellular glutathione content using flow cytometry. *Cytometry* **1994**, *15*, 9. [CrossRef] [PubMed]
33. Shiu, H.Y.; Chong, H.C.; Leung, Y.C.; Wong, M.K.; Che, C.M. A Highly Selective FRET-Based Fluorescent Probe for Detection of Cysteine and Homocysteine. *Chem. A Eur. J.* **2010**, *16*, 3308–3313. [CrossRef] [PubMed]
34. JungáKim, M. A thiol-specific fluorescent probe and its application for bioimaging. *Chem. Commun.* **2010**, *46*, 2751–2753.
35. Gu, Y.; Zhao, Z.; Niu, G.; Zhang, R.; Zhang, H.; Shan, G.-G.; Feng, H.-T.; Kwok, R.T.; Lam, J.W.; Yu, X. Ratiometric detection of mitochondrial thiol with a two-photon active AIEgen. *ACS Appl. Bio Mater.* **2019**, *2*, 3120–3127. [CrossRef]
36. Kim, G.-J.; Lee, K.; Kwon, H.; Kim, H.-J. Ratiometric fluorescence imaging of cellular glutathione. *Org. Lett.* **2011**, *13*, 2799–2801. [CrossRef] [PubMed]
37. Lin, W.; Yuan, L.; Cao, Z.; Feng, Y.; Long, L. A sensitive and selective fluorescent thiol probe in water based on the conjugate 1, 4-addition of thiols to α, β-unsaturated ketones. *Chem. A Eur. J.* **2009**, *15*, 5096–5103. [CrossRef]
38. Pang, L.; Zhou, Y.; Gao, W.; Zhang, J.; Song, H.; Wang, X.; Wang, Y.; Peng, X. Curcumin-based fluorescent and colorimetric probe for detecting cysteine in living cells and zebrafish. *Ind. Eng. Chem. Res.* **2017**, *56*, 7650–7655. [CrossRef]
39. Ahn, Y.H.; Lee, J.S.; Chang, Y.T. Combinatorial rosamine library and application to in vivo glutathione probe. *J. Am. Chem. Soc.* **2007**, *129*, 4510–4511. [CrossRef]
40. Lee, M.H.; Han, J.H.; Kwon, P.S.; Bhuniya, S.; Kim, J.Y.; Sessler, J.L.; Kang, C.; Kim, J.S. Hepatocyte-targeting single galactose-appended naphthalimide: A tool for intracellular thiol imaging in vivo. *J. Am. Chem. Soc.* **2012**, *134*, 1316–1322. [CrossRef]
41. Tang, L.; Yu, F.; Tang, B.; Yang, Z.; Fan, W.; Zhang, M.; Wang, Z.; Jacobson, O.; Zhou, Z.; Li, L. Tumor Microenvironment-Activated Ultrasensitive Nanoprobes for Specific Detection of Intratumoral Glutathione by Ratiometric Photoacoustic Imaging. *ACS Appl. Mater. Interfaces* **2019**, *11*, 27558–27567. [CrossRef]
42. Li, Y.; Yang, Y.; Guan, X. Benzofurazan sulfides for thiol imaging and quantification in live cells through fluorescence microscopy. *Anal. Chem.* **2012**, *84*, 6877–6883. [CrossRef]
43. Lee, H.Y.; Choi, Y.P.; Kim, S.; Yoon, T.; Guo, Z.; Lee, S.; Swamy, K.M.K.; Kim, G.; Lee, J.Y.; Shin, I.; et al. Selective homocysteine turn-on fluorescent probes and their bioimaging applications. *Chem. Commun.* **2014**, *50*, 6967–6969. [CrossRef]
44. Liu, Y.; Yu, D.; Ding, S.; Xiao, Q.; Guo, J.; Feng, G. Rapid and ratiometric fluorescent detection of cysteine with high selectivity and sensitivity by a simple and readily available probe. *ACS Appl. Mater. Interfaces* **2014**, *6*, 17543–17550. [CrossRef] [PubMed]
45. Yue, Y.; Huo, F.; Li, X.; Wen, Y.; Yi, T.; Salamanca, J.; Escobedo, J.O.; Strongin, R.M.; Yin, C. pH-Dependent Fluorescent Probe That Can Be Tuned for Cysteine or Homocysteine. *Org. Lett.* **2016**, *19*, 82–85. [CrossRef] [PubMed]
46. Lee, S.; Li, J.; Zhou, X.; Yin, J.; Yoon, J. Recent progress on the development of glutathione (GSH) selective fluorescent and colorimetric probes. *Coord. Chem. Rev.* **2018**, *366*, 29–68. [CrossRef]
47. Chen, X.; Zhou, Y.; Peng, X.; Yoon, J. Fluorescent and colorimetric probes for detection of thiols. *Chem. Soc. Rev.* **2010**, *39*, 2120–2135. [CrossRef]
48. Yan, F.; Sun, X.; Zu, F.; Bai, Z.; Jiang, Y.; Fan, K.; Wang, J. Fluorescent probes for detecting cysteine. *Methods Appl. Fluoresc.* **2018**, *6*, 042001. [CrossRef] [PubMed]

49. Wang, S.; Shen, S.; Zhang, Y.; Dai, X.; Zhao, B. Recent Progress in Fluorescent Probes for the Detection of Biothiols. *Chin. J. Org. Chem.* **2014**, *34*, 1717–1729. [CrossRef]
50. Liu, L.; Li, T.; Ruan, Z.; Yuan, P.; Yan, L. Reduction-sensitive polypeptide nanogel conjugated BODIPY-Br for NIR imaging-guided chem/photodynamic therapy at low light and drug dose. *Mater. Sci. Eng. C* **2018**, *92*, 745–756. [CrossRef]
51. Liu, Y.; Niu, L.Y.; Liu, X.L.; Chen, P.Z.; Yao, Y.S.; Chen, Y.Z.; Yang, Q.Z. Synthesis of N, O, B-Chelated Dipyrromethenes through an Unexpected Intramolecular Cyclisation: Enhanced Near-Infrared Emission in the Aggregate/Solid State. *Chem. A Eur. J.* **2018**, *24*, 13549–13555. [CrossRef] [PubMed]
52. Sadhu, S.S.; Wang, S.; Dachineni, R.; Averineni, R.K.; Seefeldt, T.; Xie, J.; Tummala, H.; Bhat, G.J.; Guan, X. In vitro and in vivo antimetastatic effect of glutathione disulfide liposomes. *Cancer Growth Metastasis* **2017**, *10*, 11790644–17695255. [CrossRef] [PubMed]
53. Meister, A.; Anderson, M.E. Glutathione. *Annu. Rev. Biochem.* **1983**, *52*, 711–760. [CrossRef]
54. Ghezzi, P. Protein glutathionylation in health and disease. *Biochim. Et Biophys. Acta (Bba) Gen. Subj.* **2013**, *1830*, 3165–3172. [CrossRef]
55. Balendiran, G.K.; Dabur, R.; Fraser, D. The role of glutathione in cancer. *Cell Biochem. Funct. Cell. Biochem. Its Modul. By Act. Agents Or Dis.* **2004**, *22*, 343–352. [CrossRef]
56. Wu, G.; Fang, Y.-Z.; Yang, S.; Lupton, J.R.; Turner, N.D. Glutathione metabolism and its implications for health. *J. Nutr.* **2004**, *134*, 489–492. [CrossRef] [PubMed]
57. Townsend, D.M.; Tew, K.D.; Tapiero, H. The importance of glutathione in human disease. *Biomed. Pharmacother.* **2003**, *57*, 145–155. [CrossRef]
58. Wang, K.; Peng, H.; Wang, B. Recent advances in thiol and sulfide reactive probes. *J. Cell. Biochem.* **2014**, *115*, 1007–1022. [CrossRef]
59. Yu, J.-G.; Zhao, X.-H.; Yu, L.-Y.; Yang, H.; Chen, X.-Q.; Jiang, J.-H. Fluorescent probes for selective probing thiol-containing amino acids. *Curr. Org. Synth.* **2014**, *11*, 377–402. [CrossRef]
60. Bao, B.; Liu, M.; Liu, Y.; Zhang, X.; Zang, Y.; Li, J.; Lu, W. NIR absorbing DICPO derivatives applied to wide range of pH and detection of glutathione in tumor. *Tetrahedron* **2015**, *71*, 7865–7868. [CrossRef]
61. Li, M.; Wu, X.; Wang, Y.; Li, Y.; Zhu, W.; James, T.D. A near-infrared colorimetric fluorescent chemodosimeter for the detection of glutathione in living cells. *Chem. Commun.* **2014**, *50*, 1751–1753. [CrossRef] [PubMed]
62. Xu, K.; Qiang, M.; Gao, W.; Su, R.; Li, N.; Gao, Y.; Xie, Y.; Kong, F.; Tang, B. A near-infrared reversible fluorescent probe for real-time imaging of redox status changes in vivo. *Chem. Sci.* **2013**, *4*, 1079–1086. [CrossRef]
63. Xu, G.; Tang, Y.; Lin, W. Development of a two-photon ratiometric fluorescent probe for Glutathione and its applications in living cells. *Chem. Res. Chin. Univ.* **2018**, *34*, 523–527. [CrossRef]
64. Xiong, K.; Huo, F.; Chao, J.; Zhang, Y.; Yin, C. Colorimetric and NIR fluorescence probe with multiple binding sites for distinguishing detection of Cys/Hcy and GSH in vivo. *Anal. Chem.* **2018**, *91*, 1472–1478. [CrossRef]
65. Dai, X.; Wang, Z.-Y.; Du, Z.-F.; Cui, J.-Y.; Miao, J.-Y.; Zhao, B.-X. A colorimetric, ratiometric and water-soluble fluorescent probe for simultaneously sensing glutathione and cysteine/homocysteine. *Anal. Chim. Acta* **2015**, *900*, 103–110. [CrossRef] [PubMed]
66. Li, X.; Huo, F.; Yue, Y.; Zhang, Y.; Yin, C. A coumarin-based "off-on" sensor for fluorescence selectivity discriminating GSH from Cys/Hcy and its bioimaging in living cells. *Sens. Actuators B: Chem.* **2017**, *253*, 42–49. [CrossRef]
67. Xia, X.; Qian, Y.; Shen, B. Synthesis of a BODIPY disulfonate near-infrared fluorescence-enhanced probe with high selectivity to endogenous glutathione and two-photon fluorescent turn-on through thiol-induced SN Ar substitution. *J. Mater. Chem. B* **2018**, *6*, 3023–3029. [CrossRef]
68. Hou, X.; Li, Z.; Li, B.; Liu, C.; Xu, Z. An "off-on" fluorescein-based colormetric and fluorescent probe for the detection of glutathione and cysteine over homocysteine and its application for cell imaging. *Sens. Actuators B: Chem.* **2018**, *260*, 295–302. [CrossRef]
69. Gilli, P.; Bertolasi, V.; Ferretti, V.; Gilli, G. Evidence for intramolecular N−H···O resonance-assisted hydrogen bonding in β-enaminones and related heterodienes. A combined crystal-structural, IR and NMR spectroscopic, and quantum-mechanical investigation. *J. Am. Chem. Soc.* **2000**, *122*, 10405–10417. [CrossRef]
70. Lee, K.-S.; Kim, H.-J.; Kim, G.-H.; Shin, I.; Hong, J.-I. Fluorescent chemodosimeter for selective detection of cyanide in water. *Org. Lett.* **2008**, *10*, 49–51. [CrossRef]
71. Shu, H.; Wu, X.; Zhou, B.; Han, Y.; Jin, M.; Zhu, J.; Bao, X. Synthesis and evaluation of a novel fluorescent chemosensor for glutathione based on a rhodamine B and N-[4-(carbonyl) phenyl] maleimide conjugate and its application in living cell imaging. *Dye. Pigment.* **2017**, *136*, 535–542. [CrossRef]
72. Liu, T.; Huo, F.; Yin, C.; Li, J.; Chao, J.; Zhang, Y. A triphenylamine as a fluorophore and maleimide as a bonding group selective turn-on fluorescent imaging probe for thiols. *Dye. Pigment.* **2016**, *128*, 209–214. [CrossRef]
73. Jiang, X.; Yu, Y.; Chen, J.; Zhao, M.; Chen, H.; Song, X.; Matzuk, A.J.; Carroll, S.L.; Tan, X.; Sizovs, A. Quantitative imaging of glutathione in live cells using a reversible reaction-based ratiometric fluorescent probe. *ACS Chem. Biol.* **2015**, *10*, 864–874. [CrossRef]
74. Jiang, X.; Chen, J.; Bajić, A.; Zhang, C.; Song, X.; Carroll, S.L.; Cai, Z.-L.; Tang, M.; Xue, M.; Cheng, N. Quantitative real-time imaging of glutathione. *Nat. Commun.* **2017**, *8*, 1–13. [CrossRef] [PubMed]

75. Liu, Z.; Zhou, X.; Miao, Y.; Hu, Y.; Kwon, N.; Wu, X.; Yoon, J. A reversible fluorescent probe for real-time quantitative monitoring of cellular glutathione. *Angew. Chem. Int. Ed.* **2017**, *56*, 5812–5816. [CrossRef] [PubMed]
76. Tian, M.; Yang, M.; Liu, Y.; Jiang, F.-L. Rapid and reversible reaction-based ratiometric fluorescent probe for imaging of different glutathione levels in living cells. *ACS Appl. Bio Mater.* **2019**, *2*, 4503–4514. [CrossRef]
77. Tian, M.; Liu, X.-Y.; He, H.; Ma, X.-Z.; Liang, C.; Liu, Y.; Jiang, F.-L. Real-time imaging of intracellular glutathione levels based on a ratiometric fluorescent probe with extremely fast response. *Anal. Chem.* **2020**, *92*, 10068–10075. [CrossRef]
78. Morozumi, A.; Kamiya, M.; Uno, S.-N.; Umezawa, K.; Kojima, R.; Yoshihara, T.; Tobita, S.; Urano, Y. Spontaneously blinking fluorophores based on nucleophilic addition/dissociation of intracellular glutathione for live-cell super-resolution imaging. *J. Am. Chem. Soc.* **2020**, *142*, 9625–9633. [CrossRef]
79. Becker, P.S.; Cohen, C.; Lux, S.E. The effect of mild diamide oxidation on the structure and function of human erythrocyte spectrin. *J. Biol. Chem.* **1986**, *261*, 4620–4628. [CrossRef]
80. Gao, X.; Li, X.; Li, L.; Zhou, J.; Ma, H. A simple fluorescent off–on probe for the discrimination of cysteine from glutathione. *Chem. Commun.* **2015**, *51*, 9388–9390. [CrossRef]
81. Xie, X.; Li, M.; Tang, F.; Li, Y.; Zhang, L.; Jiao, X.; Wang, X.; Tang, B. Combinatorial strategy to identify fluorescent probes for biothiol and thiophenol based on diversified pyrimidine moieties and their biological applications. *Anal. Chem.* **2017**, *89*, 3015–3020. [CrossRef] [PubMed]
82. Yang, Y.; Guan, X. Rapid and Thiol-Specific High-Throughput Assay for Simultaneous Relative Quantification of Total Thiols, Protein Thiols, and Nonprotein Thiols in Cells. *Anal. Chem.* **2014**, *87*, 649–655. [CrossRef]
83. Liu, J.; Sun, Y.-Q.; Zhang, H.; Huo, Y.; Shi, Y.; Guo, W. Simultaneous fluorescent imaging of Cys/Hcy and GSH from different emission channels. *Chem. Sci.* **2014**, *5*, 3183–3188. [CrossRef]
84. Song, L.; Tian, H.; Pei, X.; Zhang, Z.; Zhang, W.; Qian, J. Colorimetric and fluorescent detection of GSH with the assistance of CTAB micelles. *RSC Adv.* **2015**, *5*, 59056–59061. [CrossRef]
85. Yoshida, M.; Kamiya, M.; Yamasoba, T.; Urano, Y. A highly sensitive, cell-membrane-permeable fluorescent probe for glutathione. *Bioorganic Med. Chem. Lett.* **2014**, *24*, 4363–4366. [CrossRef]
86. Zhu, W.; Huang, X.; Guo, Z.; Wu, X.; Yu, H.; Tian, H. A novel NIR fluorescent turn-on sensor for the detection of pyrophosphate anion in complete water system. *Chem. Commun.* **2012**, *48*, 1784–1786. [CrossRef]
87. Guo, Z.; Zhu, W.; Tian, H. Dicyanomethylene-4H-pyran chromophores for OLED emitters, logic gates and optical chemosensors. *Chem. Commun.* **2012**, *48*, 6073–6084. [CrossRef] [PubMed]
88. Yin, J.; Kwon, Y.; Kim, D.; Lee, D.; Kim, G.; Hu, Y.; Ryu, J.-H.; Yoon, J. Cyanine-based fluorescent probe for highly selective detection of glutathione in cell cultures and live mouse tissues. *J. Am. Chem. Soc.* **2014**, *136*, 5351–5358. [CrossRef] [PubMed]
89. Han, X.; Liu, Y.; Liu, G.; Luo, J.; Liu, S.H.; Zhao, W.; Yin, J. A Versatile Naphthalimide–Sulfonamide-Coated Tetraphenylethene: Aggregation-Induced Emission Behavior, Mechanochromism, and Tracking Glutathione in Living Cells. *Chem. Asian J.* **2019**, *14*, 890–895. [CrossRef]
90. Huang, X.; Guo, Z.; Zhu, W.; Xie, Y.; Tian, H. A colorimetric and fluorescent turn-on sensor for pyrophosphate anion based on a dicyanomethylene-4 H-chromene framework. *Chem. Commun.* **2008**, *41*, 5143–5145. [CrossRef]
91. Huang, L.; Duan, R.; Li, Z.; Zhang, Y.; Zhao, J.; Han, G. BODIPY-Based Fluorescent Nanomicelles as Near-Infrared Fluorescent "Turn-On" Sensors for Biogenic Thiols. *ChemNanoMat* **2016**, *2*, 396–399. [CrossRef]
92. Zhu, B.; Zhang, X.; Li, Y.; Wang, P.; Zhang, H.; Zhuang, X. A colorimetric and ratiometric fluorescent probe for thiols and its bioimaging applications. *Chem. Commun.* **2010**, *46*, 5710–5712. [CrossRef]
93. Ye, M.; Wang, X.; Tang, J.; Guo, Z.; Shen, Y.; Tian, H.; Zhu, W.-H. Dual-channel NIR activatable theranostic prodrug for in vivo spatiotemporal tracking thiol-triggered chemotherapy. *Chem. Sci.* **2016**, *7*, 4958–4965. [CrossRef] [PubMed]
94. Yin, C.; Tang, Y.; Li, X.; Yang, Z.; Li, J.; Li, X.; Huang, W.; Fan, Q. A Single Composition Architecture-Based Nanoprobe for Ratiometric Photoacoustic Imaging of Glutathione (GSH) in Living Mice. *Small* **2018**, *14*, 1703400. [CrossRef] [PubMed]
95. Zhu, B.; Zhang, X.; Jia, H.; Li, Y.; Chen, S.; Zhang, S. The determination of thiols based using a probe that utilizes both an absorption red-shift and fluorescence enhancement. *Dye. Pigment.* **2010**, *86*, 87–92. [CrossRef]
96. Yuan, L.; Lin, W.; Xie, Y.; Zhu, S.; Zhao, S. A Native-Chemical-Ligation-Mechanism-Based Ratiometric Fluorescent Probe for Aminothiols. *Chem. A Eur. J.* **2012**, *18*, 14520–14526. [CrossRef] [PubMed]
97. Yang, X.; Huang, Q.; Zhong, Y.; Li, Z.; Li, H.; Lowry, M.; Escobedo, J.; Strongin, R. A dual emission fluorescent probe enables simultaneous detection of glutathione and cysteine/homocysteine. *Chem. Sci.* **2014**, *5*, 2177–2183. [CrossRef]
98. Zhang, Y.; Shao, X.; Wang, Y.; Pan, F.; Kang, R.; Peng, F.; Huang, Z.; Zhang, W.; Zhao, W. Dual emission channels for sensitive discrimination of Cys/Hcy and GSH in plasma and cells. *Chem. Commun.* **2015**, *51*, 4245–4248. [CrossRef]
99. Yang, X.; Guo, Y.; Strongin, R.M. Conjugate addition/cyclization sequence enables selective and simultaneous fluorescence detection of cysteine and homocysteine. *Angew. Chem.* **2011**, *123*, 10878–10881. [CrossRef]
100. Dai, X.; Wu, Q.-H.; Wang, P.-C.; Tian, J.; Xu, Y.; Wang, S.-Q.; Miao, J.-Y.; Zhao, B.-X. A simple and effective coumarin-based fluorescent probe for cysteine. *Biosens. Bioelectron.* **2014**, *59*, 35–39. [CrossRef]
101. Guo, Z.; Nam, S.; Park, S.; Yoon, J. A highly selective ratiometric near-infrared fluorescent cyanine sensor for cysteine with remarkable shift and its application in bioimaging. *Chem. Sci.* **2012**, *3*, 2760–2765. [CrossRef]
102. Liu, J.; Sun, Y.-Q.; Huo, Y.; Zhang, H.; Wang, L.; Zhang, P.; Song, D.; Shi, Y.; Guo, W. Simultaneous fluorescence sensing of Cys and GSH from different emission channels. *J. Am. Chem. Soc.* **2014**, *136*, 574–577. [CrossRef] [PubMed]

103. Liu, S.-R.; Chang, C.-Y.; Wu, S.-P. A fluorescence turn-on probe for cysteine and homocysteine based on thiol-triggered benzothiazolidine ring formation. *Anal. Chim. Acta* **2014**, *849*, 64–69. [CrossRef] [PubMed]
104. Wang, F.; Zhou, L.; Zhao, C.; Wang, R.; Fei, Q.; Luo, S.; Guo, Z.; Tian, H.; Zhu, W.-H. A dual-response BODIPY-based fluorescent probe for the discrimination of glutathione from cystein and homocystein. *Chem. Sci.* **2015**, *6*, 2584–2589. [CrossRef] [PubMed]
105. Niu, L.-Y.; Yang, Q.-Q.; Zheng, H.-R.; Chen, Y.-Z.; Wu, L.-Z.; Tung, C.-H.; Yang, Q.-Z. BODIPY-based fluorescent probe for the simultaneous detection of glutathione and cysteine/homocysteine at different excitation wavelengths. *RSC Adv.* **2015**, *5*, 3959–3964. [CrossRef]
106. Wang, P.; Liu, J.; Lv, X.; Liu, Y.; Zhao, Y.; Guo, W. A naphthalimide-based glyoxal hydrazone for selective fluorescence turn-on sensing of Cys and Hcy. *Org. Lett.* **2012**, *14*, 520–523. [CrossRef]
107. Sun, Y.-Q.; Liu, J.; Zhang, H.; Huo, Y.; Lv, X.; Shi, Y.; Guo, W. A mitochondria-targetable fluorescent probe for dual-channel NO imaging assisted by intracellular cysteine and glutathione. *J. Am. Chem. Soc.* **2014**, *136*, 12520–12523. [CrossRef]
108. Lim, S.; Escobedo, J.O.; Lowry, M.; Xu, X.; Strongin, R. Selective fluorescence detection of cysteine and N-terminal cysteine peptide residues. *Chem. Commun.* **2010**, *46*, 5707–5709. [CrossRef] [PubMed]
109. Shu, W.; Wang, Y.; Wu, L.; Wang, Z.; Duan, Q.; Gao, Y.; Liu, C.; Zhu, B.; Yan, L. Novel carbonothioate-based colorimetric and fluorescent probe for selective detection of mercury ions. *Ind. Eng. Chem. Res.* **2016**, *55*, 8713–8718. [CrossRef]
110. Yue, Y.; Yin, C.; Huo, F.; Chao, J.; Zhang, Y. The application of natural drug-curcumin in the detection hypochlorous acid of real sample and its bioimaging. *Sens. Actuators B: Chem.* **2014**, *202*, 551–556. [CrossRef]
111. Rohanizadeh, R.; Deng, Y.; Verron, E. Therapeutic actions of curcumin in bone disorders. *Bonekey Rep.* **2016**, *5*, 793. [CrossRef]
112. Niu, W.; Guo, L.; Li, Y.; Shuang, S.; Dong, C.; Wong, M.S. Highly selective two-photon fluorescent probe for ratiometric sensing and imaging cysteine in mitochondria. *Anal. Chem.* **2016**, *88*, 1908–1914. [CrossRef] [PubMed]
113. Ali, F.; Anila, H.; Taye, N.; Gonnade, R.G.; Chattopadhyay, S.; Das, A. A fluorescent probe for specific detection of cysteine in the lipid dense region of cells. *Chem. Commun.* **2015**, *51*, 16932–16935. [CrossRef] [PubMed]
114. Lee, H.; Kim, D.I.; Kwon, H.; Kim, H.-J. Bromoacetylfluorescein monoaldehyde as a fluorescence turn-on probe for cysteine over homocysteine and glutathione. *Sens. Actuators B Chem.* **2015**, *209*, 652–657. [CrossRef]
115. Xiao, H.; Li, P.; Hu, X.; Shi, X.; Zhang, W.; Tang, B. Simultaneous fluorescence imaging of hydrogen peroxide in mitochondria and endoplasmic reticulum during apoptosis. *Chem. Sci.* **2016**, *7*, 6153–6159. [CrossRef] [PubMed]
116. Loudet, A.; Burgess, K. BODIPY dyes and their derivatives: Syntheses and spectroscopic properties. *Chem. Rev.* **2007**, *107*, 4891–4932. [CrossRef]
117. Niu, L.-Y.; Guan, Y.-S.; Chen, Y.-Z.; Wu, L.-Z.; Tung, C.-H.; Yang, Q.-Z. A turn-on fluorescent sensor for the discrimination of cystein from homocystein and glutathione. *Chem. Commun.* **2013**, *49*, 1294–1296. [CrossRef]
118. Niu, L.-Y.; Zheng, H.-R.; Chen, Y.-Z.; Wu, L.-Z.; Tung, C.-H.; Yang, Q.-Z. Fluorescent sensors for selective detection of thiols: Expanding the intramolecular displacement based mechanism to new chromophores. *Analyst* **2014**, *139*, 1389–1395. [CrossRef]
119. Kand, D.; Saha, T.; Talukdar, P. Off-on type fluorescent NBD-probe for selective sensing of cysteine and homocysteine over glutathione. *Sens. Actuators B Chem.* **2014**, *196*, 440–449. [CrossRef]
120. Zhai, L.; Shi, Z.; Tu, Y.; Pu, S. A dual emission fluorescent probe enables simultaneous detection and discrimination of Cys/Hcy and GSH and its application in cell imaging. *Dye. Pigment.* **2019**, *165*, 164–171. [CrossRef]
121. Ye, Y.; Duan, C.; Hu, Q.; Zhang, Y.; Qin, C.; Zeng, L. A dual-channel responsive near-infrared fluorescent probe for multicolour imaging of cysteine in living cells. *J. Mater. Chem. B* **2017**, *5*, 3600–3606. [CrossRef]
122. Das, P.; Mandal, A.K.; Chandar, N.B.; Baidya, M.; Bhatt, H.B.; Ganguly, B.; Ghosh, S.K.; Das, A. New chemodosimetric reagents as ratiometric probes for cysteine and homocysteine and possible detection in living cells and in blood plasma. *Chem. A Eur. J.* **2012**, *18*, 15382. [CrossRef] [PubMed]
123. Wang, H.; Zhou, G.; Gai, H.; Chen, X. A fluorescein-based probe with high selectivity to cysteine over homocysteine and glutathione. *Chem. Commun.* **2012**, *48*, 8341–8343. [CrossRef]
124. Ding, S.; Feng, G. Smart probe for rapid and simultaneous detection and discrimination of hydrogen sulfide, cysteine/homocysteine, and glutathione. *Sens. Actuators B Chem.* **2016**, *235*, 691–697. [CrossRef]
125. He, L.; Yang, X.; Xu, K.; Kong, X.; Lin, W. A multi-signal fluorescent probe for simultaneously distinguishing and sequentially sensing cysteine/homocysteine, glutathione, and hydrogen sulfide in living cells. *Chem. Sci.* **2017**, *8*, 6257–6265. [CrossRef]
126. Wang, P.; Wang, Y.; Li, N.; Huang, J.; Wang, Q.; Gu, Y. A novel DCM-NBD conjugate fluorescent probe for discrimination of Cys/Hcy from GSH and its bioimaging applications in living cells and animals. *Sens. Actuators B Chem.* **2017**, *245*, 297–304. [CrossRef]
127. Lee, D.; Kim, G.; Yin, J.; Yoon, J. An aryl-thioether substituted nitrobenzothiadazole probe for the selective detection of cysteine and homocysteine. *Chem. Commun.* **2015**, *51*, 6518–6520. [CrossRef] [PubMed]
128. Ma, L.; Qian, J.; Tian, H.; Lan, M.; Zhang, W. A colorimetric and fluorescent dual probe for specific detection of cysteine based on intramolecular nucleophilic aromatic substitution. *Analyst* **2012**, *137*, 5046–5050. [CrossRef] [PubMed]
129. Jung, H.S.; Han, J.H.; Pradhan, T.; Kim, S.; Lee, S.W.; Sessler, J.L.; Kim, T.W.; Kang, C.; Kim, J.S. A cysteine-selective fluorescent probe for the cellular detection of cysteine. *Biomaterials* **2012**, *33*, 945–953. [CrossRef]
130. Zhou, X.; Jin, X.; Sun, G.; Li, D.; Wu, X. A cysteine probe with high selectivity and sensitivity promoted by response-assisted electrostatic attraction. *Chem. Commun.* **2012**, *48*, 8793–8795. [CrossRef] [PubMed]
131. Baba, S.P.; Bhatnagar, A. Role of thiols in oxidative stress. *Curr. Opin. Toxicol.* **2018**, *7*, 133–139. [CrossRef]

132. Ow, Y.-L.P.; Green, D.R.; Hao, Z.; Mak, T.W. Cytochrome c: Functions beyond respiration. *Nat. Rev. Mol. Cell Biol.* **2008**, *9*, 532–542. [CrossRef] [PubMed]
133. Hoye, A.T.; Davoren, J.E.; Wipf, P.; Fink, M.P.; Kagan, V.E. Targeting mitochondria. *Acc. Chem. Res.* **2008**, *41*, 87–97. [CrossRef]
134. Dickinson, B.C.; Srikun, D.; Chang, C.J. Mitochondrial-targeted fluorescent probes for reactive oxygen species. *Curr. Opin. Chem. Biol.* **2010**, *14*, 50–56. [CrossRef] [PubMed]
135. Su, P.; Zhu, Z.; Tian, Y.; Liang, L.; Wu, W.; Cao, J.; Cheng, B.; Liu, W.; Tang, Y. A TAT peptide-based ratiometric two-photon fluorescent probe for detecting biothiols and sequentially distinguishing GSH in mitochondria. *Talanta* **2020**, *218*, 121–127. [CrossRef]
136. Gao, P.; Pan, W.; Li, N.; Tang, B. Fluorescent probes for organelle-targeted bioactive species imaging. *Chem. Sci.* **2019**, *10*, 6035–6071. [CrossRef]
137. Yang, Y.; Zhou, T.; Jin, M.; Zhou, K.; Liu, D.; Li, X.; Huo, F.; Li, W.; Yin, C. Thiol–Chromene "Click" reaction triggered self-immolative for NIR visualization of thiol flux in physiology and pathology of living cells and mice. *J. Am. Chem. Soc.* **2019**, *142*, 1614–1620. [CrossRef]
138. Chen, J.; Jiang, X.; Zhang, C.; MacKenzie, K.R.; Stossi, F.; Palzkill, T.; Wang, M.C.; Wang, J. Reversible reaction-based fluorescent probe for real-time imaging of glutathione dynamics in mitochondria. *ACS Sens.* **2017**, *2*, 1257–1261. [CrossRef] [PubMed]
139. Umezawa, K.; Yoshida, M.; Kamiya, M.; Yamasoba, T.; Urano, Y. Rational design of reversible fluorescent probes for live-cell imaging and quantification of fast glutathione dynamics. *Nat. Chem.* **2017**, *9*, 279. [CrossRef]
140. Liu, X.-L.; Niu, L.-Y.; Chen, Y.-Z.; Zheng, M.-L.; Yang, Y.; Yang, Q.-Z. A mitochondria-targeting fluorescent probe for the selective detection of glutathione in living cells. *Org. Biomol. Chem.* **2017**, *15*, 1072–1075. [CrossRef]
141. Qi, S.; Liu, W.; Zhang, P.; Wu, J.; Zhang, H.; Ren, H.; Ge, J.; Wang, P. A colorimetric and ratiometric fluorescent probe for highly selective detection of glutathione in the mitochondria of living cells. *Sens. Actuators B Chem.* **2018**, *270*, 459–465. [CrossRef]
142. Xu, C.; Li, H.; Yin, B. A colorimetric and ratiometric fluorescent probe for selective detection and cellular imaging of glutathione. *Biosens. Bioelectron.* **2015**, *72*, 275–281. [CrossRef] [PubMed]
143. Xu, Z.; Zhang, M.-X.; Xu, Y.; Liu, S.H.; Zeng, L.; Chen, H.; Yin, J. The visualization of lysosomal and mitochondrial glutathione via near-infrared fluorophore and in vivo imaging application. *Sens. Actuators B: Chem.* **2019**, *290*, 676–683. [CrossRef]
144. Pawlicki, M.; Collins, H.A.; Denning, R.G.; Anderson, H.L. Two-photon absorption and the design of two-photon dyes. *Angew. Chem. Int. Ed.* **2009**, *48*, 3244–3266. [CrossRef]
145. Ding, D.; Goh, C.C.; Feng, G.; Zhao, Z.; Liu, J.; Liu, R.; Tomczak, N.; Geng, J.; Tang, B.Z.; Ng, L.G. Ultrabright organic dots with aggregation-induced emission characteristics for real-time two-photon intravital vasculature imaging. *Adv. Mater.* **2013**, *25*, 6083–6088. [CrossRef]
146. Dong, Y.; Lam, J.W.; Qin, A.; Sun, J.; Liu, J.; Li, Z.; Sun, J.; Sung, H.H.; Williams, I.D.; Kwok, H.S. Aggregation-induced and crystallization-enhanced emissions of 1, 2-diphenyl-3, 4-bis (diphenylmethylene)-1-cyclobutene. *Chem. Commun.* **2007**, *31*, 3255–3257. [CrossRef] [PubMed]
147. Zheng, X.; Peng, Q.; Zhu, L.; Xie, Y.; Huang, X.; Shuai, Z. Unraveling the aggregation effect on amorphous phase AIE luminogens: A computational study. *Nanoscale* **2016**, *8*, 15173–15180. [CrossRef]
148. Liang, J.; Chen, Z.; Yin, J.; Yu, G.-A.; Liu, S.H. Aggregation-induced emission (AIE) behavior and thermochromic luminescence properties of a new gold (I) complex. *Chem. Commun.* **2013**, *49*, 3567–3569. [CrossRef]
149. Kwok, R.T.; Leung, C.W.; Lam, J.W.; Tang, B.Z. Biosensing by luminogens with aggregation-induced emission characteristics. *Chem. Soc. Rev.* **2015**, *44*, 4228–4238. [CrossRef]
150. Yuan, W.Z.; Lu, P.; Chen, S.; Lam, J.W.; Wang, Z.; Liu, Y.; Kwok, H.S.; Ma, Y.; Tang, B.Z. Changing the behavior of chromophores from aggregation-caused quenching to aggregation-induced emission: Development of highly efficient light emitters in the solid state. *Adv. Mater.* **2010**, *22*, 2159–2163. [CrossRef]
151. Luo, J.; Xie, Z.; Lam, J.W.; Cheng, L.; Chen, H.; Qiu, C.; Kwok, H.S.; Zhan, X.; Liu, Y.; Zhu, D. Aggregation-induced emission of 1-methyl-1,2,3,4,5-pentaphenylsilole. *Chem. Commun.* **2001**, *18*, 1740–1741. [CrossRef]
152. Pan, L.; Cai, Y.; Wu, H.; Zhou, F.; Qin, A.; Wang, Z.; Tang, B.Z. Tetraphenylpyrazine-based luminogens with full-colour emission. *Mater. Chem. Front.* **2018**, *2*, 1310–1316. [CrossRef]
153. Yu, T.; Ou, D.; Yang, Z.; Huang, Q.; Mao, Z.; Chen, J.; Zhang, Y.; Liu, S.; Xu, J.; Bryce, M.R. The HOF structures of nitrotetraphenylethene derivatives provide new insights into the nature of AIE and a way to design mechanoluminescent materials. *Chem. Sci.* **2017**, *8*, 1163–1168. [CrossRef] [PubMed]
154. Yamaguchi, M.; Ito, S.; Hirose, A.; Tanaka, K.; Chujo, Y. Control of aggregation-induced emission versus fluorescence aggregation-caused quenching by bond existence at a single site in boron pyridinoiminate complexes. *Mater. Chem. Front.* **2017**, *1*, 1573–1579. [CrossRef]
155. Mei, J.; Leung, N.L.; Kwok, R.T.; Lam, J.W.; Tang, B.Z. Aggregation-induced emission: Together we shine, united we soar! *Chem. Rev.* **2015**, *115*, 11718–11940. [CrossRef] [PubMed]
156. Wang, Y.; Chen, M.; Alifu, N.; Li, S.; Qin, W.; Qin, A.; Tang, B.Z.; Qian, J. Aggregation-induced emission luminogen with deep-red emission for through-skull three-photon fluorescence imaging of mouse. *ACS Nano* **2017**, *11*, 10452–10461. [CrossRef]
157. Leung, N.L.; Xie, N.; Yuan, W.; Liu, Y.; Wu, Q.; Peng, Q.; Miao, Q.; Lam, J.W.; Tang, B.Z. Restriction of intramolecular motions: The general mechanism behind aggregation-induced emission. *Chem. A Eur. J.* **2014**, *20*, 15349–15353. [CrossRef] [PubMed]

158. Zhao, Z.; Gao, S.; Zheng, X.; Zhang, P.; Wu, W.; Kwok, R.T.; Xiong, Y.; Leung, N.L.; Chen, Y.; Gao, X. Rational Design of Perylenediimide-Substituted Triphenylethylene to Electron Transporting Aggregation-Induced Emission Luminogens (AIEgens) with High Mobility and Near-Infrared Emission. *Adv. Funct. Mater.* **2018**, *28*, 1705609. [CrossRef]
159. Ni, J.-S.; Liu, H.; Liu, J.; Jiang, M.; Zhao, Z.; Chen, Y.; Kwok, R.T.; Lam, J.W.; Peng, Q.; Tang, B.Z. The unusual aggregation-induced emission of coplanar organoboron isomers and their lipid droplet-specific applications. *Mater. Chem. Front.* **2018**, *2*, 1498–1507. [CrossRef]
160. Zhang, J.; Bao, X.; Zhou, J.; Peng, F.; Ren, H.; Dong, X.; Zhao, W. A mitochondria-targeted turn-on fluorescent probe for the detection of glutathione in living cells. *Biosens. Bioelectron.* **2016**, *85*, 164–170. [CrossRef] [PubMed]
161. Cui, M.; Li, W.; Wang, L.; Gong, L.; Tang, H.; Cao, D. Twisted intramolecular charge transfer and aggregation-enhanced emission characteristics based quinoxaline luminogen: Photophysical properties and a turn-on fluorescent probe for glutathione. *J. Mater. Chem. C* **2019**, *7*, 3779–3786. [CrossRef]
162. Xu, Z.; Huang, X.; Han, X.; Wu, D.; Zhang, B.; Tan, Y.; Cao, M.; Liu, S.H.; Yin, J.; Yoon, J. A visible and near-infrared, dual-channel fluorescence-on probe for selectively tracking mitochondrial glutathione. *Chem* **2018**, *4*, 1609–1628. [CrossRef]
163. Xu, Z.; Huang, X.; Zhang, M.-X.; Chen, W.; Liu, S.H.; Tan, Y.; Yin, J. Tissue imaging of glutathione-specific naphthalimide–cyanine dye with two-photon and near-infrared manners. *Anal. Chem.* **2019**, *91*, 11343–11348. [CrossRef] [PubMed]
164. Liu, Z.; Wang, Q.; Wang, H.; Su, W.; Dong, S. A FRET based two-photon fluorescent probe for visualizing mitochondrial thiols of living cells. *Sensors* **2020**, *20*, 1746. [CrossRef]
165. Fan, J.; Han, Z.; Kang, Y.; Peng, X. A Two-Photon Fluorescent Probe for Lysosomal Thiols in Live Cells and Tissues. *Sci. Rep.* **2016**, *6*. [CrossRef]
166. Zhang, W.; Huo, F.; Liu, T.; Wen, Y.; Yin, C. A rapid and highly sensitive fluorescent imaging materials for thiophenols. *Dye. Pigment.* **2016**, *133*, 248–254. [CrossRef]
167. Wang, F.-F.; Liu, Y.-J.; Wang, B.-B.; Gao, L.-X.; Jiang, F.-L.; Liu, Y. A BODIPY-based mitochondria-targeted turn-on fluorescent probe with dual response units for the rapid detection of intracellular biothiols. *Dye. Pigment.* **2018**, *152*, 29–35. [CrossRef]
168. Tekdaş, D.A.; Viswanathan, G.; Topal, S.Z.; Looi, C.Y.; Wong, W.F.; Tan, G.M.Y.; Zorlu, Y.; Gürek, A.G.; Lee, H.B.; Dumoulin, F. Antimicrobial activity of a quaternized BODIPY against Staphylococcus strains. *Org. Biomol. Chem.* **2016**, *14*, 2665–2670. [CrossRef]
169. Jiang, N.; Fan, J.; Liu, T.; Cao, J.; Qiao, B.; Wang, J.; Gao, P.; Peng, X. A near-infrared dye based on BODIPY for tracking morphology changes in mitochondria. *Chem. Commun.* **2013**, *49*, 10620–10622. [CrossRef]
170. Gao, T.; He, H.; Huang, R.; Zheng, M.; Wang, F.-F.; Hu, Y.-J.; Jiang, F.-L.; Liu, Y. BODIPY-based fluorescent probes for mitochondria-targeted cell imaging with superior brightness, low cytotoxicity and high photostability. *Dye. Pigment.* **2017**, *141*, 530–535. [CrossRef]
171. Li, Y.; Wang, K.-N.; Liu, B.; Lu, X.-R.; Li, M.-F.; Ji, L.-N.; Mao, Z.-W. Mitochondria-targeted two-photon fluorescent probe for the detection of biothiols in living cells. *Sens. Actuators B Chem.* **2018**, *255*, 193–202. [CrossRef]
172. Lin, X.; Hu, Y.; Yang, D.; Chen, B. Cyanine-coumarin composite NIR dye based instantaneous-response probe for biothiols detection and oxidative stress assessment of mitochondria. *Dye. Pigment.* **2020**, *174*, 107956. [CrossRef]
173. Lim, C.S.; Masanta, G.; Kim, H.J.; Han, J.H.; Kim, H.M.; Cho, B.R. Ratiometric Detection of Mitochondrial Thiols with a Two-Photon Fluorescent Probe. *J. Am. Chem. Soc.* **2011**, *133*, 11132–11135. [CrossRef] [PubMed]
174. Wang, L.; Wang, J.; Xia, S.; Wang, X.; Yu, Y.; Zhou, H.; Liu, H. A FRET-based near-infrared ratiometric fluorescent probe for detection of mitochondria biothiol. *Talanta* **2020**, *219*, 121296. [CrossRef]
175. Long, L.; Lin, W.; Chen, B.; Gao, W.; Yuan, L. Construction of a FRET-based ratiometric fluorescent thiol probe. *Chem. Commun.* **2011**, *47*, 893–895. [CrossRef]
176. Yuan, L.; Lin, W.; Zheng, K.; Zhu, S. FRET-based small-molecule fluorescent probes: Rational design and bioimaging applications. *Acc. Chem. Res.* **2013**, *46*, 1462–1473. [CrossRef] [PubMed]
177. Wang, J.; Xia, S.; Bi, J.; Zhang, Y.; Fang, M.; Luck, R.L.; Zeng, Y.; Chen, T.-H.; Lee, H.-M.; Liu, H. Near-infrared fluorescent probes based on TBET and FRET rhodamine acceptors with different p K a values for sensitive ratiometric visualization of pH changes in live cells. *J. Mater. Chem. B* **2019**, *7*, 198–209. [CrossRef]
178. Zeng, Q.; Guo, Q.; Yuan, Y.; Yang, Y.; Zhang, B.; Ren, L.; Zhang, X.; Luo, Q.; Liu, M.; Bouchard, L.-S. Mitochondria targeted and intracellular biothiol triggered hyperpolarized 129Xe magnetofluorescent biosensor. *Anal. Chem.* **2017**, *89*, 2288–2295. [CrossRef]
179. Lim, S.-Y.; Hong, K.-H.; Kim, D.I.; Kwon, H.; Kim, H.-J. Tunable Heptamethine–Azo Dye Conjugate as an NIR Fluorescent Probe for the Selective Detection of Mitochondrial Glutathione over Cysteine and Homocysteine. *J. Am. Chem. Soc.* **2014**, *136*, 7018–7025. [CrossRef]
180. Meng, F.; Liu, Y.; Yu, X.; Lin, W. A dual-site two-photon fluorescent probe for visualizing mitochondrial aminothiols in living cells and mouse liver tissues. *New J. Chem.* **2016**, *40*, 7399–7406. [CrossRef]
181. Zhang, H.; Wang, C.; Wang, K.; Xuan, X.; Lv, Q.; Jiang, K. Ultrasensitive fluorescent ratio imaging probe for the detection of glutathione ultratrace change in mitochondria of cancer cells. *Biosens. Bioelectron.* **2016**, *85*, 96–102. [CrossRef]
182. Bekdemir, Y.; Gediz Erturk, A.; Kutuk, H. Investigation of the acid-catalyzed hydrolysis and reaction mechanisms of N,N'-diarylsulfamides using various criteria. *J. Phys. Org. Chem.* **2014**, *27*, 94–98. [CrossRef]
183. Xia, M.-C.; Cai, L.; Zhang, S.; Zhang, X. Cell-penetrating peptide spirolactam derivative as a reversible fluorescent pH probe for live cell imaging. *Anal. Chem.* **2017**, *89*, 1238–1243. [CrossRef] [PubMed]

184. Gao, W.; Li, T.; Wang, J.; Zhao, Y.; Wu, C. Thioether-bonded fluorescent probes for deciphering thiol-mediated exchange reactions on the cell surface. *Anal. Chem.* **2017**, *89*, 937–944. [CrossRef] [PubMed]
185. Wang, F.; Wang, Y.; Zhang, X.; Zhang, W.; Guo, S.; Jin, F. Recent progress of cell-penetrating peptides as new carriers for intracellular cargo delivery. *J. Control. Release* **2014**, *174*, 126–136. [CrossRef] [PubMed]
186. Sun, C.; Du, W.; Wang, P.; Wu, Y.; Wang, B.; Wang, J.; Xie, W. A novel mitochondria-targeted two-photon fluorescent probe for dynamic and reversible detection of the redox cycles between peroxynitrite and glutathione. *Biochem. Biophys. Res. Commun.* **2017**, *494*, 518–525. [CrossRef]
187. Hanson, G.T.; Aggeler, R.; Oglesbee, D.; Cannon, M.; Capaldi, R.A.; Tsien, R.Y.; Remington, S.J. Investigating mitochondrial redox potential with redox-sensitive green fluorescent protein indicators. *J. Biol. Chem.* **2004**, *279*, 13044–13053. [CrossRef]
188. Han, C.; Yang, H.; Chen, M.; Su, Q.; Feng, W.; Li, F. Mitochondria-targeted near-infrared fluorescent off–on probe for selective detection of cysteine in living cells and in vivo. *ACS Appl. Mater. Interfaces* **2015**, *7*, 27968–27975. [CrossRef]
189. Kim, C.Y.; Kang, H.J.; Chung, S.J.; Kim, H.-K.; Na, S.-Y.; Kim, H.-J. Mitochondria-targeting chromogenic and fluorescence turn-on probe for the selective detection of cysteine by caged oxazolidinoindocyanine. *Anal. Chem.* **2016**, *88*, 7178–7182. [CrossRef]
190. Zhang, P.; Guo, Z.-Q.; Yan, C.-X.; Zhu, W.-H. Near-Infrared mitochondria-targeted fluorescent probe for cysteine based on difluoroboron curcuminoid derivatives. *Chin. Chem. Lett.* **2017**, *28*, 1952–1956. [CrossRef]
191. Tang, L.; Xu, D.; Tian, M.; Yan, X. A mitochondria-targetable far-red emissive fluorescence probe for highly selective detection of cysteine with a large Stokes shift. *J. Lumin.* **2019**, *208*, 502–508. [CrossRef]
192. Yang, M.; Fan, J.; Sun, W.; Du, J.; Peng, X. Mitochondria-anchored colorimetric and ratiometric fluorescent chemosensor for visualizing cysteine/homocysteine in living cells and daphnia magna model. *Anal. Chem.* **2019**, *91*, 12531–12537. [CrossRef] [PubMed]
193. Ji, X.; Wang, N.; Zhang, J.; Xu, S.; Si, Y.; Zhao, W. Meso-pyridinium substituted BODIPY dyes as mitochondria-targeted probes for the detection of cysteine in living cells and in vivo. *Dye. Pigment.* **2021**, *187*, 109089. [CrossRef]
194. Fan, L.; Zhang, W.; Wang, X.; Dong, W.; Tong, Y.; Dong, C.; Shuang, S. A two-photon ratiometric fluorescent probe for highly selective sensing of mitochondrial cysteine in live cells. *Analyst* **2019**, *144*, 439–447. [CrossRef]
195. Yin, K.; Yu, F.; Zhang, W.; Chen, L. A near-infrared ratiometric fluorescent probe for cysteine detection over glutathione indicating mitochondrial oxidative stress in vivo. *Biosens. Bioelectron.* **2015**, *74*, 156–164. [CrossRef]
196. Yang, X.; Liu, W.; Tang, J.; Li, P.; Weng, H.; Ye, Y.; Xian, M.; Tang, B.; Zhao, Y. A multi-signal mitochondria-targeted fluorescent probe for real-time visualization of cysteine metabolism in living cells and animals. *Chem. Commun.* **2018**, *54*, 11387–11390. [CrossRef]
197. Liu, J.; Sun, Y.-Q.; Zhang, H.; Huo, Y.; Shi, Y.; Shi, H.; Guo, W. A carboxylic acid-functionalized coumarin-hemicyanine fluorescent dye and its application to construct a fluorescent probe for selective detection of cysteine over homocysteine and glutathione. *RSC Adv.* **2014**, *4*, 64542–64550. [CrossRef]
198. Phan, U.T.; Arunachalam, B.; Cresswell, P. Gamma-Interferon-inducibleLysosomal Thiol Reductase (GILT): Maturation, activity, and mechanism of action. *J. Biol. Chem.* **2000**, *275*, 25907–25914. [CrossRef]
199. Hastings, K.T.; Cresswell, P. Disulfide reduction in the endocytic pathway: Immunological functions of gamma-interferon-inducible lysosomal thiol reductase. *Antioxid. Redox Signal.* **2011**, *15*, 657–668. [CrossRef]
200. Satoh, J.-i.; Kino, Y.; Yanaizu, M.; Ishida, T.; Saito, Y. Microglia express gamma-interferon-inducible lysosomal thiol reductase in the brains of Alzheimer's disease and Nasu-Hakola disease. *Intractable Rare Dis. Res.* **2018**, *7*, 251–257. [CrossRef]
201. Chiang, H.-S. The Regulatory Roles of Gamma Interferon Inducible Lysosomal Thiol Reductase (gilt) in Cellular Redox Homeostasis. Ph.D. Thesis, Georgetown University, Washington, DC, USA, 2011.
202. Mego, J. Role of thiols, pH and cathepsin D in the lysosomal catabolism of serum albumin. *Biochem. J.* **1984**, *218*, 775–783. [CrossRef]
203. Huang, R.; Wang, B.-B.; Si-Tu, X.-M.; Gao, T.; Wang, F.-F.; He, H.; Fan, X.-Y.; Jiang, F.-L.; Liu, Y. A lysosome-targeted fluorescent sensor for the detection of glutathione in cells with an extremely fast response. *Chem. Commun.* **2016**, *52*, 11579–11582. [CrossRef]
204. Zhang, H.; Xu, L.; Chen, W.; Huang, J.; Huang, C.; Sheng, J.; Song, X. A lysosome-targetable fluorescent probe for simultaneously sensing Cys/Hcy, GSH, and H2S from different signal patterns. *ACS Sens.* **2018**, *3*, 2513–2517. [CrossRef]
205. Song, X.; Tu, Y.; Wang, R.; Pu, S. A lysosome-targetable fluorescent probe for simultaneous detection and discrimination of Cys/Hcy and GSH by dual channels. *Dye. Pigment.* **2020**, *177*, 108270. [CrossRef]
206. Alqahtani, Y.; Wang, S.; Huang, Y.; Najmi, A.; Guan, X. Design, Synthesis, and Characterization of Bis (7-(N-(2-morpholinoethyl) sulfamoyl) benzo [c][1,2,5] oxadiazol-5-yl) sulfane for Nonprotein Thiol Imaging in Lysosomes in Live Cells. *Anal. Chem.* **2019**, *91*, 15300–15307. [CrossRef] [PubMed]
207. Alqahtani, Y.; Wang, S.; Najmi, A.; Huang, Y.; Guan, X. Thiol-specific fluorogenic agent for live cell non-protein thiol imaging in lysosomes. *Anal. Bioanal. Chem.* **2019**, *411*, 6463–6473. [CrossRef]
208. Cao, M.; Chen, H.; Chen, D.; Xu, Z.; Liu, S.H.; Chen, X.; Yin, J. Naphthalimide-based fluorescent probe for selectively and specifically detecting glutathione in the lysosomes of living cells. *Chem. Commun.* **2016**, *52*, 721–724. [CrossRef] [PubMed]
209. Gao, Q.; Zhang, W.; Song, B.; Zhang, R.; Guo, W.; Yuan, J. Development of a novel lysosome-targeted ruthenium (II) complex for phosphorescence/time-gated luminescence assay of biothiols. *Anal. Chem.* **2017**, *89*, 4517–4524. [CrossRef] [PubMed]
210. Zheng, Z.; Huyan, Y.; Li, H.; Sun, S.; Xu, Y. A lysosome-targetable near infrared fluorescent probe for glutathione sensing and live-cell imaging. *Sens. Actuators B Chem.* **2019**, *301*, 127065. [CrossRef]

211. Wang, H.; Zhang, P.; Zhang, C.; Chen, S.; Zeng, R.; Cui, J.; Chen, J. A rational design of a cancer-specific and lysosome-targeted fluorescence nanoprobe for glutathione imaging in living cells. *Mater. Adv.* **2020**, *1*, 1739–1744. [CrossRef]
212. Gao, J.; Tao, Y.; Zhang, J.; Wang, N.; Ji, X.; He, J.; Si, Y.; Zhao, W. Development of lysosome-targeted fluorescent probes for Cys by regulating the boron-dipyrromethene (BODIPY) molecular structure. *Chem. A Eur. J.* **2019**, *25*, 11246–11256. [CrossRef]
213. Xiao, H.; Zhang, R.; Wu, C.; Li, P.; Zhang, W.; Tang, B. A new pH-sensitive fluorescent probe for visualization of endoplasmic reticulum acidification during stress. *Sens. Actuators B Chem.* **2018**, *273*, 1754–1761. [CrossRef]
214. Li, P.; Shi, X.; Xiao, H.; Ding, Q.; Bai, X.; Wu, C.; Zhang, W.; Tang, B. Two-photon imaging of the endoplasmic reticulum thiol flux in the brains of mice with depression phenotypes. *Analyst* **2019**, *144*, 191–196. [CrossRef] [PubMed]
215. Jiang, C.-S.; Cheng, Z.-Q.; Ge, Y.-X.; Song, J.-L.; Zhang, J.; Zhang, H. An endoplasmic reticulum-targeting fluorescent probe for the imaging of GSH in living cells. *Anal. Methods* **2019**, *11*, 3736–3740. [CrossRef]
216. Yue, X.; Chen, J.; Chen, W.; Wang, B.; Zhang, H.; Song, X. An endoplasmic reticulum-targeting fluorescent probe for discriminatory detection of Cys, Hcy and GSH in living cells. *Spectrochim. Acta Part A Mol. Biomol. Spectrosc.* **2021**, *250*, 119347. [CrossRef] [PubMed]
217. Tian, M.; Liu, Y.; Jiang, F.-L. On the Route to Quantitative Detection and Real-Time Monitoring of Glutathione in Living Cells by Reversible Fluorescent Probes. *Anal. Chem.* **2020**, *92*, 14285–14291. [CrossRef]

Article

Investigation of Organic Acids in Saffron Stigmas (*Crocus sativus* L.) Extract by Derivatization Method and Determination by GC/MS

Laurynas Jarukas [1], Olga Mykhailenko [2], Juste Baranauskaite [3], Mindaugas Marksa [1],* and Liudas Ivanauskas [1]

1. Department of Analytical and Toxicological Chemistry, Lithuanian University of Health Sciences, A. Mickeviciaus Str. 9, LT-44307 Kaunas, Lithuania; laurynas.jarukas@lsmuni.lt (L.J.); liudas.ivanauskas@lsmuni.lt (L.I.)
2. Department of Botany, National University of Pharmacy, Valentynivska, Str. 4, 461168 Kharkov, Ukraine; mykhailenko.farm@gmail.com
3. Department of Pharmaceutical Technology, Faculty of Pharmacy, Yeditepe University Atasehir, Inonu Mah., Kayısdagı Cad., Istanbul 34755, Turkey; baranauskaite.juste@gmail.com
* Correspondence: mindaugas.marksa@lsmuni.lt; Tel.: +370-602-54-544

Received: 25 May 2020; Accepted: 25 July 2020; Published: 28 July 2020

Abstract: The beneficial health properties of organic acids make them target compounds in multiple studies. This is the reason why developing a simple and sensitive determination and investigation method of organic acids is a priority. In this study, an effective method has been established for the determination of organic (lactic, glycolic, and malic) acids in saffron stigmas. N-(*tert*-butyldimethylsilyl)-N-methyltrifluoroacetamide (MTBSTFA) was used as a derivatization reagent in gas chromatography combined with mass spectrometric detection (GC/MS). The saffron stigmas extract was evaporated to dryness with a stream of nitrogen gas. The derivatization procedure: 0.1 g of dried extract was diluted into 0.1 mL of tetrahydrofuran, then 0.1 mL MTBSTFA was orderly and successively added into a vial. Two different techniques were used to obtain the highest amount of organic acid derivatives from saffron stigmas. To the best of our knowledge, this is the first report of the quantitative and qualitative GC/MS detection of organic acids in saffron stigmas using MTBSTFA reagent, also comparing different derivatization conditions, such as time, temperature and the effect of reagent amount on derivatization process. The identification of these derivatives was performed via GC-electron impact ionization mass spectrometry in positive-ion detection mode. Under optimal conditions, excellent linearity for all organic acids was obtained with determination coefficients of $R^2 > 0.9955$. The detection limits (LODs) and quantitation limits (LOQs) ranged from 0.317 to 0.410 µg/mL and 0.085 to 1.53 µg/mL, respectively. The results showed that the highest yield of organic acids was conducted by using 0.1 mL of MTBSTFA and derivatization method with a conventional heating process at 130 °C for 90 min. This method has been successfully applied to the quantitative analysis of organic acids in saffron stigmas.

Keywords: lactic acid; malic acid; glycolic acid; GC-MS/EI

1. Introduction

Saffron is considered the world's most expensive spice and medicinal plant. Besides uses in food, saffron has attracted interest because of its health-promoting properties [1,2]. In addition, saffron stigmas have been proven to have anti-inflammatory, antioxidant, anti-allergic, and antidepressant biological functions [3,4]. These specific properties are considered to be connected to the presence of diverse compounds such as proteins, fats, minerals, sugars, and organic acids [2]. Among these ingredients, organic acids are in a prominent position because not only do they affect the flavor of saffron but also, they have various pharmacological actions [5,6].

Moreover, there is increasing interest in studies examining characteristics of organic acids, searching for positive effects of given compounds on the human body. According to the literature, the short-chain fatty acids, medium-chain fatty acids, and other organic acids have more or less pronounced antimicrobial activity, depending on the concentration of the acids and the bacterial species exposed to the acids [5,6]. It is well known that benzoic and salicylic acids exhibit antibacterial activity, hydroxycinnamic acid, and their derivatives–anti-inflammatory activity, gallic acid is an antimutagenic, anticarcinogenic, and anti-inflammatory agent [7]. Furthermore, succinic acid, acetic acid, citric acid, lactic acid, malic acid, glutamic acid, and their salts promote gastrointestinal absorption of iron [7]. Moreover, citric and malic acids have significant protective effects on the myocardium and act on ischemic lesions, according to a study by Tang et al., where supplying a patients' diet with these compounds gave significant positive results [8]. Moreover, organic acids play a principal role in maintaining the quality and nutritional value of food. These compounds can be added as acidulants or stabilizers (e.g., citric, ascorbic, benzoic, fumaric, and malic acids).

The quantitative determination of organic acids in such type of samples is of high interest in many industrial and research institutes. However, as the attention on the health benefits of organic acids is increasing, a simple and sensitive method for determining and investigating organic acids is needed [9–12]. According to the literature, organic acids are weak acids and are only partly dissociated [6]. To increase the stability and solubility of organic acids, different extraction or other chemical processing techniques are needed. The scientific literature describes that one of the most commonly used processes is derivatization, a chemical process for modifying compounds in order to generate new products with better chromatographic properties [13]. Different derivatization techniques can be used; chemical derivatization is usually used for amino acid detection, and it becomes a necessary procedure to transform analytes into derivatives that can be easily isolated, separated, and detected [14–17]. Gas chromatography is the most widely used and accepted technique for quantitative analysis of derivatization products due to its high resolution, sensitivity, great versatility, and simple sample treatment.

The derivatization parameters were systematically studied. In this study, a simple and sensitive GC/MS method for determination and investigation of organic acids in saffron stigmas after derivatization with N-(tert-butyldimethylsilyl)-N-methyltrifluoroacetamide reagent was presented. The adequacy of the proposed method was estimated in terms of accuracy, linearity, precision, and detection limit. To the best of our knowledge, this is the first report of quantitative and qualitative organic acid GC/MS detection in saffron stigmas with MTBSTFA derivatization reagent, and comparison of different conditions.

2. Results and Discussion

2.1. Derivatization Solvents and Reagents

When it comes to dealing with highly complex matrices, such as organic acids, it is advisable to use a derivatization process in order to improve parameters of separation, such as volatility, thermal stability, resolution, as well as detection parameters, when gas chromatography is used [18]. During this process, the derivatization reagent plays an important role in the separation and resolution of the analytes [18]. Among the different derivatization reagents for organic acids, silylating agents are the most popular ones, which, moreover, have been proven as excellent reagents for derivatization after extraction. Hence, two silylating agents were tested: N-methyl-N-(trimethylsilyl) trifluoroacetamide (MSTFA) and MTBSTFA [13,19,20]. This study is intended for increasing knowledge about the behavior and interest of both reagents on the efficiency of organic acid derivatization yield. However, the use of MSTFA yielded an incomplete crocus stigmas extract derivatization of the organic acid, and thus, further studies were performed with MTBSTFA. According to results, significantly 1.2 times higher amount of organic acids (expressed in percentage of the total amount of lactic, malic, and glycolic acids) has been found while using MTBSTFA reagent, and it was selected as a derivatization reagent for further study. Similar results were reported in previous studies [21,22]. Morville et al. evaluated the efficiency of the derivatization process on organic acids' (glutaric, adipic, and suberic acids) yield when MTBSTFA and MSTFA reagents were used. Results showed that glutaric, adipic and suberic acid yields significantly increased, 1.6, 1.8, and 1.3 times respectively, after using MTBSTFA reagent [21]. This might be explained by the high volatility of MTBSTFA that it did not coelute in the GC system with other peaks and improved parameters of separation, thermal stability and resolution. The MTBSTFA produces dimethyl-*tert*-butylsilyl (TBDMS) derivatives. MTBSTFA-derivatives produce characteristic fragmentation patterns presenting mainly the fragments of $[M - 15]^+$ (cleavage of methyl from the molecular ion) and $[M - 57]^+$ (cleavage of the *t*-butyl moiety), of which $[M - 57]^+$ is generally dominant in the mass spectrum. MSTFA-derivatives yielding trimethylsilyl (TMS) derivatives, mainly show the fragments $[M - 15]^+$ (cleavage of a methyl from the molecular ion) and $[M - 31]^+$ (cleavage of the trimethylsilyl ether moiety followed by cyclization involving the silyl group). The TBDMS are more stable to hydrolysis than the corresponding TMS derivatives. As demonstrated in previous studies, TBDMS derivatives are formed more easily and have, thus, higher sensitivities (10–100 times) as well as repeatabilities than the corresponding TMS derivatives [22].

The ultimate goal of extraction is the maximization of the yield and coverage of metabolites in a rapid and reproducible way while minimizing enzymatic, chemical, and physical degradation [20]. The derivatization yield of carboxylic acids with MTBSTFA depends on factors including the nature of the solvent in which the analytes are dissolved. The main factors contributing to the increase of the efficiency and the rate of the silylation reaction are the silyl donor ability of the reagent and the ease of silylation of different functional groups in the analyte. The solvent used as a medium and the compounds present or added in the silylation medium may also play a role in derivatization efficiency. The reagent excess is sometimes important for displacing the equilibrium in the desired direction, and usually, an excess up to ten times larger than stoichiometrically needed is used for silylation. The primary purpose was to determine the effect of different solvents (tetrahydrofuran (THF) and acetonitrile (ACN)) on derivatization's yield of organic acid. During this part of the study, two samples were produced by using derivatization procedure, mentioned above (sample of 0.1 g dried Saffron stigmas extract was diluted into 0.1 mL of extraction solvent, and 0.1 mL derivatization agent (MTBSTFA) was added in sequence; the vial was sealed and oscillated by vortex-mixer for 1 min, then the mixture was placed in glycerol bath allowing it to react at 50 °C for 60 min). Primary investigations revealed that between the ACN and THF samples, significant differences in the organic acid derivate yields (expressed in percentage of the total amount of lactic, malic, and glycolic acids) were obtained ($p < 0.05$; Figure 1).

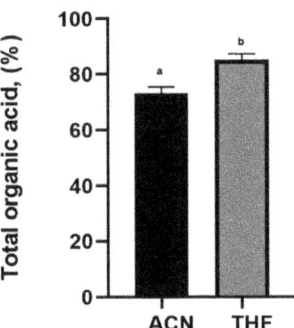

Figure 1. The effect of extraction solvents on the yield of organic acid derivatives (expressed in percentage of the total amount of lactic, malic, and glycolic acids) from Saffron stigmas extract, $n = 6$. Values within columns followed by the same lowercase letter (a, b) differed statistically at $p < 0.05$ (Tukey's test). Results are expressed as means ± standard error.

Moreover, the results showed 1.2 times higher yield of organic acids derivate by using THF as the derivatization solvent in comparison to ACN ($p < 0.05$) (Figure 1). The results could be explained by the sensitivity of the analysis. According to Wittmann et al., by using the THF as extraction solvent increased analysis sensitivity of more polar compounds (lactic acid, oxalic acid, methylcitric acid, 3-hydroxypropionic acid, 3-hydroxyisovaleric acid, kynurenic acid, glycolic acid, orotic acid and quinolinic acid) in comparison with less polar compounds (glycine conjugates) [23]. Hence, the THF was chosen for future experiments.

During this study we compared MSTFA and MTBSTFA in the derivatization efficiency of organic acids. This study is intended for increasing knowledge about the behavior and interest of both reagents. These results clearly demonstrate that solvent plays a significant role in the derivatization procedure. As a matter of fact, MTBSTFA and THF possess the most appropriate derivatization efficiency of the above-mentioned compounds, and they were selected for further studies.

2.2. Comparison of the Derivatization Parameters

Sample preparation is a critical part of every analytical procedure. The increasing demand to determine compounds at low concentrations in complex matrices requires a preliminary step. To achieve the best derivatization efficiency, a variety of important parameters, such as a derivatization temperature, time and amount of MTBSTFA, were optimized. In this study, the saffron stigmas extract was employed to optimize derivatization conditions. The concentration trends of three representative organic acids (lactic, glycolic, and malic acids) relative to different parameters are shown in Figure 2.

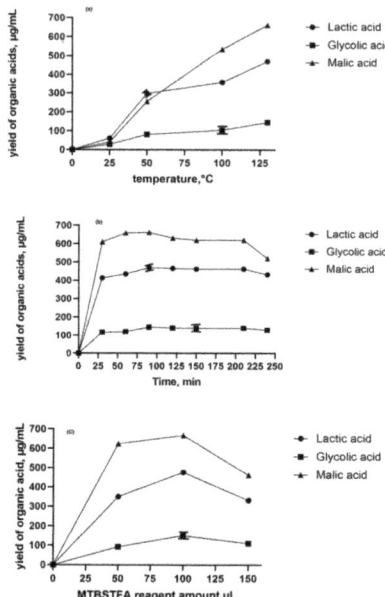

Figure 2. The effect of operation parameters on derivatization reaction: (the yield of lactic acid—62 µg/g, glycolic acid—40 µg/g and malic acid—30 µg/g in the example sample). Temperature (**a**), Time (**b**), derivatization agent (MTBSTFA) (**c**). Results are expressed as means ± standard error ($n = 6$).

The primary purpose was to determine the influence of different temperatures and extraction time on the derivatization yield of investigated organic acids. The results showed that increasing temperature significantly increased the yields of organic acids (Figure 2a; $p < 0.05$). At a derivatization temperature of 130 °C, lactic, glycolic, and malic acid derivative yields significantly increased, 7.7, 16.1, and 5 times, respectively, in comparison with samples prepared at a temperature of 25 °C

(Figure 2a). The explanation of such results could be that higher temperature is speculated to enhance derivatization efficiency, by increasing solubility of derivatization reagents and organic acid metabolites [24]. Similarly, Gulberg et al. found that increasing temperatures had an appreciable effect on derivatization efficiency [24]. Moreover, the effect of extraction time (30, 60, 90, 120, 150, 210, and 240 min) on the yield of organic acids derivatization was investigated. The derivatization procedure was carried in the same way as mentioned above. Primary investigations revealed that a prolonged extraction time of 90 min, had no significant influence on derivatization yields of lactic, glycolic and malic acids, extracted from saffron stigmas ($p > 0.05$; Figure 2b). As the reaction time was prolonged, the signal response of organic acid derivatives remained constant. Similarly, Elias and co-authors found that long term silylation-derivatization process was beneficial to stearic acid and glucose-6-phosphate [25]. Therefore, the derivatization reaction between MTBSTFA and organic acids was carried out for 90 min at 130 °C.

To ensure the complete and repeatable derivatization, the desired amount of MTBSTFA reagent was required. The influence of different volumes (50, 100, 150 μL) of MTBSTFA reagent on organic acid yields was optimized. As indicated in Figure 2c, the highest derivatization yield of organic acids was obtained when the reagent amount was 100 μL ($p < 0.05$). However, a decrease in derivatization efficiency was observed when the MTBSTFA reagent amount in the extract increased to 150 μL. According to literature, the most reported amounts of silylation agents for the silylation of polar plant extracts range between 30–125 μL [26]. This is in line with the findings of Koek, who showed that organic acids and sugars need relatively low volumes of the silylating agent [27]. It could be explained as molecular interaction because the reactivity of TMS groups is low to oxygen in organic acids, having a lower number of unshared electrons, higher steric hindrance, and transition state energy [25].

As a final conclusion of this study, when the derivatization temperature was 130 °C, silylation time was 90 min, and the MTBSTFA reagent amount was 100 μL, the highest yield of organic acid derivates in the sample, extracted from the Saffron stigmas was obtained (Table 1). Such a significant shortening of time was achieved by applying high temperature, which allowed to avoid time-consuming sorption of derivatization reagent and time-consuming desorption of analytes, which allowed to reach the high derivatization efficiency. Moreover, the rapid derivatization procedure improved parameters of separation, such as volatility, thermal stability, resolution, as well as detection parameters. For the standard (lactic, glycolic, and malic acids) and saffron stigmas extract, produced by using optimal conditions, chromatograms are shown in Figure 3.

Table 1. Linearity and sensitivity data for organic acids (lactic, glycolic, malic acids) used as a standard.

Compound	RT	Ion (m/z)	Linearity Range (μg/mL)	R^2 *	LOD [a] (μg/mL)	LOQ [a] (μg/mL)
Lactic acid	14.038	147, 73, 261	15–242	0.9977	0.156	0.317
Glycolic acid	14.265	73, 147, 247	12–379	0.9955	0.101	0.410
Malic acid	15.256	73, 147, 259	12–758	0.9986	0.085	0.339

Experimental conditions as in Section 3.4. * For R^2 the correlation coefficient. The p value was <0.0001 for all calibration curves. [a] LOD were based on S/N = 3; LOQ were based on S/N = 10.

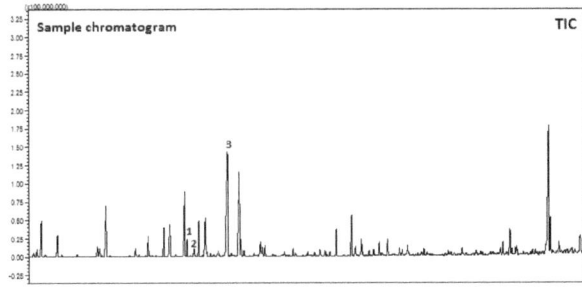

Figure 3. Cont.

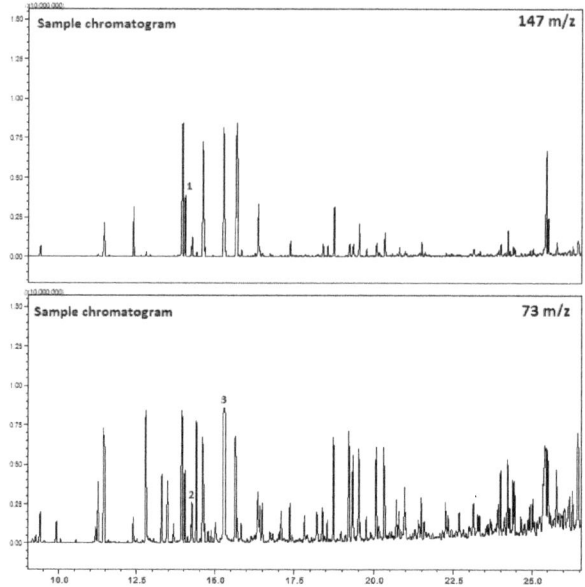

Figure 3. GC/MS chromatograms of standards (1-lactic, 2-glycolic, and 3-malic acids) and organic acid derivatives extracted from Saffron stigmas.

2.3. GC/MS Method Validation

The GC method was validated by following the ICH Q2 (R1) guidelines [28]. The developed method was evaluated via the correlation coefficient (R^2), linear range, detection limit (LOD), quantitative limit (LOQ), accuracy, and precision. The electron impact ionization of lactic, glycolic and malic acids, produced the $[M]^+$ ions at 156, 83 and 84 under positive ionization conditions. The product ion spectra ions at m/z 147, m/z 73, and m/z 73 were produced as the prominent product ions for lactic, glycolic and malic acids (Table 1). The calibration curves of the three organic acids (lactic, glycolic, and malic acids) were established by injecting the standard solutions in the range of 15–242 µg·L^{-1}, 12–379 µg·L^{-1} and 12–758 µg·L^{-1}, respectively and plotting the average peak areas versus the average concentrations of organic acids based on the data of triplicate measurements. The good linearity response over the tested concentration range was obtained with the developed method for the compounds used as lactic acid standards, having $R^2 > 0.997$, as shown in Table 1. The LOD value was 0.153 µg/mL, while the LOQ was 0.317 µg/mL (Table 1), which indicates that the method is sensitive. Moreover, R^2 values of the glycolic acid standard were higher than 0.996, thus confirming the linearity of the method (Table 1). Thus, the LOD value was 0.101 µg/mL, while the LOQ was 0.41 µg/mL (Table 1), which suggested full capacity for the quantification of the glycolic acid investigated. Furthermore, R^2 values of the malic acid standard were higher than 0.999, and the LOD value was 0.085 µg/mL, while the LOQ was 0.339 µg/mL (Table 1). According to the described data above, it can be concluded that this method is a reliable tool for the identification and quantification of organic acid in saffron stigmas, conforming to the ICH guidelines.

3. Materials and Methods

3.1. Materials and Methods

Ultrapure water was obtained in the laboratory using a Milli-Q water purification system (Millipore, Billerica, MA, USA). *N*-(t-butyldimethylsilyl)-*N*-methyltrifluoroacetamide (MTBSTFA) (>99%), *N*-methyl-*N*-(trimethylsilyl) trifluoroacetamide (MSTFA) (>98.5%) tetrahydrofuran, acetonitrile were purchased from Sigma–Aldrich (St. Louis, MO, USA). The GC-equipment was run with helium (purity 5.0) as the carrier gas was purchased from Gazchema (Lithuania). Ethanol (96%) for extraction was purchased from Vilniaus degtinė (Vilnius, Lithuania). Lactic acid (>98%), glycolic acid (>99%), and malic acid (>99%) standards were purchased from Sigma–Aldrich (Co., Birkenhead, UK).

3.2. Sample Preparation

Saffron stigma was purchased from Novin Saffron Company, Mashhad, Iran. Prior to the extract preparation, the saffron stigma was dried with a stream of nitrogen gas, then was grounded in a cross beater mill IKA A11 Basic Grinder (IKA Works, Guangzhou, China) and sieved using vibratory sieve shaker AS 200 basic (Retch, UK) equipped with a 125 μm sieve. Then, the powdered sample (1 g) was extracted with 10 mL of 70% (v/v) methanol-aqua solution in a volumetric flask using an ultrasound bath for 20 min and filtered through a 0.45 μm nylon filter.

3.3. Derivatization Procedure

0.1 g of prepared extract solution was evaporated to dryness with a stream of nitrogen gas. Briefly, to a 2 mL ampoule bottle, 0.1 g of dried extract sample was diluted into 0.1 mL of extraction solvent (tetrahydrofuran), and 0.1 mL derivatization agent (MTBSTFA) was added in sequence. The vial was sealed and oscillated by vortex-mixer for 1 min, and then, to allow the mixture to react at room temperature (25 °C) for 60 min, it was placed in glycerol bath. The subsequent solution was transferred to 200 μL insert placed autosampler vials, and 1 μL aliquot was injected into GC-MS system for analysis. Efficiency extraction parameters were evaluated and optimized, including derivatization time, extraction temperature and reagent amount on derivatization. The comparison of chromatographic responses was used to evaluate the extraction efficiency. A similar procedure was used for lactic, glycolic, and malic acid standards. Standards were diluted in the THF, producing a mixture of 150 μmol L^{-1}. The same derivatization reaction was applied to this standard mixture, with the exception of the addition of 0.1 mL of THF, because it was already present in the standard mixture.

3.4. GC/MS Method

Analyses were performed using a SHIMADZU GC/MS-QP2010nc Ultra chromatography system (coupled to an Electron Ionization (EI) ion source and a single quadrupole MS (Shimadzu Technologies, Kyoto, Japan). A robotic autosampler and a split/splitless injection port were used. Injection port temperature was kept at 250 °C until the end of the analysis. The separation of analytes was carried out on a with Rxi-5 ms (Restek Corporation, Bellefonte, PA, USA, capillary column (30 m long, 0.25 mm outer diameter and 0.25 μm liquid stationary phase thickness) with a liquid stationary phase) 5% diphenyl and 95% polydimethylsiloxane) with helium at a purity of 99.999% as the carrier gas in a constant flow of 1.49 mL/min. The oven temperature was programmed at 75 °C for 5 min, then increased to 290 °C at 10 °C/min and increased to 320 °C at 20 °C/min and kept for 10 min. The total time was 41 min. The temperatures of the MS interface and ion source were set at 280, 200 °C, respectively. The MS was operated in positive mode (electron energy 70 eV). The full-scan acquisition was performed with the mass detection range set at 35–500 *m/z* to determine retention times of analytes, optimize oven temperature gradient, and to observe characteristic mass fragments for each compound. Data acquisition and analysis were executed by LabSolution GC/MS (version 5.71) (Shimadzu Corporation). For the identification and quantification of the analytes, single-ion monitoring (SIM) mode was used.

3.5. GC/MS Method Validation

The validation of the GC/MS method was performed according to the international guidelines on analytical techniques for the quality control of pharmaceuticals (ICH guidelines) [28]. Method validation was performed to assess linearity, LOD, LOQ, and precision. The calibration curves of the organic acids (lactic, malic and glycolic acids) were established by injecting the standard solutions in the range of

12–758 µg/mL) and plotting the average peak areas versus the average concentrations of organic acids based on the data of triplicate measurements (Table 1). Analytes stock solution was prepared in THF by diluting of the analytical standards to reach a concentration of lactic acid 242 µg/mL, glycolic acid 379 µg/mL, and malic acid 758 µg/mL. Then, the subsequent dilutions were prepared with MilliQ water. The standard solutions (n = 3) were prepared at approximate concentrations of lactic acid: 242, 121, 60.5, 30.25, 15.13 µg/mL; glycolic acid: 379, 189.5, 94.75, 47.38, 23.69, 11.84 µg/mL, and malic acid: 758, 379, 189.5, 94.75, 47.38, 23.69, 11.84 µg/mL, due to the wide levels found in saffron samples. The concentrations of the lactic, glycolic and malic acids in each solution was maintained arranged as follows 15.13, 11.84, 11.84 µg/mL, respectively. The LOD and LOQ were calculated at signal-to-noise (S/N) ratios of 3 and 10, respectively. The precision of the method was evaluated by calculating the repeatability (r). The precision of the extraction technique was validated by repeating the extraction procedure with the standard mix solutions six times. An aliquot of each extract was then injected and quantified. The precision of the chromatographic system was tested by checking the %RSD of retention times and peak areas. Six injections were performed each day for three consecutive days.

3.6. Statistical Analysis

We used between five and six biological replicates for saffron stigmas extract samples and six technical replicates for standard samples. Each biological replicate consisted of a stigma of 10 plants, resulting in the isolation of 4–8 mg of stigmas. Raw data was assessed using ANOVA statistical testing (specifically one-way analysis of variance) and Tukey's multiple comparison test. For this purpose, a software package was utilized (Prism v. 5.04, GraphPad Software Inc., La Jolla, CA, USA) with statistical significance being defined as $p < 0.05$. Results were expressed as average ± standard error.

4. Conclusions

The method for organic acid analysis in saffron was developed and validated. The method consisted of sample preparation, derivatization, and chromatographic analysis. All steps were extensively studied and optimized for the derivatization procedure. To the best of our knowledge, this is the first report for the GC/MS detection of the amount and types of organic acids in saffron stigmas with MTBSTFA derivatization reagent and comparison of different conditions. The derivatization reaction was rapid, and the maximum yields of organic acids (lactic, glycolic, and malic acids) were observed by using optimal derivatization conditions. The major advantages of optimal conditions led to reach the highest derivatization efficiency of organic acids in only 90 min by using a conventional heating process. The developed method has been successfully applied to the quantification of organic acids in saffron stigmas. This research also shows the interesting agricultural potential of saffron stigmas, in relation to the preparation of certified extracts with a high content of organic acid to be used in the pharmaceutical and nutraceutical area.

Author Contributions: L.J. contributed to investigation, data analysis, and original draft preparation. M.M. and J.B. contributed to methodology, data analysis, visualization, review, and editing. L.I. and O.M. contributed to conceptualization, resources, original draft preparation, review and editing, project administration, and supervision. All authors have read and agreed to the published version of the manuscript.

Funding: This research received no external funding.

Acknowledgments: The authors are thankful for the financial support provided by the Science Foundation of Lithuanian University of Health Sciences.

Conflicts of Interest: The authors declare no conflict of interest.

References

1. Guo, M.; Shi, T.; Duan, Y.; Zhu, J.; Li, J.; Cao, Y. Investigation of amino acids in wolfberry fruit (*Lycium barbarum*) by solid-phase extraction and liquid chromatography with precolumn derivatization. *J. Food Compos. Anal.* **2015**, *42*, 84–90. [CrossRef]

2. Nerome, H.; Ito, M.; Machmudah, S.; Wahyudiono; Kanda, H.; Goto, M. Extraction of phytochemicals from saffron by supercritical carbon dioxide with water and methanol as entrainer. *J. Supercrit. Fluids* **2015**, *107*, 377–383. [CrossRef]
3. Ríos, J.L.; Recio, M.C.; Giner, R.M.; Meñez, S. An update review of saffron and its active constituents. *Phyther. Res.* **1996**, *10*, 189–193. [CrossRef]
4. Heydari, S.; Haghayegh, G.H. Extraction and Microextraction Techniques for the Determination of Compounds from Saffron. *Can. Chem. Trans.* **2014**, *2*, 221–247.
5. Khan, S.H.; Iqbal, J. Recent advances in the role of organic acids in poultry nutrition. *J. Appl. Anim. Res.* **2016**, *44*, 359–369. [CrossRef]
6. Shahidi, S.; Yahyavi, M.; Zare, D.N. Influence of Dietary Organic Acids Supplementation on Reproductive Performance of Freshwater Angelfish (*Pterophyllum scalare*). *Glob. Vet.* **2014**, *13*, 373–377.
7. Shahrzad, S.; Aoyagi, K.; Winter, A.; Koyama, A.; Bitsch, I. Pharmacokinetics of gallic acid and its relative bioavailability from tea in healthy humans. *Hum. Nutr. Metab.* **2001**, *1*, 1207–1210. [CrossRef] [PubMed]
8. Tang, X.; Liu, J.; Dong, W.; Li, P.; Li, L.; Lin, C.; Li, D. The cardioprotective effects of citric acid and L-malic acid on myocardial ischemia/reperfusion injury. *Evid. Based Complement. Altern. Med.* **2013**, *2013*, 1–11. [CrossRef] [PubMed]
9. Kaspar, H.; Dettmer, K.; Gronwald, W.; Oefner, P.J. Automated GC-MS analysis of free amino acids in biological fluids. *J. Chromatogr. B Anal. Technol. Biomed. Life Sci.* **2008**, *870*, 222–232. [CrossRef]
10. Kaźmierczak, D.; Ciesielski, W.; Zakrzewski, R.; Zuber, M. Application of iodine-azide reaction for detection of amino acids in thin-layer chromatography. *J. Chromatogr. A* **2004**, *1059*, 171–174. [CrossRef]
11. Desiderio, C.; Iavarone, F.; Rossetti, D.V.; Messana, I.; Castagnola, M. Capillary electrophoresis-mass spectrometry for the analysis of amino acids. *J. Sep. Sci.* **2010**, *33*, 2385–2393. [CrossRef] [PubMed]
12. Azilawati, M.I.; Hashim, D.M.; Jamilah, B.; Amin, I. Validation of a reverse-phase high-performance liquid chromatography method for the determination of amino acids in gelatins by application of 6-aminoquinolyl-N-hydroxysuccinimidyl carbamate reagent. *J. Chromatogr. A* **2014**, *1353*, 49–56. [CrossRef] [PubMed]
13. Schummer, C.; Delhomme, O.; Appenzeller, B.M.R.; Wennig, R.; Millet, M. Comparison of MTBSTFA and BSTFA in derivatization reactions of polar compounds prior to GC/MS analysis. *Talanta* **2009**, *77*, 1473–1482. [CrossRef] [PubMed]
14. Ilisz, I.; Robert, B.; Antal, P. Application of chiral derivatizing agents in the high-performance liquid chromatographic separation of amino acid enantiomers: A review. *J. Pharm. Biomed. Anal.* **2008**, *47*, 1–15. [CrossRef]
15. Schummer, C.; Sadiki, M.; Mirabel, P.; Millet, M. Analysis of t-butyldimethylsilyl derivatives of chlorophenols in the atmosphere of urban and rural areas in East of France. *Chromatographia* **2006**, *63*, 189–195. [CrossRef]
16. Hao, C.; Zhao, X.; Yang, P. GC-MS and HPLC-MS analysis of bioactive pharmaceuticals and personal-care products in environmental matrices. *TrAC Trends Anal. Chem.* **2007**, *26*, 569–580. [CrossRef]
17. Hong, J.E.; Pyo, H.; Park, S.J.; Lee, W. Determination of hydroxy-PCBs in urine by gas chromatography/mass spectrometry with solid-phase extraction and derivatization. *Anal. Chim. Acta* **2005**, *531*, 249–256. [CrossRef]
18. Robles, A.; Fabjanowicz, M.; Chmiel, T.; Płotka-Wasylka, J. Determination and identification of organic acids in wine samples. Problems and challenges. *TrAC Trend. Anal. Chem.* **2019**, *120*, 115630. [CrossRef]
19. Halket, J.M.; Waterman, D.; Przyborowska, A.M.; Patel, R.K.P.; Fraser, P.D.; Bramley, P.M. Chemical derivatization and mass spectral libraries in metabolic profiling by GC/MS and LC/MS/MS. *J. Exp. Bot.* **2005**, *56*, 219–243. [CrossRef]
20. Halket, J.M.; Zaikin, V.G. Derivatization in mass spectrometry—1. Silylation. *Eur. J. Mass Spectrom.* **2003**, *9*, 1–21. [CrossRef]
21. Morville, S.; Scheyer, A.; Mirabel, P.; Millet, M. A multiresidue method for the analysis of phenols and nitrophenols in the atmosphere. *J. Environ. Monit.* **2004**, *6*, 963–966. [CrossRef]
22. Orata, F. Derivatization reactions and reagents for gas chromatography analysis. In *Advanced Gas Chromatography-Progress in Agricultural, Biomedical and Industrial Applications*; InTechOpen: London, UK, 2012; pp. 83–108. [CrossRef]
23. Wittmann, G.; Karg, E.; Mühl, A.; Bodamer, O.A.; Turi, S. Comparison of tetrahydrofuran and ethyl acetate as extraction solvents for urinary organic acid analysis. *J. Inher. Met. Dis.* **2007**, *31*, 73–80. [CrossRef] [PubMed]

24. Gullberg, J.; Jonsson, P.; Nordstrom, A.; Sjostrom, M.; Moritz, T. Design of experiments: An efficient strategy to identify factors influencing extraction and derivatization of Arabidopsis thaliana samples in metabolomic studies with gas chromatography/mass spectrometry. *Anal. Biochem.* **2004**, *331*, 283–295. [CrossRef] [PubMed]
25. Bekele, E.A.; Annaratone, C.E.; Hertog, M.L.; Nicolai, B.M.; Geeraerd, A.H. Multi-response optimization of the extraction and derivatization protocol of selected polar metabolites from apple fruit tissue for GC–MS analysis. *Anal. Chim. Acta* **2014**, *824*, 42–56. [CrossRef] [PubMed]
26. Roessner, U.; Wagner, C.; Kopka, J.; Trethewey, R.N.; Willmitzer, L. Simultaneous analysis of metabolites in potato tuber by gas chromatography–mass spectrometry. *Plant. J.* **2000**, *23*, 131–142. [CrossRef]
27. Koek, M.M. Gas Chromatography Mass Spectrometry: Key Technology in Metabolomics. Ph.D. Thesis, Leiden University, Leiden, The Netherlands, 2009.
28. Validation of Analytical Procedures: Text and methodology. In *ICH Harmonization Tripartite Guideline*; European Medicines Agency: Geneva, Switzerland, 2005; pp. 1–13.

Sample Availability: Samples of the Saffron stigmas extract are available from the authors.

© 2020 by the authors. Licensee MDPI, Basel, Switzerland. This article is an open access article distributed under the terms and conditions of the Creative Commons Attribution (CC BY) license (http://creativecommons.org/licenses/by/4.0/).

Article

Cyanogenic Glycoside Analysis in American Elderberry

Michael K. Appenteng [1], Ritter Krueger [1], Mitch C. Johnson [1], Harrison Ingold [1], Richard Bell [2], Andrew L. Thomas [3] and C. Michael Greenlief [1,*]

1. Department of Chemistry, University of Missouri, Columbia, MO 65211, USA; mkappenteng@mail.missouri.edu (M.K.A.); rmkqm2@mail.umkc.edu (R.K.); mcjohnson0989@gmail.com (M.C.J.); hgidn9@mail.missouri.edu (H.I.)
2. Department of Chemistry, Truman State University, Kirksville, MO 63501, USA; rjb2318@truman.edu
3. Division of Plant Sciences, Southwest Research Center, University of Missouri, Columbia, MO 65211, USA; ThomasAL@missouri.edu
* Correspondence: greenliefm@missouri.edu; Tel.: +01-573-882-3288

Citation: Appenteng, M.K.; Krueger, R.; Johnson, M.C.; Ingold, H.; Bell, R.; Thomas, A.L.; Greenlief, C.M. Cyanogenic Glycoside Analysis in American Elderberry. *Molecules* **2021**, *26*, 1384. https://doi.org/10.3390/molecules26051384

Academic Editor: Wilfried Rozhon

Received: 28 January 2021
Accepted: 1 March 2021
Published: 4 March 2021

Publisher's Note: MDPI stays neutral with regard to jurisdictional claims in published maps and institutional affiliations.

Copyright: © 2021 by the authors. Licensee MDPI, Basel, Switzerland. This article is an open access article distributed under the terms and conditions of the Creative Commons Attribution (CC BY) license (https://creativecommons.org/licenses/by/4.0/).

Abstract: Cyanogenic glycosides (CNGs) are naturally occurring plant molecules (nitrogenous plant secondary metabolites) which consist of an aglycone and a sugar moiety. Hydrogen cyanide (HCN) is released from these compounds following enzymatic hydrolysis causing potential toxicity issues. The presence of CNGs in American elderberry (AE) fruit, *Sambucus nigra* (subsp. *canadensis*), is uncertain. A sensitive, reproducible and robust LC-MS/MS method was developed and optimized for accurate identification and quantification of the intact glycoside. A complimentary picrate paper test method was modified to determine the total cyanogenic potential (TCP). TCP analysis was performed using a camera-phone and UV-Vis spectrophotometry. A method validation was conducted and the developed methods were successfully applied to the assessment of TCP and quantification of intact CNGs in different tissues of AE samples. Results showed no quantifiable trace of CNGs in commercial AE juice. Levels of CNGs found in various fruit tissues of AE cultivars studied ranged from between 0.12 and 6.38 µg/g. In pressed juice samples, the concentration range measured was 0.29–2.36 µg/mL and in seeds the levels were 0.12–2.38 µg/g. TCP was highest in the stems and green berries. Concentration levels in all tissues were generally low and at a level that poses no threat to consumers of fresh and processed AE products.

Keywords: American elderberry; total cyanogenic potential; cyanogenic glycosides; picrate method; solid phase extraction; UHPLC-MS/MS

1. Introduction

American elderberry (AE), *Sambucus nigra* (subsp. *canadensis*) is a rapidly growing specialty crop in the United States [1]. Native to eastern and midwestern North America, AE is increasingly cultivated for its fruits and flowers that are used in a variety of foods, jellies, syrups, wines, and more importantly, dietary supplement products [2]. Elderberry is known for its nutritional and medicinal health benefits [3–5]. The fruit is rich in carbohydrates, fatty acids, organic acids, minerals, vitamins (A, B6 and C), essential oils, and is high in fiber [6,7]. Researchers have linked elderberry products to anti-inflammatory, anti-oxidant, anti-carcinogenic, anti-viral, anti-influenza, and antibacterial activities [3,8–13]. Whereas little scientific research has been conducted on AE as compared to its close relative, the European elderberry (EE), *Sambucus nigra* (subsp. *nigra*), both species are excellent sources of flavonoids, polyphenols and anthocyanins [3,8,9,14,15]. The elderberry industry is poised for major expansion and has increased significantly in sales (~0.8 to 107.6 M dollars) between 2011 and 2019 [16]. However, its competitiveness with other herbal dietary supplements [16] may by hampered in part due to uncertainty regarding the presence of cyanogenic glycosides (CNGs) and/or their putative toxicity.

Cyanogenic glycosides are naturally occurring molecules in plants (nitrogenous secondary plant metabolites) which consist of aglycone and a sugar moiety [17,18]. A generalized structure is shown in Figure 1. There are about 60 CNGs widely distributed in the plant kingdom, occurring in over 2600 plant species representing more than 130 families [19–22]. CNGs are stored in vacuoles within plant cells, separating them from plant hydrolyzing endogenous enzymes (β-1,6-glycosidases and hydroxynitrile lyases) [17,23]. Although intact CNGs are nontoxic, endogenous plant enzymes can react with CNGs, and release hydrogen cyanide (HCN) causing potential toxicity issues [17–19,24,25]. Most CNG-containing plants also produce these endogenous enzymes so when their tissues are disrupted, for example by crushing, herbivory, or disease, CNGs can react with endogenous enzymes resulting in the release of HCN [23,26]. In plants, CNGs serve as important chemical defense compounds against herbivores and pathogens [19,21,27]. Clinical trials have shown mixed results regarding the potential of amygdalin (a CNG found in *Sambucus*) in cancer treatment and as a cough suppressant in various preparations [28,29]. In humans, consumption of cyanogenic plants can cause sub-acute cyanide poisoning (depending on dose) with symptoms including anxiety, headache, vomiting, nausea, abdominal cramps diarrhea, dizziness, weakness and mental confusion. Acute cyanide toxicity in humans (0.5–3.5 mg kg^{-1} body weight) [18,19] can result in decreased consciousness, hypotension, paralysis, coma and even death [17–19,24,25,30,31]. Acute cyanide poisoning has been reported from the ingestion of apricot kernels [32], bitter almonds [33], and cassava [34].

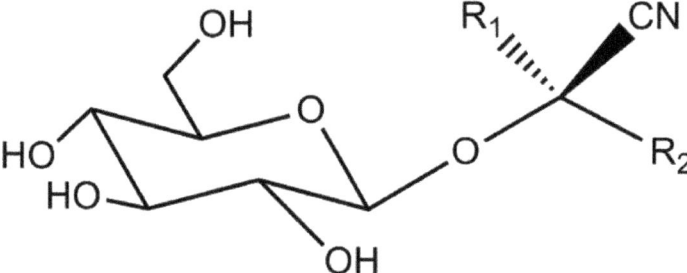

Figure 1. Generic structure for a cyanogenic glycoside, where R_1 is often methyl or a proton and R_2 is a variable organic group.

The only study previously made on *Sambucus canadensis* (American elderberry) is by Buhrmester et al. [20]. They examined the presence or absence of cyanogenic glycosides for individuals from nine populations of *Sambucus canadensis* L. (elderberry) in east-central Illinois. The study tested for cyanogenic glycosides in the leaves. Of the nine elderberry populations examined, only one population had a test positive for HCN production each of the three times tested. In another population the production of HCN was highly variable. The cyanogenic glycoside was determined to be (S)-sambunigrin by gas chromatographic separation of the TMS-derivative.

A review of the medical literature revealed no reports of elderberry juice poisoning in the past 30 years. The Centers for Disease Control and Prevention [35] did issue a bulletin about a poisoning incident on 26 August 1983 involving a group in California attributed to consumption of juice prepared from fresh wild elderberries along with leaves and stems (most likely blue elderberry, *Sambucus cerulea*) [35]. Cyanide was initially implicated in the incident, but was subsequently disproven. There remains uncertainty as to the presence of CNGs in elderberry juice and its products. Recent studies of European elderberry by Senica et al. [25,36] reported average levels of sambunigrin in fresh and processed berry products ranging between 0.8 and 18.8 µg/g [25] and higher amounts in elder leaves (27.68 and 209.61 µg/g FW) [36]. Koss-Mikolajczyk et al. [37] in similar work however recorded no quantifiable amounts of CNGs. To date, no exhaustive work has been completed on AE to conclusively ascertain the presence, forms, and levels of CNGs in ripe and unripe berries.

Traditional and modern food-processing techniques such as chopping, grinding, and heating are used to reduce the potential toxicity of plants containing CNGs [38–40]. However, the effectiveness of these techniques depends on the processing method [40], the plant tissue, and the intended processed forms. Soaking, for instance, may be effective when CNGs are soluble in the solution (discarded later) without enzymatic degradation [39–41]. Boiling can inhibit the activity of endogenous β-glucosidase due to high temperatures and halt the production of HCN [39]. However, this is only partially effective in reducing HCN because some CNGs are relatively heat stable [39] and HCN is water soluble [42]. Therefore, if CNGs do not hydrolyze due to enzyme inactivation, toxicity may still result from catabolism of these compounds in the gastrointestinal tract [43,44].

Quantification of CNGs can be made either indirectly (by determining HCN released after hydrolysis) or directly (by determining the intact glycoside) [17,19]. A very sensitive, reproducible and robust liquid chromatography/mass spectrometry-based method (LC-MS/MS) was developed and optimized for accurate identification and quantification of intact CNGs. Ultrahigh-performance liquid chromatography triple-quadrupole mass spectrometry (UHPLC-MS/MS) was used for this purpose [45,46]. A complimentary picrate paper method was modified to assess the total cyanogenic potential (TCP) by determining the total cyanide concentration following action of endogenous enzymes with CNGs. Analysis was performed using a camera-phone and UV-Vis spectrophotometry. In this study, we examine different elderberry fruit tissues. This study provides definitive and much needed answers to lingering questions regarding the safety of AE.

2. Results and Discussion

2.1. Picrate Paper Method

The TCP was first determined by adapting a picrate paper method originally developed by Bradbury and co-workers [47]. This is a colorimetric method where picrate paper changes color in the presence of HCN. It is based on the reaction of picric acid with HCN. The method is described in more detail in Section 3.5. Amygdalin was used as a CNG standard to generate HCN. Cyanide equivalent (CN^- eq.) solutions were prepared from a 1000 µg/mL KCN/NaOH stock solution to develop a calibration curve over the range of 1 to 100 µg/mL. The observed color change of the picrate paper for amygdalin improved significantly and became consistent when the adapted method [47] was modified by using minimal liquid (<0.5 mL water). This was to enhance the HCN reaction with picric acid since HCN is highly water-soluble [42]. A standard calibration curve showing the amount of CN^- eq. with its corresponding absorbance is shown in Figure 2, using amygdalin as the cyanide source. Supplemental Figure S1 shows the corresponding standard curve using a camera-phone as a detector. Supplemental Figure S2 shows the expanded UV-Vis data from 0 to 10 µg. Table 1 summarizes the LLOD, upper limit of quantification (ULOQ), and the regression coefficients (R^2-values) for camera-phone and UV-Vis analysis. UV-Vis showed better linearity compared to camera-phone method in repeated analysis.

Table 1. LLOD, ULOQ, and Pearson coefficients (R^2-values) for calibration curves from camera-phone and UV-Vis analysis using amygdalin as a CNG standard.

Method/µg CN^- eq.	ULOQ	LLOQ	R^2
UV-Vis	50	0.14	0.9971
Camera-phone	50	1.59	0.9889

Qualitative inspection of the picrate paper strips showed no visible color change for the commercial elderberry juice sample. UV-Vis analysis of the picrate paper test strips detected no quantifiable amount of cyanide (<0.14 µg CN^- eq.). Two different AE genotypes, Ozark and Ozone, were then analyzed using the picrate paper method. Sample tissues (juice, skin, stem, seeds) for each genotype showed no visible color change on qualitative assessment (Supplemental Figure S3). Generally, results obtained for lyophilized samples were comparable to fresh samples. Quantitative determination by UV-Vis revealed low levels of cyanide with average amounts ranging from (2.60–9.20 µg CN$^-$ eq.)/g of sample. TCP levels obtained were comparable for both AE genotypes and for all tissue types of Ozone and Ozark, respectively. TCP amounts increased in the order juice < seeds < skin < stem for both genotypes as shown in Figure 3 [48].

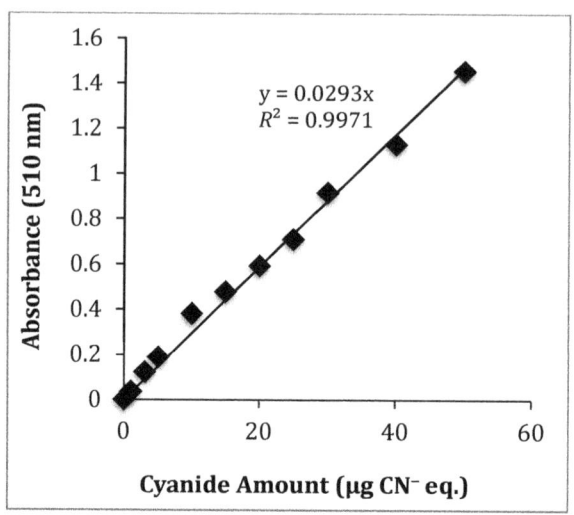

Figure 2. Calibration curve for the amount of CN$^-$ eq. produced by reaction of picric acid with HCN using amygdalin as a CNG standard measured by UV-Vis spectrophotometry at λ_{max} = 510 nm.

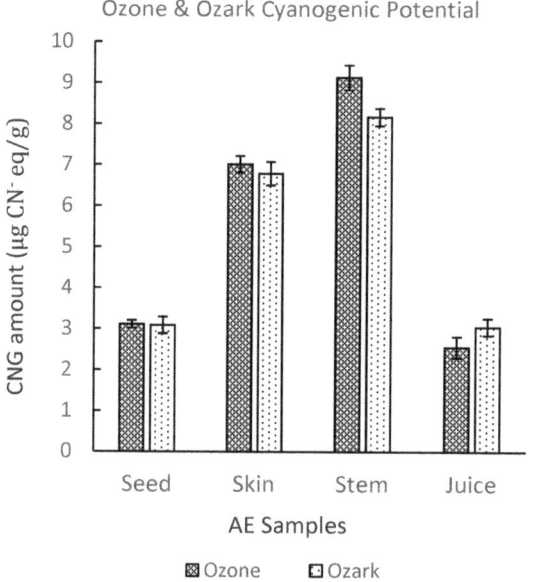

Figure 3. Total cyanogenic potential for different types of tissue of Ozone and Ozark AE genotypes. The amounts of CNGs in these genotypes were determined using UV-Vis spectrophotometry. The error bars represent the standard deviation of at least three replicate samples.

A set of pooled AE samples was generated using five AE genotypes (Ozark, York, Wyldewood, Ocoee, and Bob Gordon). The pooled samples were divided into different types of tissue. These included seeds, skin, pulp, stems, juice, and whole green, red, and ripe berries. Qualitative inspection of picrate paper test strips for pooled AE samples showed a visible faint color change for the green berries and stems (Supplemental Figure S4). UV-Vis analysis showed the highest CN^- levels for stems and green berries with lower amounts for the other tissue types (Figure 4). TCP levels in analyzed tissues increased in the order: whole ripe berries < whole red berries < juice < seeds < skin < pulp < whole green berries < stem, with highest average amounts in the stems (37.43 ± 9.19 µg CN^- eq./g) and whole green berries (25.6 ± 5.07 µg CN^- eq./g). Koss-Mikolajczyk et al. [37] in a recent EE study observed a weak and unstable signal for a peak corresponding to sambunigrin which decreased with advancing stage of ripeness in elderberry fruit. In another study, Zahmanov et al. [49] reported metabolic differences in mature and immature fruits, and plant leaves of *Sambucus ebulus*. These observations may account for the slightly higher levels recorded in green berries. Although the CNG amounts in the stems and green berries are not sufficient to pose a threat of toxicity, it is nevertheless advisable to carefully exclude green elderberries and stems during juice preparation.

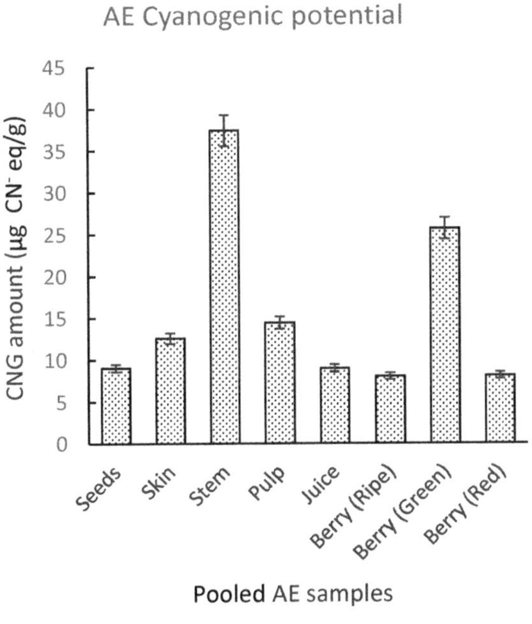

Figure 4. Total cyanogenic potential for different types of AE tissue of pooled samples made up of five different genotypes. The amounts of CNGs in pooled samples were determined using UV-Vis spectrophotometry. The error bars represent the standard deviation of at least three replicate samples.

Two different types of seeds from Gala and Granny Smith apples were obtained and prepared for analysis as discussed in Section 3.5. Apple seeds were chosen as their TCP levels are known and should be readable using the picrate paper method. Color change on the picrate paper test strip for the apple seeds occurred swiftly at room temperatures even before test strips were transferred into the oven for overnight heating (30–40 °C). A deep red color change was observed on inspection for both fresh and lyophilized samples (Supplemental Figure S5). UV-Vis analysis showed high average cyanide amounts (TCP) ranging from (497.50–603.20 µg CN^- eq.)/g of apple seeds. TCP levels in analyzed seeds were higher in Granny Smith as compared to Gala apple varieties. These results were comparable to available literature [17,50].

Results from the endogenous enzymes test made using pooled AE stems and green berries revealed higher cyanide levels in samples with added β-glucosidase than those without added β-glucosidase (Supplemental Figure S6). Approximately 77% and 33% more cyanide were measured in pooled AE stems and whole green berries, respectively, with added β-glucosidase. These findings indicated that while AE contains endogenous β-glucosidase enzymes sufficient to initiate self-hydrolysis of CNGs, it may not be sufficient for complete hydrolysis of all CNGs (55–75%) when the tissues are disrupted. This implies that not all available CNGs in elderberry may necessarily be able to transform into HCN. These observations are supported in a similar analysis by Miller et al. [51] using foliage of the tropical trees *Beilschmiedia collina* and *Mischocarpus spp*. Apple seeds however showed no appreciable change in cyanide concentration with or without addition of β-glucosidase enzymes (Supplemental Figure S7), thus indicating that the seeds of the apple varieties used contain sufficient endogenous β-glucosidase for complete hydrolysis of all CNGs when the tissues are disrupted. The picrate paper test method is quick and simple and could serve as an effective field test for elderberry producers.

2.2. UHPLC MS/MS Method of Analysis

2.2.1. Method Development and Optimization

An attempt was made to find multiple reaction monitoring (MRM) transitions for four cyanogenic standards (CNS) using both electrospray ionization (ESI) and atmospheric pressure chemical ionization (APCI) sources. Positive and negative ionization modes were performed for each standard with both ionization techniques. The only successful MRM transition identified was for amygdalin in ESI positive mode. Figure 5 shows the positive mode product ion (296 m/z, product) spectrum for amygdalin (465 m/z, precursor). All other standards readily formed sodium adducts, which did not sufficiently fragment due to their high stability. Alternative mass spectrometry scans were investigated to overcome this problem. Quantification for all four CNS's were performed using selected ion recording (SIR) mode. The developed UHPLC and MS method displayed excellent separation of the four standards and exhibited retention time repeatability and good peak shape. A chromatogram for the separation with retention times (RT) and scanning modes is shown in Figure 6. It took less than 6 min to separate and elute all 4 CNS.

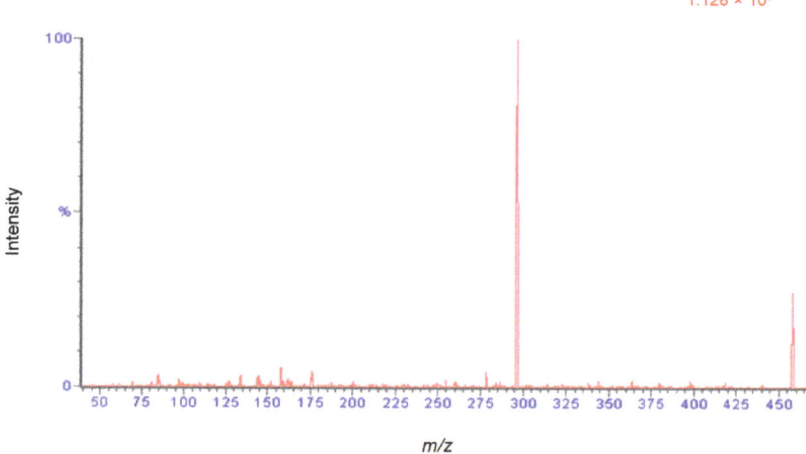

Figure 5. ESI positive mode product ion (296 m/z, product ion) spectrum for amygdalin (465 m/z, precursor ion).

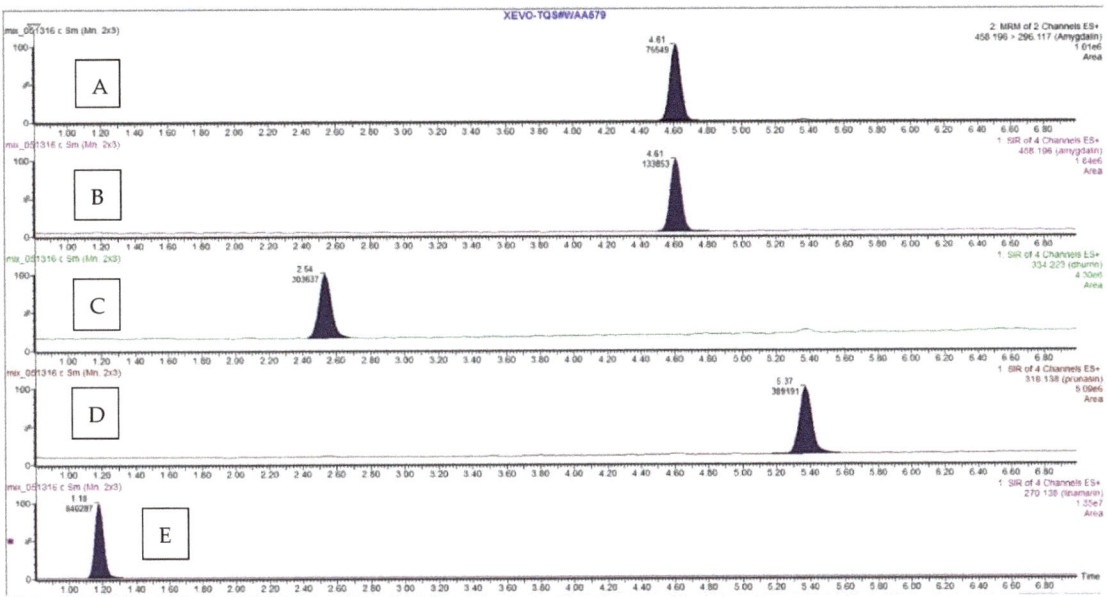

Figure 6. Ion chromatograms for (**A**) amygdalin (MRM), (**B**) amygdalin, (**C**) dhurrin, (**D**) prunasin, and (**E**) linamarin (SIR). Retention times in min. are: 4.61, 4.61, 2.54, 5.37, and 1.18, respectively.

Standard calibration curves showed good linear correlations (R^2 values) between integrated peak areas and known CNS concentrations. The lower limit of detection (LLOD) was determined based on a signal to noise ratio of three and a targeted coefficient of variation [52] (CV% ≤ 20%, for seven repeated injections) for confirmation. A linear range with lower (LLOQ, S/N =10, CV% ≤ 20%) and upper (ULOQ) limit of quantification was determined. The ULOQ was determined as the highest concentration of the linear curve beyond which the linearity breaks. Details are summarized in Table 2.

Table 2. Summary of Pearson coefficient, detection and quantification limits (ng/mL) for CNGs.

Parameters ng/mL	MRM Amygdalin	SIR			
		Amygdalin	Dhurrin	Prunasin	Linamarin
LLOD	0.3	3	3	3	1
LLOQ	1	10	10	5	5
ULOQ	8000	8000	6000	6000	2000
R^2	0.9998	0.9998	0.9983	0.9984	0.9910

2.2.2. Optimized Extraction, Recovery and Matrix Effect

Selecting the most appropriate extraction solvent was key to development of the extraction methodology. Recoveries from aqueous ethanol or methanol combinations were evaluated. Higher recoveries were obtained with 75% methanol extraction with overnight shaking (16–24 h) at room temperature and 30 min sonication at 30 °C as compared to other extraction methods and conditions. The recovery (RE) and matrix effect (ME) were evaluated by an approach based on responses from pre-extraction spike matrix (a), post-extraction spike matrix (b) and a neat spike standard (c). RE and ME were calculated using Equations (1) and (2) [53] (where +ME implies ion enhancement, −ME implies ion suppression). Table 3 below compares the mean recoveries and standard deviations for 30 min sonication at 30 °C to overnight shaking (16–24 h) at room temperature with intended spike concentration of 1000 ng/mL and 100 ng/mL CNS mixture. ME estimation was found to range between 10.20 and 18.34%. This was a negative estimation and as such indicated some degree of ion suppression. Although further dilution of sample matrix from 10 to 1000-fold reduced this value appreciably, it also decreased the sensitivity of sample detection. Hence a 10-fold dilution was used.

$$\mathbf{RE} = \frac{a}{b} \times 100 \qquad (1)$$

$$\mathbf{ME} = \left[\left(\frac{b}{c}\right) - 1\right] \times 100 \qquad (2)$$

Table 3. Mean recoveries and standard deviations for spike concentrations of 100 and 1000 ng/mL.

CNG Standards/	Conc.	MRM				SIR					
Mean Recovery%	(ng/mL)	Amygdalin		Amygdalin		Dhurrin		Prunasin		Linamarin	
		Mean RE%	SD%	Mean RE%	SD%	Mean RE%	SD%	Mean RE%	SD%	Mean RE%	SD%
Sonication	1000	85.40	1.01	86.14	1.61	95.34	2.1	91.35	0.95	95.26	1.98
(30 min at 30 °C)	100	91.43	0.92	92.14	3.11	88.54	2.04	79.14	5.1	92.1	0.96
Overnight shaking	1000	93.69	0.31	91.40	0.98	98.19	0.73	106.98	0.86	112.21	0.62
(16–24 h)	100	81.70	3.61	101.54	2.98	109.99	2.75	87.94	1.85	98.11	2.13

2.2.3. Optimized SPE Method

To assess and evaluate the elution solvent strength in the SPE method, an elution profile showing methanol and water proportions from 0:100 to 100:0 (v/v) versus peak area was made for all four standards. Figure 7 shows the elution profile for amygdalin. Evaluation of these profile diagrams revealed 30–40% methanol as the best solvent strength for elution of all four CNS. Confirmation using water methanol proportions from 0:40 to 40:0 (v/v) was made to determine the best elution solvent as 35% methanol.

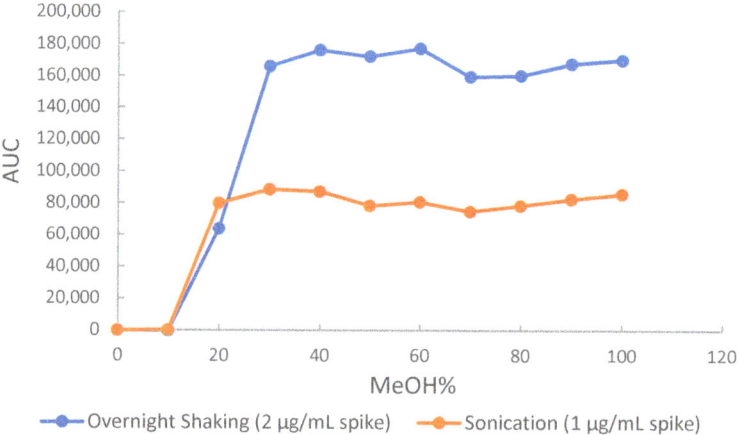

Figure 7. Elution profile for amygdalin as a function of methanol content in the extraction solvent. (AUC is the area under the curve).

2.2.4. Sample Test

The developed UHPLC-MS/MS method, as detailed in Section 3.6, was used to determine the levels of intact CNGs in different AE samples. Analysis of commercial elderberry juice showed no quantifiable amounts of CNGs. However, extracts of Ozark and Ozone elderberry tissues (seeds, juice, skin and stem) for both lyophilized and fresh samples showed low traces of CNGs (amygdalin, dhurrin, linamarin and prunasin). Levels of CNGs detected in lyophilized samples were comparable to fresh samples. The amounts (µg/g) in tissues were generally higher in Ozone compared to Ozark. Higher levels (µg/g) were recorded in the stems and skin tissues as compared to levels in the seeds and juice, respectively for the Ozone and Ozark samples. A detailed summary of amounts for each detected CNGs in AE tissues are shown in Table 4. Figures 8 and 9 show the amounts of these cyanogens in tissues for Ozone and Ozark, respectively. The levels (µg/g) of CNGs in analyzed tissues increased in the order: linamarin < dhurrin < prunasin < amygdalin, respectively, for Ozone and Ozark samples tissues. In contrast to this trend, prunasin levels were highest in the juice and stems of Ozark AE.

Table 4. Amounts (µg/g) of CNGs found in tissues of Ozone and Ozark AE samples.

Elderberry Samples		Concentration ± Standard Deviation (µg/g)			
		Amygdalin	Dhurrin	Prunasin	Linamarin
Seeds	Ozone	2.38 ± 0.09	0.27 ± 0.05	0.58 ± 0.04	0.12 ± 0.06
	Ozark	0.68 ± 0.12	0.22 ± 0.03	0.36 ± 0.05	0.13 ± 0.05
Juice	Ozone	1.57 ± 0.08	0.70 ± 0.12	1.45 ± 0.06	0.29 ± 0.03
	Ozark	0.36 ± 0.03	0.63 ± 0.04	2.36 ± 0.08	0.31 ± 0.01
Skin	Ozone	6.38 ± 0.40	0.12 ± 0.08	2.39 ± 0.04	0.75 ± 0.06
	Ozark	3.48 ± 0.14	1.46 ± 0.20	2.53 ± 0.08	0.90 ± 0.11
Stem	Ozone	5.42 ± 0.12	0.94 ± 0.06	2.84 ± 0.02	0.48 ± 0.04
	Ozark	2.15 ± 0.17	1.91 ± 0.03	3.07 ± 0.06	0.57 ± 0.06

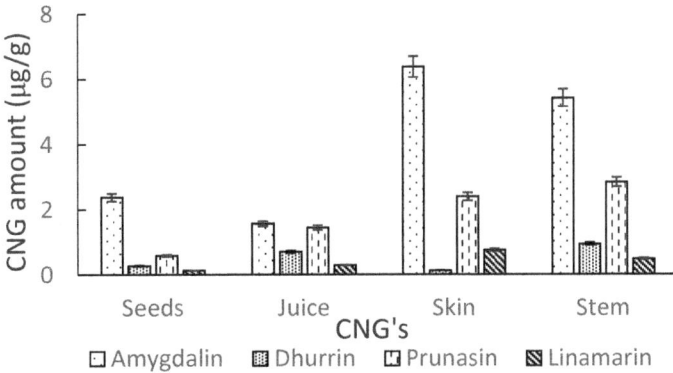

Figure 8. Amounts of CNGs (µg/g) in tissues (seeds, juice, skin and stem) of Ozone elderberry samples as measured by UHPLC-MS/MS.

In our UHPLC MS/MS, we are not able to distinguish between the two diastereomers, (R)-prunasin and (S)-sambunigrin. The two compounds were not uniquely separated by UHPLC using a C18 column. Further, their fragmentation patterns are very similar. Therefore, the prunasin concentrations should be viewed as a sum of the prunasin and sambunigrin concentrations.

A review of literature in similar areas of study found comparable results, but also revealed an interesting trend of observation. A recent study by Senica et al. [40] on the EE (subsp. *nigra*) reported average levels of sambunigrin (µg/g) in fresh berries (18.8 ± 4.3), processed juice (10.6 ± 0.7), tea (3.8 ± 1.7), spread (0.8 ± 0.19) and liqueur (0.8 ± 0.21). Our measured levels of (prunasin + sambunigrin) for AE are lower for fresh berries compared to EE. Senica et al. [36] in a similar work reported highest amounts of sambunigrin in elder leaves (27.68–209.61 µg/g FW), lower amounts in flowers (1.23–18.88 µg/g FW) and lowest amounts in berries (0.08–0.77 µg/g FW). In the work by Buhrmester et al. [20], also observed similar sambunigrin concentrations for AE leaves. Senica and co-workers concluded that the content of sambunigrin in elderberry changes depending on the growing altitudes (higher content on hill tops and lower in foothills) [36]. Another study by Koss-Mikolajczyk et al. [37] on EE (subsp. *nigra*) observed the highest signal for a peak detected as sambunigrin in the elder leaves although this peak became undetectable after one day of cold storage of extracts. It was also reported that the level of cyanogens in cassava leaves are 10 times more than in the roots [39]. Deductions from this trend of results suggest that the leaves of most cyanogenic plants may accumulate larger amounts of CNGs, with elder as no exception. The trend of results also corroborates the fact that elderberry juice, being it, AE or EE showed very low levels of CNGs.

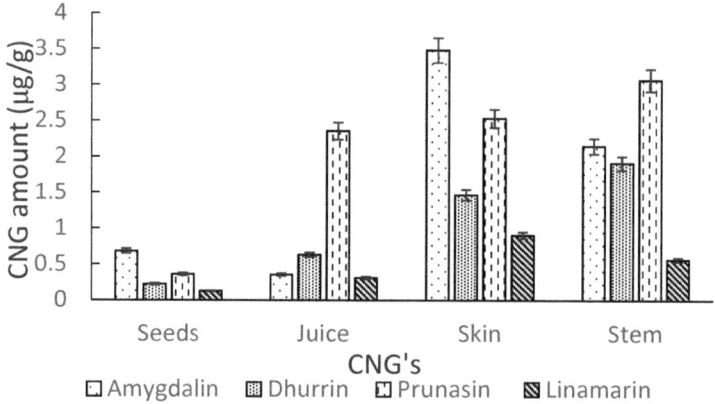

Figure 9. Amounts of CNGs (µg/g) in tissues (seeds, juice, skin and stem) of Ozark elderberry samples as measured by UHPLC-MS/MS.

The levels of CNGs detected in all tissues of AE samples were extremely low compared to levels of amygdalin detected in apple seeds (950–3910) µg/g, pressed apple juice (10–39) µg/mL and commercially available apple juice (1–7) µg/mL for 15 apple varieties [17]. Acute CN toxicity occurs at a concentration of 0.5–3.5 mg/kg of body weight. For cyanide in blood, the toxicity threshold for cyanide alone ranges from 0.5 to 1.0 mg/L, and the lethal threshold ranges from 2.5 to 3.0 mg/L [54]. Despite the high cyanide levels in apple seeds as revealed in the control picrate test (497.50–603.20 µg CN^- eq./g), signs of cyanide toxicity may occur in an average adult male of weight 82 kg, only after consuming about 14 or more apples including mastication of all associated seeds. This estimation was made considering the threshold value of 0.50 mg/kg body weight for cyanide toxicity, average TCP per seed of 550 µg/g (or 0.55 mg/g), the average weight of an apple seed (0.75 g), the average number of seeds per apple (7 or 8), and assuming maximum enzyme activity.

Different processing techniques such as chopping, grinding, soaking, fermentation, drying, roasting, boiling, and steaming have been used to remove or reduce the potential toxicity of cyanogens in plants [38–40]. The effectiveness of these processes is dependent on the specific processing method [40], the plant tissues and the intended processed forms. Boiling of juice for instance may have a different effect compared with boiling or soaking cassava chips where the associated water can easily be discarded [39–41]. This may be due to enzyme inactivation and solubilization of CNGs in discarded water [39]. A study by Montagnac et al. further indicated that the effectiveness of these techniques depends on the processing steps, the sequence utilized, and is often time-dependent [39]. They proposed that to increase the efficiency of cyanogen removal from cassava, efficient processing techniques should be combined [39]. For example, soaking, fermenting and roasting removes about 98% of cyanogens [39]. A recent study by Senica et al. [40] also showed that thermal processing, time and type of extraction solution greatly affected phenolics and cyanogenic glycosides in different elderberry products. They showed that higher processing temperatures decreased the levels of cyanogenic glycosides by 44% in elderberry juice, 80% in tea and as much as 96% in elderberry liqueur and spread [40]. It has been confirmed that pasteurization effectively decreases the levels of harmful compounds, such as cyanogenic glycosides [17,40]. Bolarinwa et al. [17] moreover observed that holding apple juice at room temperature for 120 min either before or after pasteurizing decreased the amygdalin content by about 19% compared to the original juice. These methods are very effective and can be applied to remove or further reduce the levels of CNGs in elderberry product. It is however important to establish that the types and levels of CNGs observed in AE are very low and pose no threat to consumers in the use of fresh or processed AE products.

3. Materials and Methods

3.1. Chemicals and Reagents

Water, acetonitrile (ACN), methanol, ethanol and formic acid were purchased from Fisher Scientific (Fair Lawn, NJ, USA, HPLC grade). β-glucosidase enzymes (250 mg lyophilized powder $\geq$ 6U/mg), amygdalin (1 g, $\geq$99%), dhurrin (1 mg, $\geq$98%), prunasin (1 mg, $\geq$95%) and linamarin standards (1 mg, $\geq$95%), together with picric acid (100 g moisten with water $\geq$ 98%) and potassium cyanide ($\geq$98%) were purchased from Sigma Aldrich Chemical Co. (St. Louis, MO, USA). Whatman no.1 filter paper, sodium carbonate, sodium hydroxide and pH 8 phosphate buffer (500 mL) were also purchased from Fisher Scientific (Fair Lawn, NJ, USA). Plastic backing and hobby glue (adhesive neutral pH) were purchased from the Mizzou Store (Columbia, MO, USA).

3.2. American Elderberry Samples

Plant Material. Elderberry fruit samples were harvested from experimental field plots that were previously described in detail [2,45]. Briefly, a replicated evaluation of eight American elderberry genotypes was established in Missouri (USA) in 2008. Fruits from six genotypes (Bob Gordon, Ocoee, Ozark, Ozone, Wyldewood, and York) were harvested from one of the study sites (Mt. Vernon, MO, USA) at peak ripeness in August 2016, and promptly frozen. Frozen, de-stemmed, whole berries (>400 g) from the five genotypes were provided to the laboratory, along with frozen unripe and almost-ripe berries (green and red-colored, respectively). Additionally, hundreds of individual berries from each genotype were thawed and painstakingly separated into skins (epicarp), pulp (mesocarp), seeds, juice, and small green stems (pedicels) that connect the berry to the infructescence. After dissection, samples were re-frozen. For detailed CNG analysis, tissue and juice samples from the genotypes Ozone and Ozark were analyzed separately. Further, tissue and juice from five genotypes (Bob Gordon, Ocoee, Ozark, Wyldewood, and York) were combined into pooled samples for a broader analysis. Sufficient material was dissected to produce samples exceeding 10 mg.

Commercially processed elderberry juice was purchased from River Hills Harvest, Hartsburg, MO, USA.

3.3. Sample Preparation and Extraction

Berries were transferred into small zippered plastic bags, thawed, and gently pressed. The juice was transferred to a clean vial. Seeds were separated from skin and placed into different vials. 100 g of berries produced about 60 g of juice, 20 g of seeds, 12 g of skin, and some left-over stems. Between 5–10 g of each sample tissue (excluding juice) was transferred into 15 mL Eppendorf tubes, flash frozen for about 5 min in liquid nitrogen and freeze-dried for 24 h using a Labconco FreeZone 4.5 Liter Benchtop Lyophilizer (Labconco Corp., Kansas City, MO, USA) at -105 °C. The lyophilized samples were ground using a clean mortar and pestle to obtain about 3–5 g of homogenized seeds, stem and skin samples for extraction.

To obtain an optimized sample pretreatment and extraction, equal volumes of commercially processed elderberry juice, in replicates of 4, were spiked with varying amounts of 10 µg/mL cyanogenic standard (CNS) stock mixture (amygdalin, dhurrin, prunasin and linamarin, Figure 10) to obtain intended spike concentrations of 1000, 100, and 10 ng/mL. The solutions were extracted with 1 mL of different ethanol/methanol and water proportions from 60:40 to 80:20 (v/v). Extraction was performed via sonication (10–60 min) at 30 °C, overnight shaking (16–24 h) and 2 min vortexing at room temperature on a Genie 2 Shaker Mixer (Scientific Industries, Inc., Bohemia, NY, USA) at 600 rpm. Extracts were centrifuged for 15 min, dried under nitrogen gas and reconstituted in 1 mL of mobile phase (0.1% FA in ACN) for SPE clean up. Sample extraction was performed with both fresh and lyophilized samples.

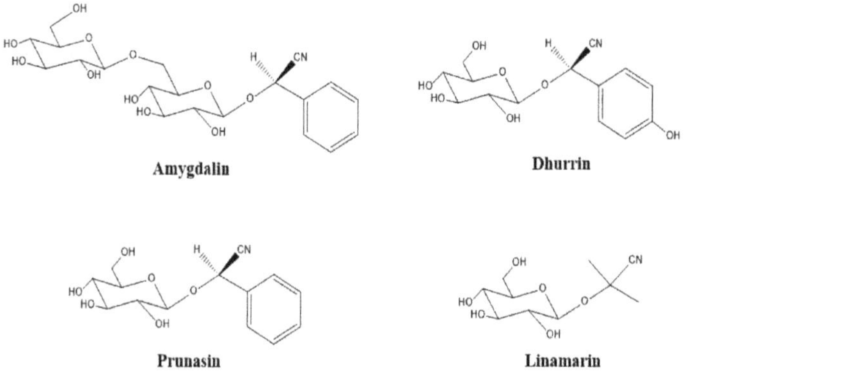

Figure 10. Cyanogenic glycosides standards used in this study: amygdalin, dhurrin, prunasin, and linamarin.

3.4. Solid Phase Extraction (SPE)

An SPE method [24] previously used for the determination of amygdalin in almonds was adapted and optimized for our use. A Supelco Visiprep™ SPE vacuum manifold (Sigma-Aldrich, St. Louis, MO, USA) was used for this purpose. A vacuum of 10 in Hg (35 kPa) and a flowrate of about 1–2 drops/s was maintained throughout the process. A SPE cartridge (Sep-Pak Vac 3 cc (500 mg) C18 cartridge, Waters, Milford, MA, USA) was conditioned with 2 mL of methanol and equilibrated with 2 mL of water. 1 mL of the sample was loaded onto the column. An additional 1 mL of 0.1% FA in water was used to remove remaining residue in the extraction tube. The column was flushed with 2 mL of 0.1% FA in water. CNGs were finally eluted with 2 mL of varying proportions of methanol/water (0, 10, 20, 30 to 100% v/v). The extracts were dried under nitrogen gas, reconstituted into 0.1% FA in water and filtered through a 0.45 µm filter prior to UHPLC-MS/MS analysis.

3.5. Picrate Paper Method of Analysis

The picrate paper method is based on the reaction of enzymes that catalyze the release of HCN gas, which reacts with picric acid on a paper test strip producing 2,6-dinitro-5-hydroxy-4-hydroxylamino-1,3 dicyclobenzene inducing a color change (Supplemental Figure S8) [46].

A previously published picrate method described by Bradbury et al. [47] for the determination of the total cyanogenic content in cassava roots was adapted and modified for use. Briefly, the picrate paper was prepared beforehand by dipping a sheet of Whatman 3 MM filter paper in a picrate solution (1.4% w/v moist picric acid diluted in 2.5% w/v Na_2CO_3 solution), allowing the paper to air dry and cutting it into 3 cm × 1 cm strips. The strips were attached using a drop of PVA hobby glue to 5 cm × 1.2 cm clear plastic strips to keep the paper clear of the liquid. They were stored at 4 °C prior to use. Cyanide equivalent (CN⁻ eq.) solutions were prepared from a 1000 μg/mL KCN stock solution. The stock solution was prepared by dissolving KCN in 0.1M NaOH as the solvent. The calibration curve covered the range of 1 to 100 μg CN⁻ eq. and the method was verified using amygdalin as a positive control. One of the most complicated portions of this analysis is the enzymatic hydrolysis of amygdalin. Enzymes are macromolecular biological catalysts whose amount for a specific enzymatic activity is measured in Units (U). One U is defined as the amount of enzyme needed to catalyze the conversion of 1 micromole of substrate per minute [48]. Enzymatic degradation of amygdalin was achieved by adding 50 μL of 3U/mL β-glucosidase.

The commercially processed elderberry juice was tested for TCP along with the AE samples. 100 μL/100 mg each of lyophilized and fresh tissues of Ozone, Ozark and pooled AE samples were measured/weighed into clean 20 mL scintillation vials. 50 μL of β-glucosidase solution (3U/mL) in pH 8 phosphate buffer was added alongside the picrate paper and the vial was immediately closed with a screw stopper. Each sample analysis was made in replicates of four. Similar set-ups were made for amygdalin standards and blank (no amygdalin added). These were left overnight (16–24 h) in an oven at 30–40 °C.

Two control experiments were performed. The first was to confirm the effectiveness of the picrate paper test method to known literature. Seeds from two apple varieties, Granny Smith (GS) and Gala (G), were prepared and tested for TCP using the same protocol for the AE samples. The second control experiment used seeds (ground) from the two apple varieties and stems and green berries from AE pooled samples. This was to test for the presence of endogenous enzymes in the samples to assess the extent of self-hydrolysis of CNGs. To accomplish this, two different picrate paper set-ups were made, a control (without an external β-glucosidase solution) and a second typical picrate set-up (with an external β-glucosidase solution). Four replicates for each set-up were performed.

A simple and quick method of analyzing the reacted picric acid is by qualitative inspection. This appears to be an effective method for quantifying CN⁻ equivalents ranging between 1 and 100 μg. However, as the amount of CN⁻ eq. increases, the ability to differentiate between the colors decreases. A color chart shown in Figure 11 can be used to relate the color change of the paper to total amount of CN⁻ evolved. A semi-quantitative approach using a camera-phone as a detector was used. An image of a concentration from Figure 11 was converted from color to greyscale. This was done using Image J software (https://imagej.nih.gov/ij/index.html (accessed on 22 December 2020), version 1.46r, National Institutes of Health, Bethesda, MD, USA). The method generated mean intensity values corresponding to each CN⁻ eq. and was used to generate a calibration curve. Quantification was confirmed using a UV-Vis spectrometer (Agilent 8454 photodiode array, Agilent Technologies, Santa Clara, CA, USA). The reacted picrate paper test strip was extracted in 3.5 mL of water in cuvettes and the resulting solution analyzed at a wavelength (λ_{max}) of 510 nm after standing for 30 min.

3.6. UHPLC-MS/MS Method of Analysis

Separation and analysis of cyanogenic glycosides were performed with a C18 column (Acquity BEH, 1.7 μm, 50 × 2.1 mm, Waters, Milford, MA, USA) using a Waters Acquity UHPLC coupled to a Xevo TQ-S triple quadrupole mass spectrometer (UHPLC-MS/MS). A previously published gradient [24] for the quantification of amygdalin in almonds was reduced from 20 min down to 10 min. The mobile phase included 0.1% formic acid in water (mobile phase A) and 0.1% formic acid in acetonitrile (mobile phase B). The gradient used was 95% A, 0−1 min; 95−80% A, 1−3 min; 80−40% A, 3−7 min; 40% B, 7−8 min and 95% A 8.1−10 min re-equilibration. The flow rate was 200 μL min^{-1} and the following conditions were used for the electrospray ionization (ESI) source: source temperature 150 °C, desolvation temperature 350 °C, capillary voltage 3.07 kV, cone voltage 21, and nebulizer gas 500 L h^{-1} N$_2$. Argon was used as the collision gas. The collision energies were optimized and ranged from 17 to 30 eV for individual analytes. The column and sample temperatures were 40 ° and 10 ° C, respectively. The ESI source was operated in the positive ion mode. Instrument control and data processing were performed by using MassLynx software (version 4.1, Waters, Milford, MA, USA). Cyanogenic standard solutions were prepared with concentrations ranging from 1 ng mL^{-1} to 10 µg mL^{-1}. All analyses were done in triplicate along with a blank. The interday and intraday precisions of the method had a CV% of less than 5%.

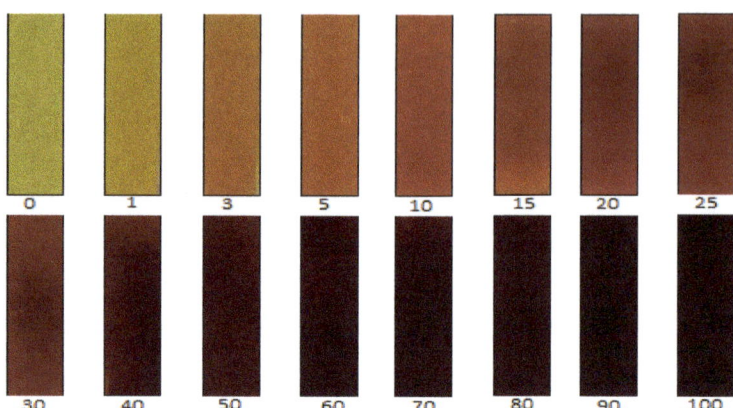

Figure 11. A picrate-paper cyanide color chart for qualitative analysis of CNGs for the range of 0–100 µg CN$^-$ eq.

3.7. Statistical Analysis

For the determination of cyanide by UV-Vis and cyanogenic glycosides by LC-MS/MS, samples were prepared in three biological and three analytical replicates for each sample. Statistical analyses were performed in Excel (Microsoft Office 2016). Results are expressed as the mean ± standard error of mean (SEM). Additionally, the coefficient of variation for six (6) repeated injections (CV% ≤ 20%) was used to confirm candidate concentrations for LLOQ and LLOD.

4. Conclusions

The UHPLC-MS/MS and picrate paper methods developed were used to reliably determine the intact CNGs and assess the TCP in various AE fruit tissue. No quantifiable trace of cyanide or CNG was detected in commercial elderberry juice. Moreover, traces of CNGs (amygdalin, dhurrin, (prunasin + sambunigrin), and linamarin) detected in tissues of AE samples were generally low with lower levels in the juice and seeds as compared to stems and skin. TCP assessed in both pure and pooled AE sample tissues were generally low with higher concentrations recorded in pooled stems and unripe (green) berries. The picrate paper method can also be used to help detect the presence of CNGs. A camera-phone and UV-Vis spectrophotometer can both be used as a detector. The camera-phone can give results easily with limits of detection that are useful for CNG analysis. Although the TCP and CNGs levels in tissues of AE pose no threat to consumers, it is advisable to separate out the stems, green berries and leaves [36] from AE ripe berries during product preparation.

Supplementary Materials: The following are available online: Figure S1: Picrate-paper results for the using a camera-phone to detect the presence of CN$^-$ using amygdalin. Figure S2: Picrate-paper results for the using a UV-Vis spectrophotometer in the low concentration range (0–10 µg) to detect the presence of CN$^-$ using amygdalin. Figure S3: Picrate Paper results for tissues of Ozone and Ozark AE samples, Figure S4: Picrate paper results for pooled AE samples, Figure S5: Picrate paper results for apple seeds and juice, Figure S6: Picrate paper results for endogenous enzymes test for pooled

AE tissues, Figure S7: Picrate paper results for endogenous enzymes test for fresh and lyophilized apple seeds, Figure S8: Cyanide reaction with picric acid.

Author Contributions: Conceptualization, C.M.G. and A.L.T.; methodology, M.K.A., M.C.J., H.I., R.B., R.K.; formal analysis, M.K.A. and C.M.G.; investigation M.K.A., M.C.J., H.I., R.B., R.K.; Writing—original draft preparation, M.K.A., C.M.G. and A.L.T.; writing—review and editing, M.K.A., C.M.G. and A.L.T.; project administration, C.M.G. and A.L.T.; funding acquisition, C.M.G. and A.L.T. All authors have read and agreed to the published version of the manuscript.

Funding: This research was supported by the Missouri Department of Agriculture through the USDA's Specialty Crop Block Grant Program; by Hatch project accession no. 1011521 from the USDA National Institute of Food and Agriculture; and by the University of Missouri Center for Agroforestry under a Cooperative Agreement (58-6020-6-001) with the USDA/ARS Dale Bumpers Small Farms Research Center.

Institutional Review Board Statement: Not applicable.

Informed Consent Statement: Not applicable.

Data Availability Statement: Original data is available upon request to the corresponding author.

Acknowledgments: We thank Jillian Boydston and Samuel Sergent for their excellent technical assistance.

Conflicts of Interest: The authors declare no conflict of interest.

Sample Availability: Not available.

Abbreviations

ACN	acetonitrile
AE	American elderberry
APCI	atmospheric pressure chemical ionization
CDC	Centers for Disease Control and Prevention
CNGs	cyanogenic glycosides
CNS	cyanogenic standards
EE	European elderberry
ESI	electrospray ionization
FA	formic acid
HCN	hydrogen cyanide
HPLC-DAD	High-performance liquid chromatography with photo diode array detectors
LLOQ	lower limit of quantification
LLOD	lower limit of detection
ME	matrix effect
MRM	multiple reaction monitoring
RE	recovery
RT	retention time
SD	standard deviation
SIR	selected ion recording
S/N	signal to noise ration
SPE	solid-phase extraction
TCP	total cyanogenic potential
UHPLC-QqQ-MS/MS	Ultra-high performance liquid chromatography triple-quadrupole mass spectrometry
ULOQ	upper limit of quantification
UV-Vis	ultraviolet visible spectrophotometry

References

1. Byers, P.L.; Thomas, A.L.; Cernusca, M.M.; Godsey, L.D.; Gold, M.A. *Growing and Marketing Elderberries*; in Missouri. Agroforestry in Action Pub, AF1016; Univ. Missouri Center for Agroforestry: Columbia, MO, USA, 2014.
2. Thomas, A.L.; Byers, P.L.; Avery, J.D., Jr.; Kaps, M.; Gu, S. Horticultural Performance of Eight American Elderberry Genotypes at Three Missouri Locations. *Acta Hortic.* **2015**, *1061*, 237–244. [CrossRef]

3. Lee, J.; Finn, C.E. Anthocyanins and other polyphenolics in American elderberry (*Sambucus canadensis*) and European elderberry (*S. nigra*) cultivars. *J. Sci. Food Agric.* **2007**, *87*, 2665–2675. [CrossRef]
4. Moerman, D.E. *Native American Ethnobotany*; Timber Press: Portland, OR, USA, 2002.
5. Thomas, A.L.; Byers, P.L.; Vincent, P.L.; Applequist, W.L. Medicinal Attributes of American Elderberry. In *Medicinal and Aromatic Plants of North America*; Máthé, Á., Ed.; Springer International Publishing: Cham, Switzerland, 2020; pp. 119–139.
6. Elderberry: Plant Profile. Available online: http://stitchandboots.com/kitchen-garden/fruit/elderberry-plant-profile/ (accessed on 22 December 2020).
7. Młynarczyk, K.; Walkowiak-Tomczak, D.; Łysiak, G.P. Bioactive properties of *Sambucus nigra* L. as a functional ingredient for food and pharmaceutical industry. *J. Funct. Foods* **2018**, *40*, 377–390. [CrossRef]
8. Barak, V.; Birkenfeld, S.; Halperin, T.; Kalickman, I. The effect of herbal remedies on the production of human inflammatory and anti-inflammatory cytokines. *Isr. Med. Assoc. J.* **2002**, *4*, 919–922.
9. Roschek, B., Jr.; Fink, R.C.; McMichael, M.D.; Li, D.; Alberte, R.S. Elderberry flavonoids bind to and prevent H1N1 infection in vitro. *Phytochemistry* **2009**, *70*, 1255–1261. [CrossRef] [PubMed]
10. Uncini Manganelli, R.E.; Zaccaro, L.; Tomei, P.E. Antiviral activity in vitro of *Urtica dioica* L., *Parietaria diffusa* M. et K. and *Sambucus nigra* L. *J. Ethnopharmacol.* **2005**, *98*, 323–327. [CrossRef] [PubMed]
11. Wu, H.; Johnson, M.C.; Lu, C.H.; Fritsche, K.L.; Thomas, A.L.; Cai, Z.; Greenlief, C.M. Determination of Anthocyanins and Total Polyphenols in a Variety of Elderberry Juices by UPLC-MS/MS and Other Methods. *Acta Hortic.* **2015**, *1061*, 43–51. [CrossRef]
12. Mohammadsadeghi, S.; Malekpour, A.; Zahedi, S.; Eskandari, F. The Antimicrobial Activity of Elderberry (*Sambucus nigra* L.) Extract Against Gram Positive Bacteria, Gram Negative Bacteria and Yeast. *Res. J. Appl. Sci.* **2013**, *8*, 240–243.
13. Antolak, H.; Czyzowska, A.; Kregiel, D. Antibacterial and Antiadhesive Activities of Extracts from Edible Plants against Soft Drink Spoilage by *Asaia* spp. *J. Food Prot.* **2017**, *80*, 25–34. [CrossRef] [PubMed]
14. Werlein, H.D.; Kütemeyer, C.; Schatton, G.; Hubbermann, E.M.; Schwarz, K. Influence of elderberry and blackcurrant concentrates on the growth of microorganisms. *Food Control* **2005**, *16*, 729–733. [CrossRef]
15. Mohebalian, P.M.; Aguilar, F.X.; Cernusca, M.M. Conjoint Analysis of U.S. Consumers' Preference for Elderberry Jelly and Juice Products. *HortScience* **2013**, *48*, 338–346. [CrossRef]
16. Smith, T.; May, G.; Eckl, V.; Reynolds, C.M. US Sales of Herbal Supplements Increase by 8.6% in 2019. *HerbalGram* **2020**, *127*, 54–69.
17. Bolarinwa, I.F.; Orfila, C.; Morgan, M.R. Determination of amygdalin in apple seeds, fresh apples and processed apple juices. *Food Chem.* **2015**, *170*, 437–442. [CrossRef] [PubMed]
18. Mazza, G.; Cottrell, T. Carotenoids and cyanogenic glucosides in saskatoon berries (*Amelanchier alnifolia* Nutt). *J. Food Compos. Anal.* **2008**, *21*, 249–254. [CrossRef]
19. Bolarinwa, I.F.; Orfila, C.; Morgan, M.R. Amygdalin content of seeds, kernels and food products commercially-available in the UK. *Food Chem.* **2014**, *152*, 133–139. [CrossRef]
20. Buhrmester, R.A.; Ebinger, J.E.; Seigler, D.S. Sambunigrin and cyanogenic variability in populations of *Sambucus canadensis* L. (*Caprifoliaceae*). *Biochem. Syst. Ecol.* **2000**, *28*, 689–695. [CrossRef]
21. Ganjewala, D. Advances in cyanogenic glycosides biosynthesis and analyses in plants: A review. *Acta Biol. Szeged.* **2010**, *54*, 1–14.
22. Vetter, J. Plant cyanogenic glycosides. *Toxicon* **2000**, *38*, 11–36. [CrossRef]
23. Abraham, K.; Buhrke, T.; Lampen, A. Bioavailability of cyanide after consumption of a single meal of foods containing high levels of cyanogenic glycosides: A crossover study in humans. *Arch. Toxicol.* **2016**, *90*, 559–574. [CrossRef]
24. Lee, J.; Zhang, G.; Wood, E.; Rogel Castillo, C.; Mitchell, A.E. Quantification of amygdalin in nonbitter, semibitter, and bitter almonds (*Prunus dulcis*) by UHPLC-(ESI)QqQ MS/MS. *J. Agric. Food Chem.* **2013**, *61*, 7754–7759. [CrossRef]
25. Senica, M.; Stampar, F.; Veberic, R.; Mikulic-Petkovsek, M. Transition of phenolics and cyanogenic glycosides from apricot and cherry fruit kernels into liqueur. *Food Chem.* **2016**, *203*, 483–490. [CrossRef] [PubMed]
26. Sánchez-Pérez, R.; Howad, W.; Garcia-Mas, J.; Arús, P.; Martínez-Gómez, P.; Dicenta, F. Molecular markers for kernel bitterness in almond. *Tree Genet. Genomes* **2010**, *6*, 237–245. [CrossRef]
27. Zagrobelny, M.; Bak, S.; Rasmussen, A.V.; Jorgensen, B.; Naumann, C.M.; Lindberg Moller, B. Cyanogenic glucosides and plant-insect interactions. *Phytochemistry* **2004**, *65*, 293–306. [CrossRef]
28. PDQ® Integrative, A., and Complementary Therapies Editorial Board PDQ Laetrile/Amygdalin. Available online: https://www.cancer.gov/about-cancer/treatment/cam/hp/laetrile-pdq (accessed on 22 December 2020).
29. Sarker, S.D.; Nahar, L. *Chemistry for Pharmacy Students General, Organic, and Natural Produce Chemistry*; John Wiley and Sons: Chichester, UK, 2007.
30. Burns, A.E.; Bradbury, J.H.; Cavagnaro, T.R.; Gleadow, R.M. Total cyanide content of cassava food products in Australia. *J. Food Compos. Anal.* **2012**, *25*, 79–82. [CrossRef]
31. Geller, R.J.; Barthold, C.; Saiers, J.A.; Hall, A.H. Pediatric cyanide poisoning: Causes, manifestations, management, and unmet needs. *Pediatrics* **2006**, *118*, 2146–2158. [CrossRef]
32. Sahin, S. Cyanide Poisoning in a Children Caused by Apricot Seeds. *J. Health Med. Informat.* **2011**, *2*, 1–2.
33. Sanchez-Verlaan, P.; Geeraerts, T.; Buys, S.; Riu-Poulenc, B.; Cabot, C.; Fourcade, O.; Megarbane, B.; Genestal, M. An unusual cause of severe lactic acidosis: Cyanide poisoning after bitter almond ingestion. *Intensive Care Med.* **2011**, *37*, 168–169. [CrossRef]
34. Akintonwa, A.; Tunwashe, O.L. Fatal cyanide poisoning from *Cassava*-based meal. *Hum. Exp. Toxicol.* **1992**, *11*, 47–49. [CrossRef]

35. MMWRCDC. Poisoning from Elderberry Juice-California. Available online: https://www.cdc.gov/mmwr/preview/mmwrhtml/00000311.htm (accessed on 22 December 2020).
36. Senica, M.; Stampar, F.; Veberic, R.; Mikulic-Petkovsek, M. The higher the better? Differences in phenolics and cyanogenic glycosides in Sambucus nigra leaves, flowers and berries from different altitudes. *J. Sci Food Agric.* **2017**, *97*, 2623–2632. [CrossRef]
37. Koss-Mikołajczyk, I.; Lewandowska, A.; Pilipczuk, T.; Kusznierewicz, B.; Bartoszek, A. Composition of bioactive secondary metabolites and mutagenicity of *Sambucus nigra L.* Fruit at different stages of ripeness. *Acta Alimentaria* **2016**, *45*, 442–451. [CrossRef]
38. Gleadow, R.M.; Moldrup, M.E.; O'Donnell, N.H.; Stuart, P.N. Drying and processing protocols affect the quantification of cyanogenic glucosides in forage sorghum. *J. Sci Food Agric.* **2012**, *92*, 2234–2238. [CrossRef] [PubMed]
39. Montagnac, J.A.; Davis, C.R.; Tanumihardjo, S.A. Processing Techniques to Reduce Toxicity and Antinutrients of *Cassava* for Use as a Staple Food. *Comp. Rev. Food Sci. Food Saf.* **2009**, *8*, 17–27. [CrossRef]
40. Senica, M.; Stampar, F.; Veberic, R.; Mikulic-Petkovsek, M. Processed elderberry (*Sambucus nigra L.*) products: A beneficial or harmful food alternative? *LWT* **2016**, *72*, 182–188. [CrossRef]
41. Akande, K.E.; Fabiyi, E.F. Effect of Processing Methods on Some Antinutritional Factors in Legume Seeds for Poultry Feeding. *Int. J. Poult. Sci.* **2010**, *9*, 996–1001. [CrossRef]
42. Brinker, A.M.; Seigeler, D.S. Determination of cyanide and cyanogenic glycosides from plants. In *Plant Toxin Analysis*; Linskens, H.F., Jackson, J.F., Eds.; Springer: Berlin, Germany, 1992; Volume 13, pp. 359–381.
43. *Analyzing Food for Nutrition Labeling and Hazardous Contaminants*, 1st ed.; Jeon, I.; Ikins, W.G. (Eds.) CRC Press: Boca Raton, FL, USA, 1994.
44. Poulton, J.E. Cyanogenic compounds in plants and their toxic effects. In *Handbook of Natural Toxins*; Keeler, R.F., Tu, W.T., Eds.; Marcel Dekker: New York, NY, USA, 1983; Volume 1, pp. 117–160.
45. Thomas, A.L.; Byers, P.L.; Gu, S.; Avery, J.D., Jr.; Kaps, M.; Datta, A.; Fernando, L.; Grossi, P.; Rottinghaus, G.E. Occurrence of Polyphenols, Organic Acids, and Sugars among Diverse Elderberry Genotypes Grown in Three Missouri (USA) Locations. *Acta Hortic.* **2015**, *1061*, 147–154. [CrossRef]
46. Douchioiou, G.; Pui, A.; Danac, R.; Basa, C.; Murariu, M. Improved Spectrophotometric Assay of Cyanide with Picric Acid. *Rev. Roum. Chim.* **2003**, *48*, 601–606.
47. Bradbury, M.G.; Egan, S.V.; Bradbury, J.H. Picrate paper kits for determination of total cyanogens in *Cassava* roots and all forms of cyanogens in *Cassava* products. *J. Sci. Food Agric.* **1999**, *79*, 593–601. [CrossRef]
48. Bergmeyer, H.U. *Methods of Enzymatic Analysis*, 2nd ed.; Elsevier Academic Press: New York, NY, USA, 1974.
49. Zahmanov, G.; Alipieva, K.; Simova, S.; Georgiev, M.I. Metabolic differentiations of dwarf elder by NMR-based metabolomics. *Phytochem. Lett.* **2015**, *11*, 404–409. [CrossRef]
50. Drochioiu, G.; Arsene, C.; Murariu, M.; Oniscu, C. Analysis of cyanogens with resorcinol and picrate. *Food Chem. Toxicol.* **2008**, *46*, 3540–3545. [CrossRef]
51. Miller, R.E.; Tuck, K.L. Reports on the distribution of aromatic cyanogenic glycosides in Australian tropical rainforest tree species of the *Lauraceae* and *Sapindaceae*. *Phytochemistry* **2013**, *92*, 146–152. [CrossRef]
52. Gonzalez, O.; Blanco, M.E.; Iriarte, G.; Bartolome, L.; Maguregui, M.I.; Alonso, R.M. Bioanalytical chromatographic method validation according to current regulations, with a special focus on the non-well defined parameters limit of quantification, robustness and matrix effect. *J. Chromatogr. A* **2014**, *1353*, 10–27. [CrossRef]
53. Matuszewski, B.K.; Constanzer, M.L.; Chavez-Eng, C.M. Strategies for the Assessment of Matrix Effect in Quantitative Bioanalytical Methods Based on HPLC−MS/MS. *Anal. Chem.* **2003**, *75*, 3019–3030. [CrossRef]
54. Borron, S.W. Recognition and treatment of acute cyanide poisoning. *J. Emerg. Nurs.* **2006**, *32* (Suppl. 4), S12–S18. [CrossRef]

Article

Subcritical Water Extraction of *Salvia miltiorrhiza*

Brahmam Kapalavavi [1], Ninad Doctor [1], Baohong Zhang [2] and Yu Yang [1,*]

[1] Department of Chemistry, East Carolina University, Greenville, NC 27858, USA; brahmam.kapalavavi@pfizer.com (B.K.); ninad11@hotmail.com (N.D.)
[2] Department of Biology, East Carolina University, Greenville, NC 27858, USA; zhangb@ecu.edu
* Correspondence: yangy@ecu.edu; Fax: +1-252-328-6210

Abstract: In this work, a green extraction technique, subcritical water extraction (SBWE), was employed to extract active pharmaceutical ingredients (APIs) from an important Chinese medicinal herb, *Salvia miltiorrhiza* (danshen), at various temperatures. The APIs included tanshinone I, tanshinone IIA, protocatechualdehyde, caffeic acid, and ferulic acid. Traditional herbal decoction (THD) of *Salvia miltiorrhiza* was also carried out for comparison purposes. Reproduction assay of herbal extracts obtained by both SBWE and THD were then conducted on *Caenorhabditis elegans* so that SBWE conditions could be optimized for the purpose of developing efficacious herbal medicine from *Salvia miltiorrhiza*. The extraction efficiency was mostly enhanced with increasing extraction temperature. The quantity of tanshinone I in the herbal extract obtained by SBWE at 150 °C was 370-fold higher than that achieved by THD extraction. Reproduction evaluation revealed that the worm reproduction rate decreased and the reproduction inhibition rate increased with elevated SBWE temperatures. Most importantly, the reproduction inhibition rate of the SBWE herbal extracts obtained at all four temperatures investigated was higher than that of traditional herbal decoction extracts. The results of this work show that there are several benefits of subcritical water extraction of medicinal herbs over other existing herbal medicine preparation techniques. Compared to THD, the thousand-year-old and yet still popular herbal preparation method used in herbal medicine, subcritical water extraction is conducted in a closed system where no loss of volatile active pharmaceutical ingredients occurs, although analyte degradation may happen at higher temperatures. Temperature optimization in SBWE makes it possible to be more efficient in extracting APIs from medicinal herbs than the THD method. Compared to other industrial processes of producing herbal medicine, subcritical water extraction eliminates toxic organic solvents. Thus, subcritical water extraction is not only environmentally friendly but also produces safer herbal medicine for patients.

Keywords: active pharmaceutical ingredients; reproduction; medicinal herbs; *Salvia miltiorrhiza*; subcritical water extraction

1. Introduction

Due to its green nature and low side effects, herbal medicine has gained greater attention in the Western world nowadays [1–3]. Both raw and preprepared herbal medicines are available in many developed countries [4,5].

The traditional way for patients to take the herbal medicine prescribed by doctors is to cook the medicinal herbs in boiling water for 60 to 90 min and then drink the "soup medicine". This herbal medicine preparation method is called traditional herbal decoction (THD). Although this herbal decoction method has been used since ancient times, there are several major drawbacks associated with it. Firstly, a large portion of the volatile active pharmaceutical ingredients (APIs) contained in medicinal herbs are lost during the cooking process with boiling water. This is because the decoction process is an open system, and volatile APIs are thus lost to the atmosphere as vapor. Secondly, some APIs contained in medicinal herbs may be degraded due to the prolonged cooking time of 60 to 90 min. In both cases, the effectiveness of THD extracts, the soup medicine, in treating diseases may be decreased due to the reduced quantity of APIs in the herbal medicine as a result of them being lost to vapor or by degradation. Concurrently, even as APIs are lost, compounds with detrimental health effects may be extracted during the lengthy THD process. The presence of such toxicants in medicinal herbal extracts may not be safe for patient use. Lastly, it would be a rare coincidence for 100 °C to be the best temperature for effective extraction of all APIs from medicinal herbs. Proper scientific investigation of other temperatures may yield more potent yet safer herbal medicine.

Several other methods have been used for extraction of herbs and plants, including Soxhlet extraction, sonication, pressurized liquid extraction, accelerated solvent extraction, microwave-assisted extraction, and sub- and supercritical fluid extraction [6–10]. Because organic solvents are used in most of these extraction techniques, such as Soxhlet and sonication extractions, they are not suitable for preparing herbal medicine due to the toxicity of organic solvents.

Herbal extracts, such as small bags of medicinal herb extracts, are prepared by large-scale THD for patients so that they can take them directly without having to cook the herbs. This preprepared herbal medicine has gained popularity due to its convenience. Other forms of preprepared herbal medicines, such as tablets, capsules, and instant beverages, are also available commercially. While these products provide convenience to the consumer, their production via commonly used industrial extraction techniques is taking its toll on the environment and perhaps even on the patients. These techniques include maceration, vertical or turbo extraction, ultrasonic extraction, percolation, and counter current extraction. Many of the organic solvents required for use in these herbal extraction methods are toxic, and some are even carcinogenic [11]. The solvents required in these herbal preparation processes are costly not only to purchase but also for its waste disposal. Overall, such harsh extraction methods carry risks for the consumer and the environment, making them principally at odds with the perceived desire of the consumer who is likely looking for natural remedies rather than pollution-causing industrial processes and persistent trace carcinogens.

A scientifically rigorous path for modernization of herbal medicine preparation techniques is of great interest. It is important to not simply mimic THD but also to improve the efficacy of herbal medicines than those prepared by THD. This leads to this research, subcritical water extraction (SBWE) of medicinal herbs. Subcritical water refers to high-temperature and high-pressure water under conditions lower than the critical point of water: 374 °C and 218 atm. Water at elevated temperatures acts like an organic solvent due to its weakened hydrogen bonds and decreased polarity [12,13]. The solubility of organic compounds such as APIs in medicinal herbs is dramatically enhanced by simply increasing the water temperature. This unique characteristic of high-temperature water makes it an alternative mobile phase solvent for reversed-phase liquid chromatography [13–16] and an excellent extraction fluid for efficient removal of organics from various sample matrices, including plants and medicinal herbs [17–23]. Because different temperatures can be employed to carry out subcritical water extractions, there will be an optimized temperature that yields the highest quantity of APIs and in turn produces the most potent herbal medicine. Ideally, the solvent for extraction of medicinal herbs should be nontoxic, and the extraction technique should be more efficient in extracting active pharmaceutical ingredients and not cause their significant loss during the extraction process. Thus, subcritical water is an excellent choice for preparing herbal medicines.

In order to evaluate and optimize the SBWE technique, *Salvia miltiorrhiza* (also known as danshen in Chinese), a popular and important herb prescribed in traditional Chinese medicine (TCM), was used in this study. *Salvia miltiorrhiza* is a perennial plant in the genus *Salvia* of the mint family. Its roots are highly valued in traditional Chinese medicine and used in the treatment of various diseases, such as blood circulation, cardiovascular, and hepatic diseases [24–26]. Researchers have isolated about 70 compounds from the extract of *Salvia miltiorrhiza* [27]. Some of the identified anticancer compounds present in *Salvia miltiorrhiza* include tanshinone I, tanshinone IIA, protocatechualdehyde, caffeic acid, and ferulic acid. These APIs have already been found to demonstrate antiproliferative effect on various cancer cells, such as colon, leukemia, lung, and breast cancers, at either pre-clinical or clinical level [28–31]. Therefore, these five APIs were investigated in this study.

The main goal of this work was to investigate a potential herbal medicine preparation technique using subcritical water to yield efficacious herbal medicine. Therefore, subcritical water extraction of *Salvia miltiorrhiza* roots was carried out at four different temperatures (75, 100, 125, and 150 °C). For comparison and evaluation purposes, traditional herbal decoction of *Salvia miltiorrhiza* was also conducted. Then, these herbal extracts were characterized using GC/MS and HPLC to identify and quantify various anticancer agents. In order to evaluate the efficacy of the SBWE herbal extracts at various temperature conditions, the reproduction assay of SBWE and THD herbal extracts were conducted on *Caenorhabditis elegans*.

Despite being a simple multicellular organism, *Caenorhabditis elegans* has been widely employed to study complex behavior and syndromes. It has been used in many recent studies to understand human diseases, including cancer, ageing, development, addiction, and neurodegenerative diseases, as well as in pharmaceutical and toxicity studies [32–36]. Research on the worm bridges the gap between in vitro systems and preclinical studies in mammalian models. Experiments using cell lines often do not represent organism-level responses. On the other hand, *Caenorhabditis elegans* is particularly useful for reverse genetic approaches due to its short life cycle, availability of strains and feasibility of customized mutants, ability to perform complex behavior, and transparent cuticle for imaging assays. In this work, we employed *Caenorhabditis elegans* as a model system to investigate the drug potency of the extracted APIs from *Salvia miltiorrhiza* using reproduction analysis.

It is a novel approach to employ subcritical water for extraction of active pharmaceutical ingredients from medicinal herbs. Compared to THD, the thousand-year-old and yet still popular herbal preparation method, subcritical water extraction is conducted in a closed system. Therefore, no loss of volatile active pharmaceutical ingredients occurs due to loss of APIs to the open environment. However, loss of analytes can still occur due to degradation at elevated temperatures. Under optimized temperature, SBWE is more efficient in extracting APIs from medicinal herbs than the THD method. Compared to other industrial processes of producing herbal medicine, subcritical water extraction eliminates toxic organic solvents. Therefore, subcritical water extraction is not only more efficient and cheaper but also environment friendly because of its green nature. Many researchers have made efforts in recent years to develop greener analytical chemistry techniques. For example, an analytical Eco-Scale has been proposed as a tool for green analysis evaluation [37]. Another new tool introduced for assessment of the green character of analytical procedures is the Green Analytical Procedure Index [38]. The work reported in this paper also contributes to the field of green analytical chemistry.

2. Results and Discussion

2.1. Subcritical Water Extraction of Salvia Miltiorrhiza

As stated later in the Materials and Methods section, the quantification of all five APIs was achieved using a standard HPLC method. The concentration of the calibration solutions ranged from 0.002 to 1.00 mg/mL. The detection limit was 0.0002 mg/mL. The correlation coefficient (r^2) ranged from 0.999 to 1.00.

A recovery study on the SBWE method was conducted using spiked samples (known amount of APIs) to validate the homemade SBWE system. The recoveries of the five APIs investigated in this work ranged from 95 to 102%, similar to that achieved in our previous study on SBWE of vanillin and coumarin [17]. This shows that the SBWE system is reliable. The subcritical extraction of *Salvia miltiorrhiza* was conducted at four different temperatures of 75, 100, 125, and 150 °C. Then, the SBWE extracts were characterized using GC/MS. Various analytes in the herbal extracts were identified by GC/MS by matching both GC retention times and mass spectra of standard samples. Among the identified analytes, five of them were anticancer agents: protocatechualdehyde, caffeic acid, ferulic acid, tanshinone

I, and tanshinone IIA. Figure 1 shows the elution of the five compounds with an internal standard on GC/MS.

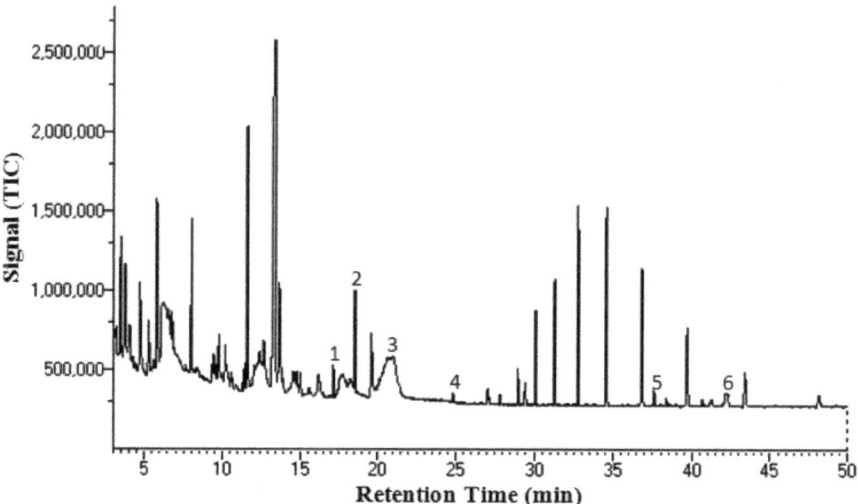

Figure 1. Total ion GC/MS chromatogram of a *Salvia miltiorrhiza* herbal extract obtained by subcritical water extraction (SBWE) at 150 °C for 30 min. Peak identification: 1, protocatechualdehyde; 2, propyl paraben (internal standard); 3, caffeic acid; 4, ferulic acid; 5, tanshinone IIA; and 6, tanshinone I.

Figure 2 shows the HPLC separation of a standard solution (Figure 2a), methylene chloride phase after liquid–liquid extraction of an herbal extract obtained by SBWE at 125 °C (Figure 2b), and water phase (methanol was added) of an herbal extract obtained by SBWE at 125 °C (Figure 2c). As one can see, all five analytes and the internal standard were well separated.

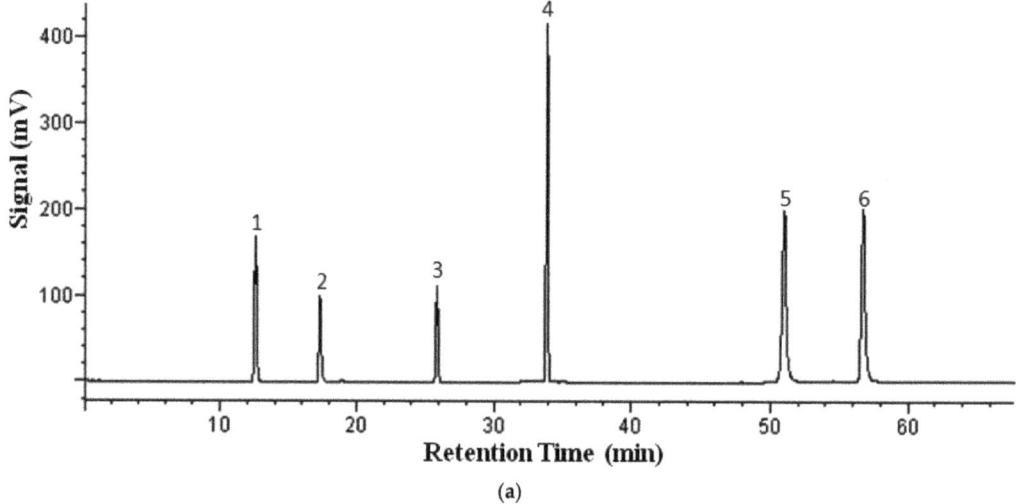

(a)

Figure 2. *Cont.*

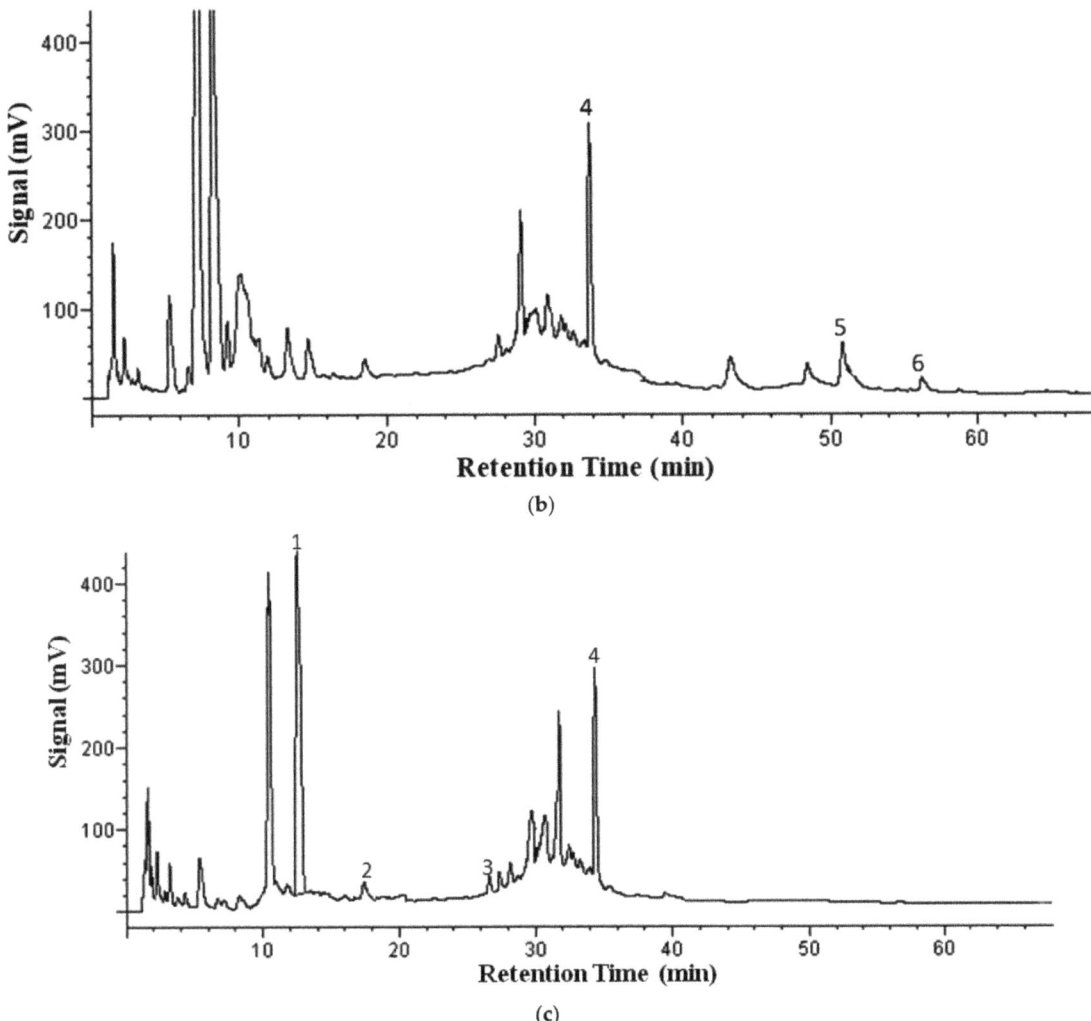

Figure 2. HPLC chromatograms of *Salvia miltiorrhiza* herbal extract obtained with 125 °C extraction temperature and evaluated in the Alltech Adsorbosil C18 column at ambient temperature. (**a**) Analyte standard solution; (**b**) methylene chloride phase; (**c**) water phase. Flow rate: 1.0 mL/min. UV detection: 254 nm. Mobile phase: A, 100 mM phosphoric acid in water; B, 100% methanol. Gradient: 0–4 min, 2% methanol; 4–8 min, 2–10% methanol; 8–23 min, 10–30% methanol; 23–32 min, 30–60% methanol; 32–43 min, 60% methanol; 43–49 min, 60–70% methanol; 49–61 min, 70–80% methanol; and 61–68 min, 80–2% methanol. Peak identification: 1, protocatechualdehyde; 2, caffeic acid; 3, ferulic acid; 4, propyl paraben; 5, tanshinone I; and 6, tanshinone IIA.

Table 1 shows the quantification results of the five analytes present in the SBWE herbal extracts obtained at four different temperatures of 75, 100, 125, and 150 °C. The quantification results indicate that the protocatechualdehyde quantity extracted increased by 2-fold with the increase of extraction temperature from 75 to 100 °C and by 24-fold with further increase of extraction temperature from 100 to 125 °C. Then, with further increase of temperature from 125 to 150 °C, the extracted protocatechualdehyde quantity was enhanced 2.5-fold. There was no significant temperature effect on extraction efficiency of caffeic acid in the temperature range of 75 to 125 °C. However, the caffeic acid quantity found in the extract decreased at 150 °C due to possible degradation at such a high temperature. Ferulic acid was not detected at 75 and 100 °C, while its quantity extracted was improved by 32-fold when the temperature increased from 125 to 150 °C. The extraction efficiency, measured by analyte concentration in herbal extracts, of the two tanshinone compounds was clearly enhanced with increasing temperature, as shown in Table 1.

Table 1. Comparison of active pharmaceutical ingredient (API) concentrations found in *Salvia miltiorrhiza* obtained by traditional herbal decoction and subcritical water extraction.

Analyte	Concentration, μg/g (%RSD)[a]				
	Traditional Herbal Decoction, 100 °C	Subcritical Water Extraction			
		75 °C	100 °C	125 °C	150 °C
Protocatechualdehyde	19.6 (12.7)	11.4 (16.9)	29.1 (9.37)	701 (9.88)	1760 (5.26)
Caffeic Acid	51.6 (15.8)	47.3 (5.35)	57.3 (10.4)	48.2 (3.12)	16.1 (1)
Ferulic Acid	ND[b]	ND	ND	1.30 (20.3)	41.6 (13.9)
Tanshinone I	0.2 (10)	4.0 (2.4)	5.8 (7.4)	19.1 (13.4)	74.0 (3.43)
Tanshinone IIA	0.8 (20.1)	3.3 (1.3)	5.0 (13)	5.18 (16.1)	15.3 (12.7)

[a] Triplicate measurements. [b] Not detected.

Table 1 also includes the quantities of the five analytes found in the THD extracts. One can easily see that tanshinone concentrations obtained by SBWE at all temperatures were much higher than those achieved by THD extractions. Specifically, tanshinone I concentration achieved by SBWE at 150 °C was 370-fold higher than that obtained by THD extraction, as demonstrated in Table 1.

We conducted the *t*-test on API concentrations in both THD and SBWE extracts. Our statistical analysis revealed that tanshinone I and tanshinone IIA concentrations obtained by THD at 100 °C and by SBWE at all four elevated temperatures were significantly different beyond the 99.9% confidence level. While there were no differences between caffeic acid concentrations achieved by THD at 100 °C and SBWE at 75–125 °C, the concentrations of caffeic acid obtained by THD and SBWE at 150 °C were significantly different at the 99.5% confidence level. Protocatechualdehyde concentrations achieved by THD at 100 °C and SBWE at all four temperatures were significantly different, mostly beyond the 99% confidence level.

2.2. Reproduction Assay of Caenorhabditis Elegans

First, we studied the impact of different concentrations (2, 10, and 50 times dilution with deionized water) of the herbal extract on *Caenorhabditis elegans* mortality. *Salvia miltiorrhiza* herbal extract was obtained by SBWE at 150 °C. After 30 h exposure to the three different diluted SBWE herbal extracts, the 10 times diluted herbal extract showed higher mortality rate than the other diluted herbal extracts. Therefore, the 10 times dilution factor was chosen for the remainder of the reproduction study. The API concentrations used for the reproduction assay are given in Table 2.

Table 2. Concentration of API in *Salvia miltiorrhiza* extracts used for reproduction study.

Analyte	Concentration, μg/mL	
	Traditional Herbal Decoction, 100 °C	Subcritical Water Extraction at 150 °C
Protocatechualdehyde	0.392	352
Caffeic Acid	1.03	3.22
Ferulic Acid	ND[a]	8.32
Tanshinone I	0.004	1.48
Tanshinone IIA	0.016	0.306

[a] Not detected.

In order to optimize the preparation conditions of efficacious herbal medication through subcritical water extraction, the reproduction inhibition of the SBWE herbal extracts obtained at four different temperatures (75, 100, 125, and 150 °C) was evaluated on *Caenorhabditis elegans*. All SBWE herbal extracts were diluted 10 times with deionized water. Table 3 shows the reproduction assay of *Caenorhabditis elegans* after 30 h exposure to the 10 times diluted SBWE water extracts obtained at 75 to 150 °C. The reproduction inhibition of the extracts increased with higher extraction temperature. The SBWE extraction temperature also influenced mortality, as shown in Table 3. In general, the worm survival rate decreased with the increase in extraction temperature except at 150 °C. The main reason for lower mortality of *Caenorhabditis elegans* with the herbal extract obtained at 150 °C may be attributed to the less intake of highly concentrated herbal extract through the skin of *Caenorhabditis elegans*. Another reason for the lower mortality of worms may be due to the degradation of compounds associated with the mortality of worms, such as caffeic acid, at 150 °C.

Table 3. Percentage reproduction inhibition and mortality of *Caenorhabditis elegans* after 30 h exposure to the 10 times diluted traditional herbal decoction and subcritical water extractions of *Salvia miltiorrhiza* at 75 to 150 °C.

Treatment NGM Plates	3-Day Average Production[a] (%RSD)[b]	%Reproduction Inhibition (%RSD)[b]	%Mortality (%RSD)[b]
Control	111 (4.13)	0	0
THD	106 (5.35)	5 (20)	10 (20)
SBWE at 75 °C	104 (3.47)	6 (40)	17 (16)
SBWE at 100 °C	89 (12)	20 (20)	33 (15)
SBWE at 125 °C	67 (23)	40 (18)	42 (14)
SBWE at 150 °C	60 (34)	46 (12)	14 (21)

[a] Total of eggs and larva average per worm over three days. [b] Five replicates.

The reproduction assay of *Caenorhabditis elegans* was also carried out using the THD extract of *Salvia miltiorrhiza* for comparison purposes. Both reproduction inhibition and mortality achieved by SBWE extracts at 100 °C and above were higher than those obtained by traditional herbal decoction, as shown in Table 3. The reproduction inhibition results indicate that SBWE is a much more efficient extraction technique than traditional herbal decoction, and it may be used to develop efficacious herbal medicine in the future.

We also carried out the *t*-test on reproduction assay results. Our statistical analysis showed that there was no difference in average reproduction and reproduction inhibition between the herbal extracts obtained by THD at 100 °C and SBWE at 75 °C, while there were significant differences between the extracts of THD 100 °C and SBWE at the other three elevated temperatures, mostly beyond the 99.9% confidence level. There were significant differences in mortality for the herbal extracts obtained by THD at 100 °C and SBWE at all four temperatures, mostly beyond the 99% confidence level.

3. Materials and Methods

3.1. Reagents and Supplies

Tanshinone I and tanshinone IIA were obtained from LKT Laboratories, Inc. (St. Paul, MN, USA). Protocatechualdehyde, caffeic acid, ferulic acid, sodium chloride, sodium hydroxide, agar, cholesterol, calcium chloride, calcium chloride dehydrate, and sodium phosphate dibasic heptahydrate were purchased from Sigma Aldrich (St. Louis, MO, USA). Sand, peptone, tryptone, magnesium sulfate, and magnesium sulfate heptahydrate were acquired from Fisher Scientific (Fair Lawn, NJ, USA). Potassium phosphate, dipotassium phosphate, yeast extract, and HPLC-grade methanol were purchased from Alfa Aesar (Ward Hill, MA, USA). Methylene chloride was obtained from Acros Organics (Fair Lawn, NJ, USA). Top Job bleaching solution was obtained from the local store. Deionized water (18 MΩ-cm) was prepared in our laboratory using a Purelab Ultra system from ELGA (Lowell, MA, USA). GD/X PVDF membrane filters (0.45 μm) were acquired from Whatman

(Florham Park, NJ, USA). Strata SPE silica-2 sample (3 mL) tubes were received from Phenomenex (Torrance, CA, USA). Petri dishes (6 cm) were obtained from BD Falcon (Franklin Lakes, NJ, USA). Alltech Adsorbosil C18 column (4.6 × 150 mm, 5 μm) was purchased from Alltech Associates, Inc. (Deerfield, IL, USA). An Empty stainless steel tube (5 × 1.00 cm I.D. with 1.27 cm O.D.) and end fittings were received from Chrom Tech, Inc. (Apple Valley, MN, USA). OP50 and *Caenorhabditis elegans* N2 Bristol wild-type worm were obtained from *Caenorhabditis* Genetics Center (University of Minnesota, Minneapolis, MN, USA).

3.2. Preparation of Solutions

Propyl paraben was used as an internal standard. This solution was prepared by adding 0.0500 g of propyl paraben to a 50 mL volumetric flask and diluted to the mark with methanol. A stock solution was prepared by adding 0.00020 g each of tanshinone I and tanshinone IIA to a 10 mL volumetric flask. Then, 4.00 mL of dichloromethane was added into the volumetric flask. The volumetric flask was vortexed to obtain a homogeneous mixture. Then, 0.0100 g each of protocatechualdehyde, caffeic acid, and ferulic acid were added to the same volumetric flask and diluted to the mark with methanol. Calibrated standard solutions were prepared using both stock and internal standard solutions. The concentrations of the calibration solutions ranged from 0.002 to 1.00 mg/mL.

3.3. Subcritical Water Extraction of Salvia Miltiorrhiza

The extraction of *Salvia miltiorrhiza* was carried out using a home-made subcritical water extraction system, as shown in Figure 3. Both end fittings of a stainless steel extraction vessel were wrapped with Teflon tape for proper sealing. One end of the vessel was sealed with an end fitting first. Approximately 2 g (the actual weight was recorded to four decimal place) of *Salvia miltiorrhiza* (finely cut to small pieces, a few millimeters in length) was added to the stainless steel vessel. The void space of extraction vessel was filled with precleaned sand. The other end of the stainless steel vessel was then sealed with another end fitting. The loaded vessel was placed in an oven (HP gas chromatograph 5890 Series 2, Hewlett Packard, Avondale, PA, USA), as shown in Figure 3.

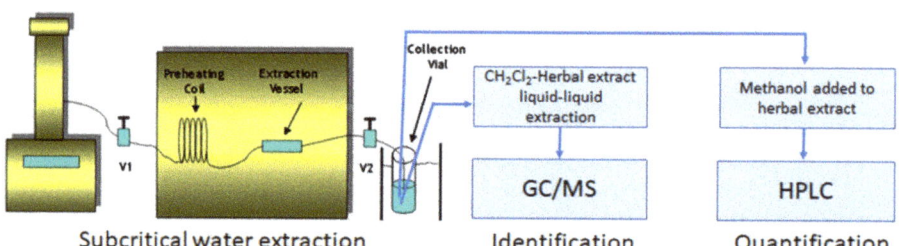

Figure 3. Block diagram of subcritical water extraction (V1 and V2 are needle valves) followed by GC/MS identification and HPLC quantification of APIs.

An ISCO model 260 D syringe pump (Lincoln, NE, USA) was used to supply 18 MΩ-cm water by opening V_1 and closing V_2 to fill the loaded vessel. Leak check of the extraction vessel was performed in the constant-pressure mode. It should be pointed out that a delay between the actual temperature of the extraction vessel and oven temperature was determined. The delay was 10 min for 75 °C, 12 min for 100 °C, 14 min for 125 °C, and 16 min for 150 °C. Static extraction was performed for 30 min after the delay time was compensated. A pressure of 15 to 25 atm was applied to keep hot water in the liquid state for all experiments. After 30 min of heating, approximately 10 mL of herbal extract was collected at 1 mL/min into a 25 mL glass vial by opening V_2. Triplicate SBWE experiments were conducted at all temperatures.

3.4. Traditional Herbal Decoction of Salvia Miltiorrhiza

Approximately 2 g (the actual weight was recorded to four decimal place) of *Salvia miltiorrhiza* (finely cut to small pieces, a few millimeters in length) was added to a 50 mL glass beaker. Then, 10.00 mL of deionized water was added to it. The beaker was covered with a watch glass and heated up to boiling on a hot plate. Then, the temperature was adjusted to ensure the water kept boiling for 30 min. Triplicate THD experiments were conducted.

3.5. Sample Treatment

For characterization of SBWE water–herbal extracts on GC/MS, solid-phase extraction (SPE) was carried out using a silica phase cartridge and methanol as the elution solvent. At first, the silica cartridge was cleaned with approximately 5 mL of methanol followed by 10 mL of water. Then, the herbal extract was run through the silica cartridge and eluted using 1.00 mL of methanol into a 2 mL glass vial. Then, 30 µL of propyl paraben internal standard solution was added.

For HPLC analysis of tanshinone I and tanshinone IIA, liquid–liquid extraction was conducted. First, 1.00 mL of methylene chloride was added to each glass vial containing SBWE water–herbal extract. These vials were then sealed with aluminum-lined caps. These vials were vortexed to effectively mix the two phases. After separation of the two phases, the methylene chloride phase was transferred into an empty 5 mL glass vial. The same liquid–liquid extraction procedure for the SBWE water–herbal extract was repeated with another 1.00 mL of fresh methylene chloride. Again, the methylene chloride layer was removed and combined with the first fraction of the methylene chloride extract in the 5 mL vial. Then, 30.00 µL of propyl paraben internal standard was added to the methylene chloride phase.

To the aqueous phase of the herbal extract sample, 1.00 mL of methanol was added. Then, 300 µL of propyl paraben internal standard was added and mixed well. This sample was then filtered through a Whatman GDX filter into a glass vial for chromatographic analysis of protocatechualdehyde, caffeic acid, and ferulic acid.

3.6. HPLC Analysis

Please note that *Salvia miltiorrhiza* contains tens if not hundreds of compounds, and the five APIs investigated in this work are mixed with all other compounds in the extract. Therefore, we needed a standard method such as HPLC to achieve separation and quantification of our analytes to ensure the quality of this research. Thus, Shimadzu Nexera UFLC was employed for separation and quantification of *Salvia miltiorrhiza* extracts on an Alltech Adsorbosil C18 column using a methanol–water mixture as the mobile phase with 1.0 mL/min at ambient temperature. The eluents were detected at 254 nm.

3.7. GC/MS Analysis

In order to separate and identify the five APIs studied in this work, Agilent Technologies 6890N Network GC System (Santa Clara, CA, USA) coupled with a JEOL Ltd. JMS-GCmate II MS System (Tokyo, Japan) was employed for the characterization of SBWE extracts of *Salvia miltiorrhiza*. The GC separations were carried out on an Agilent HP-5MS (5% phenyl)-methylpolysiloxane (30 m × 0.250 mm, 0.25 µm film thickness) capillary column with 1.0 mL/min flow of a helium carrier gas. The sample volume was 1 µL, and it was injected using split mode by keeping the injector temperature at 250 °C. The GC/MS interface and the MSD ion chamber were set at 250 °C. The MS solvent delay time was 3 min. The GC oven temperature programming was as follows. The initial temperature was held at 30 °C for 3 min. Then, it was increased at 7.4 °C/min to 250 °C and maintained at 250 °C for 16 min. TSSPro Version 3.0 (Shrader Analytical and Consulting Laboratories, Inc., Detroit, Michigan, USA) was used for data acquisition and analysis.

3.8. Reproduction Studies on Caenorhabditis Elegans

A hermaphrodite, *Caenorhabditis elegans* N2 Bristol wild-type worm, was used for the reproduction assay to determine the reproduction rate of the herbal extracts. Synchronized L1 worms were cultured on NGM with *Escherichia coli* bacteria (OP50) as food. The NGM was supplemented with SBWE herbal extracts with a certain fold of dilution, which were obtained at extraction temperatures of 75, 100, 125, and 150 °C. Each treatment contained 15 worms with five biological replicates. These plates were incubated at 20 °C. After 30 h, these worms were washed off from each plate using M9 buffer into an Eppendorf tube. The tubes were then centrifuged twice with M9 buffer to wash worms of the herbal extract. Then, from each tube, about four worms were transferred to each plate already seeded with OP50 food. These plates were continuously monitored for egg laying. When worms started laying eggs, time was noted, and the plates were labeled as day 1 plates. These plates were incubated for another 24 h. The number of laid eggs was recorded for each day for three continuous days. The following equations were used to calculate reproduction inhibition and mortality rate.

$$\text{Reproduction inhibition} = \frac{\text{Control average reproduction} - \text{Herbal extract average reproduction}}{\text{Control average reproduction}} \times 100 \quad (1)$$

$$\%\text{Mortality} = \frac{\text{Number of worms died}}{\text{Total number of worms}} \times 100 \quad (2)$$

4. Conclusions

The research described in this work is different from any other existing herbal medicine preparation techniques. Unlike traditional herbal decoction, subcritical water extraction is conducted in a closed system where no loss of volatile active pharmaceutical ingredients to open environment occurs except analyte degradation at higher temperatures. Because temperatures other than 100 °C (the condition used in traditional herbal decoction) can also be employed in SBWE, subcritical water extraction at the optimized temperature should be more efficient in extracting active pharmaceutical ingredients from medicinal herbs than traditional herbal decoction. The higher extraction efficiency of SBWE should allow the subcritical water extraction time to be shortened, thus reducing the chance for degradation of the active pharmaceutical ingredients. These three factors should assure that optimized SBWE conditions would produce herbal medicine containing higher API concentrations than traditional herbal decoction. Compared with other industrial processes of making herbal medicine, subcritical water extraction eliminates toxic organic solvents. Therefore, it is not only environment friendly but also produces safer herbal medicine for patients.

Our results showed that the API quantity obtained by subcritical water extraction of *Salvia miltiorrhiza* increased by up to 4-fold by increasing the extraction temperature from 75 to 100 °C. They were then further enhanced by up to 26-fold with the increase of temperature from 100 to 125 °C, except for caffeic acid. When the extraction temperature was raised from 125 to 150 °C, API concentrations in SBWE extracts were further increased up to 4-fold, except for caffeic acid and protocatechualdehyde. Both caffeic acid and protocatechualdehyde might be degraded at 125 °C or higher. When comparing the tanshinone concentrations achieved by SBWE of *Salvia miltiorrhiza* with that obtained by THD, the SBWE extracts contained much higher tanshinone concentrations than the THD extracts.

The extraction temperature also plays an important role in the reproduction inhibition rate of the SBWE herbal extracts collected at four different temperatures. Reproduction inhibition evaluation of *Caenorhabditis elegans* revealed that the three-day average reproduction of worms decreased with increasing extraction temperature, while the reproduction inhibition rate increased from 6 to 46% when the SBWE temperature was raised from 75 to 150 °C. Please note that the reproduction inhibition of SBWE herbal extracts obtained at all temperatures from 75 to 150 °C was higher than that of traditional herbal decoction extracts.

In closing, besides subcritical water extraction being a more efficient technique than traditional herbal decoction in extracting anticancer agents from *Salvia miltiorrhiza*, SBWE herbal extracts also have higher reproduction inhibition rate than THD extracts according to our reproduction inhibition study. These findings demonstrate the potential of employing subcritical water extraction technique to develop high API-containing herbal medicine from *Salvia miltiorrhiza*.

Author Contributions: Manuscript conception, Y.Y.; writing and original draft preparation, Y.Y. and B.Z.; subcritical water extractions, B.K. and N.D.; chromatography analysis, B.K. and N.D. All authors have read and agreed to the published version of the manuscript.

Funding: This research received no external funding.

Conflicts of Interest: The authors declare no conflict of interest.

Sample Availability: Samples of *Salvia miltiorrhiza* are available from the authors.

Abbreviations

SBWE	subcritical water extraction
APIs	active pharmaceutical ingredients
THD	traditional herbal decoction

References

1. Gibson, J.E.; Taylor, D.A. Can claims, misleading information, and manufacturing issues regarding dietary supplements be improved in the United States of America? *J. Pharmacol. Exp. Ther.* **2005**, *314*, 939–944. [CrossRef]
2. Wold, R.S.; Lopez, S.T.; Yau, C.L.; Butler, L.M.; Pareo-Tubbeh, S.L.; Waters, D.L.; Garry, P.J.; Baumgartner, R.N. Increasing trends in elderly persons use of nonvitamin, nonmineral dietary supplements and concurrent use of medicines. *J. Am. Diet. Assoc.* **2005**, *105*, 54–63. [CrossRef]
3. Huang, K.C. *The Pharmacology of Chinese Herbs*; CRC Press: Boca Raton, FL, USA, 1999.
4. Taylor, D.A. Botanical Supplements: Weeding out the health risks. *Environ. Health Perspect.* **2004**, *112*, A750–A753. [CrossRef]
5. Committee on the Framework for Evaluating the Safety of Dietary Supplements; Food and Nutrition Board; Board on Life Sciences; Institute of Medicine; National Research Council of the National Academies of Science. *Dietary Supplements: A Framework for Evaluating Safety*; National Academies Press: Washington, DC, USA, 2004.
6. Ong, E.S. Extraction methods and chemical standardization of botanicals and herbal preparations. *J. Chromatogr. B* **2004**, *812*, 23–33. [CrossRef]
7. Nile, S.H.; Nile, A.S.; Keum, Y.S. Total phenolics, antioxidant, antitumor, and enzyme inhibitory activity of Indian medicinal and aromatic plants extracted with different extraction methods. *Biotech* **2017**, *7*, 76. [CrossRef]
8. Sun, J.; Zhao, X.; Dang, J.; Sun, X.; Zheng, L.; You, J.; Wang, X. Rapid and sensitive determination of phytosterols in functional foods and medicinal herbs by using UHPLC–MS/MS with microwave-assisted derivatization combined with dual ultrasound-assisted dispersive liquid–liquid microextraction. *J. Sep. Sci.* **2017**, *40*, 725–732. [CrossRef]
9. Santos, Ê.R.M.; Oliveiraa, H.N.M.; Oliveira, E.J.; Azevedoa, S.H.G.; Jesus, A.A.; Medeiros, A.M.; Darivac, C.; Sousa, E.M.B.D. Supercritical fluid extraction of *Rumex Acetosa* L. roots: Yield, composition, kinetics, bioactive evaluation and comparison with conventional techniques. *J. Supercrit. Fluids* **2017**, *122*, 1–9. [CrossRef]
10. Sánchez-Camargo, A.P.; Ibáñez, E.; Cifuentes, A.; Herrero, M. Bioactives Obtained from Plants, Seaweeds, Microalgae and Food By-Products Using Pressurized Liquid Extraction and Supercritical Fluid Extraction. *Compr. Anal. Chem.* **2017**. [CrossRef]
11. Mukherjee, P.K. *Extraction of herbal drugs. Quality Control of Herbal Drugs, An Approach to Evaluation of Botanicals*, 1st ed.; Business Horizons Pharmaceutical Publishers: New Delhi, India, 2002; pp. 379–425.
12. Yang, Y.; Belghazi, M.; Hawthorne, S.B.; Miller, D.J. Elution of organic solutes from different polarity sorbents using subcritical water conditions. *J. Chromatogr. A* **1998**, *810*, 149–159. [CrossRef]
13. Yang, Y. Subcritical water chromatography: A green approach to high-temperature liquid chromatography. *J. Sep. Sci.* **2007**, *30*, 1131–1140. [CrossRef]
14. Miller, D.J.; Hawthorne, S.B. Subcritical water chromatography with flame ionization detection. *Anal. Chem.* **1997**, *69*, 623–627. [CrossRef]
15. Smith, R.M. Superheated water chromatography—A green technology for the future. *J. Chromatogr. A* **2008**, *1184*, 441–455. [CrossRef] [PubMed]
16. Doctor, N.; Yang, Y. Separation and Analysis of Aspirin and Metformin HCl Using Green Subcritical Water Chromatography. *Molecules* **2018**, *23*, 2258. [CrossRef]
17. Doctor, N.; Parker, G.; Vang, V.; Smith, M.; Kayan, B.; Yang, Y. Stability and Extraction of Vanillin and Coumarin under Subcritical Water Conditions. *Molecules* **2020**, *25*, 1601. [CrossRef]

18. Yang, Y.; Hawthorne, S.B.; Miller, D.J. Class-selective extraction of polar, moderately-, and nonpolar organic pollutants from solid waste using subcritical water. *Environ. Sci. Technol.* **1997**, *31*, 430–437. [CrossRef]
19. Essien, S.O.; Young, B.; Baroutian, S. Recent advances in subcritical water and supercritical carbon dioxide extraction of bioactive compounds from plant materials. *Trends Food Sci. Technol.* **2020**, *97*, 156–169. [CrossRef]
20. Yang, Y.; Kayan, B.; Bozer, N.; Pate, B.; Baker, C.; Gizir, A.M. Terpene degradation and extraction from basil and oregano leaves using subcritical water. *J. Chromatogr. A* **2007**, *1152*, 262–267. [CrossRef]
21. Abdelmoumen, B.; Jaroslava, Š.; Nataša, N.; Sofiane, G.; Daoud, H. Subcritical water extraction of polyphenols from endemic Algerian plants with medicinal properties. *Acta Period. Technol.* **2020**, *51*, 191–206.
22. Kiamahalleha, M.V.; Najafpour-Darzi, G.; Rahimnejada, M.; Moghadamnia, A.A. High performance curcumin subcritical water extraction from turmeric (*Curcuma longa* L.). *J. Chromatogr. B* **2016**, *1022*, 191–198. [CrossRef]
23. Shabkhiz, M.A.; Eikani, M.H.; Sadr, Z.B.; Golmohammad, F. Superheated water extraction of glycyrrhizic acid from licorice root. *Food Chem.* **2016**, *210*, 396–401. [CrossRef]
24. Datta, P.; Dasgupta, A. Effect of Chinese medicines Chan Su and Danshen on EMIT 2000 and Randox digoxin immunoassays: Wide variation in digoxin-like immunoreactivity and magnitude of interference in digoxin measurement by different brands of the same product. *Ther. Drug Monit.* **2002**, *24*, 637–644. [CrossRef] [PubMed]
25. Yang, M.; Liu, A.; Guan, S.; Sun, J.; Xu, M.; Guo, D. Characterization of tanshinones in the roots of *Salvia miltiorrhiza* (Dan-shen) by high-performance liquid chromatography with electrospray ionization tandem mass spectrometry. *Rapid Commun. Mass Spectrom.* **2006**, *20*, 1266–1280. [CrossRef]
26. Wang, B.Q. *Salvia miltiorrhiza*: Chemical and pharmacological review of a medicinal plant. *J. Med. Plants Res.* **2010**, *4*, 2813–2820.
27. Li, Y.G.; Song, L.; Liu, M.; Wang, Z.T. Advancement in analysis of *Salviae miltiorrhizae* Radix et Rhizoma (Danshen). *J. Chromatogr. A* **2009**, *1216*, 1941–1953. [CrossRef]
28. Nizamutdinova, I.T.; Lee, G.W.; Son, K.H.; Jeon, S.J.; Kang, S.S.; Kim, Y.S.; Lee, J.H.; Seo, H.G.; Chang, K.C.; Kim, H.J. Tanshinone I effectively induces apoptosis in estrogen receptor-positive (MCF-7) and estrogen receptor-negative (MDA-MB-231) breast cancer cells. *Int. J. Oncol.* **2008**, *33*, 485–491. [CrossRef] [PubMed]
29. Jeong, J.B.; Lee, S.H. Protocatechualdehyde possesses anti-cancer activity through downregulating cyclin D1 and HDAC2 in human colorectal cancer cells. *Biochem. Biophys. Res. Commun.* **2013**, *430*, 381–386. [CrossRef]
30. Tanaka, T.; Kojima, T.; Kawamori, T.; Wang, A.; Suzui, M.; Okamoto, K.; Mori, H. Inhibition of 4-nitroquinoline-1-oxide-induced rat tongue carcinogenesis by the naturally occurring plant phenolics caffeic, ellagic, chlorogenic and ferulic acids. *Carcinogenesis* **1993**, *14*, 1321–1325. [CrossRef]
31. Wang, X.; Wei, Y.; Yuan, S.; Liu, G.; Lu, Y.; Zhang, J.; Wang, W. Potential anticancer activity of tanshinone IIA against human breast cancer. *Int. J. Cancer* **2005**, *116*, 799–807. [CrossRef] [PubMed]
32. Kirienko, N.V.; Mani, K.; Fay, D.S. Cancer models in *Caenorhabditis elegans*. *Dev. Dyn.* **2010**, *239*, 1413–1448.
33. Honda, Y.; Honda, S.; Narici, M.; Szewczyk, N.J. Space flight and ageing: Reflecting on Caenorhabditis elegans in space. *Gerontology* **2014**, *60*, 138–142. [CrossRef]
34. Hubbard, E.J.; Korta, D.Z.; Dalfo, D. Physiological control of germline development. *Adv. Exp. Med. Biol.* **2013**, *757*, 101–131. [PubMed]
35. Feng, Z.; Li, W.; Ward, A.; Piggott, B.J.; Larkspur, E.R.; Sternberg, P.W.; Xu, X.Z. A C. elegans model of nicotine-dependent behavior: Regulation by TRP-family channels. *Cell* **2006**, *127*, 621–633. [CrossRef]
36. Ewald, C.Y.; Li, C. Understanding the molecular basis of Alzheimer's disease using a Caenorhabditis elegans model system. *Brain Struct. Funct.* **2009**, *214*, 263–283. [CrossRef]
37. Gałuszka, A.; Migaszewski, Z.M.; Konieczka, P.; Namieśnik, J. Analytical Eco-Scale for assessing the greenness of analytical procedures. *Trends Anal. Chem.* **2012**, *37*, 61–72. [CrossRef]
38. Płotka-Wasylka, J. A new tool for the evaluation of the analytical procedure: Green Analytical Procedure Index. *Talanta* **2018**, *181*, 204–209. [CrossRef]

Article

An Analytical Protocol for the Differentiation and the Potentiometric Determination of Fluorine-Containing Fractions in Bovine Milk

Nadia Spano [1,*], Sara Bortolu [2], Margherita Addis [3], Ilaria Langasco [1], Andrea Mara [1], Maria I. Pilo [1], Gavino Sanna [1,*] and Pietro P. Urgeghe [4]

1. Dipartimento di Scienze Chimiche, Fisiche, Matematiche e Naturali, Università degli Studi di Sassari, Via Vienna, 2, I-07100 Sassari, Italy
2. Istituto per la Bioeconomia, Consiglio Nazionale delle Ricerche, Traversa La Crucca, 3, I-07100 Sassari, Italy
3. Agris Sardegna, Loc. Bonassai Km.18,600, I-07040 Olmedo, Italy
4. Dipartimento di Agraria, Università degli Studi di Sassari, Viale Italia 39A, I-07100 Sassari, Italy
* Correspondence: nspano@uniss.it (N.S.); sanna@uniss.it (G.S.)

Abstract: Free fluoride ions are effective in combating caries in children, and their supplementation in milk has been widely used worldwide for this purpose. Furthermore, it is known that ionic fluoride added to milk is distributed among its components, but little is known about their quantitative relationships. This is likely due to the absence of an analytical protocol aimed at differentiating and quantifying the most important forms of fluorine present in milk. For the first time, a comprehensive protocol made up of six potentiometric methods devoted to quantifying the most important fractions of fluorine in milk (i.e., the free inorganic fluoride, the inorganic bonded fluoride, the caseins-bonded fluoride, the whey-bonded fluoride, the lipid-bonded fluoride, and the total fluorine) has been developed and tested on real samples. Four of the six methods of the procedure are original, and all have been validated in terms of limit of detection and quantification, precision, and trueness. The data obtained show that 9% of all fluorine was in ionic form, while 66.3% of total fluorine was bound to proteins and lipids, therefore unavailable for human absorption. Beyond applications in dental research, this protocol could be extended also to other foods, or used in environmental monitoring.

Keywords: fluorine; fluoride; milk; potentiometry; ion specific electrode; caries

1. Introduction

Fluorine is the 13th most common element in the Earth's crust, with a mean concentration of 600–900 mg kg^{-1} [1]. Molecular fluorine is rarely present in nature [2] whereas fluoride, its reduced form, is the nearly ubiquitous chemical form in water, soil, plants, and animals. Although the 'essentiality' of fluoride for the biosphere has been debated for many decades [3], its role for humans has only been clearly established in recent years. Based on the European Food Safety Authority (EFSA) [4], fluoride is not an essential species for humans since no health consequences of its deficiency have been observed. Conversely, excess fluoride has been found to cause dental fluorosis [5,6] or skeletal fluorosis [7], gastric and kidney disturbances [8], or even death [9,10]. On the other hand, a moderate intake of fluoride is helpful in combating the development of caries in children [11,12]. Free fluoride is the most active species in caries prophylaxis. It is incorporated in the hydroxyapatite of developing tooth enamel and dentin, forming fluorohydroxyapatite, which is more acid-resistant than hydroxyapatite. In addition, fluoride has an anticaries effect on erupting teeth by contact with enamel during ingestion, excretion in the saliva, and absorption in biofilms on teeth [13]. Even for this reason, the European Union has set the recommended daily value for fluoride intake at 3.5 mg [14].

Although ionic fluoride is the only chemical form of fluorine present in non-polluted water, this element is present in biological samples in two main forms: inorganic fluorides and "organic" (or "covalent") fluorine [15]. The inorganic fraction is represented by free ionic fluoride (FIF), i.e., the only chemical form which is detected by the fluoride ion selective electrode (FISE), and by inorganic bonded fluoride (IBF). The latter form is represented by (i) H-bonded compounds (i.e., HF), (ii) metal-bonded compounds (e.g., fluoride salts or their metal complexes), (iii) fluoride "adsorbed" by mineral/organic sediments, and (iv) inorganic fluoride incorporated into biological tissues such as teeth or bones [15]. In the past, IBF has been conveniently determined by using microdiffusion procedures [16–20]; at very low pH values, all inorganic forms of fluoride were converted to HF, which diffused into the hexamethyldisiloxane (HMDS) medium and quantitatively recovered in an alkaline solution. For this reason, the sum of the inorganic fluorides (TIF) has also been called "diffusible" fluoride.

Beyond inorganic fluorides, at least one further form of fluorine, first revealed by Taves in 1968 [21], is present in biological samples: the so-called "organic" fluorine, as previously defined by Venkateswarlu [15]. Although the nature of the bond (or interaction) between fluorine and the organic matrix has not been well established yet, the "organic" fluorine is characterized by substantial inertia towards the adsorption/exchange phenomena involving that element. Consequently, organic fluorine does not readily release fluorides, such as HF or calcium phosphate, under acidic conditions or in exchange processes with the strong adsorbent material, respectively [21].

Among the possible fluorine sources, diet is the primary one of intake for humans. Almost all foods contain this element. Unpolluted water may contain up to a few mg dm^{-3} of fluoride [7], whereas the concentration of this species in the edible part of some plants, such as grapes, spinach, elderberry, and-mainly- tomato and tea, may exceed the level of the g kg^{-1} as a function of the geochemical threshold of the soil where they are grown [8]. Since the level of fluoride in drinking water may be insufficient to achieve the outcomes in childhood caries prevention, strategies for fluoride supplementation have been adopted in previous decades [12]. Fluorination of water [22] and dietary fluoride supplementation in salt [23] or milk [24] have been widely used to carry out this activity.

Milk is the primary food during the early years of human life. The fluoride content of milk could play a role in improving the mineralization of teeth in children and preventing dental caries. Consequently, the consumption of fluoridated milk has been successfully promoted in many countries [24–32]. As a result, several authors have developed analytical methods aimed at assessing the fluoride levels in milk-based matrices. Among others, cow milk [33–52], breast milk [47,53–58], and infant milk [59,60], or formulas, [47,55,61,62] were primarily studied using fluorescence [18], molecular absorption [38,49,52] or atomic [50] spectrometry, gas-chromatography [39,42], ion chromatography [48] and, among all, potentiometric methods with ion-specific electrodes (ISE) [15,18,20,33–37,40,41,43–47,51,53–62]. In this technique, the sensing element is a crystal of LaF$_3$ doped with EuF$_2$. The lattice vacancies in this way created an increase in the mobility of fluoride ions that jump between them. Since the crystal membrane is permeable only to fluoride ions, the potential of the electrode depends only on their activity [63]. The sensor is almost specific, being that the only interfering ion is OH$^-$. To overcome this potential interference, and to avoid the protonation of the analyte, FISE measurements must be performed at pH between 5 and 6.

The methods commonly reported in the literature are aimed at the determination of free or diffusible fluoride, and total fluorine in milk samples [33–37,41,44,46,57,59], but the results regarding the measurement of other fractions are quite rare, fragmentary, and, sometimes, contradictory. Duff [36] observed that free fluoride added to milk tends to bind Ca^{2+} ions and/or unspecified organic components of milk. In particular, 72 h after the milk addition of known concentrations of fluoride ions, a fall to less than 10% of the initial free fluoride concentration is observed. Wieczorek et al. [64] investigated the tendency of α-lactalbumin, α-, β-, and κ-casein to combine with fluoride in the pH range from 6.6 to 3.9. They found that only α-lactalbumin can bind fluoride at pH = 3.9. Thus, fluoride preferentially binds the Ca^{2+} ions, that were released in solution by the proteins as a result of the pH decrease observed along the milk storage. On the contrary, Kahama et al. [46]

found that almost all fluorine in cow's milk was inorganic and the bound analyte was physically or chemically sequestered by milk proteins. Chlubek [42] reported that about 11% of fluorine in milk was bound to fat. Finally, Campus et al. [57] used an undescribed and unvalidated speciation protocol to differentiate organic and inorganic fractions of fluorine in research aimed at studying the effect of prepartum fluoride supplementation on breast milk.

From this evidence, a comprehensive and validated analytical protocol for differentiating and quantifying the most important forms of fluorine present in milk might be useful to ascertain the nature and the bioavailability of the element in this matrix. Unfortunately, this protocol is currently absent in the literature. Hence, pursuing the interest of this research group in the development and validation of original FISE methods applicable either in dental research [57,65–68] or in food analysis [69], the aim of this study is to develop for the first time (and validate, where possible) a comprehensive analytical protocol for the FISE determination of the fluorine contained in different fractions of bovine milk.

2. Results and Discussion

2.1. Classification of the Different Forms of Fluorine Present in Milk

Findings from the literature reveal that the same name has sometimes been used to indicate different forms of fluorine present in milk. To avoid any misunderstanding, it would be useful to clarify unambiguously the names of the most common forms of fluorine which have been measured in this study. In principle, inorganic fluorine is formed by free ionic fluoride (FIF) and inorganic bonded fluorine (IBF). FIF is the inorganic fraction that is directly measured in a solution (or suspension), consisting of equal volumes of the sample and a total ionic strength adjustment buffer solution (TISAB), and it can be found in the aqueous fraction of the milk. On the other hand, IBF is the fluorine fraction complexed by metal ions, or reversibly "adsorbed" by colloids, suspensions, and emulsions present in milk. The sum of FIF and IBF forms the total inorganic fluorine (TIF). The latter is the amount of analyte measured after the conversion of the free or bound inorganic forms of fluorine into HF, its diffusion in a siloxane medium, and its quantitative recovery in an alkaline solution. Non-diffusible, organic fluorine fractions in milk are mainly related to interactions of the analyte with proteins and lipids present in colloidal and emulsion particles, respectively. Proteins-bonded fluorine (PBF) is constituted by caseins-bonded fluorine (CBF) and whey protein-bonded fluorine (WBF), respectively. Furthermore, the sum of PBF and lipids-bonded fluorine (LBF) constitutes the total organic fluorine (TOF). Finally, the total fluorine (TF) is the amount of fluorine measured after a complete decomposition of the matrix (i.e., by microwave digestion or ashing).

2.2. Analytical Methods

The protocol of analytical methods proposed in this study consists of six methods, summarized in Table 1.

Table 1. Summary of the protocol aimed to measure different forms of fluorine in cow's milk.

Analytical Methods	Fluorine Fractions Measured	
M1	Free Ionic Fluoride (FIF)	FIF
M2	Total Inorganic Fluorine (TIF)	TIF
M3	Total Fluorine (TF)	TF
M4	FIF and Caseins-Bonded Fluorine (CBF)	FIF + CBF
M5	FIF and Proteins-Bonded Fluorine (PBF)	FIF + PBF
M6	Lipids-Bonded Fluorine (LBF)	LBF
Indirectly measurable fractions of fluorine		
M2-M1	Inorganic Bonded Fluorine (IBF)	IBF
M5-M4	Whey-Bonded Fluorine (WBF)	WBF

2.2.1. Diffusible (Inorganic) Fluorine

Method M1—Determination of Free Ionic Fluoride (FIF)

Despite its apparent simplicity, only a few papers have accurately described FISE potentiometric methods dedicated to measuring FIF in milk. For example, Duff [36] determined ionic fluoride in bovine milk by simply diluting the sample in potassium nitrate and sodium citrate solutions. Instead, Koparal et al. [55] measured the same analyte in breast milk by mixing the sample with equal volumes of a TISAB III solution (i.e., a solution of ammonium chloride, ammonium acetate and *trans*-1,2-diaminociclohexane-N,N,N',N'-tetraacetic acid monohydrate (CDTA) in water).

They showed that, to minimize bias, potentiometric determination of fluoride needs careful control of pH and ionic strength, as well as the elimination of any interfering species.

Therefore, pH must be buffered at 5.25, achieving in this way the best compromise in minimizing both HF and OH^- ion interferences. Similarly, high amounts of an inert electrolyte buffer the ionic strength of the final solution. Finally, small quantities of strong complexing agents are required to avoid the formation of stable complexes between the fluoride and free metal ions, such as Al(III), Fe(II), and Mg(II), whose concentrations in milk usually range between 10 μg dm^{-3} to 100 mg dm^{-3}. Among the possible strong complexing agents, CDTA is one of the most effective in removing the interferences and it allows to form stable complexes at pH below neutrality (e.g., 5.25).

For this reason, the method used by Koparal et al. has been adopted in this protocol and fully validated.

Therefore, once the solutions reached the temperature of 20.0 ± 0.5 °C, the FISE potentiometric measurement was performed after mixing 10 cm^3 of milk and 10 cm^3 of a TISAB III solution in a 50 cm^3 polyethylene beaker. After immersing both the FISE and the Ag/AgCl reference electrodes, the suspension was slowly stirred for at least 20 min until the equilibrium potential was reached.

Method M2-Total Inorganic Fluorine (TIF)

Since the methods proposed by Taves [37] and Liu [43] were insufficiently validated, the procedure used in this work for measuring TIF was an optimization of the one assessed by Kimarua et al. [44] and widely used by Kahama et al. [45,46]. First, 1 cm^3 of milk was placed in a Petri dish with a hole in the lid. A polystyrene weighing boat containing 0.050 cm^3 of 0.5 mol dm^{-3} NaOH (trap solution) was placed inside the dish. The plate was sealed around with a strip of Parafilm®, and 2 cm^3 of a 4.0 mol dm^{-3} HClO$_4$ solution in water saturated with hexamethyldisiloxane (HMDS) were added to the sample through the hole that was then closed with Parafilm®. After 18 h at 25 °C the trap solution was dissolved in 0.050 cm^3 of 0.5 mol dm^{-3} HCl aqueous solution, and ultrapure water were added up to 2 cm^3. Lastly, the potentiometric measurement was made according to the M1 method described above.

Method M3-Total Fluorine (TF)

The determination of the TF requires the conversion to fluoride of all the fluorine forms complexed or bounded with inorganic or organic compounds. To accomplish this, complete decomposition of the organic matrix is required. This is typically obtained through open incineration, fusion, or combustion procedures in an oxygen flask, oxygen bomb, or oxyhydrogen flame [16]. The determination of TF in different matrices has been thoroughly reviewed by Campbell [70]. In particular, TF in milk has been measured using incineration of the sample, followed by HDMS diffusion processes [42,46] and pyrohydrolysis (or hydrolysis with perchloric acid) [16]. Unfortunately, these methods are rather complex, time-consuming, and often not validated. Therefore, a microwave assisted milk acid digestion process is proposed in this protocol. First, 0.5 cm^3 of the sample were placed in a perfluoroalcoxy ethylene (PFA) vessel and treated with 2 cm^3 of 69% (w/v) HNO$_3$. The microwave program included two steps, the first at 300 W for 30 s and the next at 600 W for 60 s. After digestion, the vessel was cooled to 4 °C overnight. Afterward, the cold solution was quickly treated with 2 cm³ of 50% (w/v) NaOH. The neutralized solution (typical pH of 5.5 ± 0.5) was diluted to 10 cm³ with ultrapure water and 1:1 mixed with a TISAB III solution for potentiometric analysis performed according to method M1.

To carefully estimate the amount of fluoride released from the inner walls of the PFA vessels after treatment with strong oxidizing agents [71], two blanks (0.5 cm³ of ultrapure water added with 2 cm³ of 69% HNO_3) were subjected to the whole analytical procedure in the same vessel, both before and after the sample digestion. Then, the mean amount of fluoride released from each vessel was systematically subtracted from the fluoride concentration measured when the same vessel was used for milk digestion. The average concentration of fluoride for all blanks measured was 220 ± 80 µg dm^{-3} (n = 120).

2.2.2. Non-Diffusible (Organic) Fluorine

As reported by Taves [37] and Singer [72], the concentration of TF is usually two times higher than TIF. As a result, other forms of fluorine, strongly interacting with the organic phase of milk and thus not available for diffusion, were postulated to explain the difference between the two concentrations.

After the early studies by Venkateswarlu [73], other researchers have studied the interactions between fluorine and proteins [46,63] or lipids [42]. Since caseins and whey proteins showed different interactions with fluorine [73], one of the aims of this study was the differentiation of fluorine bound to these proteins. Preliminary tests performed using a literature method [55] confirmed its known drawbacks of poor reproducibility and specificity [74,75]. Therefore, two original methods, based on the fractionate precipitation of caseins and whey proteins, have been developed.

First, caseins were precipitated to the isoelectric point [76], and fluorine previously bonded to them was released in solution as fluoride ion. From this solution, a salting out procedure [77] allowed the precipitation of the whey proteins and the consequential release in the solution of the fluorine previously bonded to them. Hence, the methods allowed to measure: (i) the sum of FIF and the organic fluorine associated with caseins (CBF), after their precipitation, and (ii) the sum of FIF and the fluorine fraction released to the total amount of milk proteins (PBF), after their precipitation. Finally, the amount of fluorine associated with lipids was measured by applying the method M3 on aliquots of pure lipid extracts obtained from milk according to the Rose-Gottlieb method [78].

Method M4—Free Inorganic Fluoride and Caseins-Bonded Fluorine (FIF + CBF)

The principle of the method is based on the precipitation of caseins to their isoelectric point. When this happens, calcium and fluoride ions bonded to caseins are released into solution [37]. Then, 10 cm³ of milk was added with a 0.5 mol dm^{-3} HCl aqueous solution until pH = 4.6 and transferred into a centrifuge tube. After 20 min centrifugation, the supernatant was filtered (solution A). An aliquot of solution A was added to the same volume of TISAB III solution for the potentiometric analysis, performed according to method M1.

Method M5—Free Inorganic Fluoride and Proteins-Bonded Fluorine (FIF + PBF)

After precipitation of caseins, the whey proteins were also precipitated from the solution by a salting out procedure [77]. To 5 cm³ of solution A, 4 cm³ of a saturated aqueous solution (26.8% (w/v)) of $(NH_4)_2SO_4$ was added dropwise and under continuous stirring. The suspension was left stirring for 10 min. After 1 h centrifugation, the supernatant was filtered, and an aliquot of this solution was mixed 1:1 with TISAB III solution for the potentiometric analysis, performed according to method M1.

Method M6-Lipids-Bonded Fluorine (LBF)

The method is based on a preliminary extraction of milk lipids according to Rose-Gottlieb method [78] which are then processed with the method M3. The organic extract of lipids from milk was concentrated in a small volume (ca. 5 cm³) and quantitatively transferred in a PFA vessel. The vessel was gently heated up to 80 °C to allow the evaporation of the organic solvents. The residue was processed according to method M3.

2.2.3. Quantification

Since some exploratory measurements on real samples revealed that the concentrations of the different forms of fluoride were below the lower limit of the linearity range of the M1-M6 methods (i.e., below the concentration of 100 µg dm^{-3}), quantification has always been performed by means of internal calibration. Therefore, for each sample/blank analyzed, three known additions of a fluoride standard solution have been performed. Each addition was between 50% and 150% of the assumed amount of the analyte. The Gran's-like linearization procedure provided the concentration of the analyte [79]. Compared with external calibration, the internal calibration method is more time-consuming, as each individual sample or blank requires a calibration curve [80], and it has a higher uncertainty due to the larger width of its confidence interval [81]. Despite these disadvantages, it allows for a substantially bias-free measurement [82]. To ensure the homogeneity of the results obtained in the proposed protocol, this quantification approach was used for all methods (i.e., M1–M6). All data were blank corrected.

The procedures aimed to ensure quality assurance and quality control of the methods have been described in detail in the Supplementary material (part S1).

2.3. Validation

Validation of the methods described in this protocol has been accomplished by means of the calculation of the limit of detection (LoD), limit of quantification (LoQ), and the measurement of both precision and trueness.

2.3.1. LoD and LoQ

LoD of the potentiometric determination of free fluorides was determined using both the procedures described by the Laboratory Certification Program of the Wisconsin Department of Natural Resources (LCP-WDNR) [83] and by the International Union of Pure and Applied Chemistry (IUPAC) [84]. According to LCP-WDNR, the esteem of the standard deviation s of ten replicated measures on a standard fluoride solution at a concentration in the range between the expected LoD and 5LoD (in this case, $5 \cdot 10^{-2}$ mg dm^{-3}) was evaluated. Hence, LoD and LoQ are defined as LoD = s × t, and LoQ = 10 × s, respectively, where t is the tabulated Student's t value for 9 degrees of freedom and p = 0.99. Based on the IUPAC method, LoD has been accomplished by measuring the potential of fifteen standard solutions at concentration levels between $5 \cdot 10^{-5}$ (i.e., a concentration much below the expected LoD) and 1 mg dm^{-3}. The LoD was calculated by plotting the two linear segments and corresponds to the abscissa of the intersection point. Figure 1 shows an example of the IUPAC approach for determining the LoD for a potentiometric method.

Despite the marked differences between the approaches used in these two methods, the LoDs reported in Table 2 are similar, and both are low enough to allow quantification of the analyte in each of the methods considered. Since the amount of fluorine measured in methods M1–M6 is always obtained by means of a potentiometric determination, the LoDs and LoQs here reported are the same for all methods proposed.

Table 2. Validation parameters for the methods M1–M6.

LoD and LoQ, µg dm^{-3}	LoD [a]		LoD [b]		LoQ [a]	
	1.3		6.6		4.0	
	M1	M2	M3	M4	M5	M6
Intermediate precision, CV%	10	7	10	14	7	12
Trueness, Recovery %	110 ± 10 [c]	100 ± 10 [c]	100 ± 10 [c] 100 ± 8 [d]	88 ± 9 [c]	90 ± 9 [c]	100 ± 10 [c]

[a] according to ref. [83]; [b] according to ref. [84]; [c] measured using recovery test; [d] according to ref [46].

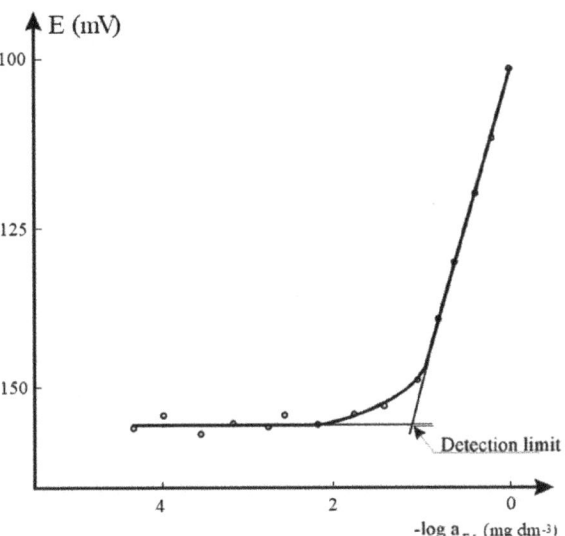

Figure 1. Calibration plot of standard fluoride solutions in the range between $5 \cdot 10^{-5}$ mg dm^{-3} and 1 mg dm^{-3}.

2.3.2. Precision

Precision was evaluated for all methods in terms of intermediate precision, whereas the evaluation of the repeatability has been performed only for method M1. Intermediate precision, measured in terms of percent coefficient of variation (CV), has been evaluated analyzing the same milk sample in five analytical sessions performed within a week. Repeatability, expressed always as CV, and evaluated by means of eight consecutive determinations of FIF performed on the same milk sample, was 3.6%. Table 2 reports the intermediate precision data, which ranged between 7% for both methods M2 and M5, and 14% for method M4. The acceptability of the precision data for a concentration interval of analyte between the tens of the µg dm^{-3} (i.e., for FIF and LBF) and the hundreds of µg dm^{-3} (for the remaining methods) has been successfully verified in terms of the Horwitz's theory [85].

2.3.3. Trueness

Since certified reference materials (CRM) were not available and no reliable and analytical methods are reported in literature (except for methods M1 and M3), trueness has been determined using recovery tests for all the methods of the protocol.

After homogenization, three aliquots of a milk sample were enriched with increasing volumes of a fresh standard fluoride solution, while no analyte was added to the fourth aliquot of milk. Then, all aliquots were subjected to the full analytical procedure aimed at determining the specific form(s) of fluoride. The whole procedure was performed in triplicate. The recoveries reported in Table 2 ranged between 88 ± 9% (method M4) and 110 ± 10% (method M1). Quantitative recoveries (t-test, α = 0.05) were observed for all the methods. Hence, the release of fluorine-containing species from mineralization vessel material (i.e., PFA), which may be a potential source of bias in the methods M3 and M6, is well compensated by the subtraction of the blanks from the measured concentrations of both TF and LBF. However, to evaluate any other possible source of bias in the determination of the TF in milk, a real sample was analyzed with both method M3 and the incineration method proposed by Kahama et al. [46]. Additionally, in these cases, Table 2 reports quantitative recoveries for these methods.

In summary, the validation data of the proposed methods confirm that this protocol has good sensitivity and accuracy.

2.4. Application of the Protocol to Cow's Milk

The whole analytical protocol has been tested on five different samples of cow's milk from markets in Sardinia. Table 3 reports the analytical data obtained both by the direct application of methods M1–M6 and indirectly by their combination, whereas Figure 2 reports the average percent distribution of the different forms of the analyte in cow's milk.

Table 3. Fluorine fractionation measured in five samples of cow's milk from the market (n = 5). Concentration of each fraction is expressed as µg dm^{-3} ± standard deviation.

	Methods	Fractions	Samples					Average	Range [b]
			1	2	3	4	5		
Direct	M1	FIF	42 ± 4	40 ± 3	43 ± 5	51 ± 4	47 ± 6	45 ± 10	40–51
	M2	TIF	180 ± 10	169 ± 4	136 ± 2	190 ± 20	160 ± 10	170 ± 20	136–182
	M3	TF	460 ± 80	450 ± 10	400 ± 60	510 ± 20	480 ± 30	500 ± 100	396–510
	M4	FIF + CBF	170 ± 10	150 ± 9	160 ± 20	180 ± 6	160 ± 10	160 ± 30	150–180
	M5	FIF + PBF	324 ± 6	335 ± 15	300 ± 20	319 ± 4	340 ± 10	320 ± 30	296–342
	M6	LBF	50 ± 10	50 ± 10	40 ± 10	60 ± 10	40 ± 10	50 ± 20	44–58
Indirect	M2−M1	IBF	140 ± 10	129 ± 5	93 ± 5	140 ± 20	110 ± 10	120 ± 30	93–140
	M4−M1	CBF	130 ± 10	110 ± 9	120 ± 20	129 ± 7	110 ± 10	120 ± 30	110–131
	M5−M1	PBF	282 ± 7	300 ± 20	250 ± 20	268 ± 6	300 ± 10	280 ± 30	268–295
	M5−M4	WBF	150 ± 10	180 ± 20	130 ± 30	139 ± 9	180 ± 20	160 ± 40	131–185
	M5−M1 + M6	TOF	330 ± 10	340 ± 20	290 ± 20	330 ± 10	340 ± 10	330 ± 30	302–343
	M5 + M6 + M2−M1	TIF + TOF	510 ± 20	510 ± 20	440 ± 20	510 ± 20	500 ± 20	500 ± 40	438–514
		(TIF + TOF) [a]/TF (%)	110	112	110	101	105	108	

FIF, Free Inorganic Fluoride; TIF, Total Inorganic Fluorine; TF, Total Fluorine; CBF, Caseins-Bonded Fluorine; PBF, Proteins-Bonded Fluorine; WBF, Whey-Bonded Fluorine; TOF, Total Organic Fluorine; TF/(TIF + TOF) (%), percent difference among total fluorine directly measured with the method M3 and the sum of both TIF and TOF fractions. Average concentrations have been rounded according to the number of significant digits of the relevant standard deviation; [a] ratio calculated based on unrounded concentrations; [b] unrounded concentrations.

The reported data clearly show that the FIF percentage is one-quarter of the TIF, being in turn minority (34%) regarding the TOF one (66%). Hence, the FIF paucity in milk (around 10% the amount of TF), already observed in the literature [36,86], has also been confirmed by these results. Furthermore, the very narrow interval of concentrations observed for FIF agrees with the range found for milk by Koparal et al. [55] and Liu et al. [43], but it is higher than those measured by Duff [36] and Bessho [87]. Additionally, TIF concentrations are comparable with those previously reported by Van Staden et al. [41], Kimarua et al. [44], and Kahama et al. [46] for milk produced in unpolluted sites.

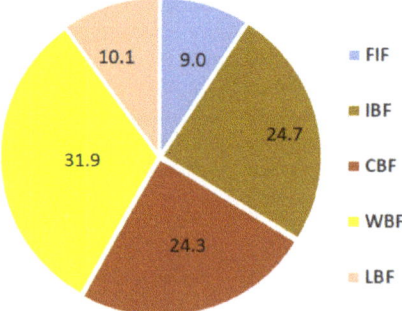

Figure 2. Average percent distribution of the different forms of fluorine in cow's milk. FIF, free inorganic fluoride; IBF, inorganic bonded fluorine; CBF, caseins-bonded fluorine; WBF, whey-bonded fluorine; LBF, lipids-bonded fluorine.

Due to the paucity and fragmentary nature of previous studies on this topic, a comparison of the data obtained here for organic fluorine with those reported in literature is rather difficult. Normally, organic fluorine has been measured indirectly as the difference between total fluorine and total inorganic fluoride; therefore, these data are affected by high uncertainty and several possible sources of bias. To the best of our knowledge, only the contributions of Venkateswarlu [88] and Chublek [42] allow direct quantification of total organic fluorine in biological fluids (other than milk) and lipid-bound fluorine in milk, respectively. As would be expected, the indirect quantification of organic fluorine in milk is derived from literature studies that produced conflicting results. As a matter of fact, in some cases the amount of organic fluorine in milk was small [46] or even negligible [37], while in others it was the most abundant [57].

The percent of the TOF measured in this research is in good agreement with that reported in breast milk by Campus et al. [57] (i.e., 62%), however the fractionation method used was poorly described and validated and did not include the determination of LBF. Hence, the procedure of fractionated precipitation of caseins and globulins described in this protocol and used for the quantification of PBF, CBF, and WBF, has been described and validated for the first time. In addition, the preliminary data obtained in this study seem to envisage no significant statistical differences in the CBF/WBF ratio as a function of the origin of milk. Indeed, in cow's milk the average ratio is 1.3, i.e., very close to the ratio of 1.27 measured by Campus et al. in the breast milk of mothers not involved in a fluoride supplementation study [57].

Furthermore, the results here obtained for the determination of LBF are in good agreement with those reported by Chlubek [42] (i.e., roughly 11% of the TF).

The TF amount measured by the method M3 ranged between 400 and 510 $\mu g\ dm^{-3}$. These values are within the range measured by Obzek and Akman [49] as well as that reported by Liu et al. [43]. On the other hand, many previous contributions reported total amounts of fluoride in milk significantly lower [33,35,37,40,47] than those measured in this study. The fluorine concentration range in milk is generally wide, however, the rough validation of the analytical methods could have led to a general underestimation of the TF in milk. Acid-assisted mineralization methods and not optimized ashing methods [15] may cause losses of analyte fractions and lead to wrong measurements, where TF was really only TIF [43,45] or even FIF [51]. In this protocol, the trueness of the method for the TF determination has been carefully validated with a recovery test either using microwave digestion or milk ashing. These results were also confirmed by the substantial consistency for all samples among the sum of TIF and TOF with respect to TF. Hence, the percent ratio (TIF + TOF)/TF is between 101% and 112%.

3. Experimental

3.1. Samples

Semi-skimmed, UHT-treated, and long-shelf-life cow's milk were purchased from local markets in North Sardinia, Italy. Sampling procedure has been described in detail in the Supplementary Materials (part S2).

3.2. Instrumentation and Labware

Potentiometric measurements were performed using a FISE (ISE Fluoride DX219, Mettler Toledo, Switzerland) connected to an Ag/AgCl reference electrode (mod. 373/SSG/6J, Amel s.r.l., Milan, Italy) and an ion analyzer (pH 1500 CyberScan, Eutech Instruments, Groningen, The Netherlands). The combined glass-electrode model LIQ-GLASS 238000/08 used for the pH measurements was from Hamilton, Bonaduz, Switzerland. A microwave digestion system Mega model MLS 1200 (Milestone, Sorisole, Italy), equipped with perfluoroalcoxy ethylene (PFA) medium-pressure digestion vessels, was used for the sample digestion. The thermostatic oven was purchased by Memmert, (Schwabach, Germany), whereas the centrifuge was by ALC 4217 MK II, ALC International s.r.l., (Cologno Monzese, MI, Italy). Fixed volume pipettes (0.500 and 1.000 cm^3, Eppendorf, Milan, Italy; 5.000 and

10.000 cm^3, Alpha, Rignano Flaminio, Italy), and 50 cm^3 Falcon tubes in polypropylene were used for milk sampling. Polystyrene Petri dishes (55 mm) were used for microdiffusion measurements. Everywhere possible, glassware was replaced with plasticware.

3.3. Reagents

All reagents and solvents were of analytical-reagent grade (Fluka, Milan, Italy) except for NaF (99.99%, Sigma-Aldrich, Milan, Italy) and CH$_3$COOH (100% extra pure, Riedel-de Haën, Milan, Italy). Type-I ultrapure water (resistivity >18 MΩ cm^{-1}), used for all methods described in the protocol, was prepared with a Milli-Q® IQ 7003 system (Millipore, Vimodrone, Italy). NaF was dried at 110 °C for two hours and cooled in a desiccator before preparation of 1000 mg dm^3 F$^-$ standard solution, which was used for preparing diluted solutions. TISAB III solution (ITW Reagents, Monza, Italy) consisted of 18 g of 1,2-diaminociclohexane N,N,N',N'-tetraacetic acid 1-hydrate, 96.65 g of ammonium chloride, 163.4 g of ammonium acetate, 0.1 g of cresol red, all dissolved in 1 dm^3 of ultrapure water. The pH of TISAB III is 5.25 ± 0.25.

4. Conclusions

For the first time, a complete protocol to quantify free ionic fluoride, total inorganic fluorine, total fluorine, caseins-bonded fluorine, proteins-bonded fluorine, lipids-bonded fluorine and, indirectly, inorganic bonded fluorine and whey-bonded fluorine has been developed, successfully validated in terms of LoD, LoQ, precision, and trueness, and tested on five different real samples of bovine milk. Among all fluorine fractions measured, free ionic fluoride is the least abundant, accounting for only 9% of the total amount of fluorine in milk. If other studies confirm this finding, there may be serious concerns about the efficacy of milk fluorination programs, since 66.3% of the total fluorine is bound to proteins or lipids, and for this reason is not available for human absorption. Beyond its novelty from a mere analytical viewpoint, this protocol might be a helpful tool for researchers engaged in studies on the bioavailability of fluoride in foods. In addition, the application of this protocol might be of interest to ascertain the authentic effectiveness of milk fluorination strategies, largely used for caries prophylaxis, and evaluating changes in the distribution of fluorine in proteins and lipids because of technological treatments. Furthermore, this protocol may be used—in whole or only for selected methods—in environmental monitoring procedures for fluorine-containing pollutants.

Supplementary Materials: The following are available online at https://www.mdpi.com/article/10.3390/molecules28031349/s1. Part S1: Quality Assurance and Quality Control; Part S2: Sampling procedure of milk.

Author Contributions: Conceptualization, N.S. and G.S.; methodology, N.S., M.A. and G.S.; validation, N.S. and G.S.; formal analysis, N.S., I.L., A.M., M.I.P. and P.P.U.; investigation, N.S., S.B. and M.A.; resources, M.A. and G.S.; data curation, N.S and G.S.; writing—original draft preparation, N.S. and G.S.; writing—review and editing, N.S:, S.B., M.A., I.L., A.M., M.I.P, G.S. and P.P.U.; visualization, I.L., A.M. and P.P.U.; supervision, N.S: and G.S.; project administration, G.S.; funding acquisition, G.S. All authors have read and agreed to the published version of the manuscript.

Funding: This research received no external funding.

Institutional Review Board Statement: Not applicable.

Informed Consent Statement: Not applicable.

Data Availability Statement: Data is contained within the article or Supplementary Materials.

Conflicts of Interest: The authors declare no conflict of interest.

Sample Availability: Samples are not more available from the authors.

References

1. Weinstein, L.H.; Davison, A.W. *Fluorides in the Environment: Effects on Plants and Animals*; CABI Digital Library: Wallingford, UK, 2004; 287p, ISBN 978-0-85199-872-5.
2. Schmedt auf der Günne, J.; Mangstl, M.; Kraus, F. Occurrence of Difluorine F_2 in Nature-In Situ Proof and Quantification by NMR Spectroscopy. *Angew. Chem. Int. Ed.* **2012**, *51*, 7847–7849. [CrossRef] [PubMed]
3. Fluorine: Essential nutrient? *Nutr. Rev.* **2009**, *12*, 156–158. [CrossRef] [PubMed]
4. EFSA Panel on Dietetic Products, Nutrition, and Allergies (NDA). Scientific Opinion on Dietary Reference Values for Fluoride. *EFSA J.* **2013**, *11*, 3332. [CrossRef]
5. Dean, H.T.; Arnold, F.A., Jr.; Elvove, E. Domestic Water and Dental Caries: V. Additional Studies of the Relation of Fluoride Domestic Waters to Dental Caries Experience in 4,425 White Children, Aged 12 to 14 Years, of 13 Cities in 4 States. *Public Health Rep.* **1942**, *57*, 1155–1179. [CrossRef]
6. Everett, E.T. Fluoride's Effects on the Formation of Teeth and Bones, and the Influence of Genetics. *J. Dent. Res.* **2010**, *90*, 552–560. [CrossRef]
7. National Research Council. *Fluoride in Drinking Water: A Scientific Review of EPA's Standards*; The National Academies Press: Washington, DC, USA, 2006; 530p. [CrossRef]
8. Johnston, N.R.; Strobel, S.A. Principles of Fluoride Toxicity and the Cellular Response: A Review. *Arch. Toxic.* **2020**, *94*, 1051–1069. [CrossRef]
9. Lech, T. Fatal Cases of Acute Suicidal Sodium and Accidental Zinc Fluorosilicate Poisoning. Review of Acute Intoxications Due to Fluoride Compounds. *Forensic Sci. Int.* **2011**, *206*, e20–e24. [CrossRef]
10. Whitford, G.M.; Pashley, D.H.; Reynolds, K.E. Fluoride Tissue Distribution: Short-Term Kinetics. *Am. J. Physiol.-Ren. Physiol.* **1979**, *236*, F141–F148. [CrossRef]
11. Hunstadbraten, K. Fluoride in caries prophylaxis at the turn of the century. *Bull. Hist. Dent.* **1982**, *30*, 117–120.
12. Sampaio, F.C.; Levy, S.M. Systemic Fluoride. In *Fluoride and the Oral Environment*; Buzalaf, M.A.R., Ed.; Karger: Basel, Switzerland, 2011; pp. 133–145. [CrossRef]
13. Buzalaf, M.A.R.; Levy, S.M. Fluoride Intake of Children: Considerations for Dental Caries and Dental Fluorosis. In *Fluoride and the Oral Environment*; Buzalaf, M.A.R., Ed.; Karger: Basel, Switzerland, 2011; pp. 1–19. [CrossRef]
14. European Union. Commission Directive 2008/100/EC of 28 October 2008 amending Council Directive 90/496/EEC on nutrition labelling for foodstuffs as regards recommended daily allowances, energy conversion factors and definitions. *Off. J. Eur. Union* **2008**, *L285*, 9–12. Available online: http://eur-lex.europa.eu/LexUriServ/LexUriServ.do?uri=OJ:L:2008:285:0009:0012:EN:PDF (accessed on 15 November 2022).
15. Venkateswarlu, P. Determination of Fluorine in Biological Materials: A Review. *Adv. Dent. Res.* **1994**, *8*, 80–86. [CrossRef]
16. Taves, D.R. Effect of Silicone Grease on Diffusion of Fluoride. *Anal. Chem.* **1968**, *40*, 204–206. [CrossRef]
17. Taves, D.R. Separation of fluoride by rapid diffusion using hexamethyldisiloxane. *Talanta* **1968**, *15*, 969–974. [CrossRef]
18. Taves, D.R. Determination of submicromolar concentrations of fluoride in biological samples. *Talanta* **1968**, *15*, 1015–1023. [CrossRef]
19. Hall, R.J. The diffusion of fluoride with hexamethyldisiloxane. *Talanta* **1969**, *16*, 129–133. [CrossRef]
20. Sara, R.; Wänninen, E. Separation and determination of fluoride by diffusion with hexamethyldisiloxane and use of a fluoride-sensitive electrode. *Talanta* **1975**, *22*, 1033–1036. [CrossRef]
21. Taves, D.R. Evidence That There Are Two Forms of Fluoride in Human Serum. *Nature* **1968**, *217*, 1050–1051. [CrossRef]
22. Maier, F.J. *Manual of Water Fluoridation Practice*; McGraw-Hill: New York, NY, USA, 1963; 234p, ISBN 978-0070397187.
23. Jones, S.; Burt, B.A.; Petersen, P.E.; Lennon, M.A. The effective use of fluorides in public health. *Bull. World Health Organ.* **2005**, *83*, 670–676. Available online: https://www.ncbi.nlm.nih.gov/pmc/articles/PMC2626340/pdf/16211158.pdf (accessed on 15 November 2022).
24. Banoczy, J.; Rugg-Gunn, A.; Woodward, M. Milk Fluoridation for the Prevention. *Acta Med. Acad.* **2013**, *42*, 156–167. [CrossRef]
25. Lights, A.E.; Smith, F.A.; Gardner, D.E.; Hodge, H.C. Effect of fluoridated milk on deciduous teeth. *J. Am. Dent. Assoc.* **1958**, *56*, 249–250. [CrossRef]
26. Rao, G.S. Dietary Intake and Bioavailability of Fluoride. *Annu. Rev. Nutr.* **1984**, *4*, 115–136. [CrossRef] [PubMed]
27. Stephen, K.W.; Boyle, I.T.; Campbell, D.; McNee, S.; Boyle, P. Five-Year Double-Blind Fluoridated Milk Study in Scotland. *Commun. Dent. Oral. Epidemiol.* **1984**, *12*, 223–229. [CrossRef] [PubMed]
28. Centers for Disease Control and Prevention. Recommendations for using fluoride to prevent and control dental caries in the United States. *MMWR* **2001**, *50*, 1–42. Available online: http://www.cdc.gov/mmwr/preview/mmwrhtml/rr5014a1.htm (accessed on 15 November 2022).
29. Mariño, R.; Villa, A.; Guerrero, S. A Community Trial of Fluoridated Powdered Milk in Chile. *Commun. Dent. Oral. Epidemiol.* **2001**, *29*, 435–442. [CrossRef]
30. Jackson, R.D.; Brizendine, E.J.; Kelly, S.A.; Hinesley, R.; Stookey, G.K.; Dunipace, A.J. The Fluoride Content of Foods and Beverages from Negligibly and Optimally Fluoridated Communities. *Commun. Dent. Oral. Epidemiol.* **2002**, *30*, 382–391. [CrossRef]
31. Mariño, R.; Traub, F.; Lekfuangfu, P.; Niyomsilp, K. Cost-Effectiveness Analysis of a School-Based Dental Caries Prevention Program Using Fluoridated Milk in Bangkok, Thailand. *BMC Oral Health* **2018**, *18*, 24. [CrossRef]

32. Petersen, P.E. Long Term Evaluation of the Clinical Effectiveness of Community Milk Fluoridation in Bulgaria. *Commun. Dent. Health* **2015**, *32*, 199–203. [CrossRef]
33. Venkateswarlu, P.; Singer, L.; Armstrong, W.D. Determination of Ionic (plus Ionizable) Fluoride in Biological Fluids. *Anal. Biochem.* **1971**, *42*, 350–359. [CrossRef]
34. Venkateswarlu, P. A Micro Method for Direct Determination of Ionic Fluoride in Body Fluids with the Hanging Drop Fluoride Electrode. *Clin. Chim. Acta* **1975**, *59*, 277–282. [CrossRef]
35. Beddows, C.G.; Kirk, D. Determination of Fluoride Ion in Bovine Milk Using a Fluoride Ion-Selective Electrode. *Analyst* **1981**, *106*, 1341–1344. [CrossRef]
36. Duff, E.J. Total and Ionic Fluoride in Milk. *Caries Res.* **1981**, *15*, 406–408. [CrossRef]
37. Taves, D.R. Dietary Intake of Fluoride Ashed (Total Fluoride) v. Unashed (Inorganic Fluoride) Analysis of Individual Foods. *Br. J. Nutr.* **1983**, *49*, 295–301. [CrossRef]
38. Takatsu, A.; Chiba, K.; Ozaki, M.; Fuwa, K.; Haraguchi, H. Direct Determination of Trace Fluorine in Milk by Aluminum Monofluoride Molecular Absorption Spectrometry Utilizing an Electrothermal Graphite Furnace. *Spectrochim. Acta B At. Spectrosc.* **1984**, *39*, 365–370. [CrossRef]
39. Tashkov, W.; Benchev, I.; Rizov, N.; Kolarska, A. Fluoride Determination in Fluorinated Milk by Headspace Gas Chromatography. *Chromatographia* **1990**, *29*, 544–546. [CrossRef]
40. Nedeljković, M.; Antonijević, B.; Matović, V. Simplified Sample Preparation for Fluoride Determination in Biological Material. *Analyst* **1991**, *116*, 477–478. [CrossRef]
41. van Staden, J.F.; van Rensburg, S.D.J. Improvement on the Microdiffusion Technique for the Determination of Ionic and Ionizable Fluoride in Cows' Milk. *Analyst* **1991**, *116*, 807–810. [CrossRef]
42. Chlubek, D. Interakcje fluorków ze składnikami mleka. *Ann. Acad. Med. Stetin.* **1993**, *39*, 23–38. [CrossRef]
43. Liu, C.; Wyborny, L.E.; Chan, J.T. Fluoride content of dairy milk from supermarket. *Fluoride* **1995**, *28*, 10–16. Available online: http://www.fluorideresearch.org/281/files/FJ1995_v28_n1_p010-016.pdf (accessed on 15 November 2022).
44. Kimarua, R.W.; Kariuki, D.N.; Njenga, L.W. Comparison of Two Microdiffusion Methods Used to Measure Ionizable Fluoride in Cows' Milk. *Analyst* **1995**, *120*, 2245–2247. [CrossRef]
45. Kahama, R.W.; Kariuki, D.N.; Kariuki, H.N.; Njenga, L.W. Fluorosis in Children and Sources of Fluoride Around Lake Elmentaita Region of Kenya. *Fluoride* **1997**, *30*, 19–25. Available online: http://www.fluorideresearch.org/301/files/FJ1997_v30_n1_p019-025.pdf (accessed on 15 November 2022).
46. Kahama, R.W.; Damen, J.J.M.; (Bob) ten Cate, J.M. Enzymatic Release of Sequestered Cows' Milk Fluoride for Analysis by the Hexamethyldisiloxane Microdiffusion Method. *Analyst* **1997**, *122*, 855–858. [CrossRef] [PubMed]
47. Peres, R.C.R.; Coppi, L.C.; Volpato, M.C.; Groppo, F.C.; Cury, J.A.; Rosalen, P.L. Cariogenic Potential of Cows', Human and Infant Formula Milks and Effect of Fluoride Supplementation. *Br. J. Nutr.* **2008**, *101*, 376–382. [CrossRef] [PubMed]
48. Yiping, H.; Caiyun, W. Ion Chromatography for Rapid and Sensitive Determination of Fluoride in Milk after Headspace Single-Drop Microextraction with in Situ Generation of Volatile Hydrogen Fluoride. *Anal. Chim. Acta* **2010**, *661*, 161–166. [CrossRef] [PubMed]
49. Ozbek, N.; Akman, S. Determination of Fluorine in Milk Samples via Calcium-Monofluoride by Electrothermal Molecular Absorption Spectrometry. *Food Chem.* **2013**, *138*, 650–654. [CrossRef]
50. Ozbek, N.; Akman, S. Determination of Fluorine in Milk and Water via Molecular Absorption of Barium Monofluoride by High-Resolution Continuum Source Atomic Absorption Spectrometer. *Microchem. J.* **2014**, *117*, 111–115. [CrossRef]
51. Gupta, P. Concentration of Fluoride in Cow's and Buffalo's Milk in Relation to Varying Levels of Fluoride Concentration in Drinking Water of Mathura City in India—A Pilot Study. *J. Clin. Diagn. Res.* **2015**, *9*, LC05–LC07. [CrossRef]
52. Akman, S.; Welz, B.; Ozbek, N.; Pereira, É.R. Chapter 5: Fluorine Determination in Milk, Tea and Water by High-Resolution, High-Temperature Molecular Absorption Spectrometry. In *Food and Nutritional Components in Focus*; Preedy, V.R., Ed.; The Royal Society of Chemistry: London, UK, 2015; pp. 75–95. [CrossRef]
53. Esala, S.; Vuori, E.; Helle, A. Effect of Maternal Fluorine Intake on Breast Milk Fluorine Content. *Br. J. Nutr.* **1982**, *48*, 201–204. [CrossRef]
54. Esala, S.; Vuori, E.; Niinistö, L. Determination of Nanogram Amounts of Fluorine in Breast Milk by Ashing-Diffusion Method and the Fluoride Electrode. *Mikrochim. Acta* **1983**, *79*, 155–165. [CrossRef]
55. Koparal, E.; Ertugrul, F.; Oztekin, K. Fluoride Levels in Breast Milk and Infant Foods. *J. Clin. Pediatr. Dent.* **2000**, *24*, 299–302. [CrossRef]
56. Şener, Y.; Tosun, G.; Kahvecioğlu, F.; Gökalp, A.; Koç, H. Fluoride levels of human plasma and breast milk. *Eur. J. Dent.* **2007**, *1*, 21–24. [CrossRef]
57. Campus, G.; Congiu, G.; Cocco, F.; Sale, S.; Cagetti, M.G.; Sanna, G.; Lingström, P.; Garcia-Godoy, F. Fluoride content in breast milk after the use of fluoridated food supplement. A randomized clinical trial. *Am. J. Dent.* **2014**, *27*, 199–202.
58. Poureslami, H.; Khazaeli, P.; Mahvi, A.H.; Poureslami, K.; Poureslami, P.; Haghani, J.; Aghaei, M. Fluoride level in the breast milk in Koohbanan, a city with endemic dental fluorosis. *Fluoride* **2016**, *49*, 485–494. Available online: https://www.fluorideresearch.org/494Pt2/files/FJ2016_v49_n4Pt2_p485-494_pq.pdf (accessed on 15 November 2022).
59. Hossein Mahvi, A.; Ghanbarian, M.; Ghanbarian, M.; Khosravi, A.; Ghanbarian, M. Determination of Fluoride Concentration in Powdered Milk in Iran 2010. *Br. J. Nutr.* **2011**, *107*, 1077–1079. [CrossRef]

60. Bussell, R.M.; Nichol, R.; Toumba, K.J. Fluoride Levels in UK Infant Milks. *Eur. Arch. Paediatr. Dent.* **2016**, *17*, 177–185. [CrossRef]
61. Noh, H.J.; Sohn, W.; Kim, B.I.; Kwon, H.K.; Choi, C.H.; Kim, H.-Y. Estimation of Fluoride Intake From Milk-Based Infant Formulas and Baby Foods. *Asia Pac. J. Public Health* **2013**, *27*, NP1300–NP1309. [CrossRef]
62. Molska, A.; Gutowska, I.; Baranowska-Bosiacka, I.; Noceń, I.; Chlubek, D. The Content of Elements in Infant Formulas and Drinks Against Mineral Requirements of Children. *Biol. Trace Elem. Res.* **2014**, *158*, 422–427. [CrossRef]
63. Frant, M.S.; Ross, J.W., Jr. Electrode for Sensing Fluoride Ion Activity in Solution. *Science* **1966**, *154*, 1553–1555. [CrossRef]
64. Wieczorek, P.; Sumujlo, D.; Chlubek, D.; Machoy, Z. Interaction of fluoride ions with milk proteins studied by gel filtration. *Fluoride* **1992**, *25*, 171–174. Available online: https://www.fluorideresearch.online/254/files/FJ1992_v25_n4_p165-210.pdf (accessed on 15 November 2022).
65. Campus, G.; Gaspa, L.; Pilo, M.; Scanu, R.; Spano, N.; Cagetti, M.G.; Sanna, G. Performance differences of two potentiometric fluoride determination methods in hard dental tissue. *Fluoride* **2007**, *40*, 111–115. Available online: https://www.fluorideresearch.org/402/files/FJ2007_v40_n2_p111-115.pdf (accessed on 15 November 2022).
66. Campus, G.; Cagetti, M.G.; Spano, N.; Denurra, S.; Cocco, F.; Bossu, M.; Pilo, M.I.; Sanna, G.; Garcia-Godoy, F. Laboratory enamel fluoride uptake from fluoride products. *Am. J. Dent.* **2012**, *25*, 13–16.
67. Campus, G.; Carta, G.; Cagetti, M.G.; Bossù, M.; Sale, S.; Cocco, F.; Conti, G.; Nardone, M.; Sanna, G.; Strohmenger, L.; et al. Fluoride Concentration from Dental Sealants. *J. Dent. Res.* **2013**, *92*, S23–S28. [CrossRef] [PubMed]
68. Cagetti, M.G.; Carta, G.; Cocco, F.; Sale, S.; Congiu, G.; Mura, A.; Strohmenger, L.; Lingstrom, P.; Campus, G.; Bossù, M.; et al. Effect of Fluoridated Sealants on Adjacent Tooth Surfaces: A 30-Mo Randomized Clinical Trial. *J. Dent. Res.* **2014**, *93* (Suppl. 1), 59–65. [CrossRef] [PubMed]
69. Spano, N.; Guccini, V.; Ciulu, M.; Floris, I.; Nurchi, V.M.; Panzanelli, A.; Pilo, M.I.; Sanna, G. Free fluoride determination in honey by ion-specific electrode potentiometry: Method assessment, validation and application to real unifloral samples. *Arab. J. Chem.* **2018**, *11*, 492–500. [CrossRef]
70. Campbell, A.D. Determination of Fluoride in Various Matrices. *Pure Appl. Chem.* **1987**, *59*, 695–702. [CrossRef]
71. Muñoz, A.; Gómez, M.; Palacios, M.A.; Camara, C. Evaluation of Nitric-Induced Teflon Degradation by Spectrochemical Fluoride Analysis and Scanning Microscopy. *Fresenius J. Anal. Chem.* **1993**, *345*, 524–526. [CrossRef]
72. Singer, L.; Ophaug, R.H. Determination of Fluoride in Foods. *J. Agric. Food Chem.* **1986**, *34*, 510–513. [CrossRef]
73. Venkateswarlu, P. Evaluation of Analytical Methods for Fluorine in Biological and Related Materials. *J. Dent. Res.* **1990**, *69*, 514–521. [CrossRef]
74. Carbonaro, M.; Cappelloni, M.; Sabbadini, S.; Carnovale, E. Disulfide Reactivity and In Vitro Protein Digestibility of Different Thermal-Treated Milk Samples and Whey Proteins. *J. Agric. Food Chem.* **1997**, *45*, 95–100. [CrossRef]
75. Kitabatake, N.; Kinekawa, Y.-I. Digestibility of Bovine Milk Whey Protein and β-Lactoglobulin in Vitro and in vivo. *J. Agric. Food Chem.* **1998**, *46*, 4917–4923. [CrossRef]
76. Fox, P.F.; McSweeney, P.L.H. (Eds.) *Dairy Chemistry and Biochemistry*; Blackie Academic Professional: London, UK, 1998; 478p.
77. Cervone, F.; Diaz Brito, J.; Di Prisco, G.; Garofano, H.; Noroña, L.G.; Traniello, S.; Zito, R. Simple Procedures for the Separation and Identification of Bovine Milk Whey Proteins. *Biochim. et Biophys. Acta (BBA)-Protein Struct.* **1973**, *295*, 555–563. [CrossRef]
78. ISO 1211:2010–IDF 1:2010; Milk—Determination of Fat Content—Gravimetric Method (Reference Method). International Organization for Standardization: Geneva, Switzerland, 2010.
79. Landry, J.-C.; Cupelin, F.; Michal, C. Potentiometric Determination of Fluoride by a Combination of Continuous-Flow Analysis and the Gran Addition Method. *Analyst* **1981**, *106*, 1275–1280. [CrossRef]
80. Goncalves, D.A.; Jones, B.T.; Donati, G.L. The Reversed-Axis Method to Estimate Precision in Standard Additions Analysis. *Microchem. J.* **2016**, *124*, 155–158. [CrossRef]
81. Andersen, J.E.T. The Standard Addition Method Revisited. *Trends Anal. Chem.* **2017**, *89*, 21–33. [CrossRef]
82. Harris, D.C. *Quantitative Chemical Analysis*, 8th ed.; W.H. Freeman: New York, NY, USA, 2010; ISBN 1429264845.
83. Wisconsin Department of Natural Resources Laboratory Certification Program. *Analytical Detection Limit Guidance & Laboratory Guide for Determining Method Detection Limits*, PUBL-TS-056-96. 1996. Available online: http://www.iatl.com/content/file/LOD%20Guidance%20Document.pdf (accessed on 15 November 2022).
84. International Union of Pure and Applied Chemistry. Recommendations for Nomenclature of ION-Selective Electrodes. *Pure Appl. Chem.* **1976**, *48*, 127–132. [CrossRef]
85. Horwitz, W. Evaluation of Analytical Methods Used for Regulation of Foods and Drugs. *Anal. Chem.* **1982**, *54*, 67–76. [CrossRef]
86. Dirks, B.; Jongeling-Eijndhoven, J.M.P.A.; Flissebaalje, T.D.; Gedalia, I. Total and Free Ionic Fluoride in Human and Cow's Milk as Determined by Gas-Liquid Chromatography and the Fluoride Electrode. *Caries Res.* **1974**, *8*, 181–186. [CrossRef]
87. Bessho, Y. Determination of Total Fluorine and Ionizable Fluorine Levels in Milk. *Showa Shigakkai Zasshi* **1987**, *7*, 154–165. [CrossRef]
88. Venkateswarlu, P. Sodium Biphenyl Method for Determination of Covalently Bound Fluorine in Organic Compounds and Biological Materials. *Anal. Chem.* **1982**, *54*, 1132–1137. [CrossRef]

Disclaimer/Publisher's Note: The statements, opinions and data contained in all publications are solely those of the individual author(s) and contributor(s) and not of MDPI and/or the editor(s). MDPI and/or the editor(s) disclaim responsibility for any injury to people or property resulting from any ideas, methods, instructions or products referred to in the content.

Article

In Situ Raman Investigation of TiO$_2$ Nanotube Array-Based Ultraviolet Photodetectors: Effects of Nanotube Length

Yanyu Ren [1], Xiumin Shi [2], Pengcheng Xia [1], Shuang Li [1], Mingyang Lv [1], Yunxin Wang [3] and Zhu Mao [1,*]

[1] School of Chemistry and Life Science, Advanced Institute of Materials Science, Changchun University of Technology, Changchun 130012, China; ryy17543012961@163.com (Y.R.); x13385955107@163.com (P.X.); lishuang1071418026@163.com (S.L.); lmy2508644298@163.com (M.L.)
[2] College of Chemical Engineering, Changchun University of Technology, Changchun 130012, China; shixiumin.edu.cn
[3] Jilin Provincial Center for Disease Control and Prevention, Changchun 130062, China; xierlian@163.com
* Correspondence: maozhu@ccut.edu.cn

Academic Editors: Gavino Sanna and Barbara Bonelli
Received: 20 March 2020; Accepted: 15 April 2020; Published: 17 April 2020

Abstract: TiO$_2$ nanotube arrays (TNAs) with tube lengths of 4, 6, and 7 µm were prepared via two-step anodization. Thereafter, ultraviolet (UV) photodetectors (PDs) with Au/TiO$_2$/Au structures were prepared using these TNAs with different tube lengths. The effects of TNA length and device area on the performance of the device were investigated using in situ Raman spectroscopy. The maximum laser/dark current ratio was achieved by using a TNA with a size of 1×1 cm^2 and a length of 7 µm, under a 532 nm laser. In addition, when the device was irradiated with a higher energy laser (325 nm), the UV Raman spectrum was found to be more sensitive than the visible Raman spectrum. At 325 nm, the laser/dark current ratio was nearly 24 times higher than that under a 532 nm laser. Six phonon modes of anatase TNAs were observed, at 144, 199, 395, 514, and 635 cm^{-1}, which were assigned to the $E_{g(1)}$, $E_{g(2)}$, $B_{1g(1)}$, $A_{1g}/B_{1g(2)}$, and $E_{g(3)}$ modes, respectively. The strong low-frequency band at 144 cm^{-1} was caused by the O-Ti-O bending vibration and is a characteristic band of anatase. The results show that the performance of TNA-based PDs is length-dependent. Surface-enhanced Raman scattering signals of 4-mercaptobenzoic acid (4-MBA) molecules were also observed on the TNA surface. This result indicates that the length-dependent performance may be derived from an increase in the specific surface area of the TNA. In addition, the strong absorption of UV light by the TNAs caused a blueshift of the $E_{g(1)}$ mode.

Keywords: TiO$_2$ nanotube arrays; UV photodetector; Raman spectroscopy; Surface-enhanced Raman scattering; SERS

1. Introduction

Highly ordered TiO$_2$ nanotube arrays (TNAs) were synthesized for the first time in 2001 [1]. Because of their controllable diameter, uniform morphology, and large specific surface area, they have been widely used in various industries, such as for gas sensors, water light solutions, dye-sensitized solar cells, and electrochromic devices [2–8]. In addition, because of the large bandgap of TiO$_2$ (3.2 eV for anatase and 3.0 eV for the rutile structure) there is no need to filter out visible or infrared light, which is ideal for ultraviolet (UV) detection applications [9–15]. Many studies have been conducted to improve the performance of TiO$_2$-based photodetectors (PDs) [16].

It is essential to understand the surface and interface structures of PDs for the further development and optimization of the performance of UV PDs [17–22]. However, the current characterization technologies for PD-related devices are mostly based on electrochemical methods. They can collect the photocurrent–voltage curves and transient photocurrent response relationships of PDs [23–25]. Thus, only limited knowledge regarding the surface and interface structures of these devices is available. Therefore, characterization techniques that can provide detailed information about the TiO_2 structure of a device in situ remain a challenge.

Raman spectroscopy utilizes a scattering spectrum that is based on the Raman scattering effect, discovered by an Indian scientist, Sir C.V. Raman [26]. It is an analytical method that can analyze the scattering spectrum at a frequency different from the incident light in order to obtain molecular vibration and rotation information; it is also used in molecular structure research. Raman spectroscopy is widely used to characterize the crystallinity and crystallographic orientation of metal oxide materials [27]. As with the detection of target molecules, solid materials can be identified by their characteristic phonon patterns. In situ Raman technology has the characteristics of micron-level spatial resolution and non-destructive detection [28,29]. Raman spectroscopy has proven itself to be a powerful characterization technique for obtaining detailed information about the molecular structure of metal oxides [30–32]. This is because each molecular state has a unique vibration spectrum related to its structure. Moreover, Raman spectroscopy is suitable for in situ research [33] and can also judge and analyze doping, defects, and small changes in the crystal structure of molecules and crystals. Its advantages include simple sample preparation, quick and easy testing, and nondestructive testing at room temperature.

In this work, a Raman laser was used as both the radiation source of the PD, and the excitation laser of the Raman spectrum, in order to characterize both the photoelectric performance and the Raman spectrum of the PD. The results of this study will help researchers to understand the effect of the TiO_2 structure on the performance of UV detectors at the phonon level.

2. Results and Discussion

An illustration of the preparation of the TNAs and PDs is shown in Figure 1. First, the TiO_2 TNAs were peeled from the Ti foil and then transferred to an Au interdigital electrode. The binder was a TiO_2 sol. The results show that the PD with a metal–semiconductor–metal (MSM) structure has a good performance because of its Schottky contacts.

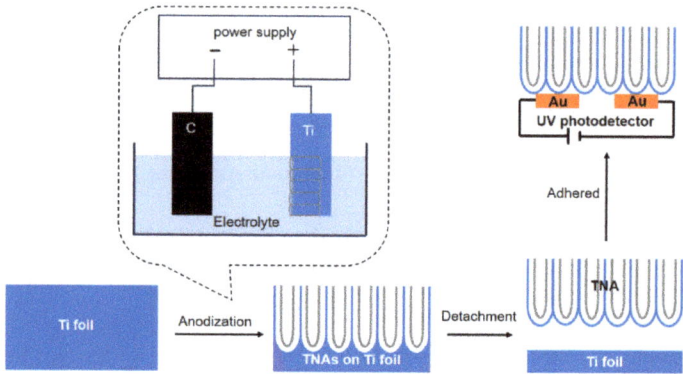

Figure 1. Preparation process of TiO_2 nanotube array (TNA)-based ultraviolet (UV) photodetectors (PDs).

In this work, the length of the tubes was adjusted by changing the time of secondary anodization, and TNAs with secondary oxidation times of 1 h, 2 h, and 3 h were prepared. The scanning electron microscopy (SEM) images shown in Figure 2 reveal that the tube lengths of the TNAs after 1, 2, and 3 h were 4, 6, and 7 μm, respectively, and the average diameter was less than 100 nm. As shown in

Figure 2, the surface of the TNAs is flat and clean. As shown in Figure 2, it is obvious that it takes 1 h to grow from 6 µm to 7 µm. As the tube becomes longer, its growth rate will be slower. In addition, it was also found that as the length of the TNA increases, it becomes easier to fracture during the transfer process, and the performance difference caused by this result is also difficult to rule out. Therefore, we have selected tube lengths of 4 µm, 6 µm, and 7 µm for discussion.

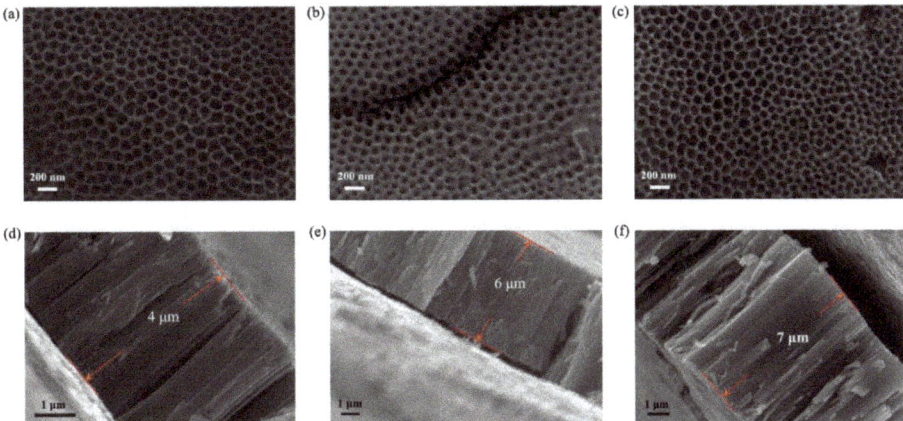

Figure 2. Scanning electron microscopy images of TNAs with different tube lengths: (**a**) top and (**d**) side views with secondary oxidation time of 1 h; (**b**) top and (**e**) side views with secondary oxidation time of 2 h; (**c**) top and (**f**) side views with 3 h oxidation time.

Figure 3 shows the X-ray diffraction (XRD) patterns of the TNAs with tube lengths of 4, 6, and 7 µm after calcination at 450 °C. As shown in Figure 3a, the seven peaks at 25.3°, 48.1°, 54.9°, 62.9°, and 70.6° are basically consistent with the corresponding sample structure No. 84-1285 in the JCPDS file and respectively correspond to the (101), (200), (211), (204), and (220) directions, proving that the TNAs are anatases. The results show that as the tube length increases, the intensity of the XRD peaks increases. The Raman spectra of the uncalcined and calcined TNA were characterized. As shown in Figure 3b, the Raman spectrum of the calcined TNA shows distinct anatase characteristic phonon modes. Furthermore, as shown in the photographs in Figure 3b, the TNA before calcination appears yellow, while after calcination, it is white. In addition, many previous publications have stated that a calcined temperature of 450 °C is sufficient to obtain the anatase structure [34–36]. The intense $E_{g(1)}$ mode (144 cm^{-1}) proves that the main phase of the TNA is anatase.

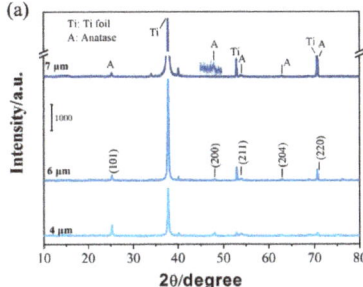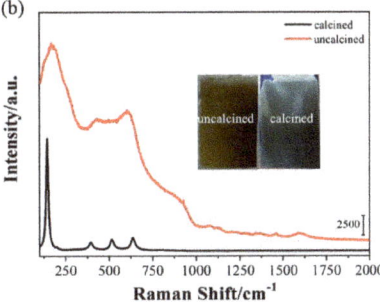

Figure 3. (**a**) X-ray diffraction patterns of TNAs with different tube lengths; (**b**) Raman spectra of uncalcined and calcined TNAs (inset: photographs of uncalcined and calcined TNAs).

Four PDs with different areas were prepared using TNAs with tube lengths of 6 μm. The device areas were 2 × 2, 1.5 × 1.26, 2 × 1, and 1 × 1 cm^2 (the width and spacing of the interdigitated Au were both 100 μm). An excitation wavelength of 325 nm was used for the irradiation light source (the laser power was 6.3 mW) in order to test the laser/dark current response performance of the devices.

To clarify the definition of the device area, a schematic view of the device area on the Au interdigitated electrode is shown in Figure 4e. On the premise of keeping the width and spacing of the interdigitated electrode at 100 μm, we changed the device area by changing the length and/or the number of the interdigitated Au. As shown in Figure 4e, the red dotted border represents the typical device area. In the preparation of large-area devices, the TNA film may inevitably break during the transfer process, resulting in reduced conductivity. This is the main reason for the serious degradation of device performance. Therefore, in this work, the main purpose of exploring the device area effect is to indirectly screen out the TNA film area that, in some cases, provides a better performance.

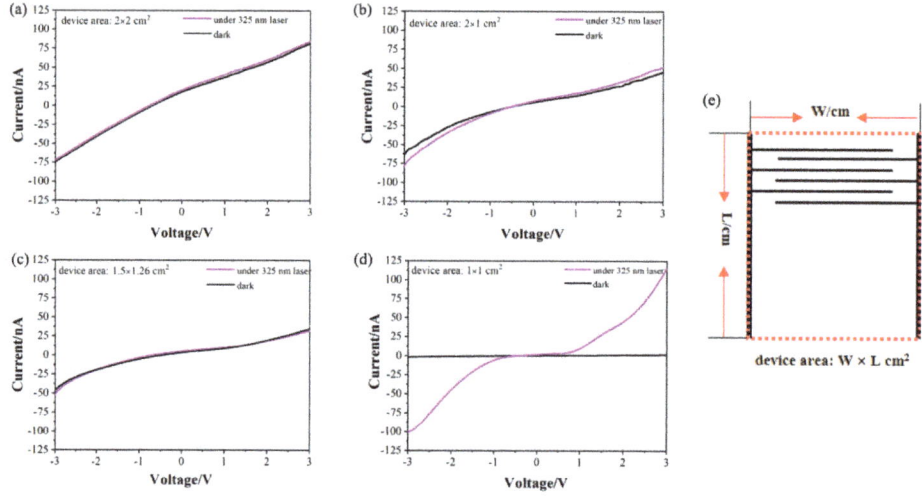

Figure 4. Current-voltage (I-V) curves of PDs with different device areas: (**a**) 2 × 2 cm^2; (**b**) 2 × 1 cm^2; (**c**) 1.5 × 1.26 cm^2; and (**d**) 1 × 1 cm^2 under 325 nm laser. (**e**) Schematic view of device area (W × L, red dotted border) on the Au interdigitated electrode.

As shown in Figure 4, the laser/dark current ratios of the PDs were calculated under a bias of 3 V. The results show that the laser/dark current ratios of the PDs with device areas of 2 × 2, 1.5 × 1.26, 1 × 2, and 1 × 1 cm^2 were 1.03, 1.07, 1.13, and 79.70, respectively. As shown in Figure 4d, the dark current of the PD with an area of 1 × 1 cm^2 was 1.447 nA, and its photocurrent was 115.30 nA. The results show that the laser/dark current ratio of the device with the area of 1 × 1 cm^2 was the largest, indicating that this device has the highest light responsivity at 325 nm. Therefore, the PD with an area of 1 × 1 cm^2 was used for subsequent experiments.

To explore the impact of laser energy on the performance of a PD, 532 nm and 325 nm lasers were used to simultaneously detect the laser/dark current ratio and Raman spectra of the PD with an area of 1 × 1 cm^2. The 532 nm laser was used with a 50 × objective lens (NA 0.5), and the 325 nm laser was used with a 15 × objective lens (NA 0.32). The spot sizes were approximately 1.30 μm for the 532 nm laser, and 1.24 μm for the 325 nm laser (spot size = 1.22 λ/NA).

Firstly, a 532 nm laser was used to study the effect of length on the PD performance. Figure 5 shows the I–V curves and the fast-response measurements of TNA-based PDs with different tube lengths. Here, when the bias was 3 V, the laser/dark current ratios of the PDs with tube lengths of 4, 6, and 7 μm were 1.89, 2.14, and 5.89, respectively. The results show that as the length of the TNA tube increases, the laser/dark current ratio of the device increases accordingly. This result indicates that as the length of the tube increases, the transfer of photogenerated charge increases—that is, as the length of the tube increases, the light response of the device also increases.

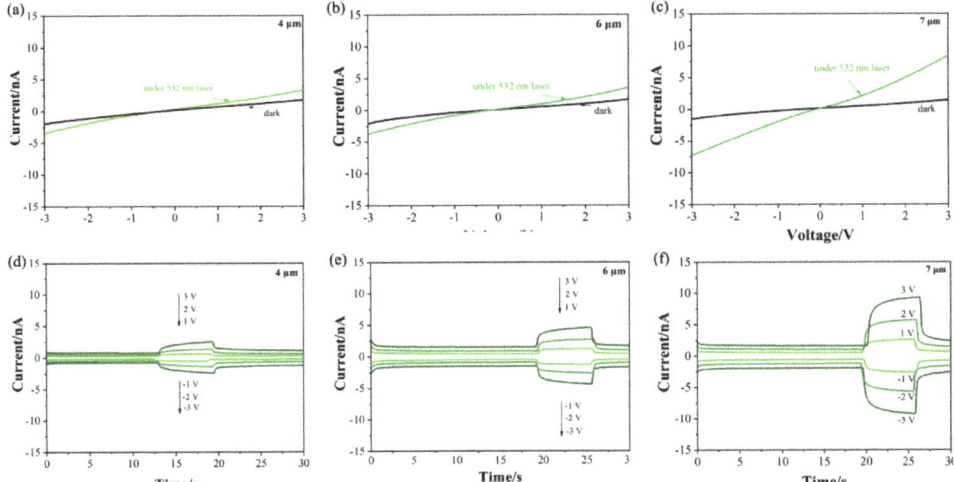

Figure 5. Performance of TNA-based PDs with different tube lengths in the dark and under a 532 nm laser: (a) 4 μm; (b) 6 μm; (c) 7 μm. Fast response measurements with different tube lengths: (d) 4 μm; (e) 6 μm; (f) 7 μm. (laser power: 5.3 mW).

Secondly, the in situ Raman spectra of TNA-based PDs with different tube lengths were collected using a 532 nm laser. The tetragonal anatase phase of a TNA has six Raman-active phonons in the vibration spectrum: $3E_g + 2B_{1g} + A_{1g}$. As shown in Figure 6, all six modes were observed at 144, 199, 395, 514, and 635 cm^{-1}, which were assigned to the $E_{g(1)}$, $E_{g(2)}$, $B_{1g(1)}$, $A_{1g}/B_{1g(2)}$, and $E_{g(3)}$ modes, respectively. The strong low-frequency band at 144 cm^{-1} was caused by the O-Ti-O bending vibration and is a characteristic band of anatase. As shown in Figure 6a, as the tube length increased, the intensity of the $E_{g(1)}$ mode increased, indicating that the $E_{g(1)}$ mode of the TNA has a tube length-dependent effect.

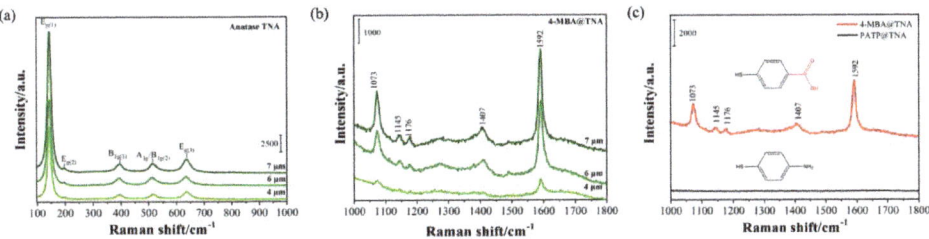

Figure 6. (a) In situ Raman spectra of TNA with different tube lengths; (b) surface-enhanced Raman scattering spectra (SERS) of 4-mercaptobenzoic acid (4-MBA)-modified TNA with different lengths; (c) Raman spectra of 4-MBA- and 4-aminothiophenol (PATP)-modified TNAs. The wavelength of excitation is 532 nm.

Moreover, as shown in Figure 6c, it is clear that the Raman signal of 4-mercaptobenzoic acid (4-MBA) molecules was enhanced on the surface of the TNA, but the Raman signal of 4-aminothiophenol (PATP) molecules was not enhanced. The results indicate that there were abundant hydroxyl groups on the surface of the TNA, which can strongly interact with the carboxyl groups of the 4-MBA molecules, thereby promoting the charge transfer resonance between the TNA and 4-MBA. In addition, the surface-enhanced Raman scattering (SERS) [37] activity of the TNA has a length-dependent effect. As shown in Figure 6b, as the length of the TNA increased, the SERS intensity of 4-MBA increased; more specifically, when the tube length was 7 μm the SERS signal was at its highest. The surface area is proportional to the volume of the tubes. However, it has been unanimously agreed in previous publications that the oxidation voltage is the main factor affecting the diameter of the TNA [38–40]. Moreover, the oxidation time is the main factor that affects the length of the TNA. In this work, to highlight the influence of the TNA length, the oxidation voltage is always constant (i.e., 60 V). In Figure 2, it is clear that there is almost no difference in the diameter of the TNAs. As the oxidation time increases, the length of the prepared TNA also increases, resulting in an increase in the specific surface area; moreover, the number of molecules adsorbed on the TNA surface increases, which ultimately leads to an increase in SERS activity. The high specific surface area of the 7 μm TNA also influences the light response performance of the TNA-based PD.

The intense bands at approximately 1592 and 1073 cm^{-1} can be assigned to totally symmetric C–C stretching and in-plane ring breathing, respectively [41,42]. From the SERS spectra of the 4-MBA-adsorbed TNA, other weak bands were also observed, corresponding to the C–H deformation modes at 1145 and 1176 cm^{-1}, respectively [41,43]. A more detailed assignment of the Raman bands of 4-MBA is shown in Table 1. It is clear from Figure 6b that when the length of the TNA was 7 μm a significantly enhanced SERS signal was obtained. In this case, the enhanced SERS signal likely originated from the charge transfer (CT) effect.

Table 1. Raman bands and assignments of 4-MBA-modified TNA.

Raman Band	Assignment
144 cm^{-1}	$E_{g(1)}$ (Anatase TiO$_2$)
199 cm^{-1}	$E_{g(2)}$ (Anatase TiO$_2$)
395 cm^{-1}	$B_{1g(1)}$ (Anatase TiO$_2$)
514 cm^{-1}	$A_{1g}/B_{1g(2)}$ (Anatase TiO$_2$)
635 cm^{-1}	$E_{g(3)}$ (Anatase TiO$_2$)
1073 cm^{-1}	in-plane ring breathing + C–S stretching (4-MBA)
1148 cm^{-1}	C–H deformation (4-MBA)
1176 cm^{-1}	C–H deformation (4-MBA)
1407 cm^{-1}	COO$^-$ stretching (4-MBA)
1592 cm^{-1}	totally symmetric C–C stretching (4-MBA)

Finally, the performance of a TNA-based PD with a tube length of 7 μm was tested using a 325 nm laser. As shown in Figure 7, the results show that when the applied bias was 3 V, the laser/dark current ratio of the device was 139.1, which is almost 24 times larger than that under a 532 nm laser.

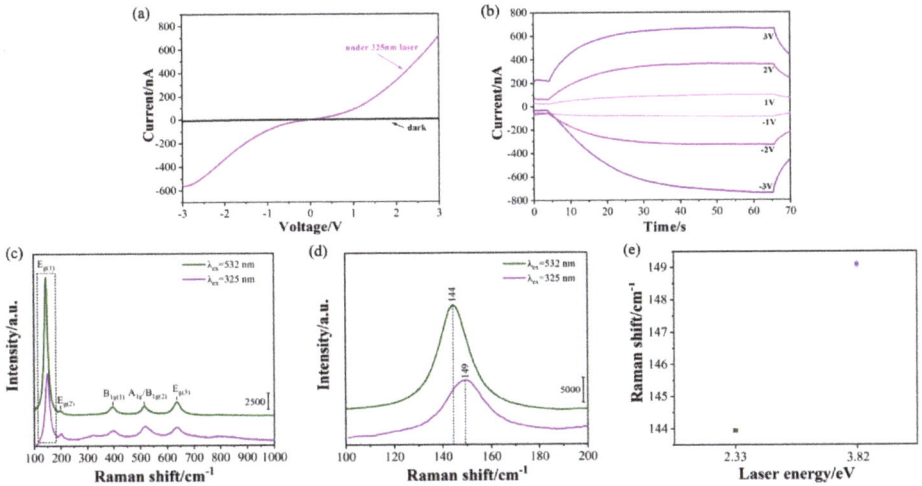

Figure 7. (**a**) *I-V* curve and (**b**) fast-response measurements of a TNA-based PD with a tube length of 7 μm under 325 nm excitation; (**c**) in situ Raman spectra of TNA-based PD collected at excitation wavelengths of 532 nm and 325 nm; (**d**) enlarged Raman spectrum in the 110–200 cm^{-1} region; (**e**) The reproducibility of Raman measurements.

With an excitation wavelength of 325 nm, the phonon modes of $E_{g(1)}$, $E_{g(2)}$, $B_{1g(1)}$, $A_{1g}/B_{1g(2)}$, and $E_{g(3)}$ were observed. It is worth mentioning that the frequency of the $E_{g(1)}$ mode shifted to 149 cm^{-1} compared with that for 532 nm excitation. This result suggests that the value of the laser energy (3.82 eV) is greater than the bandgap of the TNA, which causes photons to effectively interact with the phonons of TNA. This increase in frequency was derived from resonant Raman scattering. This work demonstrates the ability of in situ Raman technology to reveal the internal relationships between device interface structure and performance.

3. Materials and Methods

3.1. Chemicals

Ammonium fluoride, hydrogen fluoride, and n-butyl titanate were purchased from Macleans Corporation. Ethylene glycol, hydrogen peroxide, and isopropanol was purchased from Tianjin Fuyu Fine Chemical Co., Ltd. (Tianjin, China). Ethanol, hydrochloric acid, acetone, and methanol were purchased from Beijing Chemical Industry Group Co., Ltd. (Beijing, China). The abovementioned reagents were used without further purification. The water used in the experiment was ultrapure water. 4-MBA and PATP were purchased from Sigma-Aldrich Co., Ltd. (St. Louis, MO, USA) and used without further purification.

3.2. Instruments

SEM characterization was performed using a Shimadzu SSX-550 scanning electron microscope: the acceleration voltage was 3.0 kV. XRD was performed using a Japanese Rigaku Smartlab X-ray diffractometer with Cu-Kα rays (λ = 1.5418 Å) at 45 kV and 200 mA. The Raman spectra were measured using a Horiba JY LabRAM HR Evolution confocal micro-Raman spectroscopy equipped with a multichannel air-cooled charge-coupled device detector. The 532 nm and 325 nm lasers were employed as excitation laser sources. The Raman spectra for an excitation of 532 nm were measured using a 50× objective lens, 5.3 mW laser power at the sample, and 1800 gr/mm grating. The Raman spectra for the

325 nm excitation were measured using a 15× objective lens, 6.3 mW laser power at the sample, and 2400 gr/mm grating.

3.3. Preparation of TiO$_2$ Nanotube Arrays

First, the Ti foil was cut to an area of 4×5 cm^2 and ultrasonically cleaned in acetone, isopropanol, and methanol solutions for 10 min to remove impurities on the surface. Thereafter, the Ti foil was dried using an N$_2$ flow. TNA was prepared via anodic oxidation using an ethylene glycol solution of NH$_4$F and deionized water (0.3% NH$_4$F + 2% H$_2$O + 500 mL ethylene glycol) as the electrolyte, with Ti foil as the anode, and a graphite sheet as the cathode.

Anodization was performed twice, and the anodizing process is depicted in Figure 1. The first oxidation time was 3 h with a voltage of 60 V. The first TNA film obtained was peeled from the Ti foil and rinsed with a large amount of deionized water in order to obtain a regular and ordered Ti substrate, after which it was dried with N$_2$ flow. The second oxidation process was performed for either 1 h, 2 h, or 3 h at a voltage of 60 V. The TNA samples of different lengths were then calcined at 450 °C for 2 h to obtain the anatase phase.

3.4. Preparation of TiO$_2$ Sol

A set amount of 5.00 mL of n-butyl titanate and 9 mL of ethanol were mixed and stirred well to form component A. Then, 18 mL of ethanol, 1 mL of HCl, and 1 mL of H$_2$O were mixed and stirred well to obtain component B. Then, component A was added to component B and stirred for 4 h to obtain a yellow TiO$_2$ sol.

3.5. Preparation of the TiO$_2$ Nanotube Array-Based Ultraviolet Photodetector

To prepare an MSM-structured PD, the calcined TNA was stripped. Then, the TNA was oxidized at 60 V for 20 min. Thereafter, the TNA was cleaned and immersed in 40% H$_2$O$_2$ for 15 min to separate the TNA film from the Ti foil. The TiO$_2$ sol was then dropped on the Au interdigital electrode with Al$_2$O$_3$ as a substrate, and the peeled TNA film was transferred to the Au interdigital electrode. Following this, the film was calcined at 450 °C for 2 h. After annealing, the TNA film was located perpendicular to the Au interdigital electrode as an MSM structure, as shown in Figure 1.

4. Conclusions

In this work, TNA-based PDs were prepared with TNA lengths of 4, 6, and 7 μm. Firstly, the effect of the device area (2×2, 1.5×1.26, 2×1, and 1×1 cm^2) on the performance of the laser response was investigated. The results show that the device with an area of 1×1 cm^2 exhibited the best performance. Secondly, we analyzed the effect of tube length on the performance of the laser response. The results show that the device with the longest (7 μm) tube had the best laser response performance, with a laser/dark current ratio of 5.89. At 325 nm, the laser/dark current ratio was 139.1. In addition, as the tube length increased, the strength of phonon mode $E_{g(1)}$ increased, indicating that the strength of the $E_{g(1)}$ mode is length-dependent. Under irradiation of a 325 nm laser, the $E_{g(1)}$ mode shifted to a higher frequency by 5 cm^{-1} compared with that under the 532 nm laser. The SERS signals of 4-MBA molecules were also observed on the TNA surface. This result indicates that the length-dependent performance may be derived from the increase in the specific surface area of the TNA. Furthermore, in situ Raman spectroscopy successfully revealed the dependence of the phonon mode of TiO$_2$ on the laser energy and TNA length of a TNA-based PD. The results of this study may help researchers understand the effects of the TiO$_2$ structure on the performance of UV detectors from a phonon perspective.

Author Contributions: Conceptualization, Z.M.; methodology, Z.M.; validation and investigation, Y.R. and P.X.; resources, Z.M.; data curation, Z.M., Y.R., P.X., S.L., M.L., X.S., and Y.W.; writing—original draft preparation, Y.R.; writing—review and editing, Z.M.; funding acquisition, Z.M. All authors have read and agreed to the published version of the manuscript.

Funding: This research was funded by the National Natural Science Foundation of China (No. 21503021) and the Science and Technology Projects in the 13th Five-year Plan of the Education Department of Jilin Province, China (No. JJKH20200650KJ).

Acknowledgments: Z.M. thanks Haiming Lv (Shaanxi Key Laboratory of Chemical Reaction Engineering, College of Chemistry and Chemical Engineering, Yan'an University, Yan'an 716000, China) for the technical support provided for the preparation of TiO_2 nanotube arrays.

Conflicts of Interest: The authors declare no conflict of interest.

References

1. Gong, D.; Grimes, C.A.; Varghese, O.K.; Hu, W.; Singh, R.S.; Chen, Z.; Dickey, E. Titanium oxide nanotube arrays prepared by anodic oxidation. *J. Mater. Res.* **2001**, *16*, 3331–3334. [CrossRef]
2. Varghese, O.K.; Gong, D.; Paulose, M.; Ong, K.G.; Grimes, C. Hydrogen sensing using titania nanotubes. *Sens. Actuators B Chem.* **2003**, *93*, 338–344. [CrossRef]
3. Nakagawa, H.; Yamamoto, N.; Okazaki, S.; Chinzei, T.; Asakura, S. A room-temperature operated hydrogen leak sensor. *Sens. Actuators B Chem.* **2003**, *93*, 468–474. [CrossRef]
4. Grimes, C.A. Synthesis and application of highly ordered arrays of TiO_2 nanotubes. *J. Mater. Chem.* **2007**, *17*, 1451–1457. [CrossRef]
5. Allam, N.K.; Grimes, C.A. Effect of cathode material on the morphology and photoelectrochemical properties of vertically oriented TiO_2 nanotube arrays. *Sol. Energy Mater. Sol. Cells* **2008**, *92*, 1468–1475. [CrossRef]
6. Yang, L.; Yang, W.; Cai, Q. Well-dispersed Pt Au nanoparticles loaded into anodic titania nanotubes: A high antipoison and stable catalyst system for methanol oxidation in alkaline media. *J. Phys. Chem. C* **2007**, *111*, 16613–16617. [CrossRef]
7. Grimes, C.A.; Ong, K.G.; Varghese, O.K.; Yang, X.; Mor, G.; Paulose, M.; Dickey, E.; Ruan, C.; Pishko, M.V.; Kendig, J.W.; et al. A Sentinel Sensor Network for Hydrogen Sensing. *Sensors* **2003**, *3*, 69–82. [CrossRef]
8. Zhang, Y.; Fu, W.; Yang, H.; Qi, Q.; Zeng, Y.; Zhang, T.; Ge, R.; Zou, G. Synthesis and characterization of TiO_2 nanotubes for humidity sensing. *Appl. Surf. Sci.* **2008**, *254*, 5545–5547. [CrossRef]
9. Hong, S.P.; Kim, S.; Kim, N.; Yoon, J.; Kim, C. A short review on electrochemically self-doped TiO_2 nanotube arrays: Synthesis and applications. *Korean J. Chem. Eng.* **2019**, *36*, 1753–1766. [CrossRef]
10. Zhou, X.; Liu, N.; Schmuki, P. Photocatalysis with TiO_2 nanotubes:"colorful" reactivity and designing site-specific photocatalytic centers into TiO_2 nanotubes. *ACS Catal.* **2017**, *7*, 3210–3235. [CrossRef]
11. Fan, H.; Zhang, S.; Zhu, X. Nitrided TiO_2 nanoparticles/nanotube arrays for better electrochemical properties. *Chem. Phys. Lett.* **2019**, *730*, 340–344. [CrossRef]
12. Zhang, W.; Liu, Y.; Guo, F.; Liu, J.; Yang, F. Kinetic analysis of the anodic growth of TiO_2 nanotubes: Effects of voltage and temperature. *J. Mater. Chem. C* **2019**, *7*, 14098–14108. [CrossRef]
13. Zheng, Z.; Zhuge, F.; Wang, Y.; Zhang, J.; Gan, L.; Zhou, X.; Li, H.; Zhai, T. Decorating Perovskite Quantum Dots in TiO_2 Nanotubes Array for Broadband Response Photodetector. *Adv. Funct. Mater.* **2017**, *27*, 1703115. [CrossRef]
14. Motola, M.; Čaplovičová, M.; Krbal, M.; Sopha, H.; Thirunavukkarasu, G.K.; Gregor, M.; Plesch, G.; Macak, J.M. Ti^{3+} doped anodic single-wall TiO_2 nanotubes as highly efficient photocatalyst. *Electrochim. Acta* **2020**, *331*, 135374. [CrossRef]
15. Marien, C.B.D.; Cottineau, T.; Robert, D.; Drogui, P. TiO_2 Nanotube arrays: Influence of tube length on the photocatalytic degradation of Paraquat. *Appl. Catal. B Environ.* **2016**, *194*, 1–6. [CrossRef]
16. Ouyang, W.; Teng, F.; Fang, X. High Performance BiOCl Nanosheets/TiO_2 Nanotube Arrays Heterojunction UV Photodetector: The Influences of Self-Induced Inner Electric Fields in the BiOCl Nanosheets. *Adv. Funct. Mater.* **2018**, *28*, 1707178. [CrossRef]
17. Ge, M.Z.; Cao, C.Y.; Huang, J.Y.; Li, S.; Zhang, S.-N.; Deng, S.; Li, Q.; Zhang, K.-Q.; Lai, Y. Synthesis, modification, and photo/photoelectrocatalytic degradation applications of TiO_2 nanotube arrays: A review. *Nanotechnol. Rev.* **2016**, *5*, 75–112. [CrossRef]
18. Zheng, L.; Hu, K.; Teng, F.; Fang, X. Novel UV-Visible Photodetector in Photovoltaic Mode with Fast Response and Ultrahigh Photosensitivity Employing Se/TiO_2 Nanotubes Heterojunction. *Small* **2016**, *13*, 1602448. [CrossRef]

19. Chen, D.; Wei, L.; Meng, L.; Wang, D.; Chen, Y.; Tian, Y.-F.; Yan, S.; Mei, L.; Jiao, J. High-Performance Self-Powered UV Detector Based on SnO$_2$-TiO$_2$ Nanomace Arrays. *Nanoscale Res. Lett.* **2018**, *13*, 92. [CrossRef]
20. Hosseini, Z.S.; Shasti, M.; Sani, S.R.; MortezaAli, A. Photo-detector diode based on thermally oxidized TiO$_2$ nanostructures/p-Si heterojunction. *J. Appl. Phys.* **2016**, *119*, 14503. [CrossRef]
21. Li, S.; Deng, T.; Zhang, Y.; Li, Y.; Yin, W.; Chen, Q.; Liu, Z. Solar-blind ultraviolet detection based on TiO$_2$ nanoparticles decorated graphene field-effect transistors. *Nanophotonics* **2019**, *8*, 899–908. [CrossRef]
22. Zou, J.; Zhang, Q.; Huang, K.; Marzari, N. Ultraviolet Photodetectors Based on Anodic TiO$_2$ Nanotube Arrays. *J. Phys. Chem. C* **2010**, *114*, 10725–10729. [CrossRef]
23. Jahromi, H.D.; Sheikhi, M.H.; Yousefi, M.H. A numerical approach for analyzing quantum dot infrared photodetectors' parameters. *Opt. Laser Technol.* **2012**, *44*, 572–577. [CrossRef]
24. Gödel, K.C.; Steiner, U. Thin film synthesis of SbSI micro-crystals for self-powered photodetectors with rapid time response. *Nanoscale* **2016**, *8*, 15920–15925. [CrossRef] [PubMed]
25. Soci, C.; Zhang, A.; Bao, X.-Y.; Kin, H.; Lo, Y.; Wang, D. Nanowire Photodetectors. *J. Nanosci. Nanotechnol.* **2010**, *10*, 1430–1449. [CrossRef]
26. Raman, C.V.; Krishnan, K.S. A new type of secondary radiation. *Nature* **1928**, *121*, 501–502. [CrossRef]
27. Baddour-Hadjean, R.; Pereira-Ramos, J.P. Raman Microspectrometry Applied to the Study of Electrode Materials for Lithium Batteries. *Chem. Rev.* **2010**, *110*, 1278–1319. [CrossRef]
28. Ma, L.; Zhang, T.; Song, R.; Guo, L. In-situ Raman study of relation between microstructure and photoactivity of CdS@TiO$_2$ core-shell nanostructures. *Int. J. Hydrogen Energy* **2018**, *43*, 13778–13787. [CrossRef]
29. Dong, Z.; Xiao, F.; Zhao, A.; Liu, L.; Sham, T.-K.; Song, Y. Pressure induced structural transformations of anatase TiO$_2$ nanotubes probed by Raman spectroscopy and synchrotron X-ray diffraction. *RSC Adv.* **2016**, *6*, 76142–76150. [CrossRef]
30. Butler, H.J.; Ashton, L.; Bird, B.; Cinque, G.; Curtis, K.; Dorney, J.; Esmonde-White, K.; Fullwood, N.J.; Gardner, B.; Martin-Hirsch, P.L.; et al. Using Raman spectroscopy to characterize biological materials. *Nat. Protoc.* **2016**, *11*, 664–687. [CrossRef]
31. Shipp, D.W.; Sinjab, F.; Notingher, I. Raman spectroscopy: Techniques and applications in the life sciences. *Adv. Opt. Photon.* **2017**, *9*, 315–428. [CrossRef]
32. Panneerselvam, R.; Liu, G.K.; Wang, Y.H.; Liu, J.-Y.; Ding, S.-Y.; Li, J.-F.; Wu, D.-Y.; Tian, Z. Surface-enhanced Raman spectroscopy: Bottlenecks and future directions. *Chem. Commun.* **2018**, *54*, 10–25. [CrossRef]
33. Hardwick, L.J.; Holzapfel, M.; Novák, P.; Dupont, L.; Baudrin, E. Electrochemical lithium insertion into anatase-type TiO$_2$: An in situ Raman microscopy investigation. *Electrochim. Acta* **2007**, *52*, 5357–5367. [CrossRef]
34. Zhong, D.; Jiang, Q.; Huang, B.; Zhang, W.-H.; Li, C. Synthesis and characterization of anatase TiO$_2$ nanosheet arrays on FTO substrate. *J. Energy Chem.* **2015**, *24*, 626–631. [CrossRef]
35. Yao, F.U.; Wang, H.C. Preparation of transparent TiO$_2$ nanocrystalline film for UV sensor. *Chin. Sci. Bull.* **2006**, *51*, 1657–1661.
36. Mazza, T.; Barborini, E.; Piseri, P.; Milani, P.; Cattaneo, D.; Bassi, A.L.; Bottani, C.E.; Ducati, C. Raman spectroscopy characterization of TiO$_2$ rutile nanocrystals. *Phys. Rev. B Condens. Matter.* **2007**, *75*, 045416-1–045416-5. [CrossRef]
37. Campion, A.; Kambhampati, P. Surface-enhanced Raman scattering. *Chem. Soc. Rev.* **1998**, *27*, 241–250. [CrossRef]
38. Li, H.Y.; Wang, J.S.; Huang, K.L.; Sun, G.S.; Zhou, M.L. In situ preparation of multi-layer TiO$_2$ nanotube array thin films by anodic oxidation metho. *Mater. Lett.* **2011**, *65*, 1188–1190. [CrossRef]
39. Liang, F.; Zhang, J.; Zheng, L.; Tsang, C.-K.; Li, H.; Shu, S.; Cheng, H.; Li, Y.Y. Selective electrodeposition of Ni into the intertubular voids of anodic TiO$_2$ nanotubes for improved photocatalytic properties. *J. Mater. Res.* **2012**, *28*, 405–410. [CrossRef]
40. Li, Y.; Ma, Q.; Han, J.; Ji, L.; Wang, J.; Chen, J.; Wang, Y. Controllable preparation, growth mechanism and the properties research of TiO$_2$ nanotube arrays. *Appl. Surf. Sci.* **2014**, *297*, 103–108. [CrossRef]
41. Yang, L.; Jiang, X.; Ruan, W.; Zhao, B.; Xu, W.; Lombardi, J.R. Observation of Enhanced Raman Scattering for Molecules Adsorbed on TiO$_2$ Nanoparticles: Charge-Transfer Contribution. *J. Phys. Chem. C* **2008**, *112*, 20095–20098. [CrossRef]

42. Zhang, X.; Yu, Z.; Ji, W.; Sui, H.; Cong, Q.; Wang, X.; Zhao, B. Charge-Transfer Effect on Surface-Enhanced Raman Scattering (SERS) in an Ordered Ag NPs/4-Mercaptobenzoic Acid/TiO$_2$ System. *J. Phys. Chem. C* **2015**, *119*, 22439–22444. [CrossRef]
43. Ho, C.H.; Lee, S. SERS and DFT investigation of the adsorption behavior of 4-mercaptobenzoic acid on silver colloids. *Colloids Surf. A Physicochem. Eng. Asp.* **2015**, *474*, 29–35. [CrossRef]

Sample Availability: Not available.

© 2020 by the authors. Licensee MDPI, Basel, Switzerland. This article is an open access article distributed under the terms and conditions of the Creative Commons Attribution (CC BY) license (http://creativecommons.org/licenses/by/4.0/).

Article

Development and Validation of a Spectrometric Method for Cd and Pb Determination in Zeolites and Safety Evaluation

Marin Senila [1,*], Oana Cadar [1] and Ion Miu [2]

[1] National Institute for Research and Development of Optoelectronics Bucharest INOE 2000, Research Institute for Analytical Instrumentation, 67 Donath Street, 400293 Cluj-Napoca, Romania; oana.cadar@icia.ro
[2] SC UTCHIM SRL, 12 Buda Street, 240127 Ramnicu Valcea, Romania; utchim_vl@yahoo.com
* Correspondence: marin.senila@icia.ro; Tel.: +40-264-420590

Academic Editors: Gavino Sanna and Stefan Leonidov Tsakovski
Received: 9 May 2020; Accepted: 2 June 2020; Published: 2 June 2020

Abstract: An analytical method based on microwave-assisted acid digestion and atomic absorption spectrometry with graphite furnace as atomization source was developed and validated for determining trace elements (Cd and Pb) in zeolites used as dietary supplements, for their characterization and safety evaluation. The method was checked for the main performance parameters according to the legislation requirements in the field of dietary supplements. In all cases, the obtained performance parameters were satisfactory. The selectivity study showed no significant non-spectral matrix effect. The linearity study was conducted for the calibration curves in the range of 0–10 ng mL^{-1} for Cd and 0–30 ng mL^{-1} for Pb. The obtained limits of detection (LoDs) and the limits of quantification (LoQs) were sufficiently low in order to allow Pb and Cd determination in dietary supplements. For the internal quality control, certified reference materials were analysed and good recoveries were obtained. The precision study was performed in terms of repeatability and reproducibility, considering the requirements imposed by the Commission Decision (2007/333/EC) and the method fulfilled these performance parameters. Expanded measurement uncertainties were estimated to 11% for Cd and 10% for Pb. Cd and Pb content were measured in real zeolite samples and, using these data, a safety evaluation was carried out.

Keywords: zeolites; dietary supplements; cadmium; lead; GF-AAS; validation; risk assessment

1. Introduction

Natural zeolite tuffs are crystalline materials with excellent properties, for example high adsorption capacity, ion-exchange ability, catalyzing action, and thermal stability. The chemical structure of zeolites contains silicon and aluminum interlinked through one, two, or three oxygen atoms, which leads to a wide diversity of three-dimensional potential structures. Zeolites have a negative charge balanced by positively charged monovalent and divalent ions of alkali and alkali earth elements such as water molecules that can be easily substituted by other cations or molecules. The Si/Al ratio in a zeolite has influence on its negative charge and on the attraction of the foreign ions inside their pores and channels [1,2]. The natural zeolites are mostly used in water treatment, catalysis, petrochemistry, cosmetics, agronomy, and medicine [3,4].

About 70 types of natural zeolites have been identified around the world with the most common mineral that occurs in natural zeolites being clinoptilolite. The clinoptilolite possess high surface area, good adsorption capacity, chemical and thermal stability, and ion-exchange capacity, which make it a potential detoxifying agent for organisms and a support in many medical applications. Clinoptilolite was found to be efficient in the veterinary and human medicine [5]. In the last years, the clinoptilolite-based zeolite is increasingly studied for use in human medicine as dietary supplements. Basha and co-workers combined clinoptilolite with other substances such as EDTA in order to increase the detoxifying capacity [6]. Federico and co-workers used micronized clinoptilolite for adsorption of ingested ethanol in order to reduce the blood alcohol level [7].

Zeolites attract trace metals in their structure, which is a property that makes them useful to eliminate these elements from the human body [8]. However, in the natural environment, this property may cause the accumulation of toxic metals in natural zeolites. The occurrence and the concentration of toxic metals in natural zeolite minerals vary depending on the quarry [9].

Cd and Pb have no known biological functions, and are classified as non-essential elements that are toxic even at very low concentrations for human health [10–12]. Cd is associated with negative health effects on renal, skeletal, pulmonary, reproductive, and cardiovascular systems, which are classified as group I carcinogens to humans [13]. Pb may also causes serious damages for human health and may cause even death [14–16]. The presence of toxic metals in the raw ingredients used to produce dietary supplements is undesirable. Therefore, the quality of both raw materials and end products must be strictly controlled [17].

Considering the high toxicity of both metals, the maximum Cd and Pb concentrations in dietary supplements of 1.0 mg kg^{-1} and 3.0 mg kg^{-1}, respectively were established by legislation in the field [18,19]. These regulations impose to validate sensitive methods that provide trusting results even at low level of concentration, having well-established performance criteria [20,21].

Three groups of analytical techniques are the most used in the routine laboratories to measure metals in different liquid and solid samples. The main differences among these techniques consist in the detection of analyte and type of atomization source. The first group uses a mass spectrometer to measure ions of analyte produced using inductively coupled plasma (ICP-MS) [22]. In the second group of analytical techniques, the optical emission of analyte ions is measured at specific wavelengths in an inductively coupled plasma (ICP-OES) [23,24]. The third group of atomic spectrometric techniques is based on the absorption of radiation at a specific wavelength by the atoms of analyte. In this case, the atoms can be obtained by electrothermal evaporation in a graphite furnace (GF-AAS) or in a flame (F-AAS) [25–27]. Beside these, other non-conventional techniques can also be used to determine metal analytes in different samples [28–30]. For specific elements (*i.e.*, Hg) which have high volatility, the direct measurement in solid samples (including zeolites) by thermal desorption is possible [31].

Although the market growth of dietary supplements based on zeolites, no validated analytical methods for the analysis of Pb and Cd in zeolites was found in literature. Currently, the GF-AAS technique is widely used for its recognized high selectivity and sensitivity. Due to these technical features, our study focuses to perform a fully-validation of this method in view of the legislative demands for dietary supplements [32], in order to produce reliable analytical data. In addition, the concentrations of both trace metals were measured in real zeolite samples to assess if the intake of zeolites as dietary supplements represents a risk for human health.

2. Results and Discussion

2.1. Performance Parameters

The main performance parameters such as selectivity, working and linear ranges, LoD and LoQ, precision, accuracy, and measurement uncertainty were evaluated for validation of the method for determination of Cd and Pb in zeolite samples [33–35]. The targets imposed for these parameters considered the quality standards for analytical methods provided by Commission of the European Communities in Regulation No. 333/2007 [20].

A graphite furnace atomic absorption spectrometer with graphite atomizer Perkin Elmer model PinAAcle 900T (Norwalk, CT, USA) was used to determine Cd and Pb in digested samples using direct injection of samples, followed by addition of an appropriate matrix modifier, as recommended by the instrument manufacturer. The calibration standards were prepared by the instrument autosampler by diluting the highest concentrated standard solution (10 ng mL^{-1} for Cd and 30 ng mL^{-1} for Pb) with ultrapure water. Electrodeless Discharge Lamps (EDLs) were used as spectral sources. The instrument operating conditions used are presented in Table 1.

Table 1. Operation conditions for Cd and Pb determination in zeolites by graphite furnace atomic absorption spectrometry (GF-AAS).

All elements				
Signal processing	Peak area			
Read time	5 s			
Sample volume	20 µL			
Background correction	Longitudinal Zeeman-effect			
Cd				
Wavelength	228.80 nm			
EDL current	230 mA			
Calibration	0–10 ng mL^{-1} (7 points)			
Matrix modifier	50 µg NH$_4$H$_2$PO$_4$ + 3 µg Mg(NO$_3$)$_2$ (5 µL)			
Furnace Program				
Step	Temp (°C)	Ramp (s)	Hold (s)	Ar Flow-Rate (mL min^{-1})
Drying	110	1	40	250
Drying	130	15	40	250
Ashing	500	10	20	250
Vaporization	1500	0	5	0
Cleaning	2450	1	3	250
Pb				
Wavelength	283.31 nm			
EDL current	400 mA			
Calibration	0–30 ng mL^{-1} (7 points)			
Matrix modifier	50 µg NH$_4$H$_2$PO$_4$ + 50 µg Mg(NO$_3$)$_2$ (5 µL)			
Furnace Program				
Step	Temp (°C)	Ramp (s)	Hold (s)	Ar Flow-Rate (mL min^{-1})
Drying	110	1	40	250
Drying	130	15	40	250
Ashing	850	10	20	250
Vaporization	1600	0	5	0
Cleaning	2450	1	3	250

2.1.1. Selectivity

To provide trustable results, an analytical method should differentiate the analytes of interest, in this case Cd and Pb, from other compounds in the sample matrix [33]. In order to improve the selectivity, for some elements including Cd and Pb, matrix modifiers are used in the GF-AAS technique to decreases the volatility of the analytes and to prevent their loss during thermal decomposition. The estimation of the matrix effect on the analytical signal is of high importance to assure that the most suitable calibration method was selected. The digested zeolite sample contains important amounts of elements dissolved from the solid sample and high content of mineral acids (HNO$_3$, HCl, HF, and H$_3$BO$_3$) used for digestion, thus the analyte signal from the sample might differ from the signal of the same analyte concentration in aqueous calibration standards. The selectivity was evaluated based on the matrix effects produced by the high amounts of Al, Ca, Mg, K, Na, and Fe in the digested zeolite solution. In order to evaluate the matrix effects, two calibration standards were constructed for both Pb and Cd. A calibration standard of 10 µg L^{-1} Cd and a calibration standard of 30 µg L^{-1} Pb were prepared in 0.5% (*m/v*) HNO$_3$, while other standards of 5 µg L^{-1} Cd and of 20 µg L^{-1} Pb were prepared in a solution containing 2.0% (*m/v*) HNO$_3$, 3.0% (*m/v*) HCl, 0.8% (*m/v*) HF, and 1.0% (*m/v*)

H_3BO_3 as well as concentrations of 250 mg L^{-1} Al, 100 mg L^{-1} Ca, 40 mg L^{-1} Mg, 40 mg L^{-1} K, 20 mg L^{-1} Fe, and 20 mg L^{-1} Na to mimic the solution of digested samples. Ratios of the slopes for both types of calibration curves were calculated, and the value was compared with 1, to find out if positive or negative influence on the analytes signals is given by interfering components. Figure 1 presents the plots of analytical signals for Cd and Pb against their concentrations using the two sets of calibration standards.

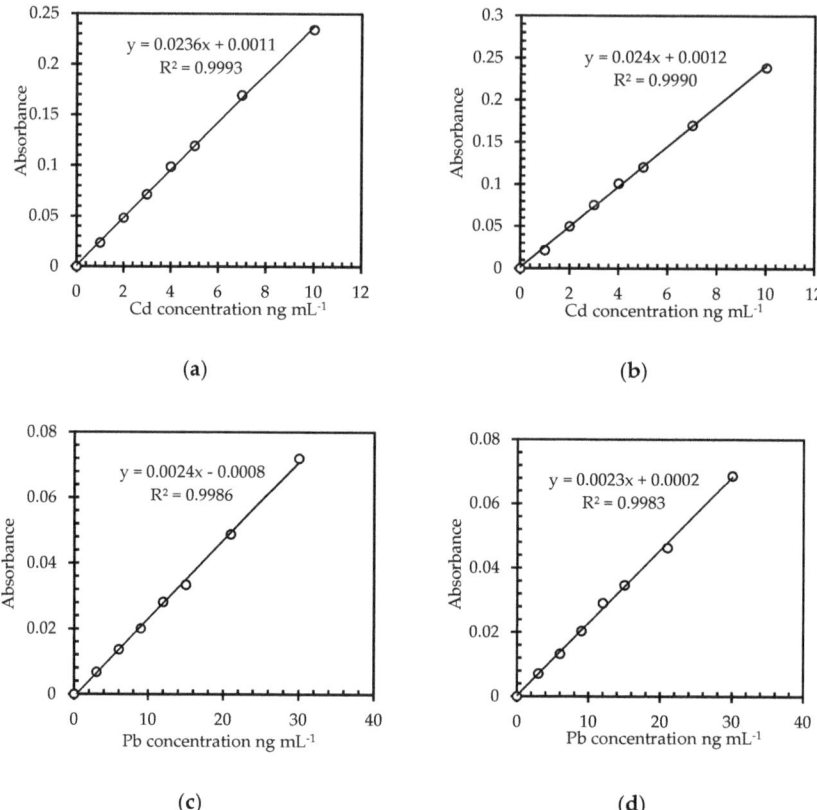

Figure 1. Calibration curves for Cd and Pb determination by graphite furnace atomic absorption spectrometry (GF-AAS): (**a**) Cd prepared in 0.5% (*m/v*) HNO_3, (**b**) Pb prepared in 0.5% (*m/v*) HNO_3, (**c**) Cd prepared in complex matrix that mimic the solution of digested samples, and (**d**) Pb prepared in complex matrix that mimic the solution of digested samples.

Comparison of the curves' slopes by means of their ratio revealed no significant influence of the matrix effect since the ratio were 1.02, and 0.96 for Cd and Pb, respectively. Consequently, calibration solutions prepared in 0.5% (*m/v*) HNO_3 were used for calibration in all experiments.

2.1.2. Linearity Study

The linear regression analysis was applied for all the calibration curves built using diluted nitric acid and complex matrix that mimic the digested zeolite solution. The calibration curves were prepared by including the allowed Residual Maximum Limit (RML) in the middle of their ranges by taking into account a factor of 200 (due to the digestion of 0.5 g sample and dilution to a volume of 100 mL), and the instrument sensitivity for each element. Thus, for Cd, the concentrations of the calibration curve were chosen in the range of 0–10 ng mL^{-1} since the RML for Cd in food supplements, according to

Decision 2008/629/EC [18], is 1.0 mg kg^{-1}. Considering the factor of 200, due to the sample digestion, this concentration of 1.0 mg kg^{-1} corresponds to a concentration in solution of 5.0 ng mL^{-1}. Similarly, for Pb, a concentration in solution of 15 ng mL^{-1} was calculated since it has a RML in food supplements of 3.0 mg kg^{-1}, and, consequently, the range of concentration was 0–30 ng mL^{-1}. Seven levels of concentration were used for the calibration of each element (0.2, 0.4, 0.6, 0.8, 1.0, 1.4, and respectively 2 times RML). The correlation coefficients (r) for Cd and Pb, respectively, for both calibration aqueous standards and standards with matrix, as presented in Figure 1, fulfill the requirements of r > 0.9950.

The test for homogeneity of variances of the linear calibration curves was done according to the requirements of ISO 8466-1 [36]. Ten replicates of the lowest (0.2 RM) and 10 of the highest (2 RML) concentrations prepared for the calibration curves were measured. The square of the standard deviations (s) for the two concentrations were calculated and the testing values $s^2_{2MRL} / s^2_{0.2MRL} = 3.31$ for Cd and $s^2_{2MRL} / s^2_{0.2MRL} = 4.40$ for Pb were compared with the Fisher–Snedecor distribution value for $n = 9$ degrees of freedom and 0.99 probability (F = 5.35). In both cases, the testing values were below than the F distribution value indicating a satisfactory homogeneity of variances for the chosen ranges of calibration curves.

The working ranges for the solid samples are limited at the lower part while, for the upper part, these concentrations are 2.0 mg kg^{-1} for Cd and 6.0 mg kg^{-1} for Pb, which are calculated from the sample preparation step and from the calibration curve ranges. Higher concentrations can be measured by dilution of samples.

2.1.3. Evaluation of LoDs and LoQs

Being the lowest content of an analyte that can be detected in a sample, LoD was calculated as the ratio between three times standard deviation of the mean resulted by measuring the blank signal ($n = 21$) and the slope obtained for calibration curves [20].

The LoD for Cd in digested liquid sample was calculated to have the value of 0.15 ng mL^{-1}, which corresponds to a concentration in a solid sample of 0.03 mg kg^{-1}. For Pb, LoD in a digested liquid sample was found to be 0.64 ng mL^{-1}, which means 0.13 mg kg^{-1} in a solid sample. These values were in agreement with the demands of Commission Regulation (EU) No 836/2011 [37] where, for RML ≥ 0.100 mg kg^{-1}, LoDs should be ≤ one tenth of the RML (1.0 mg kg^{-1} for Cd and 3.0 mg kg^{-1} for Pb). Our values were similar with those reported by Ivanova-Petropulos et al. [25] when analyzed water and brandy by GF-AAS (LoD of 0.12 ng mL^{-1} and 0.27 ng mL^{-1} for Cd, and of 0.57 ng mL^{-1} and 0.68 ng mL^{-1} for Pb). Aleluia et al. [38] reported LoDs in the same order of magnitude for Cd and Pb determination in organic pharmaceutical formulations using high-resolution continuum source GF-AAS.

LoQ is the lowest concentration of analyte that can be measured with acceptable precision and accuracy. The ten times standard deviation from the measuring the blank signal ($n = 21$) divided to the slope of calibration curves was used to calculate the LoQs [20,39]. The calculated LoQ for Cd was 0.50 ng mL^{-1}, corresponding to a concentration in a solid sample of 0.10 mg kg^{-1}, while, for Pb, the calculated LoD was 2.13 ng mL^{-1}, corresponding to a value of 0.43 mg kg^{-1} in the solid sample. These values are below the imposed limits of legislative requirements [37] for RML ≥ 0.100 mg kg^{-1}, where LoQs should be ≤ one fifth of the RML for Cd and Pb. The found values were in similar range with published data in previous studies [11,38].

Confirmation of LoQs was carried out by analysing series of ten spiked solutions with Cd and Pb content levels of 0.50 ng mL^{-1}, and, respectively, 2.00 ng mL^{-1}. The relative standard deviations of repeatability (RSDr%) were 10.4% (Cd), 12.9% (Pb), while the recoveries for that levels of concentration were 88% (Cd) and 94% (Pb). The obtained results were within the range of the imposed targets (RSDr % <20% and a recovery rate of 85–115%).

2.1.4. Trueness and Precision

The trueness was evaluated by analysing certified reference materials (CRMs) with a similar matrix to zeolite samples. Six replicates of both CRMS: potassium feldspar NIST-SRM 70b and Loam soil ERM – CC141, were analysed by GF-AAS under similar conditions as the real samples. The certified values of CRMs, measured values, and the recoveries degree (%) are offered in Table 2.

Table 2. Certified values of CRMs, measured values ($n = 6$ parallel determinations) and the recoveries degree (%).

CRM	Certified Values ± U [a] (mg kg^{-1})		Measured Values ± U [b] (mg kg^{-1})		Recovery (%)	
	Cd	Pb	Cd	Pb	Cd	Pb
NIST-SRM 70b	-	57 ± 3.0	-	55 ± 5.2	-	96
ERM–CC141	0.35 ± 0.05	41 ± 4.0	0.32 ± 0.035	43 ± 4.6	91	105

[a]: U = expanded uncertainty ($k = 2$); [b]: U = calculated expanded uncertainty ($k = 2$)

The results obtained for recoveries degree in CRMs for both Cd and Pb were situated in the acceptable range of 80–110%. This range was chosen considering the demands of the Commission Decision 2002/657/EC [32].

The precision study was performed in terms of repeatability and internal reproducibility while taking into account the requirements imposed by the legislation in the field of dietary supplements [20,37] for these performance parameters. The RSD of repeatability (RSD_r) was assessed by analyzing in parallel six aliquots of fortified samples at concentrations levels of 0.2 and respectively 1 times RML, using the same equipment on the same day. The same concentration levels of fortified samples were analysed on different days to calculate RSD of of reproducibility (RSD_R). For the acceptance criteria, the HorRat indexes were considered. HorRat index can be obtained by dividing the RSD_r or RSD_R to a predicted value of RSD (PRSD) calculated using Horvitz's equation [40]. PRSD depends on the mass fraction of the analyte in the sample

Using the Horvitz's equation, we calculated the PRSD for a mass fraction of 1000 µg kg^{-1} (RML of Cd) equal to 16%, while, and, for a mass fraction of 3000 µg kg^{-1} (RML of Pb), a PRSD equal to 14%. HorRat$_r$, HorRat$_R$ indexes, together with the PRSD calculated for 0.2 RML and 1 RML levels of concentrations for both Cd and Pb are presented in Table 3.

Table 3. Calculated PRSD, HorRat$_r$, HorRat$_R$ indexes.

Mass Fraction	Repeatability Study		Reproducibility Study		Results
	PRSD	HorRat$_r$	PRSD	HorRat$_R$	
0.2 RML Cd (200 µg kg^{-1})	20	0.55	20	0.60	Admitted
1 RML Cd (1000 µg kg^{-1})	16	0.50	16	0.44	Admitted
0.2 RML Pb (600 µg kg^{-1})	17	0.65	17	0.82	Admitted
1 RML Pb (3000 µg kg^{-1})	14	0.60	14	0.64	Admitted

In all cases, the HorRat$_r$, HorRat$_R$ indexes were below the maximum value of 2, as required in legislation in the field [37]. Thus, according to the obtained results in the repeatability and reproducibility study, the method for Cd and Pb analysis of zeolite samples using GF-AAS is sufficient precise.

2.1.5. Measurement Uncertainty

Regarding the value of measurement uncertainty for analytical methods designed for dietary supplement analysis, there are imposed limits that depend on the calculated LoD for that specific analyte [37]. The maximum standard uncertainty is calculated considering LoD, the concentration of the analyte and a numeric factor (α) that is determined by the RML of that analyte in sample (α = 0.15 for Cd RML at a mass fraction of 501–1000 µg kg^{-1}, and α = 0.12 for Pb RML at a mass fraction of 1001–10000 µg kg^{-1}).

For a RML for Cd of 1000 µg kg^{-1}, the maximum uncertainty of measurement for Cd was calculated to be 150 µg kg^{-1} (maximum relative standard uncertainty, U_{rel} = 15%). For Pb, that have RML of 3000 µg kg^{-1}, the maximum uncertainty of measurement (U_f) was found equal to 360 µg kg^{-1}, (maximum relative standard uncertainty, U_{rel} = 12%).

The measurement uncertainty is evaluated by combining different identified main sources [41]. These can be identified and quantified in the validation process and can be obtained from calibration certificates of volumetric flacks, pipettes, measuring instruments (declared uncertainty) and from the repeated measurements in the trueness and repeatability studies during the method validation.

It was assumed that the validation study comprises the total analytical procedure, and, thus, the most relevant uncertainty components that influence the expanded measurement uncertainty were assembled into only repeatability and accuracy components, that can be obtained from analysis of CRMs. Standard uncertainty is evaluated by combining the standard deviation of multiple determinations on CRM and the difference between found values and certified values of CRM.

The expanded uncertainty (U) is obtained from the standard uncertainty and a cover factor that depends on the level of confidence, P (k = 2, for p = 95%). In case of Pb, two CRMs were analysed. Thus, for the whole method a pooled expanded uncertainty (%) was calculated by combining the two expanded uncertainties.

The expanded uncertainty (%) for Cd was calculated to be 11%, while the expanded uncertainty (%) for Pb was found to be 10%. These results are within the maximum value of 15% for Cd and 12% for Pb established according to the legislative demands, and are considered satisfactory values.

2.2. Analysis of Real Samples

Five samples were used in the study including two commercial dietary supplements of micronized natural zeolite samples (S1 and S2) and three zeolite tuff samples (P1, P2, and P3) collected from different quarries from North-west Romania.

2.2.1. Zeolite Chemical and Mineralogical Characterization

All pXRD patterns of investigated zeolite samples (show the characteristic peaks of clinoptilolite at 2θ values of 9.86°, 11.16°, 22.46°, 26.03°, and 31.95° (Figure 2) [42]. According to pXRD analysis, the zeolite samples contain clinoptilolite as a major crystalline phase, which is accompanied by quartz, muscovite, feldspar, montmorillonite, and albite. The P3 sample also contains traces of calcite.

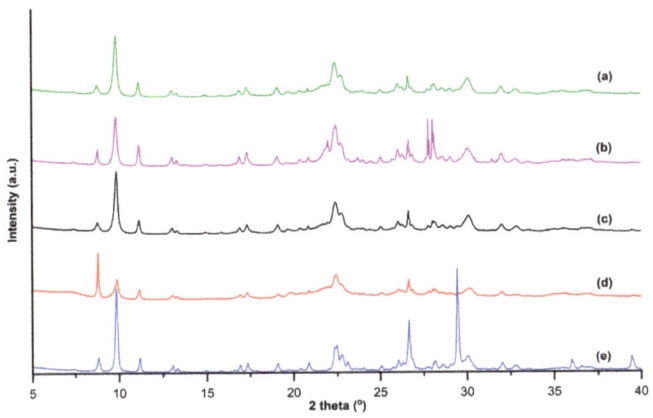

Figure 2. The pXRD patterns of (**a**) S1, (**b**) S2, (**c**) P1, (**d**) P2, and (**e**) P3 zeolites.

The samples were characterized regarding the chemical composition by measuring Al, Fe, Na, K, Ca, Mg, and Ti concentrations using ICP-OES. Then, using the atomic masses were the mass fraction the values were transformed in oxides content. SiO_2 content was obtained using the gravimetric method. The measured oxide concentrations are presented in Table 4.

Table 4. Mass fraction in oxides (%) in real samples.

Compounds	Concentrations (%)				
	S1	S2	P1	P2	P3
SiO_2	67.5	69.5	58.9	61.4	68.3
Al_2O_3	10.8	11.4	12.0	12.2	9.42
CaO	2.32	2.88	2.67	2.93	1.91
MgO	0.89	0.46	1.55	1.47	0.76
K_2O	2.34	2.47	2.20	2.03	2.96
Na_2O	1.07	0.41	0.87	0.44	0.29
TiO_2	0.20	0.18	0.13	0.12	0.19
Fe_2O_3	1.05	0.47	1.58	0.89	0.81

The chemical composition of samples confirms the pXRD analysis since clinoptilolite-type zeolite has a Si to Al ratio higher than 4 and a content of Na + K higher than Ca content [43]. Thus, according to these data, in all samples, the main mineral is clinoptilolite.

2.2.2. Cd and Pb Determination in Zeolite Samples by GF-AAS Technique after Microwave-Assisted Acid Digestion

The results obtained for Cd and Pb determination for real samples of dietary supplements and zeolites quarried from NW Romania measured using the GF-AAS technique and, for comparison, using the inductively coupled plasma mass spectrometry (ICP-MS) technique, are presented in Table 5. The Cd concentrations in all investigated samples were below the LoQ (0.1 mg kg^{-1}). Thus, from this point of view, both dietary supplements and possible use of zeolite tuffs from investigated quarries as raw material for food supplements are safe for human health. In case of Pb, in samples of dietary supplements, it was found but its concentration was lower than the maximum limit of 3.0 mg kg^{-1} imposed by Decision 2008/629/EC. In case of all samples collected from the quarries, the maximum admitted value established by legislation was exceeded. A review analysis carried out by de Vasconcelos Neto et al. [16] on the Pb contamination in food, generally showed lower concentrations than in the present study. However, for zeolites samples from Serbia, Tomasevic-Canovic [9] reported Pb concentrations in the range of 29–38 mg kg^{-1}, which is higher than in our case. Thus, there is a need to measure toxic metals in this type of samples. As presented in Table 5, the accuracy of GF-AAS technique was also proved by similarity of the results obtained using this with those obtained by ICP-MS technique.

Table 5. Concentrations of total Cd and Pb (mg kg^{-1}) measured in zeolite samples by GF-AAS and ICP-MS after sample digestion (n = 6 parallel measurements). Legend: – results presented as mean ± expanded uncertainty, <0.10 and <0.05–below limits of quantification (LoQ) for GF-AAS and ICP-MS, respectively.

Zeolite Sample	Cd (mg kg^{-1})		Pb (mg kg^{-1})	
	GF-AAS	ICP-MS	GF-AAS	ICP-MS
S1	<0.10	<0.05	2.6 ± 0.3	2.4 ± 0.3
S2	<0.10	<0.05	2.1 ± 0.3	2.0 ± 0.2
P1	<0.10	<0.05	8.2 ± 0.5	8.5 ± 0.8
P2	<0.10	<0.05	22 ± 2	23 ± 3
P3	<0.10	<0.05	15 ± 2	14 ± 2

2.2.3. Risk Exposure

The risk exposure to Pb through the consumption of zeolite dietary supplements was evaluated taking into account the provisional tolerable weekly intake (PTWI) [44,45]. The weekly intake (mg kg^{-1} b.w.) was evaluated for an usual intake of four pills containing zeolite supplements per day in which each of them are 0.35 g (a consumption of 9.8 g of zeolite supplemented per week) for a 60 kg adult. The equivalent weekly risk exposure to Pb was assessed as a percentage of PTWI considering the Pb PTWI value of 0.025 mg kg^{-1} b.w., as recommended by WHO [46].

Considering this consumption, for an average concentration of 2.4 mg kg^{-1} as found in dietary supplements, % PTWI = 1.6%, which is much lower than the imposed Pb PTWI. Even in the worst case of use of raw material zeolite containing 15 mg kg^{-1} Pb, the percentage of PTWI = 9.8% ($\cong$10%) from the recommended value for Pb of WHO. This is because the mass of the consumed supplement is relatively small and does not significantly influence the intake of Pb. However, the maximum limit of 3.0 mg kg^{-1} imposed by Decision 2008/629/EC should be considered when using zeolite as a dietary supplement. Cd was not considered in the assessment of risk exposure, since, in all cases, its concentrations in analysed samples were below LoQ.

This study is of considerable interest since: (*i*) to our knowledge, no validated analytical methods for Cd and Pb measurement in zeolites samples are presented in literature, (*ii*) provides the necessary steps to validate a method in agreement with the demands of legislation, and (*iii*) no previous data were reported on the literature on the risk exposure to Pb via consumption of zeolite dietary supplements.

3. Materials and Methods

3.1. Standard Solutions, Reagents, and CRMs

All the solutions were prepared using ultrapure water produced by a Milli-Q system Direct Q3 (Millipore, Molsheim, France). Analytical-grade reagents (65% HNO$_3$, 40% HF, 37% HCl, and 99.5–100.5% H$_3$BO$_3$) used in experiments were purchased from Merck (Darmstadt, Germany). Mono-element Cd and Pb calibration standards for GF-AAS were prepared by sequential diluting stocks solutions (Merck, Darmstadt, Germany) containing 1000 mg L^{-1} of these elements in 0.5% (*v/v*) HNO$_3$. For the ICP-OES calibration, standard solutions in the range of 0–20 mg L^{-1} were prepared from 1000 mg L^{-1} multi-elements containing elements of interest of 1000 mg L^{-1} and mono-element (Ti) standard solution 1000 mg L^{-1} (Merck, Darmstadt, Germany), diluted in 0.5% (*m/v*) HNO$_3$.

Matrix modifiers of 10% NH$_4$H$_2$PO$_4$, 1% MgNO$_3$ (Perkin Elmer Pure, Shelton, DC, USA) were used to prepare chemical modifier solution containing the mixture of 0.1% (*m/v*) Pd plus 0.05% (*m/v*) Mg in 0.5% (*v/v*) HNO$_3$. Argon (5.0 quality) from Linde Gas SRL Cluj-Napoca, Romania was used for instruments.

Potassium feldspar Standard Reference Material (NIST-SRM 70b) produced by the National Institute of Standards & Technology USA and Loam soil ERM–CC141 produced by the Institute for Reference Materials and Measurements and purchased from LGC Promochem (Wesel, Germany) were used.

3.2. Instrumentation and Methods

A GF-AAS model PinAAcle 900T (Perkin Elmer, Norwalk, CT, USA) was used for Cd and Pb determination. An inductively coupled plasma mass spectrometer ICP-MS ELAN DRC II (Perkin-Elmer, Toronto, ON, Canada) was used to measure Cd and Pb for a comparison of results with those obtained by GF-AAS. The content of major elements was determined by ICP-OES using an SPECTROFLAME FMD-07 (Spectro Analytical Instruments, Kleve, Germany). The conversion to the corresponding oxide was made by multiplying the element concentration with 1.8895 (Al$_2$O$_3$), 1.4297 (Fe$_2$O$_3$), 1.3392 (CaO), 1.6583 (MgO), 1.2046 (K$_2$O), 1.3480 (Na$_2$O), 1.2912 (MnO), and 1.6683 (TiO$_2$). SiO$_2$ content was measured by a gravimetric method.

Microwave digestion of a 0.5 g powder of sample using microwave-assisted acid digestion using a microwave system (Berghof, Eningen, Germany) was described elsewhere [31]. The heating program of the microwave system for samples digestion comprised three steps of heating at 160 °C, 200 °C, the cooling at 100 °C, in a total time of digestion of 35 min. After cooling at room temperature, 20 mL of saturated H_3BO_3 were added and the samples were heated again at 160 °C in the microwave system for 15 min.

The samples were cooled down to room temperature and filtered on cellulose filters (circles, diameter of 125 mm) (Whatman, Germany) using PTFE funnels in volumetric flasks of 50 mL and diluted to final volume using ultrapure water. The resulted solution was analyzed by GF-AAS for Cd and Pb and by ICP-OES for Na, K, Ca, Mg, Fe, Al, and Ti, respectively.

The powder X-ray diffraction (pXRD) patterns were recorded at room temperature, using a D8 Advance (Bruker, Karlsruhe, Germany) diffractometer, operating at 40 kV and 40 mA with CuK_α radiation (λ = 1.54060 Å), at room temperature.

3.3. Real Samples Preparation

Two commercial dietary supplements of micronized natural zeolite samples (S1 and S2) and three zeolite tuff samples collected from different quarries from North-West Romania (P1–Racos, Brasov County, P2–Chilioara, Salaj County, and P3–Macicas, Cluj County) were used in this study. The samples were collected from the quarries as rock, then crushed, then grounded to a fine powder. The powder of each individual sample was further micronized using a laboratory micronization unit Pilotmill-2 (FPS, Como, Italy).

3.4. Strategy for Method Validation

The GF-AAS method was validated for the analysis of Cd and Pb in zeolites used as dietary supplements in terms of matrix effect, selectivity, linearity, working range, LoD and LoQ, trueness, precision, and measurement uncertainty. The performance parameters were assessed in comparison with the target limits established by legislation in the field of food control and dietary supplements analysis [20,32].

3.5. Risk Exposure

The risk exposure to Pb through the ingestion of zeolite dietary supplements was evaluated using a methodology to assess the exposure to a contaminant based on a provisional tolerable weekly intake. The equivalent weekly risk exposure to Pb was calculated as the percentage of PTWI by considering the Pb PTWI value of 0.025 mg kg^{-1} b.w., as suggested by WHO [46].

$$\%PTWI = 100 \times \text{Estimated exposure to Pb/Pb PTWI} \quad (1)$$

where Pb PTWI = 0.025 mg kg^{-1} b.w.

4. Conclusions

In this paper, an analytical method based on the microwave-assisted acid digestion and GF-AAS technique for determining Cd and Pb in zeolites is validated, considering the demands of European legislation for the official control of trace elements in food supplements. The obtained LoDs (0.03 mg kg^{-1} for Cd and 0.13 mg kg^{-1} for Pb) and LoQ (0.10 mg kg^{-1} for Cd and 0.43 mg kg^{-1} for Pb) were lower than the imposed limits of specific legislation. The trueness was evaluated by analysing two certified reference materials, NIST-SRM 70b and ERM–CC141. Good recoveries were obtained. The precision study was performed in terms of repeatability and internal reproducibility, considering the requirements imposed by the Commission Decision (2007/333/EC). The expanded measurement uncertainties were estimated to 11% for Cd and 10% for Pb and fulfill the legislative requirements. The content of Cd and Pb was measured in several zeolite samples. Using these data, a safety evaluation was carried out. It was concluded that, considering the level of Cd and Pb concentrations and the estimated weekly intake of zeolite as dietary supplements, their consumption does not pose a risk for human health.

Author Contributions: M.S. conceived and designed the experiments. M.S., O.C., and I.M. performed the experiments and zeolite characterization. I.M. sampled zeolites and M.S. wrote the paper. All authors have read and agreed to the published version of the manuscript.

Funding: The Competitiveness Operational Programme of the Ministry of European Funds, contract no. 7/01.09.2016, code MY SMIS 105654, funded this research. The Romanian Research and Innovation Ministry under the PROINSTITUTIO project, contract no. 19PFE/17.10.2018, funded the article processing charge (APC).

Conflicts of Interest: The authors declare no conflict of interest. The funders had no role in the design of the study, in the collection, analyses, or interpretation of data, in the writing of the manuscript, or in the decision to publish the results.

References

1. Canli, M.; Abali, Y.S.; Bayca, U. Removal of methelyne blue by natural and Ca and K-exchanged zeolite treated with hydrogen peroxide. *Physicochem. Probl. Miner. Process.* **2013**, *49*, 481–496.
2. Hong, M.; Yu, L.; Wang, Y.; Zhang, J.; Chen, Z.; Dong, L.; Zan, Q.; Li, R. Heavy metal adsorption with zeolites: The role of hierarchical pore architecture. *Chem. Eng. J.* **2019**, *359*, 363–372. [CrossRef]
3. Kraljevic Pavelic, S.; Simovic Medica, J.; Gumbarevic, D.; Filosevic, A.; Przulj, N.; Pavelic, K. Critical review on zeolite clinoptilolite safety and medical applications in vivo. *Front. Pharmacol.* **2018**, *9*, 1350. [CrossRef] [PubMed]
4. Eroglu, N.; Emekci, M.; Athanassiou, C.G. Applications of natural zeolites on agriculture and food production. *J. Sci. Food Agric.* **2017**, *97*, 3487–3499. [CrossRef] [PubMed]
5. Laurino, C.; Palmieri, B. "Zeolite the magic stone"; main nutritional, environmental, experimental and clinical fields of application. *Nutr. Hosp.* **2015**, *32*, 573–581. [PubMed]
6. Basha, M.P.; Begum, S.; Mir, B.A. Neuroprotective actions of clinoptilolite and ethylenediaminetetraacetic acid against lead-introduced toxicity in mice Mus musculus. *Toxicol. Int.* **2013**, *20*, 201–207. [CrossRef] [PubMed]
7. Federico, A.; Dallio, M.; Gravina, A.G.; Iannotta, C.; Romano, M.; Rossetti, G.; Somalvico, F.; Tucillo, C.; Loguerico, C. A pilot study on the ability of clinoptilolite to absorb ethanol in vivo in healthy drinkers: Effect of gender. *J. Physiol. Pharmacol.* **2015**, *66*, 441–447.
8. Mastinu, A.; Kumar, A.; Maccarinelli, G.; Bonini, S.A.; Premoli, M.; Aria, F.; Gianoncelli, A.; Memo, M. Zeolite Clinoptilolite: Therapeutic Virtues of an Ancient Mineral. *Molecules* **2019**, *27*, 1517. [CrossRef]
9. Tomasevic-Canovic, M. Purification of natural zeolite-clinoptilolite for medical application-extraction of lead. *J. Serb. Chem. Soc.* **2005**, *70*, 1335–1345. [CrossRef]
10. Berar Sur, I.M.; Micle, V.; Avram, S.; Senila, M.; Oros, V. Bioleaching of some heavy metals from polluted soils. *Environ. Eng. Manag. J.* **2012**, *11*, 1389–1393.
11. Balaram, V. Recent advances in the determination of elemental impurities in pharmaceuticals—Status, challenges and moving frontiers. *Trends Anal. Chem.* **2016**, *80*, 83–95. [CrossRef]
12. Damian, G.E.; Micle, V.; Sur, M.I. Mobilization of Cu and Pb from multi-metal contaminated soils by dissolved humic substances extracted from leonardite and factors affecting the process. *J. Soil. Sediment.* **2019**, *19*, 2869–2881. [CrossRef]
13. Butaciu, S.; Frentiu, T.; Senila, M.; Darvasi, E.; Cadar, S.; Ponta, M.; Petreus, D.; Etz, R.; Frentiu, M. Determination of Cd in food using an electrothermal vaporization capacitively coupled plasma microtorch optical emission microspectrometer: Compliance with European legislation and comparison with graphite furnace atomic absorption spectrometry. *Food Control* **2016**, *61*, 227–234. [CrossRef]
14. Senila, M. Real and simulated bioavailability of lead in contaminated and uncontaminated soils. *J. Environ. Health Sci. Eng.* **2014**, *12*, 108. [CrossRef] [PubMed]
15. Geraldes, V.; Carvalho, M.; Goncalves-Rosa, N.; Tavares, C.; Laranjo, S.; Rocha, I. Lead toxicity promotes autonomic dysfunction with increased chemoreceptor sensitivity. *Neurotoxicology* **2016**, *54*, 170–177. [CrossRef]

16. De Vasconcelos Neto, M.C.; Castano Silva, T.B.; de Araujo, V.E.; de Souza, S.V.C. Lead contamination in food consumed and produced in Brazil: Systematic review and meta-analysis. *Food Res. Int.* **2019**, *126*, 108671. [CrossRef]
17. Amelin, V.G.; Lavrukhina, O.I. Food safety assurance using methods of chemical analysis. *J. Anal. Chem.* **2017**, *72*, 1–46. [CrossRef]
18. Commission Regulation (EC) No 629/2008 of 2 July 2008 Amending Regulation (EC) No 1881/2006 Setting Maximum Levels for Certain Contaminants in Foodstuffs. Available online: https://eur-lex.europa.eu/legal-content/EN/TXT/PDF/?uri=CELEX:32008R0629&from=EN (accessed on 4 May 2020).
19. Commission Regulation (2006/1881/EC) Setting Maximum Levels for Certain Contaminants in Foodstuffs. Available online: https://eur-lex.europa.eu/LexUriServ/LexUriServ.do?uri=OJ:L:2006:364:0005:0024:EN:PDF (accessed on 4 May 2020).
20. Commission Decision (2007/333/EC) Laying down the Methods of Sampling and Analysis for the Official Control of the Levels of Lead, Cadmium, Mercury, Inorganic Tin, 3-MCPD and benzo(a)-pyrene in Foodstuffs. Available online: https://eur-lex.europa.eu/LexUriServ/LexUriServ.do?uri=OJ:L:2007:088:0029:0038:EN:PDF (accessed on 4 May 2020).
21. Rzymski, P.; Budzulak, J.; Niedzielski, P.; Klimaszyk, P.; Proch, J.; Kozak, L.; Poniedzialek, B. Essential and toxic elements in commercial microalgal food supplements. *J. Appl. Phycol.* **2019**, *31*, 3567–3579. [CrossRef]
22. Carter, J.A.; Barros, A.I.; Nobrega, J.A.; Donati, G.L. Traditional calibration methods in atomic sprectrometry and calibration strategies for inductively coupled plasma mass spectrometry. *Front. Chem.* **2018**, *6*, 504. [CrossRef]
23. Miclean, M.; Levei, E.A.; Senila, M.; Roman, C.; Cordos, E. Assessment of Cu, Pb, Zn and Cd availability to vegetable species grown in the vicinity of tailing deposits from Baia Mare Area. *Rev. Chim. (Bucurest.)* **2009**, *60*, 1–4.
24. Elgammal, S.M.; Khorshed, M.A.; Ismai, E.H. Determination of heavy metal content in whey protein samples from markets in Giza, Egypt, using inductively coupled plasma optical emission spectrometry and graphite furnace atomic absorption spectrometry: A probabilistic risk assessment study. *J. Food. Compost. Anal.* **2019**, *84*, 103300. [CrossRef]
25. Ivanova-Petropulos, V.; Balabanova, B.; Bogeva, E.; Frentiu, T.; Ponta, M.; Senila, M.; Gulaboski, R.; Irimie, F.D. Rapid determination of trace elements in Macedonian grape brandies for their characterization and safety evaluation. *Food. Anal. Methods* **2017**, *10*, 459–468. [CrossRef]
26. de Oliveira, T.M.; Peres, J.A.; Felsner, M.L.; Justi, K.C. Direct determination of Pb in raw milk by graphite furnace atomic absorption spectrometry (GF AAS) with electrothermal atomization sampling from slurries. *Food Chem.* **2017**, *229*, 721–725. [CrossRef] [PubMed]
27. Ferreira, S.L.C.; Bezerra, M.A.; Santos, A.S.; dos Santos, W.N.L.; Novaes, C.G.; de Oliveira, O.M.C.; Oliveira, M.L.; Garcia, R.L. Atomic absorption spectrometry-A multi element technique. *Trends Anal. Chem.* **2018**, *100*, 1–6. [CrossRef]
28. Frentiu, T.; Ponta, M.; Senila, M.; Mihaltan, A.I.; Darvasi, E.; Frentiu, M.; Cordos, E. Evaluation of figures of merit for Zn determination in environmental and biological samples using EDL excited AFS in a new radiofrequency capacitively coupled plasma. *J. Anal. Atomic Spectrom.* **2010**, *25*, 739–742. [CrossRef]
29. Frentiu, T.; Butaciu, S.; Darvasi, E.; Ponta, M.; Senila, M.; Petreus, D.; Frentiu, M. Analytical characterization of a method for mercury determination in food using cold vapour capacitively coupled plasma microtorch optical emission spectrometry – compliance with European legislation requirements. *Anal. Methods* **2015**, *7*, 747–752. [CrossRef]
30. Hornackova, M.; Plavcan, J.; Hornacek, M.; Hudec, P.; Veis, P. Heavy metals detection in zeolites using the LIBS Method. *Atoms* **2019**, *7*, 98. [CrossRef]
31. Senila, M.; Cadar, O.; Senila, L.; Hoaghia, A.; Miu, I. Mercury determination in natural zeolites samples by thermal decomposition atomic absorption spectrometry: Method validation in compliance with requirements for use as dietary supplements. *Molecules* **2019**, *24*, 4023. [CrossRef]
32. Commission Decision of 12 August 2002 2002/657/EC Implementing Council Directive 96/23/EC Concerning the Performance of Analytical Methods and the Interpretation of Results. Available online: https://eur-lex.europa.eu/legal-content/EN/TXT/PDF/?uri=CELEX:32002D0657&from=EN (accessed on 4 May 2020).

33. Drolc, A.; Pintar, A. Measurement uncertainty evaluation and inhouse method validation of the herbicide iodosulfuron-methylsodium in water samples by using HPLC analysis. *Accred. Qual. Assur.* **2011**, *16*, 21–29. [CrossRef]
34. Herrero Fernandez, Z.; Valcarcel Rojas, L.A.; Montero Alvarez, A.; Estevez Alvarez, J.R.; dos Santos Junior, J.A.; Pupo Gonzalez, I.; Rodriguez Gonzalez, M.; Alberro Macias, N.; Lopez Sanchez, D.; Hernandez Torres, D. Application of Cold Vapor-Atomic Absorption (CVAAS) Spectrophotometry and Inductively Coupled Plasma-Atomic Emission Spectrometry methods for cadmium, mercury and lead analyses of fish samples. Validation of the method of CVAAS. *Food Control* **2015**, *48*, 37–42.
35. Tudorache, A.; Ionita, D.E.; Marin, N.M.; Marin, C.; Badea, I.A. Inhouse validation of a UV spectrometric method for measurement of nitrate concentration in natural groundwater samples. *Accred. Qual. Assur.* **2017**, *22*, 29–35. [CrossRef]
36. ISO 8466-1 (1990) Water quality—Calibration and Evaluation of Analytical Methods and Estimation of Performance Characteristics, Part 1: Statistical Evaluation of the Linear Calibration Function. International Organization for Standardization. Available online: https://www.iso.org/standard/15664.html (accessed on 4 May 2020).
37. Commission Regulation (EU) No 836/2011 of 19 August 2011 Amending Regulation (EC) No 333/2007 Laying down the Methods of Sampling and Analysis for the Official Control of the Levels of Lead, Cadmium, Mercury, Inorganic Tin, 3-MCPD and benzo(a)pyrene in Foodstuffs. Available online: https://eur-lex.europa.eu/LexUriServ/LexUriServ.do?uri=OJ:L:2011:215:0009:0016:EN:PDF (accessed on 4 May 2020).
38. Aleluia, A.C.M.; de Santana, F.A.; Brandao, G.C.; Ferreira, S.L.C. Sequential determination of cadmium and lead in organic pharmaceutical formulations using high-resolution continuum source graphite furnace atomic absorption spectrometry. *Microchem. J.* **2017**, *130*, 157–161. [CrossRef]
39. Adolfo, F.R.; do Nascimento, P.C.; Leal, G.C.; Bohrer, D.; Viana, C.; de Carvalho, L.M.; Colim, A.N. Simultaneous determination of iron and nickel as contaminants in multimineral and multivitamin supplements by solid sampling HR-CS GF AAS. *Talanta* **2019**, *195*, 745–751. [CrossRef] [PubMed]
40. Horwitz, W.; Kamps, L.R.; Boyer, R.W. Quality assurance in the analysis of foods and trace constituents. *J. Assoc. Off. Anal. Chem.* **1980**, *63*, 1344–1354. [CrossRef]
41. ISO 11352 (2012) Water Quality-Estimation of Measurement Uncertainty Based on Validation and Quality Control Data. International Organization for Standardization (ISO): Geneva, Switzerland. Available online: https://www.iso.org/standard/50399.html (accessed on 4 May 2020).
42. Cadar, O.; Hoaghia, M.A.; Kovacs, E.; Senila, M.; Miu, I. Behavior of some clinoptilolite rich natural zeolites from Romania in simulated biological fluids. *Int. Multidiscip. Sci. GeoConf.* **2019**, *19*, 59–66.
43. Bish, D.; Boak, J. Clinoptilolite-heulandite nomenclature. *Rev. Min. Geochem.* **2001**, *45*, 207–216. [CrossRef]
44. Senila, M.; Covaci, E.; Cadar, O.; Ponta, M.; Frentiu, M.; Frentiu, T. Mercury speciation in fish tissue by eco-scale thermal decomposition atomic absorption spectrometry: Method validation and risk exposure to methylmercury. *Chem. Pap.* **2018**, *72*, 441–448. [CrossRef]
45. Covaci, E.; Senila, M.; Ponta, M.; Darvasi, E.; Frentiu, M.; Frentiu, T. Mercury speciation in seafood using non-chromatographic chemical vapor generation capacitively coupled plasma microtorch optical emission spectrometry method-Evaluation of methylmercury exposure. *Food Control* **2017**, *82*, 266–273. [CrossRef]
46. *FAO/WHO Food Additives Series: 64 SAFETY Evaluation of Certain Food Additives and Contaminants: Sixty-Third Report of the Joint FAO/WHO Expert Committee on Food Additives*; WHO: Geneva, Switzerland, 2011.

Sample Availability: Samples of the zeolites are available from the authors.

 © 2020 by the authors. Licensee MDPI, Basel, Switzerland. This article is an open access article distributed under the terms and conditions of the Creative Commons Attribution (CC BY) license (http://creativecommons.org/licenses/by/4.0/).

Article

In Situ Determination of Nitrate in Water Using Fourier Transform Mid-Infrared Attenuated Total Reflectance Spectroscopy Coupled with Deconvolution Algorithm

Fangqun Gan [1,2], Ke Wu [2], Fei Ma [1] and Changwen Du [1,3,*]

[1] The State Key Laboratory of Soil and Sustainable Agriculture, Institute of Soil Science Chinese Academy of Sciences, Nanjing 210008, China; qunfanggan@163.com (F.G.); fma@issas.ac.cn (F.M.)
[2] College of Environment and Ecology, Jiangsu Open University, Nanjing 210017, China; kwu@issas.ac.cn
[3] College of Advanced Agricultural Sciences, University of Chinese Academy of Sciences, Beijing 100049, China
* Correspondence: chwdu@issas.ac.cn; Tel.: +86-25-8688-1565

Academic Editor: Derek J. McPhee
Received: 11 November 2020; Accepted: 8 December 2020; Published: 10 December 2020

Abstract: Fourier transform infrared attenuated total reflectance (FTIR-ATR) spectroscopy has been used to determine the nitrate content in aqueous solutions. However, the conventional water deduction algorithm indicated considerable limits in the analysis of samples with low nitrate concentration. In this study, FTIR-ATR spectra of nitrate solution samples with high and low concentrations were obtained, and the spectra were then pre-processed with deconvolution curve-fitting (without water deduction) combined with partial least squares regression (PLSR) to predict the nitrate content. The results show that the typical absorption of nitrate (1200–1500 cm^{-1}) did not clearly align with the conventional algorithm of water deduction, while this absorption was obviously observed through the deconvolution algorithm. The first principal component of the spectra, which explained more than 95% variance, was linearly related to the nitrate content; the correlation coefficient (R^2) of the PLSR model for the high-concentration group was 0.9578, and the ratio of the standard deviation of the prediction set to that of the calibration set (*RPD*) was 4.22, indicating excellent prediction performance. For the low-concentration group model, R^2 and *RPD* were 0.9865 and 3.15, respectively, which also demonstrated significantly improved prediction capability. Therefore, FTIR-ATR spectroscopy combined with deconvolution curve-fitting can be conducted to determine the nitrate content in aqueous solutions, thus facilitating rapid determination of nitrate in water bodies with varied concentrations.

Keywords: nitrate; water bodies; Fourier transform attenuated total reflection; deconvolution; curve-fitting; partial least squares

1. Introduction

The main forms of nitrogen in aquatic ecosystems are total nitrogen, ammonium nitrogen (NH_4^+–N), and nitrate nitrogen (NO_3^-–N and NO_2^-–N) [1]. In recent years, excess nitrogen in water bodies, especially NO_3^-–N, has led to major ecological problems [2]. In addition, NO_3^-–N in drinking water can be converted to NO_2^-–N by the commensal bacteria in the mouth and digestive tract, which is harmful to the health of adults and children. Their long-term consumption can cause, for example, cancer of the digestive and excretory systems [3,4]. Therefore, there is a need to develop techniques that rapidly detect NO_3^-–N in water bodies to prevent water eutrophication and promote human health management.

Conventional methods for measuring NO_3^--N in water include reduction distillation, colorimetry, and the use of ion-specific electrodes [5,6], which are time-consuming and tend to produce secondary pollution. As a fast and nondestructive analysis method, infrared spectroscopy has many advantages, such as a simple analytical process, low cost, high efficiency, and no chemical reagent consumption [7–9]. It has recently been used to rapidly determine nitrate nitrogen levels in water. Previous studies have indicated that mid-infrared attenuated total reflection (FTIR-ATR) spectroscopy can be conducted for the rapid quantitative analysis of nitrate in solutions. The results show that the intensity of the characteristic absorption peak of N-O vibration in nitrate (1200–1500 cm^{-1}) was proportional to the NO_3^--N concentration. They used this relationship to establish a partial least squares (PLS) model that predicted the nitrate nitrogen content [10]. Shaviv et al. used FTIR-ATR to determine NO_3^--N in deionized water and in soil solutions [11]. Although these studies used FTIR-ATR to detect nitrate nitrogen in water, it was problematic to determine NO_3^--N with low concentrations (such as lower than 20 mg L^{-1}) in aqueous solutions due to the significant interference from strong water absorption.

In previous studies, for soil solution and vegetable samples with high concentrations of NO_3^--N, direct water deduction was generally used to remove the interference [12–14]. However, for the spectral analysis of low-concentration nitrate samples, water deduction causes large errors. Therefore, the spectral data must be pre-processed effectively to obtain useful information. Deconvolution is a mathematical procedure and a signal processing method typically conducted in many fields such as pattern recognition, seismology, system identification, electromagnetic scattering, and tomography [15]. The application of deconvolution in spectral processing has also proven to be effective. Deconvolution techniques can be used to enhance the resolution beyond the instrumental limit and significantly improve the signal-to-noise ratio [16,17]. In addition, to obtain useful, accurate and reliable information, spectral deconvolution could be associated with the Gaussian fit of the absorption spectra to adjust the Gaussian mathematical curves and obtain the corresponding characteristic absorption from overlapped peaks in a complex spectrum [18,19].

Thus, the objective of this study was to use FTIR-ATR to rapidly determine both high and low concentrations of nitrate in aqueous solutions through the spectra pretreatment of deconvolution curve-fitting, combined with principal component analysis (PCA) and partial least squares regression (PLSR), which could provide a new alternative option for the rapid determination of varied nitrate concentrations in water.

2. Materials and Methods

2.1. Materials

The test reagents were KNO_3 (analytical reagent grade, AR, purchased from Nanjing Ronghua Apparatus Co., Ltd., Nanjing, China) and deionized water. High and low nitrate concentrations (NO_3^--N) were prepared separately, in which the high-concentration group included concentrations of 0, 5, 10, 20, 30, 40, 50, 60, 70, 80, 90, and 100 mg L^{-1} and the low-concentration group included 0, 1, 2, 4, 6, 8, 10, 12, 14, 16, 18, and 20 mg L^{-1}.

2.2. Spectra Recording

An FTIR-ATR instrument (Nicolet 6700) was used (Thermo Fisher Scientific, Waltham, MA, USA), with a DTG detector, and the attenuated total reflection accessory was a 45 °C ZnSe ATR (Bruker, Karlsruhe, Germany). When recording the FTIR-ATR spectra, the nitrate solutions were directly added to the ATR crystal tank and the nitrate solution of each concentration was measured four times. The spectral scan range was set to 500–4000 cm^{-1} and 32 repeated scans were continuously recorded, with a resolution of 4 cm^{-1} and a mirror velocity of 0.4747 cm s^{-1}.

2.3. Pretreatment of Spectral Data

2.3.1. Water Deduction

The FTIR-ATR spectra were pre-processed with a Savitzky–Golay filter to reduce noise and improve the signal-to-noise ratio [20,21]. MATLAB 2016a (The MathWorks, Natick, MA, USA) was used to deduct the absorption peaks of water with the reference band (wavenumber range of 1500–2200 cm^{-1}); then, PCA and PLS analysis were subsequently conducted.

2.3.2. Deconvolution Curve-Fitting (without Water Deduction)

For all solutions, smoothing, baseline correction, and deconvolution curve-fitting (Gaussian) of the spectra within the range 1200–1500 cm^{-1} were performed through the Peakfit 4.12 software (SeaSolve Software Inc., San Jose, CA, USA). The objective of deconvolution was to separate each peak from the comprehensive information in the spectrum [22,23]. The details of the deconvolution process are demonstrated in the supplementary materials (Figures S1–S3). Briefly, the principles are as follows:

$$Y(x) = \Sigma F_i(x) \tag{1}$$

where Y is the spectrum; x is the wavenumber; i (1, 2, 3, ... n) is the number of isolated peaks; F is the expansion function or the kernel function of deconvolution. The Gaussian function is used as the kernel function:

$$y = \frac{a_0}{\pi \sqrt{\pi a_2}} exp\left[-\frac{1}{2}\left(\frac{x-a_1}{a_2}\right)^2\right] \tag{2}$$

where a_0, a_1, and a_2 represent the peak amplitude, position, and width, respectively, and x and y are the wavenumber and absorption intensity, respectively.

2.4. Model Evaluation

The following equations were used to calculate $RMSE$, RPD, and R^2 in order to evaluate the performance of the models in the validation set as following:

$$RMSE = \sqrt{\frac{1}{n}\sum_{i=1}^{n}(y_i - \hat{y}_i)^2} \tag{3}$$

$$RPD = \frac{SD}{RMSE} \tag{4}$$

$$R^2 = 1 - \frac{\sum_{i=1}^{n}(y_i - \hat{y}_i)^2}{\sum_{i=1}^{n}(y_i - \bar{y})^2} \tag{5}$$

where y_i and $\hat{y}_i$ are the measured and predicted nitrate levels of ith samples, respectively, $\bar{y}$ is the mean of the measured nitrate, and n is the number of samples. High values of R^2 and RPD along with a low $RMSE$ value indicated a robust and accurate model. RPD_V values of <1.4 were poor; ≥1.4 and <1.8 were fair and allowed the model prediction to be used for assessment and correlation; ≥1.8 and <2.0 were good, in which case quantitative predictions were possible; ≥2.0 and <2.5 were very good for quantitative analysis; and ≥2.5 were excellent [24,25].

3. Results and Discussion

3.1. FTIR-ATR Spectra of Nitrate

The FTIR-ATR spectra of the high- and low-concentration groups of nitrates showed the same spectral appearance (Figure 1). Two strong absorption peaks appeared in the range of 3000–3800 and 1500–1800 cm^{-1}, which are characteristic absorptions of water, indicating that absorptions by water greatly interfered with the absorptions of nitrate in the spectra. The characteristic absorptions of nitrate appeared in the range 1200–1500 cm^{-1}, but it was difficult to observe directly because its intensity was much weaker than that of water.

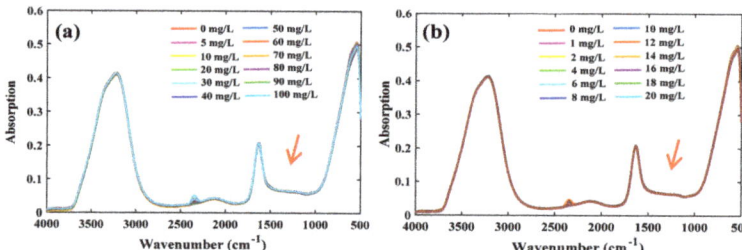

Figure 1. Fourier transforms mid-infrared attenuated total reflectance (FTIR-ATR) spectra of nitrate solutions. (**a**) High-concentration group; (**b**) low-concentration group.

The spectra of the nitrate solutions of the two concentration groups, after deducting the signal arising from water, are shown in Figure 2a,b. For both the groups, the characteristic peak intensities at different nitrate concentrations did not follow a consistent trend, which mainly resulted from the interference of water absorption. The spectra ranging from 1200 to 1500 cm^{-1} (Figure 2c,d) were then deconvoluted, and the absorption intensity of NO_3^- was visually proportional to the NO_3^-–N concentration; therefore, the characteristic peaks within this range could be used for the quantitative analysis of NO_3^-–N in solutions. Comparing the nitrate spectra obtained with the water deduction and deconvolution (without water deduction), it showed that deducting water could not effectively reduce the signal interference, while deconvolution could significantly extract the characteristic peaks of nitrate.

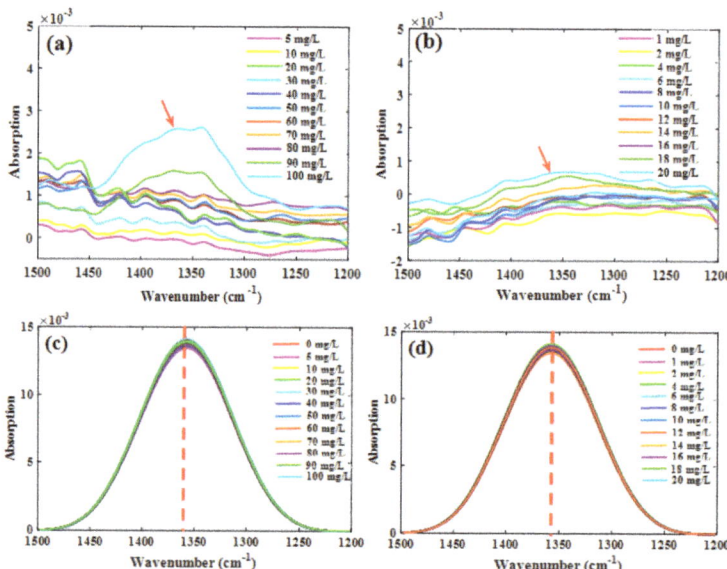

Figure 2. Characteristic absorption bands of nitrate solutions through water deduction ((**a**) high-concentration group; (**b**) low-concentration group; $n = 44$) and deconvolution from raw spectra ((**c**) high-concentration group; (**d**) low-concentration group; $n = 48$).

3.2. Principal Component Analysis

PCA was conducted on the spectra within the range 1200–1500 cm^{-1}. For high- and low-concentration groups, the first two principal components of both the concentration groups accounted for more than 80% of the spectral information within the range 1200–1500 cm^{-1}. Thus, PC1 and PC2 can be used to represent variations in the spectra. However, the scores of these two principal components did not show an obvious and consistent trend. This may have been caused by interference from water or the systematic environment. This might have also occurred because the scores of each component used for mapping only contained information about the original independent variables, without taking into account the relationship between independent and dependent variables [26], which reduced the model's robustness and prediction capability [27–29]. A second PCA of the spectra, within the range 1200–1500 cm^{-1}, was conducted after deconvolution. The results showed that PC1 of the high-concentration group reached 99.52% and that of the low-concentration group was 99.39% (Figure 3a,b, respectively). The NO_3^-–N concentration showed a regular distribution in the PC1 area, wherein the plot shifted to positive values of PC1 as the NO_3^-–N concentration of the solution increased.

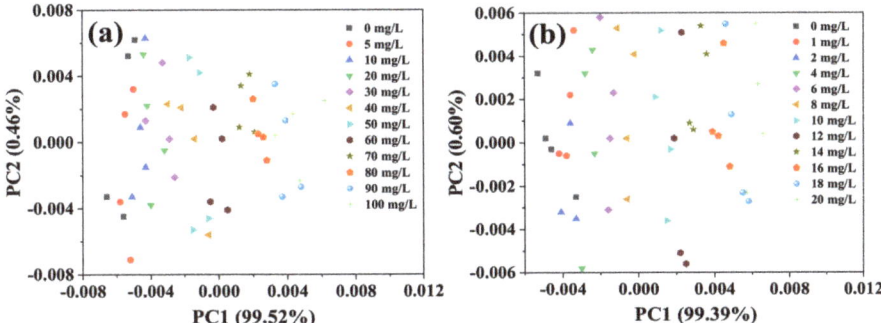

Figure 3. Principal component distribution of FTIR-ATR spectra of the nitrate solution after deconvolution ((**a**) high-concentration group; (**b**) low-concentration group).

3.3. Prediction of Nitrate Nitrogen in Water with Water Deduction

PLSR was used to model the 1200–1500 cm^{-1} region of the spectra, and the overall dataset was divided into a training set (75% of the overall set) and a testing set (25% of the overall set) using random division. The cross-validation method was used to determine the optimal number of PLS factors. As shown in Figure 4a, the optimal number of PLS factors in the high-concentration group was 7, which corresponds to the minimum of *RMSECV* [13]; therefore, the first seven PLS factors were used to construct the PLSR model. Figure 4b,c show the distributions of the real and predicted values of the training and testing sets, respectively. The linear regression coefficient (R^2) of the measured and predicted values of NO_3^-–N in the training set was 0.9756, representing a significant correlation. The R^2 of the prediction set was 0.8325, and the *RPD* value was 1.86 (Table 1).

RPD is an important model evaluation parameter in infrared spectrum analysis; it is the ratio of the standard deviation (*SD*) of a sample to the root mean square error (*RMSE*). Generally, when *RPD* > 1.8, quantitative detection can be conducted. An *RPD* between 2 and 2.5 indicates a good quantitative prediction model, while one higher than 3 suggests excellent model prediction performance [30,31]. Therefore, this model can be applied for the rapid quantitative determination of high NO_3^-–N concentrations in water bodies. The optimal number of PLS factors in the low-concentration group was 4 (Figure 4d); thus, the PLSR model was built using the first four factors. The distributions of the true and predicted values of the training and testing sets are shown in Figure 4e,f, respectively. The evaluation index (R^2) of the training set was 0.9221, suggesting a significant correlation. However, the R^2 of the testing set was much lower at only 0.7932 and RPD also decreased to 1.75, which is lower than the minimum standard of 1.8 for quantitative detection. These results show that the predictive performance of this model is poor and that the model is not suitable for detecting low concentrations of NO_3^-–N in water.

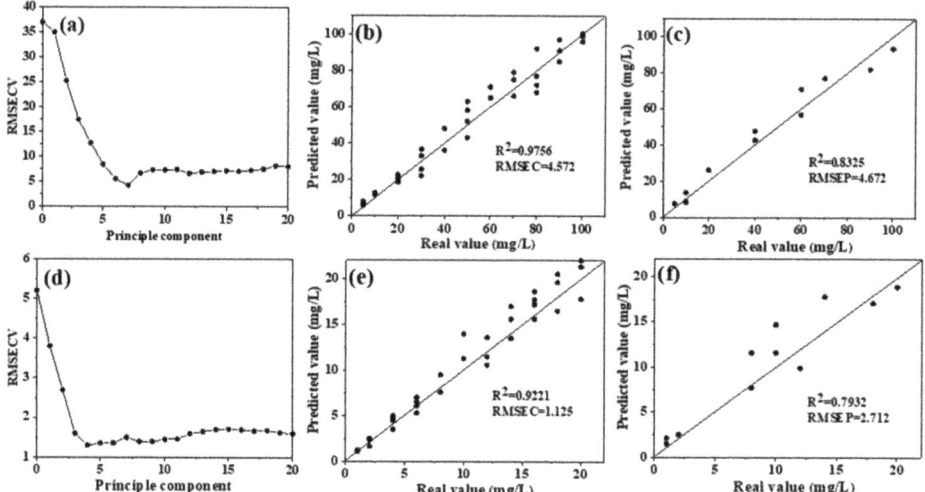

Figure 4. Distribution and model evaluation of the partial least squares (PLS) factor (**a,d**), training set ((**b,e**); $n = 33$), and testing set ((**c,f**); $n = 11$) of the partial least squares regression (PLSR) prediction model (without water deduction) for high nitrate solutions (**a–c**) and low nitrate solutions (**d–f**). Note: RMSECV, root mean square error in cross validation; RMSEC, root mean square error of calibration; RMSEP, root mean square error of prediction.

Table 1. Number of optimal PLS factors and model evaluation results of the PLSR model with and without water deduction.

Treatment	Statistical Parameters	High-Concentration Group	Low-Concentration Group
Water deduction	Correlation coefficient (R^2)	0.8325	0.7932
	RPD	1.86	1.75
	Optimized PLS factor	7	4
Without water deduction	Correlation coefficient (R^2)	0.9578	0.9865
	RPD	4.22	3.15
	Optimized PLS factor	5	3

3.4. Prediction of Nitrate Nitrogen in Water with Deconvolution (without Water Deduction)

Similarly, PLSR was also used to model the characteristic bands of NO_3^-–N, within the 1200–1500 cm^{-1} region, obtained by deconvolution curve-fitting. The overall dataset was divided into a training set (75% of the overall set) and a testing set (25% of the overall set) using random division. Cross-validation was used to obtain the optimal number of principal components in the high- and low-concentration groups and then, to establish PLSR models. For the high-concentration group, the optimal number of PLS factors was 5 (Figure 5a). The R^2 of real and predicted values of nitrate nitrogen in the training set and testing set were 0.9723 and 0.9578 (Figure 5b,c), respectively, implying a significant correlation. The RPD value was 4.22 (Table 1), which was higher than 3, suggesting that the model had an excellent predictive capability. The optimal number of principal components in the low-concentration group model was 3 (Figure 5d), the correlation coefficients (R^2) in the training set and testing set were 0.9853 and 0.9865 (Figure 5e,f), respectively, and the RPD was 3.15 (Table 1), indicating an excellent predictive performance. The above results showed that in both concentration groups, the PLSR model established based on spectra deconvolution (without water deduction) achieved better performance than the model established with water deduction, which indicated that deconvolution peak-fitting could effectively reduce water interference to extract useful spectral information.

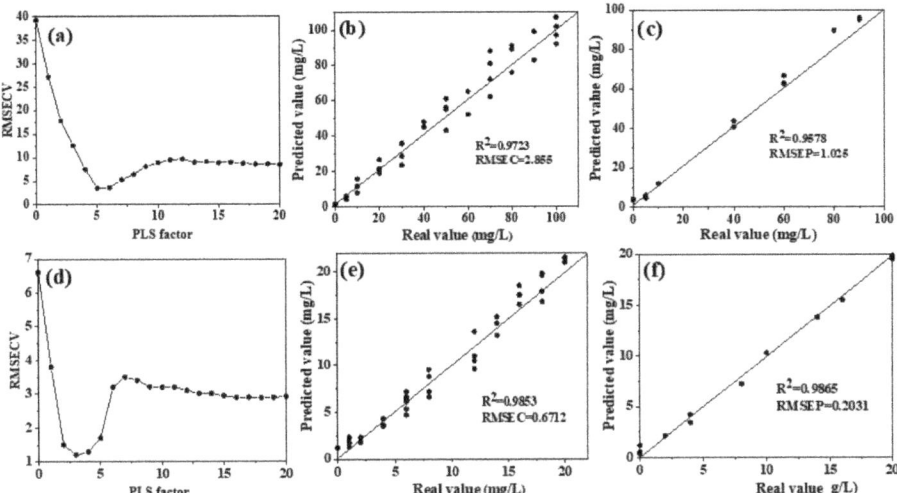

Figure 5. Distribution and model evaluation of the PLS factor (**a**,**d**), training set ((**b**,**e**); $n = 36$), and testing set ((**c**,**f**); $n = 12$) of the PLSR prediction model with deconvolution (without water deduction) for high nitrate solutions (**a**–**c**) and low nitrate solutions (**d**–**f**). Note: *RMSECV*, root mean square error in cross validation; *RMSEC*, root mean square error of calibration; *RMSEP*, root mean square error of prediction.

In the process of linear multivariate calibration analysis, the limit of detection (LOD) could be estimated by 3σ or $3\sigma/m$ [32]. σ was the standard deviation of the predicted concentration, which could be replaced with *RMSE*, and m was the fitting-curve slope of the model (using the real value as the X-axis and the predicted value as the Y-axis). The *m* value of the high concentration group model was 0.6236, and the *RMSEP* was 1.025 (Figure 5c), so the 3σ was 3.075 and the $3\sigma/m$ was 4.931. In the low concentration group, the *m* was 0.7122, and *RMSEP* was 0.2031 (Figure 4f); therefore, the 3σ and $3\sigma/m$ were 0.6039 and 0.8491, respectively.

4. Conclusions

In this study, combined with the PLSR model, FTIR-ATR spectroscopy was applied to detect nitrate in high- and low-concentration solutions, with deconvolution algorithm (without water deduction) comparing with conventional water deduction algorithms. In both the high- and low-concentration groups, the PLSR model based on the non-deduction of water (deconvolution curve-fitting) performed significantly higher prediction accuracy than the model established by deducting water to quantitatively predict nitrate nitrogen, which provided a more effective analysis method for the rapid determination of different concentrations of nitrate in water bodies.

Supplementary Materials: The following are available online. Figure S1: The process of deconvolution, Figure S2: Second derivative of nitrate absorption. (a), high-concentration group with the range of 0–100 mg L^{-1}; (b), low-concentration group with the range of 0–20 mg L^{-1}. Figure S3: The equilibrium of electron clouds in nitrate.

Author Contributions: C.D. designed the framework; F.G. measured FTIR-ATR and analyzed the data; F.G. and K.W. wrote original draft; F.M. and C.D. wrote and reviewed the manuscript. All authors have read and agreed to the published version of the manuscript.

Funding: This work was supported by the Strategic Priority Research Program of Chinese Academy of Sciences (XDA23030107), the National Natural Science Foundation of China (41907154, 42077019), the National Natural Science Foundation of Jiangsu Province (BK20191110) and the "Green blue project" of Jiangsu University.

Conflicts of Interest: The authors declare no conflict of interest.

References

1. Rabalais, N.N. Nitrogen in aquatic ecosystems. *AMBIO J. Hum. Environ.* **2002**, *31*, 102–112. [CrossRef] [PubMed]
2. Verma, A.; Rawat, A.K.; More, N. Extent of nitrate and nitrite pollution in ground water of rural areas of Lucknow, U.P. India. *Curr. World Environ.* **2014**, *9*, 114–122. [CrossRef]
3. Inoue-Choi, M.; Virk-Baker, M.K.; Aschebrook-Kilfoy, B.; Cross, A.J.; Subar, A.F.; Thompson, F.E.; Sinha, R.; Ward, M.H. Development and calibration of a dietary nitrate and nitrite database in the NIH–AARP Diet and Health Study. *Public Health Nutr.* **2016**, *19*, 1934–1943. [CrossRef] [PubMed]
4. Kalaycıoğlu, Z.; Erim, F.B. Nitrate and nitrites in foods: Worldwide regional distribution in view of their risks and benefits. *J. Agric. Food Chem.* **2019**, *67*, 7205–7222. [CrossRef]
5. Killham, K. *Soil Ecology*; Cambridge University Press: Cambridge, UK, 1994; pp. 108–141.
6. China Soil Society Agrochemistry Speciality Committee. *Agrochemical Convention Analytical*; Science Press: Beijing, China, 1989; pp. 91–93.
7. Balabin, R.M.; Safieva, R.Z. Biodiesel classification by base stock Type (vegetable oil) using near infrared spectroscopy data. *Anal. Chim. Acta* **2011**, *689*, 190–197. [CrossRef]
8. Smola, N.; Urleb, U. Qualitative and quantitative analysis of oxytetracycline by near-infrared spectroscopy. *Anal. Chim. Acta* **2000**, *410*, 203–210. [CrossRef]
9. Zhang, M.; Sheng, G.; Mu, Y.; Li, W.; Yu, H.; Harada, H.; Li, Y. Rapid and accurate determination of VFA s and ethanol in the effluent of an anaerobic H_2-producing bioreactor using near-infrared spectroscopy. *Water Res.* **2009**, *43*, 1823–1830. [CrossRef]
10. Shao, Y.; Du, C.; Shen, Y.; Ma, F.; Zhou, J. Rapid determination of nitrogen isotope labeled nitrate using Mid-infrared attenuated Total reflectance spectroscopy. *Chin. J. Anal. Chem.* **2014**, *42*, 747–752. [CrossRef]
11. Shaviv, A.; Kenny, A.; Shmulevitch, I.; Singher, L.; Raichlin, Y.; Katzir, A. Direct monitoring of soil and water nitrate by FTIR based FEWS or membrane systems. *Environ. Sci. Technol.* **2003**, *37*, 2807–2812. [CrossRef]
12. Shao, Y.; Du, C.; Shen, Y.; Ma, F.; Zhou, J. Evaluation of net nitrification rates in paddy soil using mid-infrared attenuated total reflectance spectroscopy. *Anal. Methods* **2017**, *9*, 748–755. [CrossRef]
13. Shao, Y.; Du, C.; Zhou, J.; Ma, F.; Zhu, Y.; Yang, K.; Tian, C. Quantitative analysis of different nitrogen isotope labelled nitrate in paddy soil using mid-infrared attenuated total reflectance spectroscopy. *Anal. Methods* **2017**, *9*, 5388–5394. [CrossRef]
14. Yang, J.; Du, C.; Shen, Y.; Zhou, J. Rapid Determination of Nitrate in Chinese Cabbage Using Fourier Transforms Mid-infrared Spectroscopy. *Chin. J. Anal. Chem.* **2013**, *41*, 1264–1268. [CrossRef]
15. He, M. A parallel-distributed processing for time-domain deconvolution coefficients. *Signal Process.* **1999**, *74*, 309–315. [CrossRef]
16. Wang, D.; Kong, X.; Dong, L.; Chen, L.; Wang, Y.; Wang, X. A predictive deconvolution method for non-white-noise reflectivity. *Appl. Geophys.* **2019**, *16*, 101–115. [CrossRef]
17. Du, H.; Yi, R.; Dong, L.; Liu, M.; Jia, W.; Zhao, Y.; Liu, X.; Hui, M.; Kong, L.; Chen, X. Rotating asymmetrical phase mask method for improving signal-to-noise ratio in wave front coding systems. *Appl. Opt.* **2019**, *58*, 6157–6164.
18. Zou, M.; Unbehauen, R. A deconvolution method for spectroscopy. *Meas. Sci. Technol.* **1995**, *6*, 482–487.
19. Barth, A.; Haris, P.I. *Biological and Biomedical Infrared Spectroscopy*; IOS Press: Amsterdam, The Netherlands, 2009.
20. Du, C.; Zhou, J.; Wang, H.; Chen, X.; Zhang, A.; Zhang, J. Determination of soil properties using Fourier transform mid-infrared photoacoustic spectroscopy. *Vib. Spectrosc.* **2009**, *49*, 32–37. [CrossRef]
21. Savitzky, A.; Golay, M.J.E. Smoothing and differentiation of data by simplified least squares procedures. *Anal. Chem.* **1964**, *8*, 1627–1639. [CrossRef]
22. Buslov, D.K.; Nikonenko, N.A. Regularized method of spectral curve deconvolution. *Appl. Spectrosc.* **1997**, *51*, 666–672. [CrossRef]

23. Buslov, D.K.; Nikonenko, N.A.; Sushko, N.I.; Zhbankov, R.G. Analysis of the structure of the bands in the IR spectrum of β-D glucose by the regularized method of deconvolution. *J. Appl. Spectrosc.* **2002**, *69*, 817–824. [CrossRef]
24. Viscarra Rossel, R.A.; McGlynn, R.N.; McBratney, A.B. Determining the composition of mineral-organic mixes using UV-vis-NIR diffuse reflectance spectroscopy. *Geoderma* **2006**, *137*, 70–82. [CrossRef]
25. Yuzhen, L.; Changwen, D.; Changbing, Y.; Jianmin, Z. Use of FTIR-PAS combined with chemometrics to quantify nutritional information in rapeseeds (*Brassica napus*). *J. Plant Nutr. Soil Sci.* **2014**, *177*, 927–933.
26. Godoy, J.L.; Vega, J.R.; Marchetti, J.L. Relationships between PCA and PLS-regression. *Chemom. Intell. Lab. Syst.* **2014**, *130*, 182–191. [CrossRef]
27. Jie, D.; Xie, L.; Fu, X.; Rao, X.; Ying, Y. Variable selection for partial least squares analysis of soluble solids content in watermelon using near-infrared diffuse transmission technique. *J. Food Eng.* **2013**, *118*, 387–392. [CrossRef]
28. Mehmood, T.; Liland, K.H.; Snipen, L.; Sæbø, S. A review of variable selection methods in partial least squares regression. *Chemom. Intell. Lab. Syst.* **2012**, *118*, 62–69. [CrossRef]
29. Wu, D.; He, Y.; Nie, P.; Cao, F.; Bao, Y. Hybrid variable selection in visible and near-infrared spectral analysis for non-invasive quality determination of grape juice. *Anal. Chim. Acta* **2010**, *659*, 229–237. [CrossRef]
30. Viscarra, R.A.; Walvoort, D.J.J.; McBratney, A.B.; Janik, L.J.; Skjemstad, J.O. Visible, near infrared, mid infrared or combined diffuse reflectance spectroscopy for simultaneous assessment of various soil properties. *Geoderma* **2006**, *131*, 59–75. [CrossRef]
31. Du, C.; Zhou, J. Prediction of soil available phosphorus using Fourier transform infrared photoacoustic spectroscopy. *Chin. J. Anal. Chem.* **2007**, *35*, 119–122.
32. Doran, E.M.; Yost, M.G.; Fenske, R.M. Measuring dermal exposure to pesticide residues with attenuated total reflectance Fourier transform infrared (ATR-FTIR) spectroscopy. *Bull. Environ. Contam. Toxicol.* **2000**, *64*, 666–672. [CrossRef]

Sample Availability: Samples of the compounds are available from the authors.

Publisher's Note: MDPI stays neutral with regard to jurisdictional claims in published maps and institutional affiliations.

© 2020 by the authors. Licensee MDPI, Basel, Switzerland. This article is an open access article distributed under the terms and conditions of the Creative Commons Attribution (CC BY) license (http://creativecommons.org/licenses/by/4.0/).

Article

Simple Derivatization–Gas Chromatography–Mass Spectrometry for Fatty Acids Profiling in Soil Dissolved Organic Matter

Neil Yohan Musadji [1,2] and Claude Geffroy-Rodier [1,*]

1. Institut de Chimie des Milieux et Matériaux de Poitiers (IC2MP), Université de Poitiers, UMR CNRS 7285, Equipe EBiCOM, 4 rue Michel Brunet, 86076 Poitiers, France; nyohan1@gmail.com
2. Institut National Supérieur d'Agronomie et de Biotechnologies (INSAB), Université des Sciences et Techniques de Masuku (USTM), 941 Franceville, Gabon
* Correspondence: claude.geffroy@univ-poitiers.fr; Tel.: +33-549453590

Academic Editors: Gavino Sanna and Stefan Leonidov Tsakovski
Received: 9 September 2020; Accepted: 10 November 2020; Published: 12 November 2020

Abstract: Dissolved organic matter is an important component of the global carbon cycle that allows the distribution of carbon and nutrients. Therefore, analysis of soil dissolved organic matter helps us to better understand climate change impacts as it is the most dynamic and reactive fraction in terrestrial ecosystems. Its characterization at the molecular level is still challenging due to complex mixtures of hundreds of compounds at low concentration levels in percolating water. This work presents simple methods, such as thermochemolysis– or derivatization–gas chromatography, as an alternative for the analysis of fatty acids in dissolved organic matter without any purification step. The variables of the protocols were examined to optimize the processing conditions for the C_9–C_{18} range. As a proof of concept, fatty acid distributions of soil percolating water samples from a long-term field experiment were successfully assessed. The variability of dissolved organic acid distributions was pronounced through depth profile and soil treatment but no major change in composition was observed. However, although the optimization was done from C_9 to C_{18}, detection within the C_6-C_{32} fatty acids range was performed for all samples.

Keywords: soil dissolved organic matter; fatty acids; methylation

1. Introduction

Soil dissolved organic matter (SDOM, OM dissolved in soil solution which passes a 0.45 µm filter), which represents a small portion of soil organic matter (<2%), is the most dynamic and reactive fraction in terrestrial ecosystems [1,2]. Its characterization is an important issue for soil environmental studies as DOM is a useful indicator for SOM quality [1,3,4], pollution [5] and soil management impact [1,6–8]. Although quantifying SDOM amount (commonly by measuring dissolved organic carbon concentration, DOC) is necessary, it is also important to characterize its composition at the molecular level, as the components drive its reactivity and properties [9].

From an analytical perspective, SDOM characterization at the molecular level poses a special challenge. The complex and diverse soil dissolved organic matter requires a multi-method analytical characterization [10]. A single sample can be composed of tens of thousands of individual molecules that together rarely exceed 1 mg C/L. Actual molecular-level structural characterization, which requires detailed extraction and separation protocols, is limited to a few compounds or compound classes, and thus a very small portion of the total DOM pool [11]. Recently, studies have focused on targeted molecules relevant for particular biogeochemical processes but, as far as we know, none were dedicated to biomarkers analysis. Biomarkers preserving molecular fingerprints are indicative of past vegetation cover, soil organic matter input and microbial diversity. Their identification would be of major interest to an investigation of soil functioning. Among biomarkers, carboxylic acids, which represent, for the low molecular weight ones, a significant part of the water-soluble fractions of organic molecules released in the rhizosphere [12], exhibit important roles in soil nutrient availability, ecology and productivity. Generally, they are weak acids ranging from 0.46 to 1 million Dalton (Da). Although numerous techniques are devoted to volatile fatty acids (gas, liquid or ion chromatography, titration, mid-infrared spectroscopy) in aqueous samples, none or only few are used for soil leachates characterization [13–17] and are useful for fast total fatty acids monitoring. Most of the above methods either require sophisticated instruments, or are time-consuming or not useful for higher molecular weight fatty acids, making them unsuitable for fatty acids fingerprints. Therefore, the optimization of in situ characterization of the latter makes it possible to eliminate any pretreatment of aqueous samples.

To study DOM dynamics, numerous laboratory experiments using disturbed soil samples extracted with aqueous solutions have been conducted, but few have been performed on samples collected in field experiments [18,19]. Experiments at field scales are now needed to characterize undisturbed samples [20,21]. However, field experiments mean time consuming experiments with many samples to handle to take into account time and scale soil variabilities. Developing a rapid screening of fatty acids in SDOM from field experiments will allow the selection of representative samples for further investigations.

The main objectives of this study were (i) to develop a simple, fast and accurate strategy to detect fatty acids in water solutions with few pretreatments to preserve the native molecular structures and (ii) to validate it on percolating soil solutions from a long-term experiment performed on an amended soil and its reference [21,22]. Two analytical strategies have been evaluated in this study for the identification of fatty acids in SDOM. Thermochemolysis, with the alkaline reagent tetramethylammonium hydroxide (TMAH) which is efficient in hydrolyzing and derivatizing all fatty acids in soil humic substances, would allow the detection of total fatty acids (both free and covalently bound acids) [21,22]. Derivatization, commonly used for extracted organic solvent fatty acids, will allow to detect SDOM free fatty acids [11,23,24].

2. Results and Discussion

2.1. Analysis of SDOM Fatty Acids from Evaporated to Dryness Samples

2.1.1. Thermochemolysis: Fingerprints of Total Fatty Acids in SDOM

Thermochemolysis in the C_9–C_{18} Fatty Acids Range

To have an overview of the SDOM fatty acids, optimal analytical conditions of thermochemolysis in the presence of TMAH were first determined on a standard mixture of C_9 to C_{18} fatty acids. An amount of 25%TMAH in methanol at 600 °C was found to be the best conditions as it gave the highest peak intensity and a good overall molecular response (Figure 1).

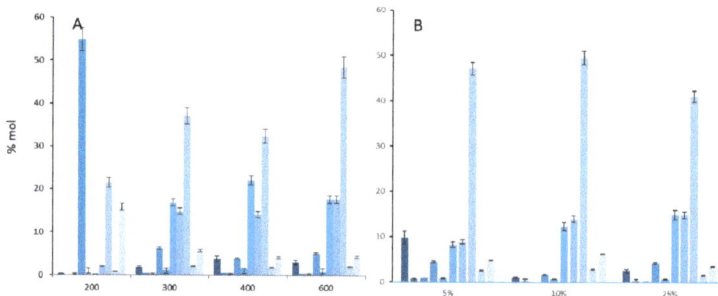

Figure 1. %mol (Ai/∑Ai) of fatty acid methyl esters (C_9 to C_{18}) in function of (**A**) the temperature of thermochemolysis (with 25% TMAH in MeOH) and (**B**) the percentage of TMAH in methanol (at 600 °C, 15 µL TMAH solution).

SDOM Analysis by Thermochemolysis

The fatty acid distribution of percolating waters from urban green waste amended soil and its non-amended reference was investigated to validate thermochemolysis for SDOM analysis. In total, 500 mL of each sample (ranging from 8 to 61 mgC/L, Figure S1) was evaporated at 40 °C to dryness and optimal thermochemolysis was performed on the resulting residues.

Fatty acids, observed as methyl esters, ranged in all samples from C_6 to C_{30}. The bimodal distribution was dominated by C_{16} for short mode (C_6–C_{18}) and C_{28} and C_{30} for long-chain acids (Figure 2). In reference soil solutions, even-over-odd predominance in the C_{24}–C_{32} range was a marker of higher plant vegetation (Figure 2A). In amended soil solutions, branched isomers of C_{15} and C_{17} were as abundant as long-chain compounds in the first 30 cm depth, showing higher microbial activity than in reference soil solutions (Figure 2B). The higher input of vegetation was at 30 cm in amended soil whereas it was at 100 cm in reference soil solutions. The influence of amendment performed seven years before the sampling was thus still significant at 30 cm depth.

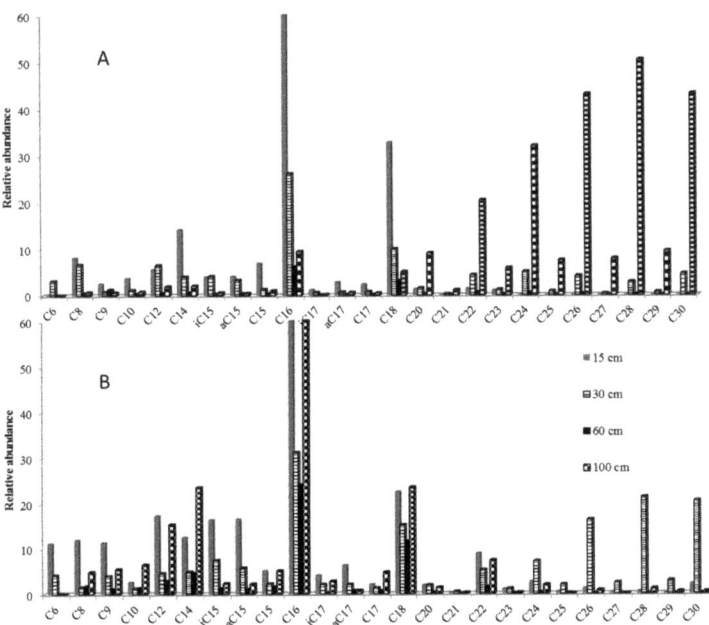

Figure 2. Thermochemolysis of 500 mL reference (**A**) and amended (**B**) soil solutions from 15 to 100 cm depth after a 40 °C evaporation step (TMAH 25%, 600 °C).

2.1.2. Boron Trifluoride-Methanol Complex (BF$_3$/MeOH) Derivatization: Fingerprints of Free Fatty Acids in SDOM

To identify only the free fatty acids, a derivatization with BF$_3$/MeOH was then evaluated.

When performing BF$_3$/MeOH derivatization on evaporated samples, the resulting distribution was only due to free fatty acids (Figure 3). The bimodal distribution ranged from C_{12} to C_{32}. High concentrations in 30 cm reference soil solutions compared to thermochemolysis results showed that high concentrations of fatty acids at 100 cm were rather associated to macromolecules such as humic substances than to free molecules. This procedure is, however, not suitable for analyzing carboxylic acids of low molecular weight (<C_{12}) since their increased volatility after derivatization can lead to unquantifiable losses related to evaporation.

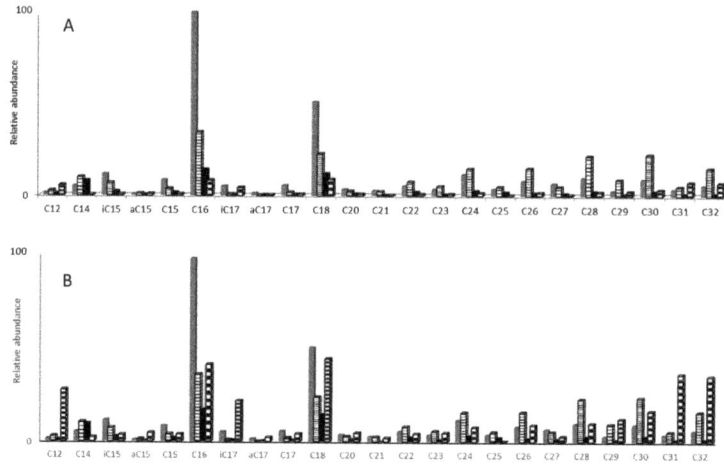

Figure 3. BF$_3$/MeOH derivatization on 500 mL reference (**A**) and amended (**B**) soil solutions from 15 to 100 cm depth after a 40 °C evaporation step.

To have a rapid overview of low molecular weight compounds, we have developed an in situ derivatization on a 2 mL sample.

2.2. In Situ Analysis of SDOM Fatty Acids

2.2.1. Optimal In Situ Derivatization

Before optimization of the derivatization step, derivatives recovery was estimated. The simple and low cost dichloromethane CH$_2$Cl$_2$ liquid/liquid extraction of the derivatized compounds was selected as the optimized direct immersion solid phase microextraction (DI-SPME) method that showed a peak area decreasing inversely according to the molecular weight of the compounds (Figures 4 and S2).

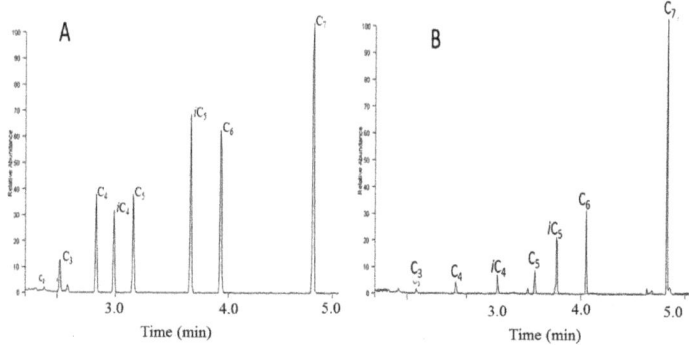

Figure 4. Spectral ionic mass (SIM) chromatogram of liquid/liquid (**A**) and direct immersion solid phase microextraction (SPME) (**B**) optimal recoveries of iso molar low molecular weight fatty acid methyl esters.

The derivatization involving direct methylation of fatty acids with boron trifluoride (BF$_3$) in methanol, used on evaporated samples, was then optimized for the direct methylation in water samples to study low molecular weight compounds. To perform in situ derivatization, different ratios of BF$_3$/MeOH and MeOH (*v/v*) were studied (Figure 5). BF3-MeOH, MeOH 45/45 (%/%), that showed the best recoveries, was used to analyze water samples.

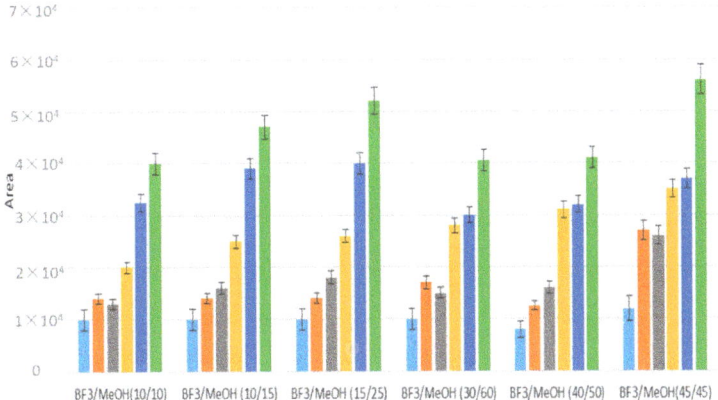

Figure 5. Optimization of BF_3-MeOH and MeOH ratios (%/%) for the derivatization of a 2 mL water solution of low molecular weight fatty acids (C_4 blue, isoC_4 orange, C_5 grey, iso C_5 yellow, C_6 dark blue, C_7 green) with CH_2Cl_2 liquid/liquid extraction. Analyses were performed in triplicates.

2.2.2. SDOM Analysis by In Situ Derivatization

C_6-C_{18} fatty acids were detected in all samples, when single ion monitoring analyses were performed (m/z 74, Figure 6) in 2 mL samples without any pretreament. This protocol will allow us to discriminate samples so that further analyses will be performed only on relevant samples.

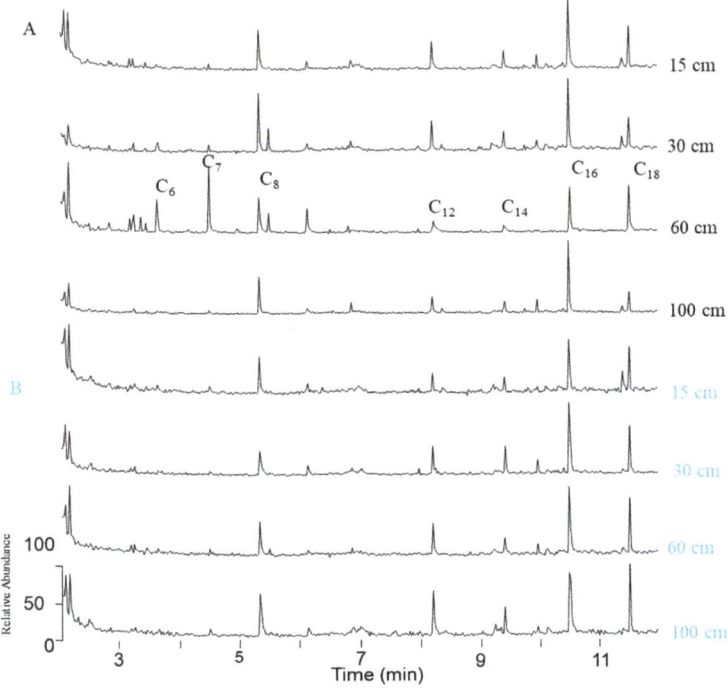

Figure 6. SIM chromatograms of volatile fatty acid methyl esters after BF_3/MeOH derivatization and CH_2Cl_2 liquid/liquid extraction of 2 mL reference (**A**) and amended (**B**) soil solutions.

3. Materials and Methods

3.1. Experimental Site

The study was conducted in the Hydrogeological Experimental Site which is a part of the Network of National Hydrogeological Sites (SNO H+) and of the French network of Critical Zone Observatories: Research and Applications (OZCAR) of the University of Poitiers located in the Regional Observatory of Deffend (OR, Mignaloux-Beauvoir, France). The field, previously managed as cereal cropping rotation (maize-wheat), is currently under grassland. The soil is a luvic cambisol fully characterized in a previous study [22]. The experimental field has been divided into 6 plots (3 reference and 3 amended, all equipped with suction cups). An amount of 150 t/ha of bio and green wastes (from the composting plant of La Villedieu du Clain, France) have been amended to follow the long-term effects of exogenous OM in the soil.

3.2. Percolating Water Samples and Pretreatments

The sampling was performed with 12 ceramic suction cups (31 mm, SDEC, Reignac sur Indre, France) to collect soil solutions disposed in triplicates at 5, 30, 60 or 100 cm depth. Then, the samples were filtered at 0.45 µm with a Whatman filters to separate particular to DOM. Before the storage at 4 °C for their conservation, DOC was measured. Thereafter, molecular study of dissolved fatty acids was accessible after evaporation step at 40 °C; the resulting extract was further dried in a desiccator.

3.3. Chemicals and Reagents

Methanol, chloroform, dichloromethane of analytical standard grade and SPME fibers were purchased from Supelco (Bellefonte, PA, USA). The fibers of carboxen/polydimethylsiloxane (CAR/PDMS), polydimethylsiloxane (PDMS), carbowax/polydimethylsiloxane divinylbenzene (CAR/PDMS/DVB) were tested. BF_3-MeOH and tretramethylamonnium hydroxyde (TMAH) used for derivatization were purchased from Sigma-Aldrich (Darmstadt, Hessen, Germany). C_9–C_{18} standard solution was obtained from separate compounds provided by Sigma Aldrich. Solutions were performed in methanol at 2 mM, mixed and further diluted before the experiment. Standards of low molecular weight compounds for SPME (certified reference material CRM46975) were purchased from Supelco.

3.4. Derivatization Process

For in situ analysis, 50 µL MeOH and 200 µL 12.5% (w/v) BF_3-MeOH were added to 2 mL water sample and were heated during 30 min at 70 °C. Resulting fatty methyl esters (FAMES) were extracted with 200 µL CH_2Cl_2; 1 µL was injected for GC-MS analyses.

For dried samples, 50 µL of methanol and 200 µL of 12.5% (w/v) BF_3-methanol were added to the vials containing the dried fatty acids (20 mg). After tightly capping the vials, they were heated for 30 min at 70 °C. Pure water (ca. 100 µL) was added and after tightly capping again, the vials were vigorously shaken and then 150 µL of dichloromethane was added. Fatty acid methyl esters were removed in the dichloromethane layer. The organic layer was transferred to a 1 mL conical vial. This procedure was repeated three times; 1 µL was injected for GC-MS analyses.

An amount of 10 mg of residues resulting from evaporation was moistened with 10 µL of a 25% (w/w) methanol solution of tetramethylamonium hydroxide (TMAH) in a cup, heated at 40 °C, 10 min and introduced in the pyrolyzer.

3.5. GC/MS and Pyr-GC/MS Analyses

The analyses were perform using a Thermo Finnigan Trace gas chromatograph interfaced to a Thermo Finnigan Automass mass spectrometer (Thermo Fisher Scientific, Waltham, MA, USA, 70 eV) operated in fullscan or SIM mode with a fused silica capillary column (HP-5% phenylmethylpolysiloxane 30 m × 0.25 mm, 0.25 µm film thickness) and helium as carrier gas. Line transfer and source were held, respectively, at 280 °C and 220 °C. The GC oven temperature program was from 60 °C, at 5 °C·min^{-1} up to 300 °C, and the temperature was finally held for 10 min at 300 °C. The temperatures of the injector and the detector were 250 and 300 °C, respectively.

Py-GC/MS experiments used a EGA-PY 3030D Pyrolyzer (Frontier Lab, Japan) connected to a gas chromatography instrument (GC2010 Pro, Shimadzu, Kyoto, Japan) with capillary column (SLB-5% phenylmethylsiloxan 30 m × 0.25 mm, 0.25 µm film thickness) and coupled with mass spectrometry (Ultra QP 2010). The GC and MS conditions were the same as for GC-MS analysis.

Supplementary Materials: The following are available online, Figure S1: SPME optimal recoveries of methyl fatty esters: influence of stationary phase, time and temperature. Figure S2: (A) Mean SPME recoveries on PDMS/DVB (blue), PDMS (red) and PDMS/CAR/DVB (grey) fibers of methyl esters (10 mL, room temperature, from 5 to 60 min. (B) Histograms from room temperature to 80 °C for C_6 (blue) and C_5 (orange) methyl ester recoveries on PDMS/CAR/DVB fiber from 5 to 60 min exposure. * recoveries at 60 °C. (C) extraction profiles at 60 °C of C_6 (blue) and C_5 (orange) methyl ester on PDMS/CAR/DVB fiber from 5 to 60 min.

Author Contributions: Conceptualization, C.G.-R.; methodology, C.G.-R. and N.Y.M.; experiments and data analysis, N.Y.M.; writing—original draft preparation, C.G.-R. and N.Y.M.; funding acquisition, C.G.-R. All authors have read and agreed to the published version of the manuscript.

Funding: The authors wish to thanks the CPER/FEDER and région Poitou-Charentes for funding.

Conflicts of Interest: The authors declare no conflict of interest.

References

1. Haynes, R.J. Labile organic matter fractions as central components of the quality of agricultural soils: An overview. In *Advances in Agronomy*; Academic Press: Cambridge, MA, USA, 2005; Volume 85, pp. 221–268, ISBN 0065-2113.
2. Marschner, B.; Kalbitz, K. Controls of bioavailability and biodegradability of dissolved organic matter in soils. *Geoderma* **2003**, *113*, 211–235. [CrossRef]
3. Filep, T.; Draskovits, E.; Szabó, J.; Koós, S.; László, P.; Szalai, Z. The dissolved organic matter as a potential soil quality indicator in arable soils of Hungary. *Environ. Monit. Assess.* **2015**, *187*. [CrossRef] [PubMed]
4. Jones, D.L.; Simfukwe, P.; Hill, P.W.; Mills, R.T.E.; Emmett, B.A. Evaluation of dissolved organic carbon as a soil quality indicator in national monitoring schemes. *PLoS ONE* **2014**, *9*, e90882. [CrossRef] [PubMed]
5. Amorello, D.; Barreca, S.; Gambacurta, S.; Gulotta, M.G.; Orecchio, S.; Pace, A. An analytical method for monitoring micro-traces of landfill leachate in groundwater using fluorescence excitation–emission matrix spectroscopy. *Anal. Methods* **2016**, *8*, 3475–3480. [CrossRef]
6. Seifert, A.-G.; Roth, V.-N.; Dittmar, T.; Gleixner, G.; Breuer, L.; Houska, T.; Marxsen, J. Comparing molecular composition of dissolved organic matter in soil and stream water: Influence of land use and chemical characteristics. *Sci. Total Environ.* **2016**, *571*, 142–152. [CrossRef] [PubMed]
7. Sharma, P.; Laor, Y.; Raviv, M.; Medina, S.; Saadi, I.; Krasnovsky, A.; Vager, M.; Levy, G.J.; Bar-Tal, A.; Borisover, M. Compositional characteristics of organic matter and its water-extractable components across a profile of organically managed soil. *Geoderma* **2017**, *286*, 73–82. [CrossRef]
8. Sun, H.Y.; Koal, P.; Gerl, G.; Schroll, R.; Joergensen, R.G.; Munch, J.C. Water-extractable organic matter and its fluorescence fractions in response to minimum tillage and organic farming in a Cambisol. *Chem. Biol. Technol. Agric.* **2017**, *4*. [CrossRef]
9. Roth, V.-N.; Lange, M.; Simon, C.; Hertkorn, N.; Bucher, S.; Goodall, T.; Griffiths, R.I.; Mellado-Vázquez, P.G.; Mommer, L.; Oram, N.J.; et al. Persistence of dissolved organic matter explained by molecular changes during its passage through soil. *Nat. Geosci.* **2019**, *12*, 755–761. [CrossRef]
10. Herbert, B.E.; Bertsch, P.M. Characterization of dissolved and colloidal organic matter in soil solution: A review. In *Carbon Forms and Functions in Forest Soils*; McFee, W.W., Kelly, J.M., Eds.; Soil Science Society of America: Madison, WI, USA, 2006; pp. 63–88. ISBN 978-0-89118-869-8.
11. Minor, E.C.; Swenson, M.M.; Mattson, B.M.; Oyler, A.R. Structural characterization of dissolved organic matter: A review of current techniques for isolation and analysis. *Environ. Sci. Process. Impacts* **2014**, *16*, 2064–2079. [CrossRef]
12. Strobel, B.W. Influence of vegetation on low-molecular-weight carboxylic acids in soil solution—A review. *Geoderma* **2001**, *99*, 169–198. [CrossRef]

13. Jobling Purser, B.J.; Thai, S.-M.; Fritz, T.; Esteves, S.R.; Dinsdale, R.M.; Guwy, A.J. An improved titration model reducing over estimation of total volatile fatty acids in anaerobic digestion of energy crop, animal slurry and food waste. *Water Res.* **2014**, *61*, 162–170. [CrossRef] [PubMed]
14. Boe, K.; Batstone, D.J.; Angelidaki, I. An innovative online VFA monitoring system for the anaerobic process, based on headspace gas chromatography. *Biotechnol. Bioeng.* **2007**, *96*, 712–721. [CrossRef] [PubMed]
15. Raposo, F.; Borja, R.; Cacho, J.A.; Mumme, J.; Orupõld, K.; Esteves, S.; Noguerol-Arias, J.; Picard, S.; Nielfa, A.; Scherer, P.; et al. First international comparative study of volatile fatty acids in aqueous samples by chromatographic techniques: Evaluating sources of error. *TrAC Trends Anal. Chem.* **2013**, *51*, 127–143. [CrossRef]
16. Falk, H.M.; Reichling, P.; Andersen, C.; Benz, R. Online monitoring of concentration and dynamics of volatile fatty acids in anaerobic digestion processes with mid-infrared spectroscopy. *Bioprocess Biosyst. Eng.* **2015**, *38*, 237–249. [CrossRef] [PubMed]
17. Robert-Peillard, F.; Palacio-Barco, E.; Dudal, Y.; Coulomb, B.; Boudenne, J.-L. Alternative spectrofluorimetric determination of short-chain volatile fatty acids in aqueous samples. *Anal. Chem.* **2009**, *81*, 3063–3070. [CrossRef]
18. Chantigny, M.H. Dissolved and water-extractable organic matter in soils: A review on the influence of land use and management practices. *Geoderma* **2003**, *113*, 357–380. [CrossRef]
19. Embacher, A.; Zsolnay, A.; Gattinger, A.; Munch, J.C. The dynamics of water extractable organic matter (WEOM) in common arable topsoils: I. Quantity, quality and function over a three year period. *Geoderma* **2007**, *139*, 11–22. [CrossRef]
20. Deb, S.K.; Shukla, M.K. A review of dissolved organic matter transport processes affecting soil and environmental quality. *J. Environ. Anal. Toxicol.* **2011**, *1*. [CrossRef]
21. Musadji, N.Y.; Lemée, L.; Caner, L.; Porel, G.; Poinot, P.; Geffroy-Rodier, C. Spectral characteristics of soil dissolved organic matter: Long-term effects of exogenous organic matter on soil organic matter and spatial-temporal changes. *Chemosphere* **2019**, 124808. [CrossRef]
22. Celerier, J.; Rodier, C.; Favetta, P.; Lemee, L.; Ambles, A. Depth-related variations in organic matter at the molecular level in a loamy soil: Reference data for a long-term experiment devoted to the carbon sequestration research field. *Eur. J. Soil Sci.* **2009**, *60*, 33–43. [CrossRef]
23. Webster, R. Soil sampling and methods of analysis–Edited by M.R. Carter & E.G. Gregorich. *Eur. J. Soil Sci.* **2008**, *59*, 1010–1011. [CrossRef]
24. Shadkami, F.; Helleur, R. Recent applications in analytical thermochemolysis. *J. Anal. Appl. Pyrolysis* **2010**, *89*, 2–16. [CrossRef]

Sample Availability: Samples of the compounds are not available from the authors.

Publisher's Note: MDPI stays neutral with regard to jurisdictional claims in published maps and institutional affiliations.

© 2020 by the authors. Licensee MDPI, Basel, Switzerland. This article is an open access article distributed under the terms and conditions of the Creative Commons Attribution (CC BY) license (http://creativecommons.org/licenses/by/4.0/).

Article

The Potential Transformation Mechanisms of the Marker Components of Schizonepetae Spica and Its Charred Product

Xindan Liu, Ying Zhang *, Menghua Wu, Zhiguo Ma and Hui Cao *

Research Center for Traditional Chinese Medicine of Lingnan (Southern China), Jinan University, Guangzhou 510632, China; liuxindan66@stu2018.jnu.edu.cn (X.L.); zyfxwmh@163.com (M.W.); mzg79@hotmail.com (Z.M.)
* Correspondence: zhangying@jnu.edu.cn (Y.Z.); huicao@stu2015.jnu.edu.cn (H.C.)

Academic Editor: Derek J. McPhee
Received: 15 July 2020; Accepted: 17 August 2020; Published: 17 August 2020

Abstract: Schizonepetae Spica (SS) is commonly used for treating colds, fevers, bloody stool and metrorrhagia in China. To treat colds and fevers, traditional Chinese medicine doctors often use raw SS, while to treat bloody stool and metrorrhagia, they usually use Schizonepetae Spica Carbonisata (SSC; raw SS processed by stir-frying until carbonization). However, there have been limited investigations designed to uncover the mechanism of stir-fry processing. In the present study, a method combining gas chromatography-mass spectrometry (GC-MS) and high-performance liquid chromatography (HPLC) was developed for the comprehensive analysis of the chemical profiles of SS and SSC samples. Principal component analysis of the GC-MS data demonstrated that there were 16 significant differences in volatile compounds between the SS and SSC samples. The simultaneous quantification of six nonvolatile compounds was also established based on HPLC, and remarkable differences were found between the two products. These changes were probably responsible for the various pharmacological effects of SS and SSC as well as the observed hepatotoxicity. Finally, the mechanisms could be rationalized by deducing possible reactions involved in the transformation of these marker components. This work reports a new strategy to reveal the chemical transformation of SS during stir-fry processing.

Keywords: Chinese medicinal material; Schizonepetae Spica; Schizonepetae Spica Carbonisata; stir-frying; marker component; transformation mechanism

1. Introduction

Chinese medicinal materials (CMMs) often have to be processed by using physical or chemical methods before prescription or clinical usage. The aims of processing are to alter the clinical efficacy and/or reduce the toxicity of the CMMs [1]. The classic Chinese medicine literature *Lei Gong Pao Zhi Lun* (Lei Gong Processing Handbook, 500 A.D.) emphasized the importance of processing for the first time as a pharmaceutical technique to fulfill the different requirements of therapy [2]. Modern research has indicated that the processing of CMMs has a large influence on the quality and quantity of chemicals in medicinal materials, which consequently affects the bioactivities, pharmacokinetics, and/or safety of the medicinal materials [3–6].

Schizonepetae Spica (SS) is the dried spike of *Schizonepeta tenuifolia* Briq. (Chinese Pharmacopoeia, 2015 edition) [7]. It was first recorded in *Shen Nong Ben Cao Jing* (Shen Nong's herbal classic), a book written 2000 years ago [8]. The herb is commonly used in traditional Chinese medicine (TCM) prescriptions to treat colds, fevers, bloody stool and metrorrhagia [9]. To treat colds and fevers, TCM practitioners often prescribe raw SS, while to treat bloody stool and metrorrhagia, they usually use Schizonepetae Spica Carbonisata (SSC; raw SS processed by stir-frying until carbonization). Pharmacological analyses have shown that SS had three main biological and pharmaceutical properties, including anti-inflammatory [10], antiviral [11,12] and hemostatic activity [13–15]. Chemical studies revealed that SS contains volatile oils, flavonoids, organic acids, etc. [16–18]. The essential oils accumulated by SS are recognized as the major constituents responsible for its anti-inflammatory and antiviral effects [10,19]. One of the major pharmacological volatile components in SS is pulegone, which is also used as a biomarker for SS in the Chinese pharmacopoeia (2015 edition) [7]. However, evidence has demonstrated that pulegone can cause severe hepatotoxicity [20]. Previous reports have shown that changes in the quantity and quality of the components in the volatile oils of Schizonepetae Herba (SH, the aerial part of *S. tenuifolia* Briq.) and its charred product Schizonepetae Herba Carbonisata (SHC; raw SH processed by stir-frying until carbonization) have been observed: nine new components were formed and eight components disappeared in the volatile oils after processing while the contents of seven original constituents decreased and the concentration of nine other constituents increased [21]. The changes in the contents of some constituents between the raw drugs and processed products are mostly due to the transformation of chemical structures and/or volatilization of volatile compounds during processing. However, the possible mechanism involved in the transformation of marker compounds induced by processing by stir-frying SS has not been fully elucidated. The influence of processing on SS is not only due to the change in volatile oils, and further study on the change in other components is needed.

In an attempt to uncover the potential mechanism of stir-fry processing on SS, a combination of gas chromatography-mass spectrometry (GC-MS) and high-performance liquid chromatography-diode-array detector (HPLC-DAD) was carried out, thus obtaining an overall characterization for both volatile and nonvolatile compounds. The information obtained from GC-MS was then analyzed by chromatographic fingerprinting to compare the chemical profiles of SS and SSC. Statistical analyses of principal component analysis (PCA) from GC-MS and Student's two-tailed t-tests from HPLC were employed to find marker compounds to differentiate SS from SSC scientifically and reliably. The processing mechanism could be rationalized by deducing possible reactions involved in the transformation of these marker compounds.

2. Results and Discussion

2.1. Comparison of SS and SSC Volatile Compounds

2.1.1. GC-MS Fingerprint of the Essential Oils from SS and SSC

The essential oils from the SS and SSC samples were extracted by hydrodistillation. The essential oil yield of SS markedly decreased after stir-fry processing, and the essential oil yield from SSC was too low to be detected. It has been reported that stir-frying can dramatically reduce the contents of volatile components from SS through heat volatilization [22]. Nonetheless, the essential oil can still be detected after stir-frying, which supports the theory of CMM processing "raw medicinal herbs processed by stir-frying until carbonized to avoid scorching" [23,24]. All samples were analyzed by GC-MS, and the chromatograms are shown in Figure 1. The correlation coefficient of similarity between each chromatographic profile of SS, SSC and their reference chromatogram, the representative standard fingerprint/chromatogram for a group of chromatograms, was calculated (Table 1). The correlation coefficients from each chromatogram for 11 SS batches were found to be 0.969–0.997, which was in agreement with previous studies [25,26], while those from the 11 batches of SSC were 0.955–0.995 (Table 1). These results demonstrated that the chromatographic fingerprints of SS and SSC were consistent to some extent despite their slightly different chemical compositions, which indicated that the developed processing method for investigating the specific variations between SS and SSC was satisfactory. After stir-fry processing, the GC-MS fingerprints of SS and SSC showed differences in their chromatographic profiling and relative contents (Figure 1). The circled peaks in section I and section III had higher contents in SS, whereas section II was relatively higher in SSC compared with in SS samples. This phenomenon reminded us that stir-fry processing could cause chemical profile changes in SS.

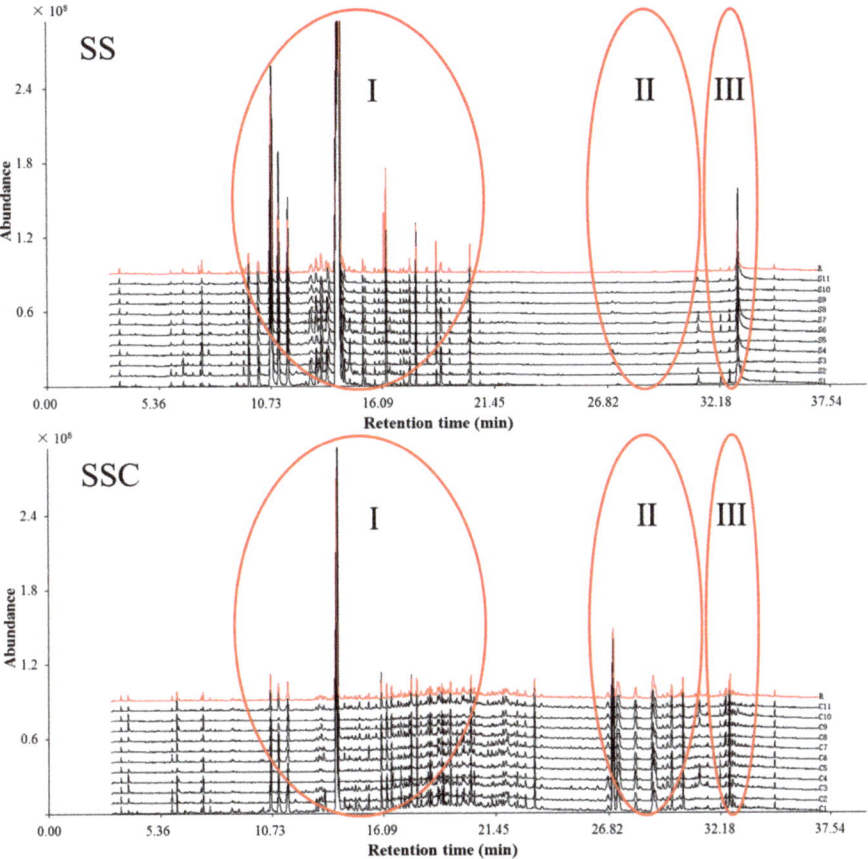

Figure 1. GC-MS fingerprints of Schizonepetae Spica (SS) and Schizonepetae Spica Carbonisata (SSC) samples and their reference chromatograms (R).

Table 1. Details of the 11 batches SS and SSC samples and their fingerprint similarities.

No.	Voucher Specimen	Place of Collection	Date of Collection	Similarity SS	Similarity SSC
1	YP132581301	Henan province	29 November 2018	0.982	0.974
2	YP132581401	Henan province	17 January 2019	0.969	0.995
3	YP132581501	Guangdong province	8 January 2019	0.997	0.972
4	YP132581601	Hubei province	9 January 2019	0.995	0.962
5	YP132581701	Hubei province	10 January 2019	0.984	0.955
6	YP132581801	Shanxi province	9 January 2019	0.990	0.989
7	YP132580101	Hebei province	7 June 2018	0.989	0.978
8	YP132582301	Hebei province	9 January 2019	0.994	0.976
9	YP132582601	Jiangsu province	11 January 2019	0.991	0.995
10	YP132582901	Jiangxi province	11 January 2019	0.983	0.984
11	YP132583001	Jiangxi province	12 January 2019	0.994	0.977

2.1.2. Identification of Volatile Components from SS and SSC

As shown in Table 2, the compositions of the volatile components of SS and SSC were analyzed. The number of identified peaks in SS and SSC were 39 and 62, accounting for 93.90% and 74.56% of their total volatile content, respectively. The main volatile constituents (>1%) of SS included V18 (*l*-menthone, 13.05%), V19 (menthofuran, 3.18%), V20 (*trans*-5-methyl-2-(1-methylvinyl)cyclohexan-1-one, 2.84%), V25 (2-allyl-4-methylphenol, 1.11%), V26 (pulegone, 58.70%), V37 (3-methyl-6-(1-methylethylidene)cyclohex-2-en-1-one, 3.21%), V48 (caryophyllene, 1.28%) and V83 (linolenic acid, 2.67%), a result that was in accordance with the findings of recent studies [27,28]. The volatile composition and relative contents of SS notably changed after processing by stir-frying, as V18 (*l*-menthone, 1.84%), V19 (menthofuran, 1.46%), V20 (*trans*-5-methyl-2-(1-methylvinyl)cyclohexan-1-one, 1.75%), V26 (pulegone, 27.44%), V35 (2,5,6-trimethylbenzimidazole, 1.01%), V61 (caryophyllene oxide, 1.40%), V72 (3,7,11-trimethyl-1-dodecanol, 1.49%), V73 (neophytadiene, 10.11%), V75 (phytol acetate, 1.58%), V77 (3,7,11,15-tetramethyl-2-hexadecen-1-ol, 3.95%), V80 (methyl palmitate, 1.56%) and V82 (methyl linolenate, 1.28%) were determined to be the main volatile constituents of SSC. In addition, 22 compounds found in SS samples disappeared in SSC samples, while 45 compounds were newly generated and identified in SSC samples. However, in a previous study, it was found that the major compounds limonene and menthone detected in SH were obviously decreased in SHC, whereas the relative contents of another four components, isomenthone, isopulegone, pulegone and piperitone, were higher in SHC than in SH [21]. Different sample preparation methods, different medicinal parts of SH and different sources of SH specimens were most likely the reasons for the variance in the results. In our study, the differences among these variables may lead to efficacy differences in the SS and SSC samples.

2.1.3. Multivariate Statistical Analyses

To further investigate the change in chemical compositions between SS and SSC, the GC-MS data (84 identified peaks) were subjected to PCA analysis. The PCA ($R^2X = 0.853$, $Q^2 = 0.788$) score plot showed that the 22 samples were obviously separated from the two groups (Figure 2). The first two PCs explained 80.59% of the data variance (PC1 = 72.06% and PC2 = 8.53%); of these, PC1 played a significant role in discriminating SS and SSC samples. SS was located on the negative side of PC1, while SSC was located on the positive side of PC1. The corresponding loading plot of PC1 (Figure 3) was used to find the components responsible for the separation between SS and SSC. The signals giving a positive effect in PC1 demonstrated that the corresponding ingredients were higher in SSC than in SS. In contrast, the signals with negative values indicated that the level of related components was higher in SS. The signals of major constituents V18 (*l*-menthone), V19 (menthofuran), V20 (*trans*-5-methyl-2-(1-methylvinyl)cyclohexan-1-one), V25 (2-allyl-4-methylphenol), V26 (pulegone), V37 (3-methyl-6-(1-methylethylidene)cyclohex-2-en-1-one), V48 (caryophyllene) and V83 (linolenic acid) gave a negative contribution to PC1. The signals with positive PC1 values included major components V35 (2,5,6-trimethylbenzimidazole), V61 (caryophyllene oxide), V72 (3,7,11-trimethyl-1-dodecanol), V73 (neophytadiene), V75 (phytol acetate), V77 (3,7,11,15-tetramethyl-2-hexadecen-1-ol), V80 (methyl palmitate) and V82 (methyl linolenate). The volatile components of SS have been recognized as the major constituents responsible for its biological effects. For example, pulegone, which is known for its pleasant odor, analgesia, and anti-inflammatory and antiviral properties [19,29], is a chemical indicator of SS in the Chinese Pharmacopoeia (2015 edition) [7]. *L*-Menthone also presents analgesia and antiviral effects [19]. Caryophyllene is a functional cannabinoid receptor type 2 agonist [30]. Menthofuran is widely used in flavorings and fragrances [31]. On the other hand, neophytadiene is a dominant metabolite in *Urtica dioica* L., a folk medicine that is commonly used as a hemostatic agent [32]. Therefore, these compounds might be the discriminant marker compounds when distinguishing SS from SSC, which are characterized by different medicinal properties.

Table 2. Volatile compounds and their relative contents in SS and SSC.

No.	Compound	Molecular Formula	RI	CAS	Similarity Indices	Relative Content (%) SS	Relative Content (%) SSC
V1	Butyl acetate	$C_6H_{12}O_2$	818.86	123-86-4	81.07%	-	0.24 ± 0.06
V2	(R)-(+)-3-Methylcyclohexanone	$C_7H_{12}O$	952.53	13368-65-5	82.43%	0.10 ± 0.04	0.23 ± 0.16
V3	Benzaldehyde	C_7H_6O	965.57	100-52-7	93.42%	-	0.52 ± 0.10
V4	1-Octen-3-ol	$C_8H_{16}O$	983.39	3391-86-4	78.46%	0.18 ± 0.06	-
V5	(1R,2S,5S)-2-Methyl-5-(3-oxoprop-1-en-2-yl)cyclopentane-1-carbaldehyde	$C_{10}H_{14}O_2$	993.51	5951-57-5	85.63%	-	0.07 ± 0.02
V6	1,3,8-p-Menthatriene	$C_{10}H_{14}$	1005.59	18368-95-1	82.08%	-	-
V7	2-Ethyl-1,4-dimethylbenzene	$C_{10}H_{14}$	1022.65	1758-88-9	91.27%	0.04 ± 0.02	0.15 ± 0.03
V8	E,E-2,6-Dimethyl-1,3,5,7-octatetraene	$C_{10}H_{14}$	1023.77	460-01-5	88.83%	0.08 ± 0.03	-
V9	D-Limonene	$C_{10}H_{16}$	1026.40	5989-27-5	95.44%	0.38 ± 0.35	0.42 ± 0.17
V10	3,5-Dimethyl-2-cyclohexen-1-one	$C_8H_{12}O$	1042.58	1123-09-7	83.79%	-	0.15 ± 0.05
V11	Benzeneacetaldehyde	C_8H_8O	1043.37	122-78-1	91.68%	0.11 ± 0.05	-
V12	1-Ethenyl-3,5-dimethyl-benzene	$C_{10}H_{12}$	1088.07	5379-20-4	88.24%	-	0.15 ± 0.04
V13	2,5-Dimethylstyrene	$C_{10}H_{12}$	1089.74	2039-89-6	80.65%	0.07 ± 0.03	-
V14	Linalool	$C_{10}H_{18}O$	1100.42	78-70-6	83.43%	0.10 ± 0.02	-
V15	trans-1-Methyl-4-(1-methylvinyl)cyclohex-2-en-1-ol	$C_{10}H_{16}O$	1118.90	7212-40-0	89.17%	0.90 ± 0.17	-
V16	1-Phenyl-1-butene	$C_{10}H_{12}$	1122.33	824-90-8	83.12%	-	0.06 ± 0.02
V17	cis-p-Mentha-2,8-dien-1-ol	$C_{10}H_{16}O$	1133.26	3886-78-0	91.38%	0.81 ± 0.12	-
V18	l-Menthone	$C_{10}H_{18}O$	1148.46	14073-97-3	90.28%	13.05 ± 6.27	1.84 ± 1.75
V19	Menthofuran	$C_{10}H_{14}O$	1159.04	494-90-6	81.68%	3.18 ± 2.34	1.46 ± 1.11
V20	trans-5-Methyl-2-(1-methylvinyl)cyclohexan-1-one	$C_{10}H_{16}O$	1171.24	29606-79-9	88.11%	2.84 ± 0.61	1.75 ± 0.56
V21	(−)-cis-Isopiperitenol	$C_{10}H_{16}O$	1202.38	96555-02-1	91.15%	0.73 ± 0.28	-
V22	3-Methyl-2-(2-methyl-2-butenyl)-furan	$C_{10}H_{14}O$	1208.98	15186-51-3	86.28%	-	0.25 ± 0.12
V23	4,7-Dimethylbenzofuran	$C_{10}H_{10}O$	1214.35	28715-26-6	84.75%	-	0.43 ± 0.23
V24	trans-Pulegone oxide	$C_{10}H_{16}O_2$	1218.24	13080-28-9	79.80%	-	0.40 ± 0.10
V25	2-Allyl-4-methylphenol	$C_{10}H_{12}O$	1222.89	6628-06-4	92.37%	1.11 ± 0.56	0.43 ± 0.37
V26	(+)-Pulegone	$C_{10}H_{16}O$	1245.58	89-82-7	91.91%	58.70 ± 11.53	27.44 ± 7.85
V27	Piperitone	$C_{10}H_{16}O$	1258.51	89-81-6	86.72%	0.35 ± 0.19	0.22 ± 0.14
V28	5-Isopropenyl-2-methyl-2-cyclohexen-1-yl pivalate	$C_{15}H_{24}O_2$	1276.52	1000124-59-2	79.10%	-	0.20 ± 0.07
V29	4-Ethyl-2-methoxyphenol	$C_9H_{12}O_2$	1283.03	2785-89-9	83.34%	-	0.36 ± 0.14
V30	(7-Hydroxy-3,3-dimethyl-4-oxo-7-vinylbicyclo [3.2.0]hept-1-yl)acetaldehyde	$C_{13}H_{18}O_3$	1290.48	1000156-78-3	79.14%	0.40 ± 0.23	-
V31	(1S,3S,5S)-1-Isopropyl-4-methylenebicyclo[3.1.0]hexan-3-yl acetate	$C_{12}H_{18}O_2$	1294.44	139757-62-3	83.38%	0.20 ± 0.03	-
V32	Thymol	$C_{10}H_{14}O$	1297.47	89-83-8	84.05%	-	0.30 ± 0.15
V33	Carveol	$C_{10}H_{16}O$	1312.88	99-48-9	85.39%	0.13 ± 0.03	-
V34	(4-Methoxymethoxy-hex-5-ynylidene)-cyclohexane	$C_{14}H_{22}O_2$	1315.74	1000186-16-6	80.43%	-	0.15 ± 0.08
V35	2,5,6-Trimethylbenzimidazole	$C_{10}H_{12}N_2$	1328.07	3363-56-2	88.88%	-	1.01 ± 0.61
V36	Carvyl acetate	$C_{12}H_{18}O_2$	1335.28	97-42-7	84.83%	0.10 ± 0.04	-
V37	3-Methyl-6-(1-methylethylidene)cyclohex-2-en-1-one	$C_{10}H_{14}O$	1342.11	491-09-8	90.88%	3.21 ± 0.55	0.84 ± 0.28
V38	1,1,5-Trimethyl-1,2-dihydronaphthalene	$C_{13}H_{16}$	1354.07	1000357-25-8	92.67%	-	0.61 ± 0.24
V39	1,2,3,4-Tetrahydro-1,1,6-trimethylnaphthalene	$C_{13}H_{18}$	1357.25	475-03-6	83.46%	-	0.17 ± 0.10
V40	α-Ethyl-4-methoxybenzenemethanol	$C_{10}H_{14}O_2$	1371.02	5349-60-0	87.85%	-	0.17 ± 0.08
V41	α-Copaene	$C_{15}H_{24}$	1376.65	1000360-33-0	90.42%	0.11 ± 0.04	0.27 ± 0.19
V42	(−)-β-Bourbonene	$C_{15}H_{24}$	1385.02	5208-59-3	92.16%	0.13 ± 0.03	-
V43	Falcarinol	$C_{17}H_{24}O$	1390.30	21852-80-2	88.33%	-	0.12 ± 0.06
V44	β-Elemene	$C_{15}H_{24}$	1393.37	515-13-9	79.54%	0.08 ± 0.04	-

Table 2. Cont.

No.	Compound	Molecular Formula	RI	CAS	Similarity Indices	Relative Content (%) SS	SSC
V45	Jasmone	$C_{11}H_{16}O$	1399.03	488-10-8	87.93%	0.07 ± 0.01	-
V46	2-[(2Z)-2-Buten-1-yl]-4-hydroxy-3-methyl-2-cyclopenten-1-one	$C_{10}H_{14}O_2$	1401.35	17190-74-8	85.22%	0.39 ± 0.18	-
V47	1-Methyl-4-[(2-methyl-3-butyn-2-yl)oxy]benzene	$C_{12}H_{14}O$	1403.18	82719-54-8	87.66%	-	0.90 ± 0.50
V48	β-Caryophyllene	$C_{15}H_{24}$	1420.14	87-44-5	94.76%	1.28 ± 0.78	0.89 ± 0.25
V49	(3β,5α)-2-Methylenecholestan-3-ol	$C_{28}H_{48}O$	1455.41	22599-96-8	80.11%	-	0.60 ± 0.17
V50	Humulene	$C_{15}H_{24}$	1458.11	6753-98-6	86.35%	0.15 ± 0.10	-
V51	β-Guaiene	$C_{15}H_{24}$	1479.26	88-84-6	85.83%	-	0.11 ± 0.07
V52	4-(2,4,4-Trimethyl-cyclohexa-1,5-dienyl)-but-3-en-2-one	$C_{13}H_{18}O$	1484.88	1000187-51-9	84.20%	-	0.52 ± 0.16
V53	Germacrene D	$C_{15}H_{24}$	1485.14	23986-74-5	90.97%	0.80 ± 0.56	-
V54	1-Chlorooctadecane	$C_{18}H_{37}Cl$	1496.85	3386-33-2	84.04%	-	0.61 ± 0.17
V55	2-Hydroxy-2,6-dimethyl-1-[(1E)-3-methyl-1,3-butadien-1-yl]bicyclo[4.1.0]hept-3-yl acetate	$C_{16}H_{24}O_3$	1507.78	1000196-25-1	83.45%	-	0.25 ± 0.08
V56	(+)-δ-Cadinene	$C_{15}H_{24}$	1523.45	483-76-1	91.52%	0.13 ± 0.05	0.96 ± 0.30
V57	4,5,9,10-Dehydroisolongifolene	$C_{15}H_{20}$	1542.96	156747-45-4	88.25%	-	0.51 ± 0.25
V58	3-(2-Methyl-propenyl)-1H-indene	$C_{13}H_{14}$	1559.02	1000187-78-5	82.46%	-	0.53 ± 0.27
V59	4,4-Dimethyl-3-(3-methylbut-3-enylidene)-2-methylenebicyclo[4.1.0]heptane	$C_{15}H_{22}$	1563.49	79718-83-5	82.83%	-	0.49 ± 0.28
V60	(−)-Spathulenol	$C_{15}H_{24}O$	1579.60	77171-55-2	81.56%	0.14 ± 0.04	0.48 ± 0.19
V61	Caryophyllene oxide	$C_{15}H_{24}O$	1584.80	1139-30-6	85.95%	0.87 ± 0.66	1.40 ± 0.66
V62	Geranyl isovalerate	$C_{15}H_{26}O_2$	1595.20	109-20-6	81.47%	-	0.56 ± 0.17
V63	(1R,3E,7E,11R)-1,5,5,8-Tetramethyl-12-oxabicyclo[9.1.0]dodeca-3,7-diene	$C_{15}H_{24}O$	1612.71	19988-34-7	83.83%	0.06 ± 0.05	-
V64	(8S)-1-Methyl-4-isopropyl-7,8-dihydroxy-spiro[tricyclo[4.4.0.0(5,9)]-decane-10,2′-oxirane]	$C_{15}H_{24}O_3$	1613.74	1000193-77-2	81.05%	-	0.15 ± 0.07
V65	α-Santonin	$C_{15}H_{18}O_3$	1628.24	481-06-1	81.28%	-	0.19 ± 0.16
V66	10,10-Dimethyl-2,6-dimethylenebicyclo-[7.2.0]undecan-5β-ol	$C_{15}H_{24}O_2$	1637.26	19431-80-2	84.47%	-	0.29 ± 0.11
V67	2-[4-methyl-6-(2,6,6-trimethylcyclohex-1-enyl)hexa-1,3,5-trienyl]cyclohex-1-en-1-carboxaldehyde	$C_{23}H_{32}O$	1646.83	1000216-09-2	79.63%	-	0.32 ± 0.16
V68	13-Heptadecyn-1-ol	$C_{17}H_{32}O$	1662.98	56554-77-9	81.41%	-	0.58 ± 0.19
V69	cis-5,8,11,14,17-Eicosapentaenoic acid	$C_{20}H_{30}O_2$	1668.82	10417-94-4	82.21%	-	0.95 ± 0.29
V70	(1R,7S,E)-7-Isopropyl-4,10-dimethylenecyclodec-5-enol	$C_{15}H_{24}O$	1687.47	81968-62-9	81.29%	0.03 ± 0.02	0.39 ± 0.15
V71	3,4′-Diethyl-1,1′-biphenyl	$C_{16}H_{18}$	1708.74	61141-66-0	88.07%	-	0.45 ± 0.09
V72	3,7,11-Trimethyl-1-dodecanol	$C_{15}H_{32}O$	1723.32	6750-34-1	80.42%	-	1.49 ± 0.38
V73	Neophytadiene	$C_{20}H_{38}$	1834.08	504-96-1	90.35%	-	10.11 ± 3.20
V74	Fitone	$C_{18}H_{36}O$	1842.16	502-69-2	87.33%	0.05 ± 0.02	-
V75	Phytol acetate	$C_{22}H_{42}O_2$	1860.42	1000375-01-4	85.59%	-	1.58 ± 0.60
V76	2,4,7,14-Tetramethyl-4-vinyl-tricyclo[5.4.3.0(1,8)]tetradecan-6-ol	$C_{20}H_{34}O$	1865.55	1000193-31-2	84.74%	-	0.26 ± 0.09
V77	3,7,11,15-Tetramethyl-2-hexadecen-1-ol	$C_{20}H_{40}O$	1879.81	102608-53-7	85.05%	-	3.95 ± 1.53
V78	Hexadecanenitrile	$C_{16}H_{31}N$	1895.78	629-79-8	81.46%	-	0.63 ± 0.19
V79	1-Heptatriacotanol	$C_{37}H_{76}O$	1914.84	105794-58-9	81.06%	-	0.21 ± 0.08
V80	Methyl palmitate	$C_{17}H_{34}O_2$	1931.28	112-39-0	88.36%	-	1.56 ± 0.48
V81	Methyl linoleate	$C_{19}H_{34}O_2$	2091.82	112-63-0	92.97%	-	0.70 ± 0.26
V82	Methyl linolenate	$C_{19}H_{32}O_2$	2097.76	301-00-8	90.66%	0.08 ± 0.06	1.28 ± 0.45
V83	Linolenic acid	$C_{18}H_{30}O_2$	2149.50	463-40-1	94.44%	2.67 ± 2.41	-
V84	2,2′-Methylenebis(6-tert-butyl-4-methylphenol)	$C_{23}H_{32}O_2$	2428.07	119-47-1	87.78%	0.11 ± 0.02	0.28 ± 0.10

Relative content (%) in the last two columns represents the mean ± SD ($n = 11$). RI, retention index. CAS, Chemical Abstracts Service. Similarity indices obtained from direct searching with the National Institute of Standards and Technology (NIST) MS database.

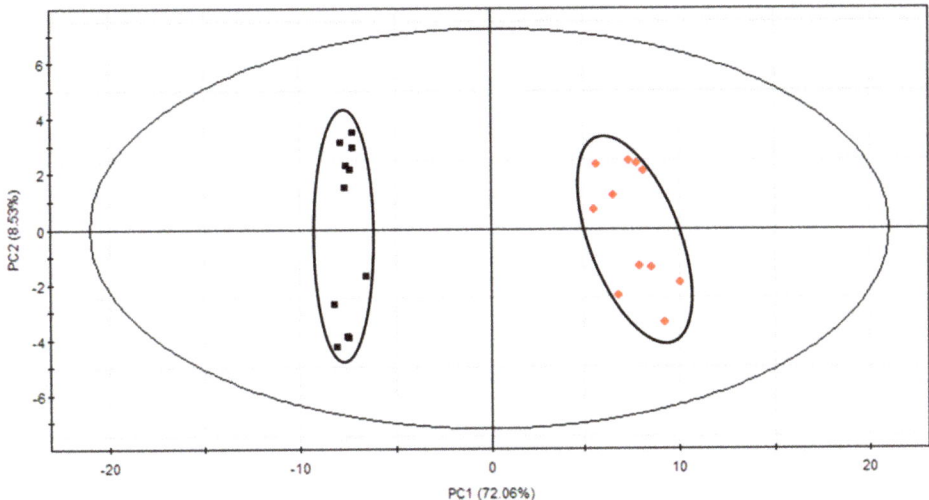

Figure 2. Principal component analysis (PCA) score plots based on gas chromatography-mass spectrometry (GC-MS) data of SS (black) and SSC (red).

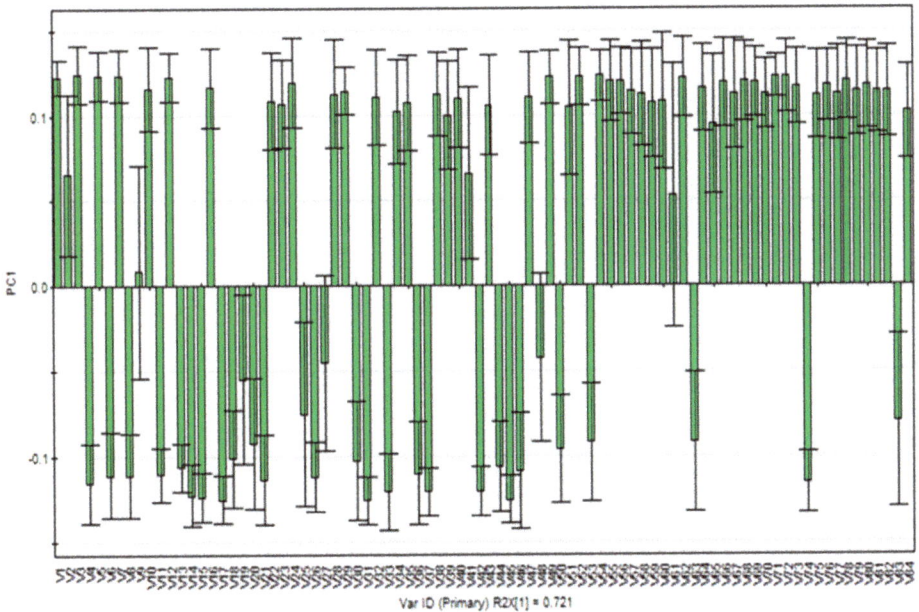

Figure 3. PC1 loading plot of the PCA results obtained from the GC-MS spectra.

2.1.4. Possible Mechanisms Involved in the Transformation of Volatile Compounds between SS and SSC

Building on the chemical marker analysis described above and the extensive organic synthesis knowledge accumulated by literature data [33–38], the possible mechanisms involved in the transformation of volatile compounds during stir-frying SS are demonstrated in Figure 4. For example, the significant decline in pulegone content in Table 2 revealed that the chemical reaction involved might generate an epoxidation reaction between the original pulegone (V26) and oxygen

219

during stir-frying [33]. According to previous studies [19,29], pulegone has been demonstrated to have good biological activity in terms of anti-inflammatory and antiviral effects. However, evidence has proven that pulegone and its oxidative product *p*-cresol can cause severe hepatotoxicity [27]. Thus, the epoxidation reaction may occur in the process of preparing SSC by stir-frying raw SS, allowing SSC to contain less toxic pulegone. On the other hand, it has been found that β-Caryophyllene exhibits potent anti-inflammatory activity [39]. Stir-fry processing may also reduce the SS's anti-inflammatory property because of the changing of β-Caryophyllene (V48) to caryophyllene oxide (V61) and 10,10-dimethyl-2,6-dimethylenebicyclo-[7.2.0]undecan-5β-ol (V66) [34,35]. Meanwhile, neophytadiene (V73) seemed to be generated from the dehydration reaction of 3,7,11,15-tetramethyl-2-hexadecen-1-ol (V77), a product that was generated from fitone (V74) by hydroxylation during stir-frying [35,36]. Neophytadiene is a major constituent in *U. dioica*, a folk medicine that is commonly used as a hemostatic agent [32]. Hence, the traditional use of SSC is in line with the observed increase in the amount of neophytadiene with plausible hemostatic activity. Likewise, linolenic acid (V83) was assumed to be transformed into methyl linolenate (V82) by esterification reaction during stir-frying [38]. Previous study has reported that linolenic acid shows bioactivities such as anti-inflammatory and antioxidant effects [40]. Therefore, it might be also related to the disappearance or significant decrease of linolenic acid (Table 2) and efficacy changing under the stir-fry processing of SS.

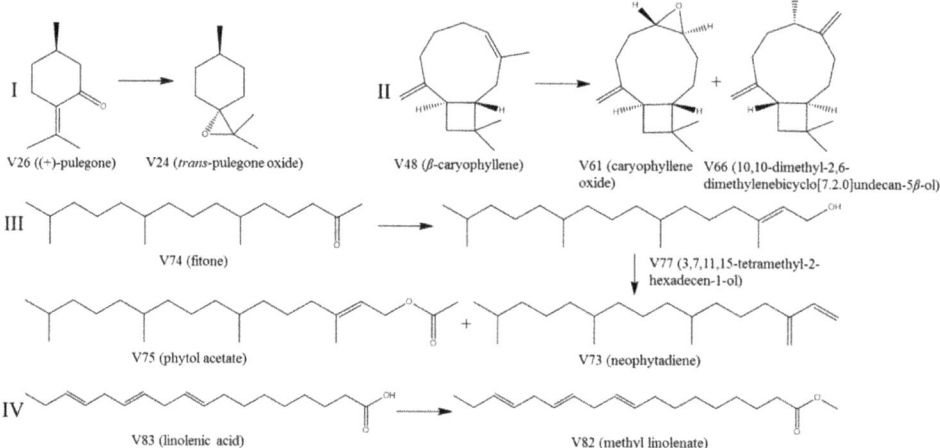

Figure 4. Possible mechanisms involved in the transformation of some main volatile components during stir-fry processing of SS.

2.2. Comparison of the Nonvolatile Components of SS and SSC

2.2.1. Determination of Nonvolatile Compound Contents of SS and SSC

Representative HPLC chromatograms of the SS and SSC samples are shown in Figure 5. The six nonvolatile compounds were identified using reference standards of luteolin-7-O-β-D-glucoside, apigenin-7-O-β-D-glucoside, hesperidin, rosmarinic acid, luteolin and apigenin, as well as comparing with literature [41–43]. Comparison of the chromatograms of SS and SSC revealed that stir-fry processing mainly caused quantitative changes among the nonvolatile compounds. The simultaneous determination of the six compounds in SS and SSC was analyzed (Table 3). Stir-fry processing significantly ($p < 0.001$) decreased the total nonvolatile compound contents of SS from 5.29 ± 0.42 mg/g to 1.56 ± 0.06 mg/g. Compared with SS, the contents of four major ingredients (luteolin-7-O-β-D-glucoside, apigenin-7-O-β-D-glucoside, hesperidin and rosmarinic acid) decreased significantly ($p < 0.001$), whereas another two main components (luteolin and apigenin) increased significantly in SSC samples ($p < 0.001$). However, the intensity of some relatively polar compounds with shorter retention times on reversed-phase chromatography that were detected in SSC was markedly higher than those in SS (Figure 5). The identities of these compounds and the mechanisms involved in their transformation need to be further studied. Previous studies have demonstrated that stir-fry processing causes a significant increase in the total flavonoid and tannin contents of SH and SS [44–46]. Total flavonoids [47], tannins [48] and carbon dots [49] were proven to be active principles with hemostatic properties in SHC. In our research, the contents of total flavonoids (luteolin-7-O-β-D-glucoside, apigenin-7-O-β-D-glucoside, hesperidin, luteolin and apigenin) decreased remarkably after the SS samples were processed. A reasonable explanation for this result is that few flavonoids were detected in the ethanol extract of SSC.

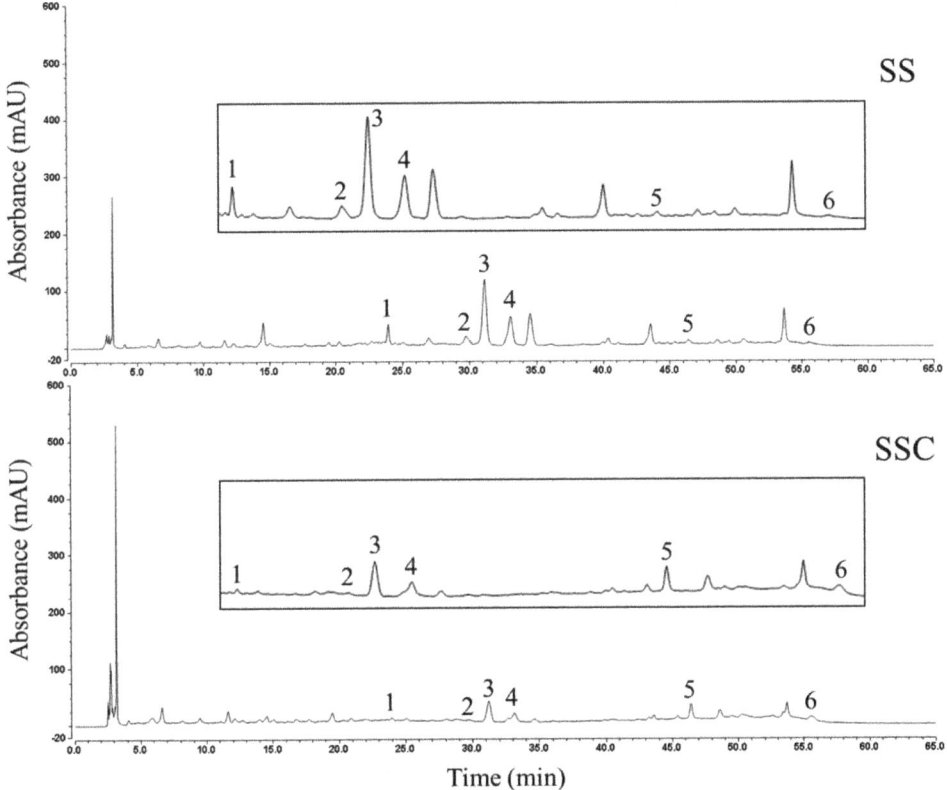

Figure 5. Representative HPLC chromatograms of SS and SSC samples. 1, luteolin-7-O-β-D-glucoside; 2, apigenin-7-O-β-D-glucoside; 3, hesperidin; 4, rosmarinic acid; 5, luteolin; and 6, apigenin.

Table 3. Nonvolatile compound contents of SS and SSC.

	Contents of Nonvolatile Compounds (mg/g)						
	Luteolin-7-O-β-D-glucoside	Apigenin-7-O-β-D-glucoside	Hesperidin	Rosmarinic Acid	Luteolin	Apigenin	Total
SS	0.40 ± 0.06	0.25 ± 0.09	2.97 ± 0.99	1.62 ± 0.67	0.04 ± 0.04	0.01 ± 0.01	5.29 ± 0.42
SSC	0.05 ± 0.01 ***	0.02 ± 0.01 ***	0.64 ± 0.16 ***	0.40 ± 0.11 ***	0.32 ± 0.05 ***	0.13 ± 0.03 ***	1.56 ± 0.06 ***

Data are presented as the mean ± SD ($n = 11$). *** $p < 0.001$ vs. SS samples. The total nonvolatile compound contents were calculated as the sum of the contents of the 6 individual nonvolatile compounds.

2.2.2. Possible Mechanisms Involved in the Transformation of the Nonvolatile Compounds between SS and SSC

In the present study, the changes in the contents of luteolin-7-O-β-D-glucoside and apigenin-7-O-β-D-glucoside probably contributed to the increases in luteolin and apigenin, respectively. The possible mechanisms involved in the transformation of these compounds are assumed in Figure 6. The results were consistent with previous reports in that glucoside could be hydrolyzed to aglycone under various conditions, including high temperature during stir-fry processing [43,50,51]. In addition, luteolin-7-O-β-D-glucoside has been documented to possess significant anti-inflammatory and antiviral effects [52,53]. Luteolin has been shown to be a potent hemostatic drug candidate [54]. Therefore, the increased contents of luteolin might result from the degradation of the glucosidic bond of luteolin-7-O-β-D-glucoside, which could be a main mechanism of stir-fry processing in SS.

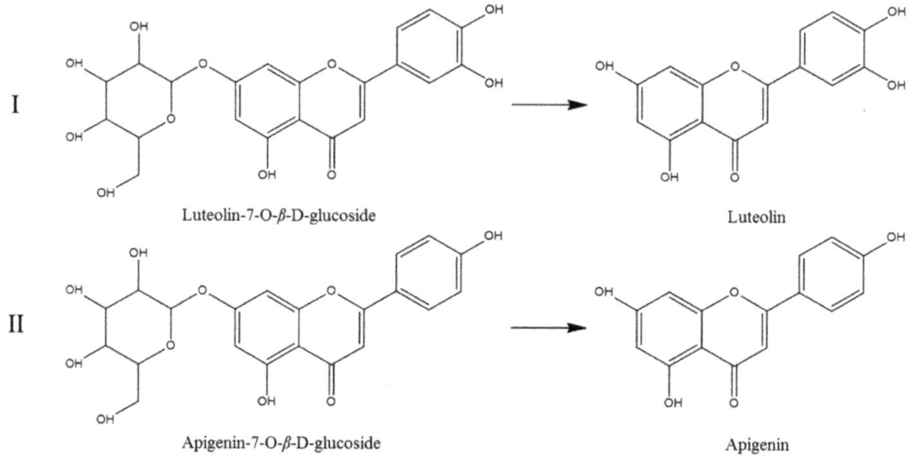

Figure 6. Possible mechanisms involved in the transformation of some main nonvolatile components during the stir-fry processing of SS.

3. Materials and Methods

3.1. Materials and Reagents

SS and SSC samples came from the same path (Table 1). SSC was processed in accordance with the Chinese Pharmacopoeia (2015 edition) [7]: dried SS slices (10–15 mm in length) were put in a pan and stir-fried until the surface became black and the interior became brown to avoid scorching. The identities of SS and SSC were confirmed by Dr. Ying Zhang, Jinan University, P. R. China. Voucher specimens were deposited at the Research Center for Traditional Chinese Medicine of Lingnan (Guangzhou, Southern China), Jinan University.

Ethyl acetate (analytical grade) was purchased from Aladdin (Aladdin, Shanghai, China). Methanol (HPLC grade), acetonitrile (HPLC grade) and formic acid (HPLC grade) were purchased from Fisher Scientific (Fair Lawn, NJ, USA). The chemical standards luteolin-7-O-β-D-glucoside, hesperidin, rosmarinic acid, luteolin and apigenin were purchased from Chengdu Ruifensi Biotechnology (Sichuan, China). apigenin-7-O-β-D-glucoside was purchased from Tianjun Biotechnology (Guangzhou, China). The purity of the compounds was greater than 95% as determined by HPLC. A mix of alkane analogues (GC grade, purity > 97%), which was used as an internal quality standard for GC-MS analysis, was purchased from o2si (Charleston, SC, USA). Deionized water was obtained by passing distilled water through a Milli-Q purification system (Millipore, Bedford, MA, USA). All other reagents were of analytical grade.

3.2. Sample Preparation

For GC-MS analysis, the steam distillation method was chosen according to the Chinese Pharmacopoeia (2015 edition) for the extraction of essential oils [7]. All samples were smashed and filtered through a 24-mesh sieve. The dried powder (50 g) was accurately weighed and transferred to a 1000 mL flask and soaked in 300 mL of redistilled water for 1 h. Redistilled water was added from the top of the volatile oil determination apparatus until the water spilled into the flask, and 2 mL of ethyl acetate was added to the water layer. Then, the essential oils were extracted by water distillation for 4 h. The volatile oil was separated from the water layer and leached into the ethyl acetate layer, and then the ethyl acetate layer was dried over anhydrous sodium sulfate for GC-MS analysis. The anhydrous essential oils were stored in dark glass vials at −20 °C until use.

For HPLC analysis, 2 g of the pulverized sample (filtered through a 24-mesh sieve) was weighed accurately and macerated in 20 mL of 75% ethanol. The sample was then extracted for 60 min by hot reflux extraction in a water bath. The supernatant of the extracts was filtered through a 0.45-μm membrane before injection. The samples were stored in a refrigerator at 4 °C until use.

3.3. Standard Solution Preparation

A mixed standard stock solution containing the 6 reference compounds was prepared in methanol. Working standard solutions were prepared by diluting the mixed standard stock solution with methanol to give different concentrations within the following ranges for calibration curves: luteolin-7-O-β-D-glucoside, 0.796–159.2 μg/mL; apigenin-7-O-β-D-glucoside, 0.48–48 μg/mL; hesperidin, 2.084–625.2 μg/mL; rosmarinic acid, 0.924–277.2 μg/mL, luteolin, 0.472–141.6 μg/mL and apigenin, 0.4–120 μg/mL. All standard solutions were filtered through a 0.45-μm membrane before injection. The solutions were stored in a refrigerator at 4 °C before analysis.

3.4. Apparatus and Chromatographic Conditions

The GC-MS instrument used was an Agilent 7890B GC system coupled to an Agilent 7000C GC/MS Triple Quad mass spectrometer (Agilent, Santa Clara, CA, USA). Initial chromatographic separations of 1 μL of sample were performed on a 15 m × 250 μm i.d. × 0.25 μm film thickness HP-5 (Agilent) capillary column with a He flow rate of 1.0 mL/min and an injection port temperature of 250 °C with a split ratio of 1:10. The oven temperature ramp was 3 min at 50 °C, then 10 °C/min to 90 °C where the temperature was held for 5 min, then ramped at the rate of 10 °C/min to 160 °C where the temperature was maintained for 10 min, then a 20 °C/min ramp to 260 °C where the temperature was held for 3 min. The detector was operated at an ionization energy of 70 eV, and the m/z values were recorded in the range 50–600 amu with a scan rate of 3.6 scan/s and a solvent delay of 3 min. Components were identified using the National Institute of Standards and Technology (NIST) 2.2L Mass Spectra Database containing approximately 189,000 compounds, as well as comparing with literature [27,55–57].

HPLC analysis was performed on an UltiMate 3000 liquid chromatography system (Thermo Scientific, Bremen, Germany). Chromatographic separation was performed with an ACE Excel 5 C18 column (4.6 mm × 250 mm, 5 μm; ACE, Scotland, UK), and the column temperature was maintained at 30 °C. The mobile phase was 0.1% formic acid aqueous solution (A) and acetonitrile (B) with a gradient program as follows: 0–20 min, linear gradient from 8–20% B; 20–30 min, isocratic elution at 20% B; 30–50 min, linear gradient from 20–30% B; 50–60 min, linear gradient from 30–8% B. The flow rate was 1 mL/min, and the injection volume was 10 μL. The detection wavelength was set at 283 nm.

3.5. Data Analysis

The GC-MS fingerprint was performed by the professional software Similarity Evaluation System for Chromatographic Fingerprint of Traditional Chinese Medicine (Version 2004 A) composed by the Chinese Pharmacopoeia Committee. PCA was performed by SIMCA-P version 11.5 software (Umetrics, Umea, Sweden). Statistical significance was assessed by Student's two-tailed t-tests with GraphPad

Prism v. 5.0 software (San Diego, CA, USA). Values of $p < 0.05$, $p < 0.01$ or $p < 0.001$ were considered to be statistically significant.

4. Conclusions

This work reported, for the first time, the application of the combination of GC-MS and HPLC methods coupled with a strategy of corresponding data processing and statistical analysis to qualitatively and quantitatively distinguish SS and SSC and to identify marker compounds with significantly changed structures or contents during stir-fry processing. The GC-MS comparative analysis results of SS and SSC showed that 16 major constituents had a large contribution to the discrimination. HPLC analyses revealed that stir-fry processing remarkably increased the contents of two main ingredients but significantly reduced the contents of another four major constituents from SS. The change in the type and amount of these marker components is probably responsible for the different functions and pharmacological effects of SS and SSC as well as the observed hepatotoxicity. Finally, we speculated how some of these marker chemical changes occurred upon stir-fry processing. Hence, the potential mechanisms of stir-fry processing could be justified. The proposed strategy provided new clues for the investigation of the stir-frying-induced chemical transformation of SS and provided useful references for understanding the potential mechanisms of other processing methods.

Author Contributions: Conceptualization, H.C. and X.L.; Funding acquisition, H.C.; Investigation, X.L.; Formal analysis, X.L.; Resources, Y.Z.; Data curation, X.L.; Methodology, M.W. and Z.M.; Writing—original draft preparation, X.L.; Writing—review and editing, H.C., Y.Z., M.W. and Z.M. All authors have read and agreed to the published version of the manuscript.

Funding: This work was supported by the 6th National Academic Experience Inheritance Program of Famous Chinese Medicine Experts (Hui Cao) (No. 176-2017-XMZC-0166-01).

Conflicts of Interest: The authors declare no conflict of interest.

Abbreviations

SS	Schizonepetae Spica
SSC	Schizonepetae Spica Carbonisata
GC-MS	Gas chromatography-mass spectrometry
HPLC	High-performance liquid chromatography
CMM	Chinese medicinal material
TCM	Traditional Chinese medicine
SH	Schizonepetae Herba
SHC	Schizonepetae Herba Carbonisata
PCA	Principal component analysis

References

1. Zhao, Z.Z.; Liang, Z.T.; Chan, K.; Lu, G.H.; Lee, E.L.M.; Chen, H.B.; Li, L. A unique issue in the standardization of Chinese materia medica: Processing. *Planta Med.* **2010**, *76*, 1975–1986. [CrossRef] [PubMed]
2. Lei, F. *Lei Gong Pao Zhi Lun*; Phoenix Science Press: Nanjing, China, 1985.
3. Su, T.; Zhang, W.W.; Zhang, Y.M.; Cheng, B.C.Y.; Fu, X.Q.; Li, T.; Guo, H.; Li, Y.X.; Zhu, P.L.; Cao, H.; et al. Standardization of the manufacturing procedure for Pinelliae Rhizoma Praeparatum cum Zingibere et Alumine. *J. Ethnopharmacol.* **2016**, *193*, 663–669. [CrossRef]
4. Yang, C.Y.; Guo, F.Q.; Zang, C.; Li, C.; Cao, H.; Zhang, B.X. The effect of ginger juice processing on the chemical profiles of Rhizoma Coptidis. *Molecules* **2018**, *23*, 380. [CrossRef] [PubMed]
5. Zhang, X.; Wang, Y.; Li, S.J.; Dai, Y.J.; Li, X.Q.; Wang, Q.H.; Wang, G.Y.; Ma, Y.L.; Gu, X.Z.; Zhang, C. The potential antipyretic mechanism of Gardeniae Fructus and its heat-processed products with plasma metabolomics using rats with yeast-induced fever. *Front. Pharmacol.* **2019**, *10*, 491. [CrossRef]
6. Zhang, X.; Wang, Y.; Li, X.Q.; Dai, Y.J.; Wang, Q.H.; Wang, G.Y.; Liu, D.P.; Gu, X.Z.; Yu, D.R.; Ma, Y.L.; et al. Treatment mechanism of Gardeniae Fructus and its carbonized product against ethanol-induced gastric lesions in rats. *Front. Pharmacol.* **2019**, *10*, 750. [CrossRef]

7. State Commission of Chinese Pharmacopoeia. *Pharmacopoeia of the People's Republic of China*, 2015th ed.; China Medical Science Press: Beijing, China, 2015.
8. Wu, P. *Shen Nong Ben Cao Jing*; Sun, X.Y., Sun, F.J., Eds.; People's Medical Publishing House: Beijing, China, 1984.
9. Fung, D.; Lau, C.B.S. *Schizonepeta tenuifolia*: Chemistry, pharmacology, and clinical applications. *J. Clin. Pharmacol.* **2002**, *42*, 30–36. [CrossRef]
10. Shan, M.Q.; Qian, Y.; Yu, S.; Guo, S.C.; Zhang, L.; Ding, A.W.; Wu, Q.N. Anti-inflammatory effect of volatile oil from *Schizonepeta tenuifolia* on carrageenin-induced pleurisy in rats and its application to study of appropriate harvesting time coupled with multi-attribute comprehensive index method. *J. Ethnopharmacol.* **2016**, *194*, 580–586. [CrossRef]
11. Chen, S.G.; Cheng, M.L.; Chen, K.H.; Horng, J.T.; Liu, C.C.; Wang, S.M.; Sakurai, H.; Leu, Y.L.; Wang, S.D.; Ho, H.Y. Antiviral activities of *Schizonepeta tenuifolia* Briq. against enterovirus 71 in vitro and in vivo. *Sci. Rep.* **2017**, *7*, 1–15. [CrossRef]
12. Ng, Y.C.; Kim, Y.W.; Lee, J.S.; Lee, S.J.; Song, M.J. Antiviral activity of *Schizonepeta tenuifolia* Briquet against noroviruses via induction of antiviral interferons. *J. Microbiol.* **2018**, *56*, 683–689. [CrossRef]
13. Ding, A.W.; Kong, L.D.; Wu, H.; Wang, S.L.; Long, Q.J.; Yao, Z.; Chen, J. Research on hemostatic constituents in carbonized *Schizonepeta tenuifolia* Briq. *China J. Chin. Mater. Med.* **1993**, *18*, 535–538.
14. Ding, A.W.; Wu, H.; Kong, L.D.; Wang, S.L.; Gao, Z.Z.; Zhao, M.X.; Tan, M. Research on hemostatic mechanism of extracts from carbonized *Schizonepeta tenuifolia* Briq. *China J. Chin. Mater. Med.* **1993**, *18*, 598–600.
15. Zhang, Y. Application of jingjiesui in treating gynecological diseases. *Clin. J. Chin. Med.* **2016**, *8*, 69–70.
16. Du, C.Z.; Qin, J.P.; Chen, Y.P.; Cai, Y. GC-MS analysis of volatile oil components in *Schizonepeta tenuifolia* Briq. from various habitats. *Hubei Agric. Sci.* **2014**, *53*, 188–190.
17. Hu, J.; Shi, R.B.; Zhang, Y.H. Content determination of luteolin and hesperidin in the effective fractions of Spica Schizonepeta by HPLC. *J. Beijing Univ. Tradit. Chin. Med.* **2005**, *28*, 52–54.
18. Jiang, Y.H.; Jiang, D.Q.; Wu, K.J.; Peng, L.L. Contents determination of ursolic acid and oleanolic acid in different parts of Jiangxi Herba Schizonepetae and carbonized Herba Schizonepetae. *China J. Tradit. Chin. Med. Pharm.* **2016**, *31*, 1068–1070.
19. He, T.; Tang, Q.; Zeng, N.; Gong, X.P. Study on effect and mechanism of volatile oil of Schizonepetae Herba and its essential components against influenza virus. *China J. Chin. Mater. Med.* **2013**, *38*, 1772–1777.
20. Zhou, S.F.; Xue, C.C.; Yu, X.Q.; Wang, G. Metabolic activation of herbal and dietary constituents and its clinical and toxicological implications: An update. *Curr. Drug Metab.* **2007**, *8*, 526–553. [CrossRef]
21. Ye, D.J.; Ding, A.W.; Yu, L.; Cao, G.X.; Wu, M. Study on the composition of volatile oil in different medicinal parts of Schizonepetae Herba and its charred products. *Bull. Chin. Mater. Med.* **1985**, *10*, 19–21.
22. Zuo, M.L. *Schizonepeta* health products and processed products of different quality inspection. *China Pharm.* **2009**, *18*, 26–27.
23. Miao, X.Y. *Pao Zhi Da Fa*; Cao, H., Wu, M.H., Eds.; China Medical Science Press: Beijing, China, 2018.
24. Huang, Q.; Wu, D.L.; Wang, Y.; Ma, Y.L.; Yu, D.R.; Zhang, C. GC-MS analysis of volatile components in Scutellariae Radix before and after being charred. *Chin. J. Exp. Tradit. Med. Form.* **2016**, *22*, 9–12.
25. Zhang, L.; Shao, X.; Yu, S.; Bao, B.H.; Ding, A.W. HSGC fingerprint of volatile oil in Schizonepetae Spica. *Chin. Tradit. Herb. Drugs* **2012**, *43*, 1767–1769.
26. Wang, Q.; Wang, L.X.; Wang, X.Y.; Wang, Y.F. Chromatographic fingerprint for *Schizonepeta tenuisfolia* Briq. *Chin. Tradit. Pat. Med.* **2007**, *29*, 941–945.
27. Yu, S.; Chen, Y.W.; Zhang, L.; Shan, M.Q.; Tang, Y.P.; Ding, A.W. Quantitative comparative analysis of the bio-active and toxic constituents of leaves and spikes of *Schizonepeta tenuifolia* at different harvesting times. *Int. J. Mol. Sci.* **2011**, *12*, 6635–6644. [CrossRef]
28. Liu, X.; Zhang, Y.; Wu, M.; Ma, Z.; Cao, H. Color discrimination and gas chromatography-mass spectrometry fingerprint based on chemometrics analysis for the quality evaluation of Schizonepetae Spica. *PLoS ONE* **2020**, *15*, 1–15. [CrossRef]
29. Yamahara, J.; Matsuda, H.; Watanabe, H.; Sawada, T. Biologically active principles of crude drug, analgesic and anti-inflammatory effects of "Keigai (*Schizonepeta tenuifolia* Briq.)". *Yakugaku Zasshi* **1980**, *100*, 713–717. [CrossRef]

30. Oliveira, G.L.D.; Machado, K.C.; Machado, K.C.; Da Silva, A.P.D.C.L.; Feitosa, C.M.; Almeida, F.R.D. Non-clinical toxicity of β-caryophyllene, a dietary cannabinoid: Absence of adverse effects in female Swiss mice. *Regul. Toxicol. Pharm.* **2018**, *92*, 338–346. [CrossRef]
31. Wang, Y.; Zhang, N. Review on the safety and fine processing technology of natural aromatic materials. *Flavour Fragrance Cosmet.* **2016**, *4*, 53–58.
32. Smoylovska, G.P. Identification of phytosterins in *Urtica dioica* L. (overground part). *Zaporozhye Med. J.* **2017**, *1*, 90–93. [CrossRef]
33. Riss, B.; Garreau, M.; Fricero, P.; Podsiadly, P.; Berton, N.; Buchter, S. Total synthesis of TMS-ent-bisabolangelone. *Tetrahedron* **2017**, *73*, 3202–3212. [CrossRef]
34. Uchida, T.; Matsubara, Y.; Koyama, Y. Structures of two novel sesquiterpenoids formed by the lead tetraacetate oxidation of β-caryophyllene. *Agric. Biol. Chem.* **1989**, *53*, 3011–3015.
35. Yamazaki, S. An effective procedure for the synthesis of acid-sensitive epoxides: Use of 1-methylimidazole as the additive on methyltrioxorhenium-catalyzed epoxidation of alkenes with hydrogen peroxide. *Org. Biomol. Chem.* **2010**, *8*, 2377–2385. [CrossRef]
36. Duhamel, L.; Ancel, J.E. Utilization of functional vinylic organometallics for convergent syntheses of phytol from 6-methylhept-5-en-2-one. *J. Chem. Res.* **1990**, *5*, 154–155.
37. Grossi, V.; Rontani, J.F. Photosensitized oxygenation of phytadienes. *Tetrahedron Lett.* **1995**, *36*, 3141–3144. [CrossRef]
38. Damm, M.; Kappe, C.O. High-throughput experimentation platform: Parallel microwave chemistry in HPLC/GC vials. *J. Comb. Chem.* **2009**, *11*, 460–468. [CrossRef] [PubMed]
39. Li, H.; Wang, D.F.; Chen, Y.J.; Yang, M.S. β-Caryophyllene inhibits high glucose-induced oxidative stress, inflammation and extracellular matrix accumulation in mesangial cells. *Int. Immunopharmacol.* **2020**, *84*, 1–9. [CrossRef]
40. Ok, F.; Kaplan, H.M.; Kizilgok, B.; Demir, E. Protective effect of α-linolenic acid on lipopolysaccharide-induced orchitis in mice. *Andrologia* **2020**, *00*, 1–6. [CrossRef]
41. Fan, J.X.; Wang, S.; Meng, X.S.; Bao, Y.R.; Li, T.J. Determination of six flavonoids in *Schizonepeta tenuifolia* from different areas by HPLC. *Chin. Tradit. Herb. Drugs* **2017**, *48*, 2292–2295.
42. Hu, J.H.; Liu, L.L.; Zhang, Y.J.; Xiao, W. Determination of cafferic acid and rosmarinic acid in *Perilla frutescens* leaves and *Schizonepata tenuifolia* by HPLC. *Chin. Tradit. Herb. Drugs* **2015**, *46*, 2155–2159.
43. Bolzon, L.B.; Dos Santos, J.S.; Silva, D.B.; Crevelin, E.J.; Moraes, L.A.B.; Lopes, N.P.; Assis, M.D. Apigenin-7-O-glucoside oxidation catalyzed by P450-bioinspired systems. *J. Inorg. Biochem.* **2017**, *170*, 117–124. [CrossRef]
44. Bao, B.H.; Yang, J.P.; Zhang, L.; Ding, A.W. Quantitative analysis of flavonoid in Schizonepetae Spica. *J. Nanjing Univ. Tradit. Chin. Med.* **2004**, *20*, 124–125.
45. Bao, B.H.; Zhang, L.; Ding, A.W. Quantitative analysis of flavonoid in *Schizonepeta tenuifolia* Briq. *Lishizhen Med. Mater. Med. Res.* **2004**, *15*, 264–265.
46. Dong, C.H.; Liu, B.; Zhu, X. Study best processing degree of carbonized catnip. *Qilu Pharm. Aff.* **2006**, *25*, 560–599.
47. Zhang, L.; Bao, B.H.; Sun, L.; Ding, A.W. Optimum processing technique of Herba Schizonepetae Carbonisatus by orthogonal design. *Chin. Tradit. Herb. Drugs* **2005**, *36*, 370–372.
48. Cao, L.L.; Li, X.; Zhang, L. Experimental study on the effect of Schizonepetae Spica Carbonisata and its effective fractions on rat coagulation system. *Chin. Tradit. Pat. Med.* **2010**, *32*, 611–613.
49. Zhang, M.L.; Zhao, Y.; Cheng, J.J.; Liu, X.M.; Wang, Y.Z.; Yan, X.; Zhang, Y.; Lu, F.; Wang, Q.G.; Qu, H.H. Novel carbon dots derived from Schizonepetae Herba Carbonisata and investigation of their haemostatic efficacy. *Artif. Cell Nanomed. B* **2018**, *46*, 1562–1571. [CrossRef]
50. Kurkin, V.A.; Kharisova, A.V. Flavonoids of *Carthamus tinctorius* flowers. *Chem. Nat. Compd.* **2014**, *50*, 446–448. [CrossRef]
51. Li, S.L.; Yan, R.; Tam, Y.K.; Lim, G. Post-harvest alteration of the main chemical ingredients in *Ligusticum chuanxiong* hort. (Rhizoma chuanxiong). *Chem. Pharm. Bull.* **2007**, *55*, 140–144. [CrossRef]
52. Ma, S.C.; Liu, Y.; Paul, B.P.H.; Yang, Y.; Vincent, O.E.C.; Spencer, H.S.L.; Song, F.L.; Lu, J.; Lin, R.C. RP-HPLC determination of hederagenin and oleanolic acid in Flos Lonicerae Japonicae. *Chin. J. Pharm. Anal.* **2006**, *26*, 885–887.

53. Ma, S.C.; Liu, Y.; Paul, B.P.H.; Yang, Y.; Vincent, O.E.C.; Spencer, H.S.L.; Song, F.L.; Lu, J.; Lin, R.C. Antiviral activities of flavonoids isolated from *Lonicera japonica* Thunb. *Chin. J. Pharm. Anal.* **2006**, *26*, 426–430.
54. Lin, Z.C.; Fang, Y.J.; Huang, A.Y.; Chen, L.Y.; Guo, S.H.; Chen, J.W. Chemical constituents from *Sedum aizoon* and their hemostatic activity. *Pharm. Biol.* **2014**, *52*, 1429–1434. [CrossRef]
55. Chun, M.H.; Kim, E.K.; Lee, K.R.; Jung, J.H.; Hong, J. Quality control of *Schizonepeta tenuifolia* Briq by solid-phase microextraction gas chromatography/mass spectrometry and principal component analysis. *Microchem. J.* **2010**, *95*, 25–31. [CrossRef]
56. Liu, C.C.; Srividya, N.; Parrish, A.N.; Yue, W.; Shan, M.Q.; Wu, Q.N.; Lange, B.M. Morphology of glandular trichomes of Japanese catnip (*Schizonepeta tenuifolia* Briquet) and developmental dynamics of their secretory activity. *Phytochemistry* **2018**, *150*, 23–30. [CrossRef]
57. Duan, S.L.; Zeng, W.X.; Sun, L.L. Chemical constituents of volatile oils in drug pair of menthae Haplocalycis herba and Schizonepetae spica by GC-MS. *Chin. J. Exp. Tradit. Med. Form.* **2015**, *21*, 50–54.

Sample Availability: Samples of the compounds are available from the authors.

© 2020 by the authors. Licensee MDPI, Basel, Switzerland. This article is an open access article distributed under the terms and conditions of the Creative Commons Attribution (CC BY) license (http://creativecommons.org/licenses/by/4.0/).

Article

Three Extraction Methods in Combination with GC×GC-TOFMS for the Detailed Investigation of Volatiles in Chinese Herbaceous Aroma-Type Baijiu

Lulu Wang, Mengxin Gao, Zhipeng Liu, Shuang Chen * and Yan Xu *

State Key Laboratory of Food Science & Technology, Key Laboratory of Industrial Biotechnology of Ministry of Education & School of Biotechnology, Jiangnan University, Wuxi 214122, China; 15761632279@163.com (L.W.); mxgao97@163.com (M.G.); lzp940918@163.com (Z.L.)
* Correspondence: shuangchen@jiangnan.edu.cn (S.C.); yxu@jiangnan.edu.cn (Y.X.); Tel.: +86-510-85964112 (Y.X.); Fax: +86-510-85918201 (Y.X.)

Academic Editors: Gavino Sanna and Stefan Leonidov Tsakovski
Received: 10 September 2020; Accepted: 24 September 2020; Published: 27 September 2020

Abstract: In this study, the detailed volatile compositions of Chinese herbaceous aroma-type Baijiu (HAB) were characterized by comprehensive two-dimensional gas chromatography-time of flight mass spectrometry (GC×GC-TOFMS). A total of 606 compounds were tentatively identified by similarity, mass spectral data, and retention indices, among which 247 compounds were positively verified by authentic standards. Esters were present in higher numbers (179), followed by aldehydes and ketones (111), and alcohols (81). In addition, there were also many terpenes (82), sulfides (37), furans (29), nitrogenous compounds (29), lactones (17), and so on. Meanwhile, the extraction effects of volatile components from different sample pretreatment methods (headspace solid-phase microextraction (HS-SPME), solid phase extraction (SPE), and stir bar sorptive extraction (SBSE)) for HAB were also revealed. The results indicated that HS-SPME has a better extraction effect on easily volatile compounds, such as alcohols and sulfides, especially for terpenes. SPE was particularly beneficial for the analysis of nitrogen-containing compounds; SBSE showed medium extraction ability for most types of compounds and was more suitable for the target analysis of trace content substances.

Keywords: GC×GC-TOFMS; HS-SPME; SPE; SBSE; Chinese herbaceous aroma-type Baijiu

1. Introduction

As a traditional indigenous spirit and the most distilled liquor produced globally [1], Baijiu plays an important role in the Chinese traditional food industry, with nearly 8 million kiloliters of production in 2019 [2]. Baijiu is made from sorghum as the main raw material, produced by a solid-state spontaneous fermentation process, which accumulates a complex community of microorganisms contributing to the generation of complex layers of flavor [3,4]. Due to the differences in production technology and aroma characteristics, Baijiu can be divided into different aroma-type categories, including soy sauce aroma type, light aroma-type, strong aroma-type, and herbaceous aroma-type Baijiu (HAB), etc. Among them, HAB is produced from sorghum mixed with more than 100 Chinese herbs [5], which gives the distillate a distinctive flavor and creates more aroma active substances.

Volatile compositions are often regarded as the main characteristics to determine Baijiu quality and are essential for consumers' criteria; therefore, most studies focus on the identification of volatile components in Baijiu. Gas chromatography-flame ionization detection (GC-FID), gas chromatography-mass spectrometry (GC-MS), and other techniques have been gradually applied to the study of Baijiu, and hundreds of volatile components have been identified so far. Nevertheless, because the resolution of one-dimensional gas chromatography (1DGC) has a limit to separating mixtures including hundreds or even thousands of components, it is often necessary to combine normal phase chromatography and other complex pre-separation methods to assist in the separation and identification of volatile components in Baijiu samples [6,7]. Therefore, to meet the requirement of stronger separation energy, comprehensive two-dimensional gas chromatography (GC×GC) has emerged. GC×GC involves the combination of two capillary columns with different separation mechanisms through a single modulator. With properly selected orthogonal separation mechanisms, GC×GC allows for the separation of a large number of compounds in a single chromatographic run due to the added selectivity of the second column and inherently high peak capacity [8]. GC×GC has been a powerful technique for analyzing volatile components in highly complex samples [9,10], such as petroleum [11], environmental samples [12], essential oils [13], and wines [14]. Through acquisition of a large amount of data from samples based on GC×GC, significantly different compounds from different regions or varieties were recognized by means of multivariate analysis [15–17]. GC×GC has also been used to create sample-specific fingerprints for sample differentiation [18]; for example, Cordero et al. used GC×GC for the creation of two-dimensional (2D) fingerprints for roasted hazelnuts from different cultivars, varieties, and geographical origins [19]. In addition, Gracka et al. used GC×GC to monitor the changes in volatile compounds related to roasting conditions [20].

Despite the significant benefits offered by GC×GC for the separation and identification of volatiles, sample preparation is also a critical step when characterizing the volatile compositions in such a complex matrix. Headspace solid-phase microextraction (HS-SPME) has been by far the most applied sample preparation method in GC×GC, followed by solid phase extraction (SPE), and stir bar sorptive extraction (SBSE) [21]. A series of validated HS-SPME methods have been proposed for targeted analyses of volatile compounds in a variety of samples [22,23], while research found that HS-SPME has limited application for some influential high-boiling compounds [8]. Accordingly, to compensate for these shortcomings, SPE combined with GC×GC was utilized for the detailed investigation of particularly low-level semi-volatiles and obtained a satisfactory result in wine [24].

In the recent years, GC×GC has been gradually applied to the component analysis of soy sauce aroma-type Baijiu [25] and strong aroma-type Baijiu [26], and more than 1000 volatile components have been identified. However, there are few studies of HAB [27,28], and no systematic analysis using GC×GC has been performed so far. Hence, we analyzed HAB by means of comprehensive two-dimensional gas chromatography-time of flight mass spectrometry (GC×GC-TOFMS) for the purpose of overall characterization of volatiles and revealed the volatile compound profile, and potentially key aromatic compounds. Meanwhile, three pretreatment methods (HS-SPME, SPE, and SBSE) in combination with GC×GC-TOFMS were used for the first time to compare the extraction ability on a complex Baijiu sample and determine the biased analysis of some methods for certain groups of compounds.

2. Results and Discussion

2.1. GC×GC-TOFMS Separation and Identification of Volatile Components in HAB

GC×GC-TOFMS was used for the overall characterization of volatile components in HAB in this study based on the higher capacity, significantly enhanced resolving power, and spectral deconvolution function. As Baijiu mainly consists of polar compounds, such as esters, alcohols, aldehydes, and acids, a (polar × medium-polar) column combination can be more beneficial to the separation [25]. Figure 1A is the 2D chromatogram obtained for HAB analyzed by HS-SPME combined with GC×GC-TOFMS. Ordered chromatograms of homologous series in HAB are observed using the above reversed-type column set. Baijiu has a complex matrix consisting of a large number of volatiles with wide-ranging physicochemical properties, and abundant coelution is observed on conventional 1DGC. This restriction is overcome in GC×GC by subjecting the sample to separation based on two different mechanisms, e.g., polarity in 1D and mid-polarity in 2D. Figure 1B,C illustrates the effectiveness of this approach for HAB analysis. Obviously, the number of compounds shown in this figure could not be separated using conventional 1DGC methods. In fact, only approximately 1/5 of the compounds detected here can be fully resolved in one dimension. As shown in Figure 2A, acetic acid 2-phenylethyl ester, 3-pyridinecarboxylic acid ethyl ester, hexyl octanoate, and β-damascenone were coeluted during 1DGC separation, and this overlapping peak makes qualitative and quantitative analysis difficult. Figure 2A shows that these four compounds are easily separated in the 2D plot. Analytes flowing from the 1st column were sequentially separated by 2nd columns with different retention mechanisms, and the interference of coeluted components was efficiently reduced. Figure 2B shows the mass spectra of the four compounds compared to the mass spectra in the NIST library, and the results indicate that identification of the compound is accurate.

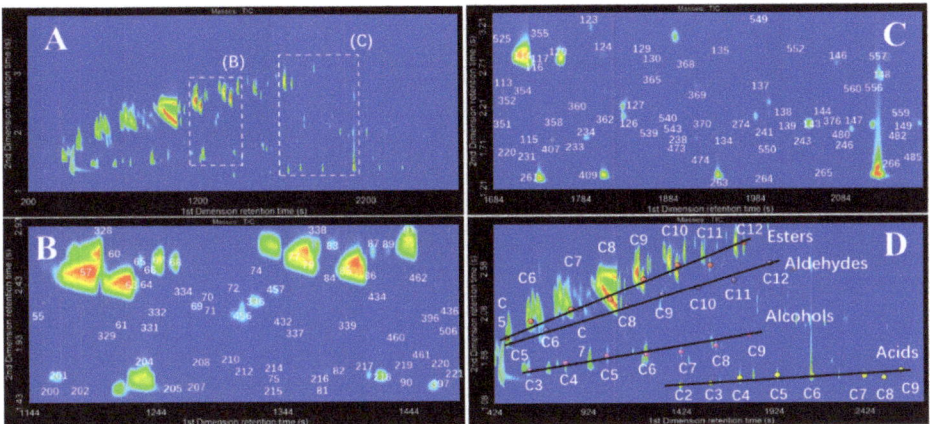

Figure 1. (**A**) Complete 2D contour plot; (**B,C**) present detailed portions of the contour plot to illustrate some of the identified compounds. Compound numbers correspond to Table S1. (**D**) GC × GC distribution of homologous series. Esters: ethyl propionate, ethyl butanoate, ethyl valerate, ethyl hexanoate, ethyl heptanoate, ethyl octanoate, ethyl nonanoate, and ethyl decanoate. Aldehydes: pentanal, hexanal, heptanal, octanal, nonanal, decanal, undecanal, and dodecanal. Alcohols: propanol, butanol, pentanol, hexanol, heptanol, octanol, and nonanol. Acids: acetic acid, propanoic acid, butanoic acid, pentanoic acid, hexanoic acid, octanoic acid, nonanoic acid, and decanoic acid.

Another advantage of GC×GC is the generation of structured chromatograms. The compounds with similar chemical structures will be grouped in a 2D plot. As shown in Figure 1D, the presence of four homologous series compounds was observed for some straight-chain esters, alcohols, acids, and aldehydes. The lines described in the graph show a tendency of organized distribution of these components in the 2D space, and labels represent the carbon atom number of the molecule. The organized distribution of homologous members can be predicted or confirmed, so it is very useful for reliable identification. Esters and alcohols have lower polarity, so they eluted early on the 1st column. Polar acid compounds were retained better in the 1D column, because of their strong polarity; they eluted at higher temperatures, so they eluted earlier on the second dimension and appeared at the bottom. However, apparent 2D tailing was observed for acid compounds, which may cause more coelutions and influence the accurate identification, especially of minor constituents. Tailing in the 2nd dimension was related to the incompatibility of the polar compounds with the mid-polar stationary phase used in 2D.

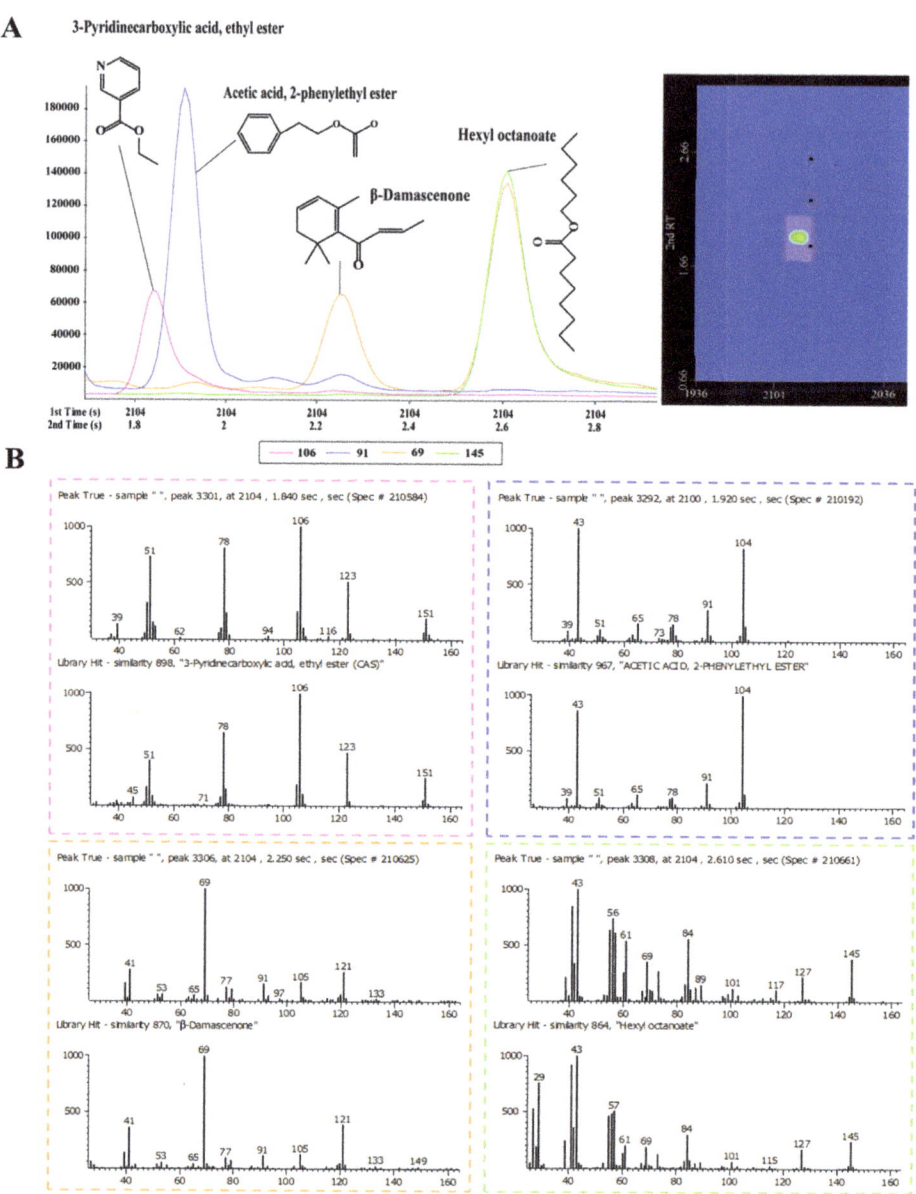

Figure 2. (**A**) Four peaks shown in the two-dimensional chromatogram and modulated peaks of four compounds found in Chinese herbaceous aroma-type Baijiu. (**B**) Deconvoluted mass spectra of compounds.

Nontargeted analysis is performed when it is desirable to have knowledge of all the components in a mixture. In this study, more than 3000 chromatographic peaks with signal-to-noise (S/N) ratios greater than 100 were recognized by deconvolution. Then, the deconvoluted mass spectra were compared with NIST 2014 and Wiley 9 spectral libraries using Chroma TOF4.61.1.0 software with a match value of 70% as the minimum requirement, and 1266 compounds were retained. Next, 472 unwanted search results were eliminated. Finally, a total of 606 compounds were verified by comparing retention indices and mass spectra to those of reference standards. Among them, 247 compounds were positively verified by authentic standards.

2.2. Comparison of Pretreatment Methods

Figure 3 shows the 2D plot of HAB extracts obtained by three different pretreatment methods (HS-SPME, SPE, and SBSE). Table S1 lists the 606 compounds identified in this study, grouped according to different chemical classes. This result showed the most detailed characterization of volatile constituents in HAB for the first time. Figure 4 presents the correlation of HAB analytes identified by three pretreatment methods, and only 205 compounds were commonly identified, which shows the great difference among the three pretreatment methods. Table 1 compares the number of compounds identified in each class using HS-SPME, SPE, and SBSE.

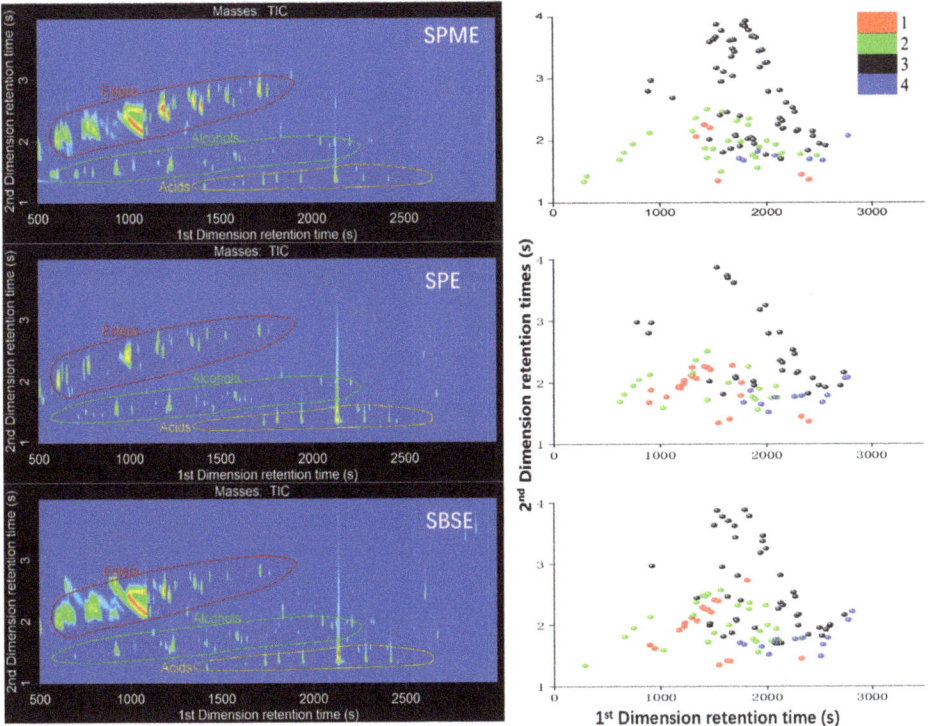

Figure 3. Total ion chromatogram (TIC) contour plot obtained from the HS-SPME-GC×GC-TOFMS, SPE-GC×GC-TOFMS, and SBSE-GC×GC-TOFMS analysis of herbaceous aroma-type Baijiu, and 4 classes of compounds distributed in contour plot (red balls are nitrogenous compounds, green balls are sulfides, gray balls are terpenes, and blue balls are lactone compounds).

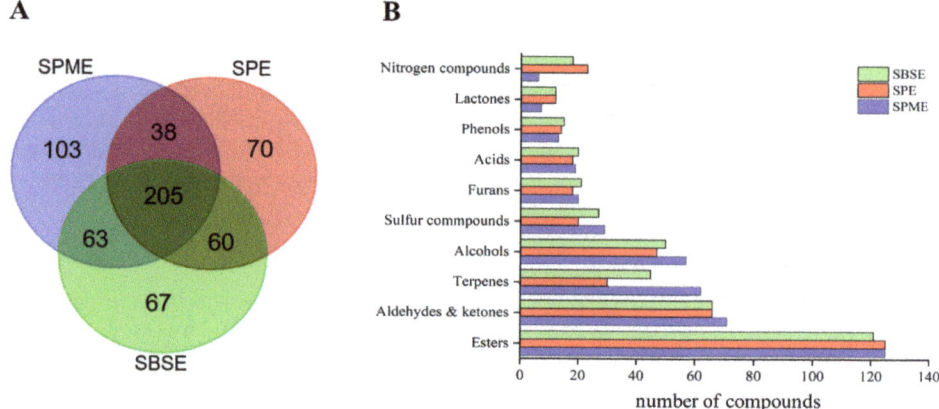

Figure 4. Comparison of identification compounds obtained by HS-SPME-GC×GC-TOFMS, SPE-GC×GC-TOFMS, and SBSE-GC×GC-TOFMS. (**A**) (Venn diagram) and (**B**) (Bar plot graph displaying compound distribution according to chemical class).

Table 1. Comparison of volatile compounds detected in Chinese herbaceous aroma-type Baijiu using HS-SPME-GC×GC-TOFMS, SPE-GC×GC-TOFMS, and SBSE-GC×GC-TOFMS.

Class	Number of Compounds			
	SPME	SPE	SBSE	Total
Esters	125	125	121	179
Aldehydes & Ketones	71	66	66	111
Terpenes	62	30	45	82
Alcohols	57	47	50	81
Sulfides	29	20	27	37
Furans	20	18	21	29
Nitrogenous compounds	6	23	18	29
Acids	19	18	20	23
Phenols	13	14	15	18
Lactones	7	12	12	17
Total	409	373	395	606

HS-SPME is based on the establishment of partition equilibrium of the analytes between the polymeric stationary phase, which covers a fused silica fiber, and the matrix of the sample. It is a simple, rapid, and inexpensive technique in which the extraction and concentration processes are performed simultaneously; furthermore, only small sample volumes are required [29]. A total of 409 volatile compounds were identified by HS-SPME-GC×GC-TOFMS, including esters, alcohols, sulfides, and terpenes; however, lactones and nitrogenous compounds were poorly detected. The results indicated that HS-SPME was particularly beneficial for the analysis of volatile compounds, but defective for the extraction of some high-boiling volatile compounds.

SPE has also been reported to have the functions of extraction, enrichment, and rinsing. Abundant adsorbent material can provide high extraction capacity when properly optimized [21]. SPE is supposed to be the complementary nature of the extraction techniques for HS-SPME, which is best exploited for the analysis of semi-volatiles. Several peaks present at the end of the contour plot by SPE were not detected when using HS-SPME, such as γ-dodecalactone, ethyl *cis*-9-octadecenoate, ethyl linoleate, and ethyl vanillate. In addition, more nitrogenous compounds were detected when using SPE protocol. However, some volatile compounds may be lost, including terpenes and volatile sulfides, because the Baijiu sample was exposed to atmosphere during extraction.

SBSE was initially introduced in 1999 as a miniaturized and solvent-free extraction technique for aqueous samples. Compared with HS-SPME, SBSE provides greater analytical sensitivity and reaches much lower detection and quantification limits. The reason is that the enrichment factor for SBSE is higher than that of HS-SPME using the same stationary phase, because of the 50–250 times larger volume of extraction phase on the stir bar [30]. However, large amounts of stationary phase extracted excessive amounts of solute, thus overloading the GC×GC system (particularly the 2nd column). After split injection (20:1), more terpenes and sulfides were obtained than by the SPE method, due to the apolar adsorbent material PDMS. On the other hand, more high-boiling compounds, including higher fatty acid esters, acids, nitrogenous compounds, and lactones, were obtained.

2.3. Volatile Components in HAB

A total of 606 volatile compounds were identified in this study, among which esters were present in the highest number (179), followed by aldehydes and ketones (111), terpenes (82), alcohols (81), sulfides (37), furans (29), nitrogenous compounds (29), acids (23), phenols (18), and lactones (17).

2.3.1. Skeleton Components

Esters, alcohols, acids, aldehydes, and ketones are the skeletal components of Chinese Baijiu. A total of 179 esters were detected in this study. Among them, ethyl esters were the most representative, and the ethyl esters homologous C2–C12 and C14–C18 were all detected. Ethyl acetate, ethyl butanoate, and ethyl hexanoate are the key aroma compounds in Baijiu, which mainly contribute to its fruity and sweet aroma. In addition, some compounds with very low odor thresholds that cannot be found in the previous literature also contribute to the overall aroma, for example (Figure 5), ethyl 3-methylvalerate has an odor threshold of 8 ng/L and contributes to a strawberry flavor [31]; ethyl cyclohexanoate has an odor threshold of 1 ng/L and reveals strawberry and anise aroma [32]; ethyl cinnamate is known for its honey and cinnamon flavor with an odor threshold of 1 µg/L [33].

Aldehydes and ketones were the second major categories of identified volatiles, amounting to a total of 111 compounds, and the homologous series of straight-chain aldehydes from C2–C12 were all detected. (E, Z)-2,6-nonadienal is the strongest aroma aldehyde compound in HAB and is described as having a strong cucumber aroma [27]; 3-methyl-butanal presents cocoa and almond aroma with an odor threshold of 0.5 µg/L [27]; 2,3-butanedione has been described as contributing a butter aroma with an odor threshold of 100 µg/L [33]; (E)-2-octenal is a key aroma odorant of Chinese chixiang aroma-type Baijiu, which has a fatty flavor with an odor threshold of 15.1 µg/L [27]. 3-Methyl-butanal, 2,3-butanedione and (E)-2-octenal were identified in this study for the first time.

Alcohols are highly volatile constituents of alcoholic beverages transformed from sugar during fermentation. Among them, 39 out of the 81 alcohols were confirmed using authentic standards, and the homologous series of saturated straight-chain primary alcohols from C3-C12 were all detected. 2-Butanol, 1-butanol, and 3-methyl-1-butanol have odor thresholds of 50, 2.73, and 179 mg/L, respectively, in Baijiu. They are also important aroma compounds in HAB, contributing to the fruit or mellow flavor [27].

A total of 23 acid compounds were identified in this study, most of which are saturated monocarboxylic fatty acids. The homologous series of straight-chain monocarboxylic fatty acids from C1–C10, C14, and C16 were all detected, among which butanoic acid, pentanoic acid, and hexanoic acid play an important role in the flavor of Baijiu, which contributes to rancid and cheesy odors [27]; 2-methyl butyric acid and phenylacetic acid have relatively low odor thresholds of 5.9 and 1.4 mg/L, respectively, and are the key food odorants (KFO) [34].

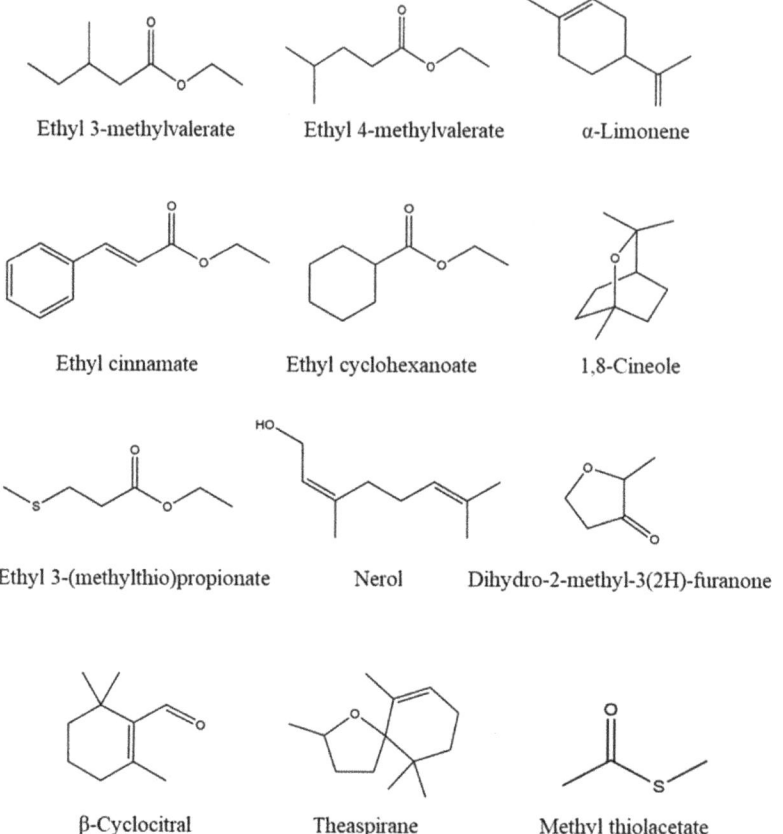

Figure 5. Chemical structure of some aroma compounds first reported in Chinese herbaceous aroma-type Baijiu.

2.3.2. Terpenes

Many terpenoids have important physiological activities, although they exist with low contents in Baijiu. A total of 82 terpenoid compounds (Table 2) were detected in this study, including mono- and polyterpene hydrocarbons, alcohols, carbonyls, and esters. This represents a significant improvement in the number of terpenoids detected by GC×GC-TOFMS compared to a previous report (only 41 compounds) [28]. Terpenes are well-known varietal compounds of Vitis vinifera, and raw materials are important sources of terpenoids; furthermore, most terpenes exist in grapes, and their contribution to wine aroma is significant [35]. During the production process of HAB, more than 100 Chinese herbs are mixed in the raw materials, which create more terpenoid compounds and give Baijiu a distinctive flavor. In addition, substantial evidence also exists to show the formation of terpene-related compounds during fermentation and aging.

Table 2. A total of 82 terpene compounds in Chinese herbaceous aroma-type Baijiu.

NO.	Compounds	RT1 [a]	RT2 [b]	Similarity	LRIcal [c]	LRIlit [d]	Identification [e]
1	δ-3-Carene	784	2.98	842	1139	1166	RI, MS, Tent
2	α-Limonene	892	2.8	913	1193	1200	RI, MS, STD
3	1,8-Cineole	920	2.97	862	1207	1211	RI, MS, Tent
4	Terpinolene	1124	2.69	886	1306	1280	RI, MS, STD
5	α-Thujone	1340	2.44	790	1414	1431	RI, MS, Tent
6	trans-Linalool oxide	1460	1.99	925	1476	1483	RI, MS, Tent
7	cis-Linalool oxide	1464	2.02	921	1478	1454	RI, MS, Tent
8	α-Longipinene	1472	3.59	828	1483	1482	RI, MS, Tent
9	α-Copaene	1504	3.63	831	1500	1497	RI, MS, Tent
10	Daucene	1508	3.63	879	1502	1495	RI, MS, STD
11	Longicyclene	1528	3.67	889	1513	1497	RI, MS, Tent
12	Theaspirane B	1532	3.17	717	1515	1522	RI, MS, Tent
13	Camphor	1532	3.87	758	1515	1540	RI, MS, STD
14	(−)-Camphor	1564	2.41	949	1532	1532	RI, MS, Tent
15	Vitispirane	1576	2.95	853	1539	1527	RI, MS, Tent
16	α-Gurjunene	1580	3.77	903	1541	1529	RI, MS, Tent
17	Linalool	1588	1.81	947	1545	1552	RI, MS, STD
18	Theaspirane	1600	3.11	838	1552	1523	RI, MS, Tent
19	α-Cedrene	1628	3.74	877	1568	1571	RI, MS, STD
20	Carvomenthone	1628	2.46	753	1567	1552	RI, MS, Tent
21	β-Funebrene	1636	3.7	860	1572	1588	RI, MS, Tent
22	Junipene	1656	3.61	926	1583	1583	RI, MS, Tent
23	D-Fenchyl alcohol	1664	1.87	942	1586	1588	RI, MS, Tent
24	α-trans-Bergamotene	1672	3.35	903	1591	1583	RI, MS, Tent
25	α-Guaiene	1684	3.48	860	1598	1598	RI, MS, Tent
26	β-Elemene	1684	3.04	908	1598	1586	RI, MS, Tent
27	Calarene	1692	3.62	916	1602	1604	RI, MS, STD
28	trans-Caryophyllene	1700	3.43	949	1607	1581	RI, MS, STD
29	Terpinen-4-ol	1704	2.09	940	1608	1628	RI, MS, STD
30	Isophorone	1708	2.07	920	1610	1600	RI, MS, STD
31	trans-Edulan	1720	2.8	748	1617	1620	RI, MS, Tent
32	β-Terpineol	1748	1.91	862	1632	1616	RI, MS, Tent
33	β-Cyclocitral	1748	2.4	824	1632	1613	RI, MS, STD
34	α-Patchoulene	1776	3.86	819	1648	1640	RI, MS, Tent
35	Alloaromadendrene	1788	3.88	884	1655	1644	RI, MS, Tent
36	β-Barbatene	1800	3.84	746	1662	1667	RI, MS, Tent
37	γ-Gurjunene	1804	3.93	919	1664	1674	RI, MS, Tent
38	Isoborneol	1820	2.03	803	1671	1672	RI, MS, Tent
39	α-Humulene	1832	3.78	919	1679	1680	RI, MS, Tent
40	L-Borneol	1852	2.05	730	1689	1675	RI, MS, Tent
41	α-Terpineol	1872	2.02	958	1700	1700	RI, MS, STD
42	γ-Amorphene	1864	3.68	895	1696	1724	RI, MS, Tent
43	Ledene	1880	3.68	902	1705	1701	RI, MS, Tent
44	trans-Borneol	1880	1.95	924	1704	1679	RI, MS, Tent
45	β-Chamigrene	1900	3.66	864	1716	1702	RI, MS, Tent
46	Valencene	1928	3.44	899	1731	1726	RI, MS, Tent
47	α-bisabolene	1936	3.18	878	1735	1720	RI, MS, STD
48	Germacrene A	1956	3.37	839	1747	1743	RI, MS, Tent
49	α-Chamigrene	1960	3.46	851	1749	1753	RI, MS, Tent
50	δ-Cadinene	1988	3.25	932	1764	1753	RI, MS, STD
51	β-Citronellol	1992	1.77	889	1765	1771	RI, MS, STD
52	7 epi-a-Selinene	2008	3.26	873	1775	1772	RI, MS, Tent
53	α-Curcumene	2016	2.79	881	1779	1788	RI, MS, Tent

Table 2. Cont.

NO.	Compounds	RT1 [a]	RT2 [b]	Similarity	LRIcal [c]	LRIlit [d]	Identification [e]
54	Nerol	2072	1.7	845	1811	1821	RI, MS, Tent
55	Isogeraniol	2096	1.69	832	1827	1818	RI, MS, Tent
56	β-Damascenone	2104	2.26	910	1832	1827	RI, MS, STD
57	Dihydro-β-ionone	2124	2.36	835	1845	1854	RI, MS, Tent
58	L-calamenene	2124	2.81	946	1846	1838	RI, MS, STD
59	Geraniol	2132	1.7	872	1850	1851	RI, MS, STD
60	trans-Geranylacetone	2148	2.19	877	1861	1862	RI, MS, STD
61	Geosmin	2148	2.32	902	1861	1858	RI, MS, STD
62	α-Ionone	2156	2.2	846	1866	1866	RI, MS, STD
63	α-Dehydro-himachalene	2184	2.61	836	1885	1882	RI, MS, Tent
64	α-Calacorene	2248	2.53	898	1930	1904	RI, MS, Tent
65	Palustrol	2264	2.46	899	1941	1938	RI, MS, Tent
66	trans-β-Ionone	2280	2.15	854	1952	1953	RI, MS, STD
67	cis-Jasmone	2292	1.99	859	1961	1955	RI, MS, STD
68	β-Caryophyllene oxide	2296	2.17	792	1964	1990	RI, MS, Tent
69	D-Nerolidol	2388	1.84	921	2036	2010	RI, MS, Tent
70	E-Nerolidol	2392	1.82	926	2040	2054	RI, MS, Tent
71	Epicubenol	2436	2.07	765	2077	2078	RI, MS, Tent
72	α-Corocalene	2436	2.15	863	2077	2083	RI, MS, Tent
73	Cubenol	2436	2.07	787	2077	2071	RI, MS, Tent
74	6-Isocedrol	2496	1.95	894	2135	2162	RI, MS, Tent
75	α-Cedrol	2496	1.95	877	2135	2127	RI, MS, Tent
76	β-Bisabolol	2520	1.82	728	2160	2151	RI, MS, Tent
77	Torreyol	2556	1.92	815	2197	2197	RI, MS, Tent
78	α-Cadinol	2556	1.92	810	2197	2217	RI, MS, STD
79	α-Eudesmol	2592	1.98	719	2237	2223	RI, MS, Tent
80	β-Eudesmol	2600	2	821	2246	2246	RI, MS, Tent
81	Farnesol	2700	1.95	846	2353	2351	RI, MS, Tent
82	9H-Fluorene	2732	2.16	907	2386	2374	RI, MS, Tent

[a] RT1: retention time on the primary column. [b] RT2: retention time on the secondary column. [c] LRIcal: calculated linear retention indices. [d] LRIlit: literature linear retention indices obtained from the NIST library (https://webbook.nist.gov/chemistry/). [e] Identification: tentative identification (Tent.) based on retention indices (RI) and mass spectra (MS), positive identification based on retention times of authentic standards (STD).

Linalool, geraniol, and β-citronellol are common terpenes in Baijiu and present floral aroma properties. The odor perception thresholds of these compounds are 13.1 µg/L, 120 µg/L, and 300 µg/L [36], respectively. Geosmin is known for its beet, earth aroma as an off-flavor compound, whose odor threshold is 0.1 µg/L in 46%vol ethanol aqueous solution [37]. β-Ionone is believed to be responsible for the characteristic violet and floral aroma, whose concentration in Baijiu is generally 0.3–2.2 µg/L, and the odor threshold in 46%vol ethanol aqueous solution is 1.3 µg/L [36]. In addition, a series of terpenes with low thresholds were also found in this study for the first time (as shown in Figure 5); for example, 1,8-cineole presents a fresh odor with an odor threshold of 0.26 µg/L, and β-cyclocitral has an odor threshold of 5 µg/L [38].

2.3.3. Sulfides

Volatile sulfides play a remarkable role in the aroma of food and beverages, even when present at low concentrations [39,40]. Recently, Wang et al. found that the imbalance of sulfides will lead to the off-odor in soy sauce aroma-type Baijiu; however, these compounds might contribute to the overall aroma of Baijiu at relatively low concentrations [41]. In this study, a total of 37 sulfides (Table 3) were identified, of which 28 compounds have not been reported before in Baijiu. Sulfides usually have a relatively low odor threshold and contribute to the onion, cabbage, and sulfur aroma. Dimethyl trisulfide is an important sulfide in Baijiu, and it is well known for the onion and cabbage aroma with an odor threshold of 0.36 µg/L [42]. Methional is characterized by cooked potatoes and has a perception threshold of 7.1 µg/L [42]. Several sulfides were first reported in HAB, and s-methyl ester butanethioic acid shows a sulfurous and cheesy aroma [43], 5-Methyl-2-formylthiophene is described as moldy odor, and 1,2,4-trithiolane contributes the roasted beef and sulfurous aroma.

Table 3. A total of 37 sulfides in Chinese herbaceous aroma-type Baijiu.

No	Compounds	RT1 [a]	RT2 [b]	Similarity	LRIcal [c]	LRIlit [d]	Identification [e]
1	Methanethiol	292	1.34	985	669	643	RI, MS, STD
2	Dimethyl sulfide	316	1.43	895	750	774	RI, MS, STD
3	Methyl thiolacetate	628	1.69	814	1054	1052	RI, MS, Tent
4	Dimethyl disulfide	668	1.81	960	1077	1078	RI, MS, STD
5	S-Methyl propanethioate	752	1.95	749	1122	1131	RI, MS, STD
6	Methyl ethyl disulfide	804	2.05	736	1149	1141	RI, MS, Tent
7	S-Methyl ester butanethioic acid	908	2.13	835	1201	1198	RI, MS, STD
8	Thiazole	1032	1.59	907	1261	1259	RI, MS, STD
9	Dimethyl trisulphide	1312	2.16	966	1399	1400	RI, MS, STD
10	S-Methyl hexanethioate	1340	2.37	895	1414	1412	RI, MS, Tent
11	Methyl pentyl disulfide	1400	2.48	764	1445	1445	RI, MS, Tent
12	4,5-Dimethyl-2-isopropyl-thiazole	1424	2.47	747	1457	1436	RI, MS, Tent
13	Ethyl 2-(methylthio)acetate	1428	1.88	902	1459	1484	RI, MS, Tent
14	Methional	1448	1.72	826	1470	1480	RI, MS, STD
15	2-Pentyl-thiophene	1448	2.51	893	1470	1452	RI, MS, Tent
16	Furfuryl methyl sulfide	1504	1.87	913	1499	1492	RI, MS, Tent
17	4,5-Dimethyl-2-isobutylthiazole	1568	2.57	709	1534	1514	RI, MS, Tent
18	2-(Methylthio)ethanol	1576	1.5	725	1538	1520	RI, MS, Tent
19	Methyl propyl trisulfide	1588	2.47	752	1545	1529	RI, MS, Tent
20	Ethyl 3-(methylthio)propionate	1644	2	961	1575	1580	RI, MS, STD
21	2,5-Dimethyl-1,3,4-trithiolane	1724	2.32	865	1619	1618	RI, MS, Tent
22	3-(Methylthio)propyl acetate	1760	1.99	752	1639	1627	RI, MS, Tent
23	2,4,5-Trithiahexane	1828	2.26	895	1676	1662	RI, MS, Tent
24	Methyl benzyl sulfide	1836	2.36	932	1680	1665	RI, MS, Tent
25	3-Thiophenecarboxaldehyde	1868	1.77	711	1697	1687	RI, MS, Tent
26	2-Thiophenecarboxaldehyde	1896	1.73	920	1713	1722	RI, MS, STD
27	Methionol	1916	1.56	914	1724	1721	RI, MS, STD
28	5-Methyl-2-formylthiophene	1932	1.9	814	1733	1759	RI, MS, Tent
29	Dimethyl tetrasulphide	1988	2.32	727	1763	1750	RI, MS, Tent
30	1,2,4-Trithiolane	2004	2	866	1772	1760	RI, MS, Tent
31	3-Acetylthiophene	2044	1.75	752	1794	1772	RI, MS, Tent
32	2-Acetylthiophen	2044	1.74	717	1794	1785	RI, MS, STD
33	Furfuryl methyl disulfide	2088	1.94	846	1822	1813	RI, MS, Tent
34	3-Methyl-2-thiophenecarbaldehyde	2104	1.76	798	1832	1815	RI, MS, Tent
35	1-(2-Thienyl) propanone	2144	1.8	714	1858	1840	RI, MS, Tent
36	Benzothiazole	2320	1.78	835	1981	1958	RI, MS, STD
37	2-Phenylthiophene	2476	1.76	780	2114	2124	RI, MS, STD

[a] RT1: retention time on the primary column. [b] RT2: retention time on the secondary column. [c] LRIcal: calculated linear retention indices. [d] LRIlit: literature linear retention indices obtained from the NIST library (https://webbook.nist.gov/chemistry/). [e] Identification: tentative identification (Tent.) based on retention indices (RI) and mass spectra (MS), positive identification based on retention times of authentic standards (STD).

2.3.4. Cyclic Components

Furans, phenols, and lactones were classified under this group. In this research, a total of 29 furans were detected. Compounds of 1-(2-furanyl)-ethanone, 5-methyl-furfural, 2-acetyl-5-methylfuran, and 2-furanmethanol have been described as contributing a honey, caramel odor [27]. Dihydro-2-methyl-3(2H)-furanone reveals sweet, bread and buttery aroma with a sensory threshold of 5 ng/L [44]. This kind of compound mainly contributes to the sweet aroma of Baijiu.

Eight out of the 18 phenols are first detected in HAB. 2-Methoxy-4-ethylphenol, 2-methoxy-4-methylphenol and 2-methoxy phenol were reported to be important aroma compounds in HAB, revealing smoke, sweet, and spice aroma with odor thresholds of 123 µg/L, 315 µg/L, and 13 µg/L, respectively. 2-Methyl-phenol and 3-ethyl-phenol are key food odorants [34], and they have not been reported in HAB before.

A total of 17 lactone compounds were detected in the current study, of which 12 were identified in Baijiu for the first time. γ-Caprolactone and γ-nonalactone present weak sweet, fruity aroma in

HAB [27]. γ-Decalactone has been reported to contribute to peachy and fatty aroma with a threshold of 5000 µg/L [33]. γ-Dodecalactone is responsible for the sweet and floral aroma with a threshold of 7 µg/L [33]. γ-6-(Z)-dodecenolactone was first detected in Baijiu and is associated with sweet and fruity aroma, and its odor threshold is 700 ng/L [45].

2.3.5. Nitrogenous Components

In this study, a total of 29 nitrogenous compounds were detected in HAB, mainly consisting of pyrazines, pyrroles, and pyridines. Tetramethyl-pyrazine was previously described as a baked flavor in HAB [27]. Isopropylpyrazine was identified in Baijiu for the first time.

3. Materials and Methods

3.1. Reagents and Chemicals

A commercially available Dongjiu Baijiu (54% ethanol) was used, which was produced by Guizhou Dongjiu Co., Ltd. (Guizhou, China) according to the National Standard of Herbaceous Aroma-Type Baijiu (DB52/T550). All chemical standards with high-purity grade (GC grade) and C7-C30 n-alkane mixture were obtained from Sigma-Aldrich Co., Ltd. (Shanghai, China). Organic solvents of methanol (HPLC grade), ethanol (HPLC grade), and dichloromethane (HPLC grade) were purchased from J&K Scientific Co., Ltd. (Beijing, China). Sodium chloride (NaCl) and anhydrous sodium sulfate (Na_2SO4) were purchased from China National Pharmaceutical Group Corp. Ultrapure water was obtained from a Milli-Q purification system (Millipore, Bedford, MA, USA).

3.2. Sample Extraction Methods

3.2.1. HS-SPME

An automatic headspace sampling system (MultiPurpose Sample MPS 2 with a SPME adaptor, from Gerstel Inc., Muülheim, Ruhr, Germany) with a 50/30 µm divinylbenzene/carboxen/polydimethylsiloxane (DVB/CAR/PDMS) fiber (2 cm, Supelco Inc., Bellefonte, PA, USA) was used to extract the volatile compounds. Following a method presented in the literature [41], the Baijiu sample was diluted with ultrapure water to a final concentration of 10% ethanol by volume. A total of 10 mL diluted Baijiu sample was transferred into a 20 mL screw-capped vial and saturated with 3 g of NaCl. Then, the sample was equilibrated at 40 °C for 5 min and extracted for 40 min at the same temperature under stirring at a rotation speed of 250 rpm. The extracts were desorbed in a GC splitless injector port at 250 °C for 5 min.

3.2.2. SPE

SPE was based on a slightly modified method described by Chen et al. [46]. A total of 10 mL HAB was diluted with ultrapure water to 50 mL and saturated with 15 g of NaCl. The SPE cartridge (0.8 cm internal diameter, 12 mL internal volume, Sigma Aldrich, Shanghai, China) was consecutively conditioned using 20 mL of dichloromethane, 20 mL of methanol, and 30 mL of ultrapure water. A total of 50 mL diluted sample was passed through the Lichrolut EN cartridge at a flow rate of 2 mL/min. After the sample had been loaded, 30 mL ultrapure water was used to rinse the sorbent. Then, the sorbent was dried by letting the air pass through it (−0.6 bar, 10 min). Extracts were recovered by elution with 10 mL of dichloromethane, concentrated under a gentle stream of nitrogen to a final volume of 500 µL, and stored at −20 °C until analysis. Finally, 1 µL extracts were injected into the GC splitless injector port at 250 °C with 450 s acquisition delay.

3.2.3. SBSE

In this study, SBSE was carried out according to the description in the literature [47]. Stir bars (Twister) coated with PDMS (10 mm length × 1.0 mm thickness) were obtained from GERSTEL. Prior to use, the stir bar was conditioned for 30 min at 280 °C in a flow of helium. The Baijiu sample was diluted with ultrapure water to a final concentration of 10% ethanol by volume. A 10 mL diluted sample was saturated with 3 g NaCl in a 20 mL glass vial, and a stir bar was immersed in the sample for enriching the substance. Then, the sample was placed in an agitation plate at 25 °C and extracted at 800 rpm for 90 min. After extraction, the stir bar was removed with a magnetic rod (twister taking tool) and forceps, rinsed briefly with ultrapure water to remove ethanol, and dried with lint-free tissue, followed by placement in a sample holder for GC×GC-TOFMS analysis.

An automatic headspace sampling system was used to analyze the extracts in this study. The stir bar was placed in a glass thermal desorption liner and thermally desorbed by programming the twister desorption unit (TDU) from 35 °C heated at a rate of 700 °C/min to a final temperature of 270 °C and held for 5 min. TDU injection was in split ratio of 20:1 mode during thermal desorption. A cooled injection system (CIS4) was used in the GC×GC-TOFMS system. CIS4 was in solvent vent mode with a venting flow of 60 mL/min for 4.7 min at a venting pressure of 80 kPa. The initial temperature of CIS4 was kept at −60 °C for 0.2 min, ramped at a rate of 10 °C/s to a final temperature of 250 °C, and held for 3 min.

3.3. GC×GC-TOFMS Instrumentation

Experiments were performed on a LECO Pegasus® 4D GC×GC-TOFMS system (LECO Corp., St. Joseph, MI, USA). This instrument consisted of an Agilent 7890B GC (Agilent Technologies, Palo Alto, CA, USA) incorporating LECO's thermal modulator (dual-stage quad-jet) and a secondary oven mounted inside the primary GC oven. The column set consisted of a 60 m × 0.25 mm × 0.25 μm DB-FFAP (Agilent Technologies, Palo Alto, CA, USA) primary column and a 1.5 m × 0.25 mm × 0.25 μm Rxi-17Sil MS secondary column (Restek, Bellefonte, PA, USA). Ultrahigh purity helium was used as the carrier gas at a constant flow of 1.00 mL/min. The separation was performed using the following temperature program: 45 °C kept for 3 min, ramped at 4 °C/min to 150 °C and held for 2 min; reaching 200 °C at 6 °C/min and 230 °C at 10 °C/min for 20 min. The secondary oven was operated at 5 °C higher than the primary oven throughout. The modulator was offset by +20 °C in relation to the primary oven. A modulation period of 4 s (hot pulse of 0.80 s) was used.

The TOFMS parameters included ion source of 230 °C and transfer line of 240 °C, electron energy of -70 volts, acquisition of 1430, mass range of 35–400 amu, and acquisition rate of 100 spectra/s.

3.4. Data Processing

ChromaTOF version 4.61.1.0 software (LECO Corp., St. Joseph, MI, USA) was used for peak finding, mass spectral deconvolution, peak area integration, and library searching. Automated peak finding and spectral deconvolution with a baseline offset of 0.5 and S/N of 100 after evaluating serval options (i.e., 25, 50, 100, 150, and 200). All analyses were performed in triplicate for each extraction method. The existence of the compound is considered reliable only when the number of detections is greater than 2 at the same retention time. Tentative identification was based on the comparison of mass spectra with the NIST 2014 and Weliy 9 databases using a minimum similarity value of 700 as the criterion, as well as experimentally determined linear retention indices compared to NIST library values. A series of n-alkanes were analyzed under the same conditions to determine first dimension linear retention indices (LRIs) for each compound. A maximum deviation of 30 between the experimental and literature RI values was used as the criterion. Some identification results may be consistent with MS and RI identification, but the 2nd dimensional retention time may not meet the linear distributions of homologous series. The ordered chromatograms of homologous series can also be used for identification. In addition, positive verification of 247 compounds (~41% of the total number) was based on comparison of retention time with authentic standards.

4. Conclusions

The combination of HS-SPME, SPE, and SBSE sample preparation methods coupled with GC×GC-TOFMS analysis enabled us to (tentatively and positively) identify 606 compounds in HAB. Many low content compounds that have never been reported before were identified for the first time. Especially for terpenes, 41 more compounds were identified than previously reported, which are important physiologically active substances in HAB. Furthermore, the contributions of some important compounds were studied in terms of aroma characteristics and odor thresholds. Meanwhile, the three extraction methods show distinct differences and biases for specific analytes. HS-SPME preferred the analysis of alcohols, sulfur-containing, and terpenes compounds; SPE generally revealed more high-boiling compounds, such as lactones and nitrogenous compounds; SBSE showed general extraction ability for all types of compounds, but too much adsorption led to chromatogram overload, making it suitable for the identification of trace content substances. Therefore, the analysis of volatiles in such a complex sample requires multiple pretreatment methods coupled with GC×GC-TOFMS. Importantly, the method can be applied to other alcoholic beverage systems for the determination of the specific kinds of volatile compounds. This approach proved beneficial for the analysis of terpenes, lactones, and sulfur containing compounds, which are important flavor contributors of Baijiu. In addition, the development of this technique laid a foundation for the quantitative determination of the content substances at very low levels (in the region of μg L^{-1} and lower). Based on the feasibility of accurate quantification, it is hoped that this method can be used to monitor the formation of key aroma substances in the production process of Baijiu.

Supplementary Materials: The following are available online, Table S1: Volatile compounds identified in Chinese herbaceous aroma-type Baijiu by HS-SPME-GC×GC-TOFMS, SPE-GC×GC-TOFMS and SBSE-GC×GC-TOFMS, Table S2: Peak area of 606 volatile compounds identified in Chinese herbaceous aroma-type Baijiu.

Author Contributions: L.W., Conceptualization, methodology, formal analysis, visualization, writing—review & editing; M.G., Validation, data curation; Z.L., investigation, writing—original draft; S.C., Conceptualization, validation, supervision, review & editing; Y.X., Supervision, project administration, funding acquisition. All authors have read and agreed to the published version of the manuscript.

Funding: This research was funded by the National Natural Science Foundation of China (no. 31530055), National Key R&D Program of China (no. 2018YFC1604100), Project funded by China Postdoctoral Science Foundation (no.2018M631971), National First-class Discipline Program of Light Industry Technology and Engineering (no. LITE2018-12), 111 Program of Introducing Talents (no. 111-2-06), and Sichuan Science and Technology Program (no. 2018JZ0033).

Conflicts of Interest: The authors declare no conflict of interest.

References

1. Liu, H.L.; Sun, B.G. Effect of Fermentation Processing on the Flavor of Baijiu. *J. Agric. Food Chem.* **2018**, *66*, 5425. [CrossRef] [PubMed]
2. Analysis of National Baijiu Production and Growth from January to December 2019. Available online: https://bg.qianzhan.com/report/detail/459/200207-e8183426.html (accessed on 31 December 2019).
3. Jin, G.Y.; Zhu, Y.; Xu, Y. Mystery behind Chinese liquor fermentation. *Trends Food Sci. Technol.* **2017**, *63*, 18–28. [CrossRef]
4. Wu, J.H.; Zheng, Y.; Sun, B.G.; Sun, X.T.; Sun, J.Y.; Zheng, F.P.; Huang, M.Q. The Occurrence of Propyl Lactate in Chinese Baijius (Chinese Liquors) Detected by Direct Injection Coupled with Gas Chromatography-Mass Spectrometry. *Molecules* **2015**, *20*, 19002–19013. [CrossRef]
5. Zheng, X.W.; Han, B.Z. Baijiu, Chinese liquor: History, classification and manufacture. *J. Ethnic Foods* **2016**, *3*, 19–25. [CrossRef]

6. Fan, W.L.; Xu, Y.; Qian, M.C. Identification of aroma compounds in Chinese "Moutai" and "Langjiu" liquors by normal phase liquid chromatography fractionation followed by gas chromatography/olfactometry. In *Flavor Chemistry of Wine and Other Alcoholic Beverages*; ACS Publications: Washington, WA, USA, 2012; pp. 303–338.
7. Fan, W.L.; Qian, M.C. Characterization of aroma compounds of Chinese "Wuliangye" and "Jiannanchun" liquors by aroma extract dilution analysis. *J. Agric. Food Chem.* **2006**, *54*, 2695–2704. [CrossRef]
8. Weldegergis, B.T.; de Villiers, A.; McNeish, C.; Seethapathy, S.; Mostafa, A.; Górecki, T.; Crouch, A.M. Characterisation of volatile components of Pinotage wines using comprehensive two-dimensional gas chromatography coupled to time-of-flight mass spectrometry (GC x GC-TOFMS). *Food Chem.* **2011**, *129*, 188–199. [CrossRef]
9. Amaral, M.S.S.; Marriott, P.J. The Blossoming of Technology for the Analysis of Complex Aroma Bouquets—A Review on Flavour and Odorant Multidimensional and Comprehensive Gas Chromatography Applications. *Molecules* **2019**, *24*, 2080. [CrossRef] [PubMed]
10. Amaral, M.S.S.; Nolvachai, Y.; Marriott, P.J. Comprehensive Two-Dimensional Gas Chromatography Advances in Technology and Applications: Biennial Update. *Anal. Chem.* **2020**, *92*, 85–104. [CrossRef]
11. Frysinger, G.S.; Gaines, R.B. Separation and identification of petroleum biomarkers by comprehensive two-dimensional gas chromatography. *J. Sep. Sci.* **2001**, *24*, 87–96. [CrossRef]
12. Focant, J.F.; Sjodin, A.; Patterson, D.G. Improved separation of the 209 polychlorinated biphenyl congeners using comprehensive two-dimensional gas chromatography-time-of-flight mass spectrometry. *J. Chromatogr. A* **2004**, *1040*, 227–238. [CrossRef]
13. Marriott, P.J.; Shellie, R.; Fergeus, J.; Ong, R.; Morrison, P. High resolution essential oil analysis by using comprehensive gas chromatographic methodology. *Flavour Fragr. J.* **2000**, *15*, 225–239. [CrossRef]
14. Robinson, A.L.; Boss, P.K.; Heymann, H.; Solomon, P.S.; Trengove, R.D. Development of a sensitive non-targeted method for characterizing the wine volatile profile using headspace solid-phase microextraction comprehensive two-dimensional gas chromatography time-of-flight mass spectrometry. *J. Chromatogr. A* **2011**, *1218*, 504–517. [CrossRef] [PubMed]
15. He, Y.X.; Liu, Z.P.; Qian, M.C.; Yu, X.W.; Xu, Y.; Chen, S. Unraveling the chemosensory characteristics of strong-aroma type Baijiu from different regions using comprehensive two-dimensional gas chromatography–time-of-flight mass spectrometry and descriptive sensory analysis. *Food Chem.* **2020**, *331*, 127335. [CrossRef] [PubMed]
16. Zhang, L.; Zeng, Z.D.; Zhao, C.X.; Kong, H.W.; Lu, X.; Xu, G.W. A comparative study of volatile components in green, oolong and black teas by using comprehensive two-dimensional gas chromatography-time-of-flight mass spectrometry and multivariate data analysis. *J. Chromatogr. A* **2013**, *1313*, 245–252. [CrossRef] [PubMed]
17. Welke, J.E.; Zanus, M.; Lazzarotto, M.; Pulgati, F.H.; Zini, C.A. Main differences between volatiles of sparkling and base wines accessed through comprehensive two-dimensional gas chromatography with time-of-flight mass spectrometric detection and chemometric tools. *Food Chem.* **2014**, *164*, 427–437. [CrossRef]
18. Stilo, F.; Tredici, G.; Bicchi, C.; Robbat, A.J.; Morimoto, J.; Cordero, C. Climate and Processing Effects on Tea (Camellia sinensis L. Kuntze) Metabolome: Accurate Profiling and Fingerprinting by Comprehensive Two-Dimensional Gas Chromatography/Time-of-Flight Mass Spectrometry. *Molecules* **2020**, *25*, 2447. [CrossRef]
19. Cordero, C.; Liberto, E.; Bicchi, C.; Rubiolo, P.; Schieberle, P.; Reichenbach, S.E.; Tao, Q. Profiling food volatiles by comprehensive two-dimensional ga schromatography coupled with mass spectrometry: Advanced fingerprinting approaches for comparative analysis of the volatile fraction of roasted hazelnuts (Corylus avellana L.) from different orig. *J. Chromatogr. A* **2010**, *1217*, 5848–5858. [CrossRef]
20. Gracka, A.; Jeleń, H.H.; Majcher, M.; Siger, A.; Kaczmarek, A. Flavoromics approach in monitoring changes in volatile compoundsof virgin rapeseed oil caused by seed roasting. *J. Chromatogr. A* **2016**, *1428*, 292–304. [CrossRef]
21. Tranchida, P.Q.; Maimone, M.; Purcaro, G.; Dugo, P.; Mondello, L. The penetration of green sample-preparation techniques in comprehensive two-dimensional gas chromatography. *Trac-Trends Anal Chem.* **2015**, *71*, 74–84. [CrossRef]

22. Schurek, J.; Portoles, T.; Hajslova, J.; Riddellova, K.; Hernandez, F. Application of head-space solid-phase microextraction coupled to comprehensive two-dimensional gas chromatography-time-of-flight mass spectrometry for the determination of multiple pesticide residues in tea samples. *Anal. Chim. Acta* **2008**, *611*, 163–172. [CrossRef]
23. Purcaro, G.; Tranchida, P.Q.; Conte, L.; Obiedzinska, A.; Dugo, P.; Dugo, G.; Mondello, L. Performance evaluation of a rapid-scanning quadrupole mass spectrometer in the comprehensive two-dimensional gas chromatography analysis of pesticides in water. *J. Sep. Sci.* **2011**, *34*, 2411–2417. [CrossRef] [PubMed]
24. Weldegergis, B.T.; Crouch, A.M.; Gorecki, T.; de Villiers, A. Solid phase extraction in combination with comprehensive two-dimensional gas chromatography coupled to time-of-flight mass spectrometry for the detailed investigation of volatiles in South African red wines. *Anal. Chim. Acta.* **2011**, *701*, 98–111. [CrossRef] [PubMed]
25. Zhu, S.K.; Lu, X.; Ji, K.H.; Guo, K.F.; Li, Y.L.; Wu, C.Y.; Xu, G.W. Characterization of flavor compounds in Chinese liquor Moutai by comprehensive two-dimensional gas chromatography/time-of-flight mass spectrometry. *Anal. Chim. Acta.* **2007**, *597*, 340–348. [CrossRef] [PubMed]
26. Yao, F.; Yi, B.; Shen, C.; Tao, F.; Liu, Y.; Lin, Z.; Xu, P. Chemical analysis of the Chinese liquor Luzhoulaojiao by comprehensive two-dimensional gas chromatography/time-of-flight mass spectrometry. *Sci. Rep.* **2015**, *5*, 9553. [CrossRef]
27. Fan, W.L.; Hu, G.Y.; Yan, X.U.; Jia, Q.Y.; Ran, X.H. Analysis of Aroma Components in Chinese Herbaceous Aroma Type Liquor. *J. Food Sci. Biotechnol.* **2012**, *31*, 8.
28. Fan, W.L.; Hu, G.Y.; Xu, Y. Quantification of Volatile Terpenoids in Chinese Medicinal Liquor Using Headspace-Solid Phase Microextraction Coupled with Gas Chromatography-Mass Spectrometry. *Food Sci.* **2012**, *33*, 110–116.
29. Castro, R.; Natera, R.; Duran, E.; Garcia-Barroso, C. Application of solid phase extraction techniques to analyse volatile compounds in wines and other enological products. *Eur. Food Res. Technol.* **2008**, *228*, 1–18. [CrossRef]
30. Ochiai, N.; Sasamoto, K.; Ieda, T.; David, F.; Sandra, P. Multi-stir bar sorptive extraction for analysis of odor compounds in aqueous samples. *J. Chromatogr. A* **2013**, *1315*, 70–79. [CrossRef]
31. Campo, E.; Ferreira, V.; Lopez, R.; Escudero, A.; Cacho, J. Identification of three novel compounds in wine by means of a laboratory-constructed multidimensional gas chromatographic system. *J. Chromatogr. A* **2006**, *1122*, 202–208. [CrossRef]
32. Campo, E.; Cacho, J.; Ferreira, V. Multidimensional chromatographic approach applied to the identification of novel aroma compounds in wine-Identification of ethyl cyclohexanoate, ethyl 2-hydroxy-3-methylbutyrate and ethyl 2-hydroxy-4-methylpentanoate. *J. Chromatogr. A* **2006**, *1137*, 223–230. [CrossRef]
33. Francis, I.L.; Newton, J.L. Determining wine aroma from compositional data. *Aust. J. Grape Wine Res.* **2005**, *11*, 114–126.
34. Dunkel, A.; Steinhaus, M.; Kotthoff, M.; Nowak, B.; Krautwurst, D.; Schieberle, P.; Hofmann, T. Nature's Chemical Signatures in Human Olfaction: A Foodborne Perspective for Future Biotechnology. *Angew. Chem. Int. Ed.* **2014**, *53*, 7124–7143.
35. Coelho, E.; Rocha, S.M.; Delgadillo, I.; Coimbra, M.A. Headspace-SPME applied to varietal volatile components evolution during Vitis vinifera L. cv. 'Baga' ripening. *Anal. Chim. Acta.* **2006**, *563*, 204–214.
36. Wang, L.; Hu, G.Y.; Lei, L.B.; Lin, L.; Wang, D.Q.; Wu, J.X. Identification and Aroma Impact of Volatile Terpenes in Moutai Liquor. *Int. J. Food Prop.* **2016**, *19*, 1335–1352.
37. Du, H.; Fan, W.L.; Xu, Y. Characterization of Geosmin as Source of Earthy Odor in Different Aroma Type Chinese Liquors. *J. Agric. Food Chem.* **2011**, *59*, 8331–8337.
38. Buttery, R.G.; Teranishi, R.; Ling, L.C.; Turnbaugh, J.G. Quantitative and sensory studies on tomato paste volatiles. *J. Agric. Food Chem.* **1990**, *38*, 336–340.
39. Roland, A.; Schneider, R.; Razungles, A.; Cavelier, F. Varietal Thiols in Wine: Discovery, Analysis and Applications. *Chem. Rev.* **2011**, *111*, 7355–7376.
40. Song, X.B.; Zhu, L.; Jing, S.; Li, Q.; Ji, J.; Zheng, F.P.; Zhao, Q.Z.; Sun, J.Y.; Chen, F.; Zhao, M.M.; et al. Insights into the Role of 2-Methyl-3-furanthiol and 2-Furfurylthiol as Markers for the Differentiation of Chinese Light, Strong, and Soy Sauce Aroma Types of Baijiu. *J. Agric. Food Chem.* **2020**, *68*, 7946–7954.

41. Wang, L.L.; Fan, S.S.; Yan, Y.; Yang, L.; Chen, S.; Xu, Y. Characterization of Potent Odorants Causing a Pickle-like Off-Odor in Moutai-Aroma Type Baijiu by Comparative Aroma Extract Dilution Analysis, Quantitative Measurements, Aroma Addition, and Omission Studies. *J. Agric. Food Chem.* **2020**, *68*, 1666–1677.
42. Sha, S.; Chen, S.; Qian, M.; Wang, C.; Xu, Y. Characterization of the Typical Potent Odorants in Chinese Roasted Sesame-like Flavor Type Liquor by Headspace Solid Phase Microextraction–Aroma Extract Dilution Analysis, with Special Emphasis on Sulfur-Containing Odorants. *J. Agric. Food Chem.* **2017**, *65*, 123–131.
43. Du, X.F.; Plotto, A.; Baldwin, E.; Rouseff, R. Evaluation of Volatiles from Two Subtropical Strawberry Cultivars Using GC-Olfactometry, GC-MS Odor Activity Values, and Sensory Analysis. *J. Agric. Food Chem.* **2011**, *59*, 12569–12577. [CrossRef] [PubMed]
44. Sunarharum, W.B.; Williams, D.J.; Smyth, H.E. Complexity of coffee flavor: A compositional and sensory perspective. *Food Res. Int.* **2014**, *62*, 315–325. [CrossRef]
45. Siebert, T.E.; Barker, A.; Barter, S.R.; Lopes, M.A.D.B.; Herderich, M.J.; Francis, I.L. Analysis, potency and occurrence of (Z)-6-dodeceno-γ-lactone in white wine. *Food Chem.* **2018**, *256*, 85–90. [CrossRef] [PubMed]
46. Chen, S.; Xu, Y.; Qian, M.C. Aroma Characterization of Chinese Rice Wine by Gas Chromatography-Olfactometry, Chemical Quantitative Analysis, and Aroma Reconstitution. *J. Agric. Food Chem.* **2013**, *61*, 11295–11302. [CrossRef]
47. Fan, W.L.; Shen, H.Y.; Xu, Y. Quantification of volatile compounds in Chinese soy sauce aroma type liquor by stir bar sorptive extraction and gas chromatography-mass spectrometry. *J. Sci. Food Agric.* **2011**, *91*, 1187–1198. [CrossRef]

Sample Availability: Samples of all chemicals used in this study are available from the authors.

© 2020 by the authors. Licensee MDPI, Basel, Switzerland. This article is an open access article distributed under the terms and conditions of the Creative Commons Attribution (CC BY) license (http://creativecommons.org/licenses/by/4.0/).

Article

Simultaneous Measurement of Urinary Trimethylamine (TMA) and Trimethylamine *N*-Oxide (TMAO) by Liquid Chromatography–Mass Spectrometry

Xun Jia [1], Lucas J. Osborn [1,2] and Zeneng Wang [1,2,*]

1. Department of Cardiovascular & Metabolic Sciences, Lerner Research Institute, Cleveland Clinic, 9500 Euclid Ave, Cleveland, OH 44195, USA; jiax2@ccf.org (X.J.); osbornl@ccf.org (L.J.O.)
2. Department of Molecular Medicine, Cleveland Clinic Lerner College of Medicine, Case Western Reserve University, Cleveland, OH 44106, USA
* Correspondence: wangz2@ccf.org; Tel.: +1-216-445-2484

Academic Editors: Gavino Sanna and Stefan Leonidov Tsakovski
Received: 30 March 2020; Accepted: 16 April 2020; Published: 17 April 2020

Abstract: Trimethylamine (TMA) is a gut microbial metabolite—rendered by the enzymatic cleavage of nutrients containing a TMA moiety in their chemical structure. TMA can be oxidized as trimethylamine *N*-oxide (TMAO) catalyzed by hepatic flavin monooxygenases. Circulating TMAO has been demonstrated to portend a pro-inflammatory state, contributing to chronic diseases such as cardiovascular disease and chronic kidney disease. Consequently, TMAO serves as an excellent candidate biomarker for a variety of chronic inflammatory disorders. The highly positive correlation between plasma TMAO and urine TMAO suggests that urine TMAO has the potential to serve as a less invasive biomarker for chronic disease compared to plasma TMAO. In this study, we validated a method to simultaneously measure urine TMA and TMAO concentrations by liquid chromatography–mass spectrometry (LC/MS). Urine TMA and TMAO can be extracted by hexane/butanol under alkaline pH and transferred to the aqueous phase following acidification for LC/MS quantitation. Importantly, during sample processing, none of the nutrients with a chemical structure containing a TMA moiety were spontaneously cleaved to yield TMA. Moreover, we demonstrated that the acidification of urine prevents an increase of TMA after prolonged storage as was observed in non-acidified urine. Finally, here we demonstrated that TMAO can spontaneously degrade to TMA at a very slow rate.

Keywords: trimethylamine (TMA); trimethylamine *N*-oxide (TMAO); biomarker; urine; liquid chromatography–mass spectrometry (LC/MS)

1. Introduction

Trimethylamine *N*-oxide (TMAO), the oxidative product of trimethylamine (TMA), which is dependent on gut microbiota, has gained wide interest due to its pro-atherogenic and pro-thrombotic properties [1–4]. Circulating levels of TMAO can predict future risk for major adverse cardiac events, myocardial infarction, stroke or death [5]. Apart from the clinical relevance to cardiovascular disease, TMAO has also been reported to contribute to chronic kidney disease progression and has been linked to type II diabetes mellitus, a key feature of the human metabolic syndrome [6–9]. As evidenced by these studies, plasma TMAO has become an important biomarker for determining the human health status in a variety of disease states. Recently, we reported a positive correlation between plasma TMAO and urine TMAO concentrations, either in spot urine or in 24 h urine collections [10]. These data suggest that urine TMAO can also be highly indicative of health status similar to plasma TMAO. Importantly, when compared to serum or plasma, urine is much easier to collect as it does not require venipuncture.

To date, little has been reported on the clinical relevance of TMA due to nearly non-detectable concentrations in plasma, with the exception of patients with fish odor syndrome resulting from a deficiency in hepatic flavin monooxygenase 3 [11]. The non-detectable TMA in healthy controls may be due to the rapid oxidation of TMA in liver or excretion via urine and it is reported that more than 90% of the TMA produced from diet can be oxidized as TMAO [12,13]. Here we demonstrated that TMA can be detected in human urine, a finding not limited to patients with fish odor syndrome [14]. Previously, different methods have been used to measure urine TMA such as proton nuclear magnetic resonance (^{1}H NMR), gas chromatography–mass spectrometry (GC/MS), liquid chromatography–mass spectrometry (LC/MS) and fast atom bombardment mass spectrometry [15–20]. For GC/MS, headspace and microfiber extraction were used to introduce samples. Urine TMAO can be measured simultaneously with TMA by LC/MS and fast atom bombardment mass spectrometry [14,19,21,22]. However, fast atom bombardment mass spectrometry is not widely used as an analytical instrument compared to LC/triple quadrupole MS. Additionally, in healthy humans, the average urine concentration of TMAO is about 160 times higher than that of TMA [22], suggesting that when the mass spectrometer's TMAO signal is saturated, the TMA signal may remain undetectable. Consequently, there exists an unmet need to relatively increase the TMA recovery rate while simultaneously decreasing the TMAO recovery rate before transferring to a mass spectrometry vial for LC/MS. Such a method would promote favorable conditions that allow for the quantification of urine TMA and TMAO with a large concentration range as is observed in human urine samples. In this study, we developed a method by the hexane/butanol extraction of TMAO and TMA under alkaline pH in urine to decrease the relative recovery rate of TMAO to TMA to attain this aim.

2. Results

2.1. Validation of the LC/MS/MS Quantitation of TMA and TMAO in Urine

Plasma/serum and urine TMAO can be simultaneously measured with other TMA-related metabolites such as choline, carnitine, γ-butyrobetaine and crotonobetaine by LC/MS/MS after precipitation of protein with methanol [10]. Using the same method, urinary TMAO yielded a S/N > 10, yet the TMA peak was unacceptable due to a high baseline and multiple peaks in the same channel (Figure 1A). Adjusting the injection volume failed to yield an acceptable TMA peak since the background signal changed accordingly, which may be related to other compounds containing a TMA moiety in urine. Furthermore, an increased injection volume can lead to TMAO signal saturation, which is not conducive to linearity throughout the detectable range. Although we can monitor another daughter ion of TMAO with a lower sensitivity or change the MS parameters to monitor ions at a sub-optimal status to avoid signal saturation, the larger injection volume required to do so would lead to a contamination of the MS source and quadrupoles thus adding to the cost of instrument maintenance. Alternatively, we utilized a method to extract the TMAO and TMA with hexane/butanol under alkaline pH followed by acidification to transfer to the aqueous phase. This approach allowed for the simultaneous measurement of urine TMAO and TMA by LC/MS/MS (Figure 1B), where we observed acceptable peaks for both TMA and TMAO.

In an effort to optimize the recovery rate, TMA, d9-TMA, TMAO, and d9-TMAO, at a concentration of 100 µM each, were extracted by the procedure described above with variable volumes of butanol and variable concentrations of 1 mL NaOH (Figure 2). Without the addition of NaOH, TMA and d9-TMA, recovery was not observed. Varying the concentration of NaOH from 0.1 M to 0.5 M had little effect on the recovery rate whereas the addition of butanol greatly improved the recovery rate of TMA and d9-TMA. For TMAO and d9-TMAO, the addition of 0.2 M NaOH significantly improved the recovery rate as did the addition of 1 mL butanol. Based on the recovery curves for TMA, d9-TMA, TMAO, and d9-TMAO, we observed that the addition of butanol uniformly improved the recovery rate. However, due to the high boiling point of butanol (bp 116 °C), butanol may not be compatible with mass spectrometry. Moreover, the relatively higher solubility of butanol in water compared to hexane may lead to excessively high pressures in the subsequent liquid chromatography (LC) system. Nevertheless, we prepared human urine samples by adding 1 mL butanol and 1 mL 0.5 M NaOH for the quantitation of urine TMA and TMAO. Under such conditions, the recovery rates of TMA, d9-TMA, TMAO, and d9-TMAO were 75.3%, 78.5%, 0.50%, and 0.53%, respectively. The recovery rate ratio of TMA to TMAO was 150, which normalized the concentration difference between endogenous levels of urinary TMAO and TMA. The end result was an extracted urine sample ready for electrospray ionization (ESI) LC/MS analysis with TMA and TMAO concentrations of the same magnitude.

The linearity of the standard curve, which was a plot of the peak area ratio of TMA to d9-TMA and TMAO to d9-TMAO, versus TMA and TMAO concentrations, respectively, was related to the volume of butanol and NaOH added. Furthermore, the linearity was more sensitive to butanol than NaOH (Figure 3A–F). The standard curves for TMA were uniformly linear with $r^2 > 0.99$ except when extracting with 1 mL butanol and 1 mL 0.2 N NaOH ($r^2 = 0.9897$). The slopes were very different with respect to the added butanol volume and NaOH concentration with a coefficient of variation (CV) of 13.8% if the effects of butanol and NaOH were ignored (Figure 3A–C). For TMAO (Figure 3D–F), the addition of 0.2 mL butanol or no butanol did not yield a linear curve ($r^2 < 0.99$), which was due to the low recovery rate. The linearity was largely affected by the addition of 1 mL butanol yet the addition of NaOH at different concentrations did not significantly change the linearity of either TMA or TMAO—suggesting that butanol showed discrimination in the recovery rates between native TMA or TMAO and deuterium-labeled TMA/TMAO, respectively. This finding was further confirmed by the recovery rate comparison (Figure 2). Isotope-labeled standard usually has the same physicochemical property as non-labeled standard, which is widely used to quantify metabolites by mass spectrometry with very high accuracy. Here, we observed a modest difference between the TMA and TMAO and their respective deuterium-labeled standards, d9-TMA and d9-TMAO. Compared to deuterium-labeled compounds, ^{13}C or ^{15}N-labeled compounds should be more similar to the naturally abundant compound in terms of their physicochemical properties. In order to confirm this, we prepared standard curves using $[^{13}C_3,^{15}N]$TMA and $[^{13}C_3]$TMAO as internal standards. Of note, the slope difference with the addition of the different volume of butanol was still observed (Figure 4), but with a trivial CV% of 2.9% for the TMA and 6.0% for the TMAO. Consequently, the effects of butanol volume were ignored during the quantitation as the CV%s were drastically smaller than the CV%s observed using d9-TMA and d9-TMAO as internal standards. However, due to cost, the deuterium-labeled compounds d9-TMA and d9-TMAO were used as internal standards in the subsequent study while strictly controlling the volume of butanol and the concentration of NaOH added. Using different known concentrations of TMA and TMAO standards to spike urine, the slopes of the spike-in standard curves had CV%s of 0.8% and 1.2% for TMA and TMAO, respectively (Figure 5). The TMA slope was 9.2% different from the non-spike-in standard curve and the difference was less than 1.5% for the TMAO (Figure 5). As such, the use of a spike-in standard curve represented a reliable method to measure the TMA in urine whereas either a spike-in standard curve or a non-spike-in curve may be used to measure TMAO in urine. The TMA and TMAO standard curves had a wide range of linearity, with concentrations extending to 100 µM for the TMA and 10,000 µM for the TMAO; all the while maintaining slopes of linearity with less than 3% variation (data not shown).

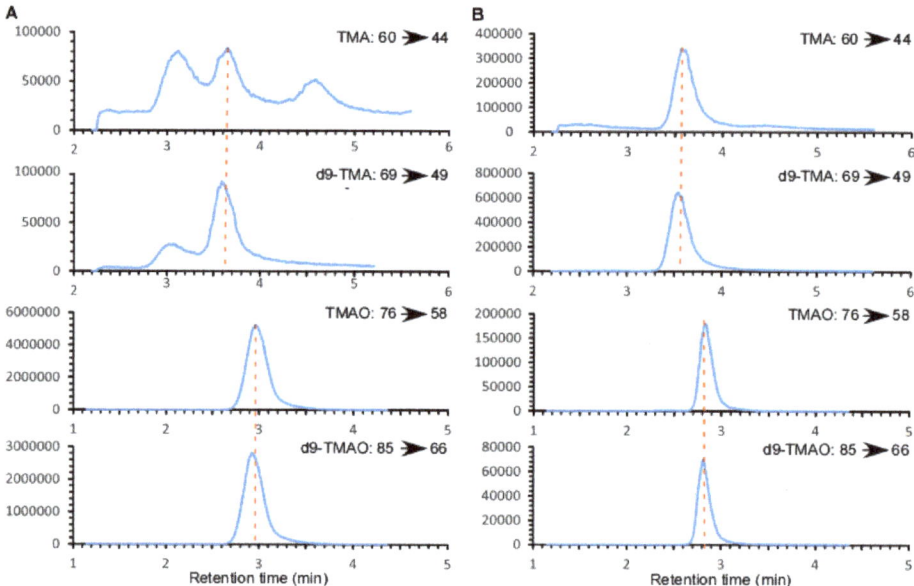

Figure 1. Extracted-ion LC chromatograms from the multiple reaction monitoring (MRM) in positive-ion mode of trimethylamine (TMA), deuterated trimethylamine (d9-TMA), trimethylamine-*N*-oxide (TMAO) and deuterated trimethylamine-*N*-oxide (d9-TMAO) in a typical human urine sample by two different processing methods. (**A**): Mixed with 4 volumes of methanol and internal standards d9-TMA and d9-TMAO; (**B**): Extracted with hexane/butanol after spiking with internal standards d9-TMA and d9-TMAO, under alkaline pH followed by acidification to transfer to the aqueous phase. Then, 5 µL sample was injected onto a Silica column followed by a solvent elution as described in the methods. The eluate was monitored for TMA, TMAO, d9-TMA and d9-TMAO in a Thermo Quantiva mass spectrometer. The precursor-to-product transitions were m/z 60 → 44, m/z 69 → 49, m/z 76 → 58 and m/z 85 → 66 for TMA, d9-TMA, TMAO and d9-TMAO, respectively. The concentrations of TMAO and TMA in this urine sample were measured by the hexane/butanol extraction method as 314 µM and 2.37 µM, respectively. The data were acquired in Thermo Quantiva mass spectrometer with a Vanquish auto-sampler and LC pump system.

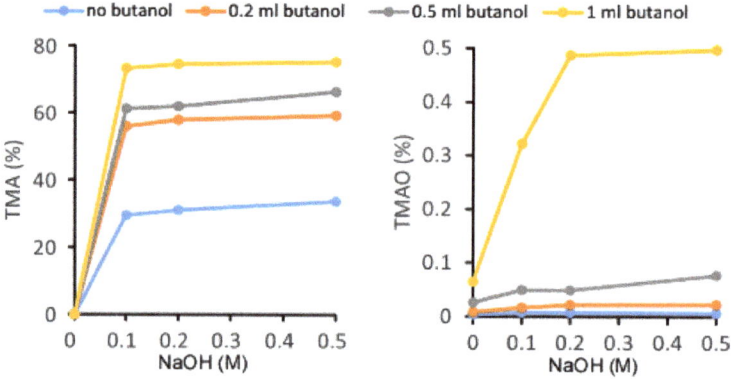

Figure 2. *Cont.*

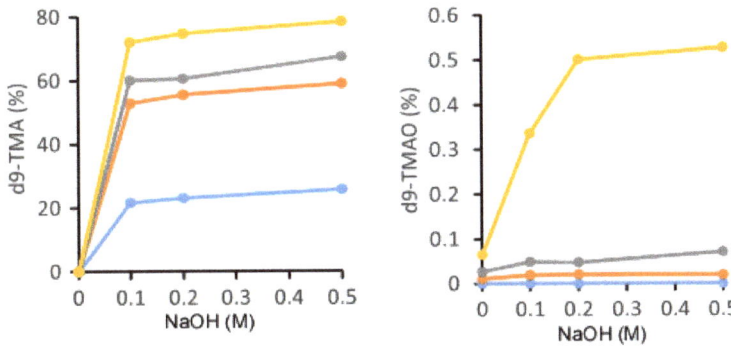

Figure 2. Recovery rates of TMA, TMAO, d9-TMA and d9-TMAO. Here, 500 µL of 40 µM (TMA + TMAO) was mixed with 20 µL of 1 mM (d9-TMA + d9-TMAO) and extracted with 2 mL hexane and different volumes of butanol in the presence of 1 mL NaOH at varying concentrations. This step was followed by the transfer to the aqueous phase by adding 0.2 mL of 0.2 N formic acid. Then, 5 µL was injected into the liquid chromatography–mass spectrometry (LC/MS). The recovery rate (%) was calculated by the peak area of the standard after extraction preparation divided by the peak area of the standard, 100 µM (TMA + TMAO + d9-TMA + d9-TMAO), without undergoing extraction. Data were acquired in a Thermo Quantiva mass spectrometer with a Vanquish auto-sampler and a LC pump system.

The detection limit is related to the instrument used and the injection volume. For the Shimadzu 8050 LC/MS with a 2 µL injection, the detection limit for TMA was 0.26 µM and 0.57 µM for TMAO. The limit of quantitation for TMA was 0.40 µM and 5.0 µM for TMAO. The accuracy of TMA and TMAO measurements are listed in Table 1, where we can see that the accuracy is very close to 100%, with 96.2% and 98.4% accuracy maintained at a concentration very close to the lower limit of quantitation for TMA and TMAO, respectively.

Table 1. Characteristics of the method for the TMAO and the TMA determination by LC/MS/MS.

Characteristic	TMA		TMAO	
LLOD	0.26 µM		0.57 µM	
LLOQ	0.40 µM		5.0 µM	
ULOQ	>50 µM		>10 mM	
Accuracy (%)	0.5 µM	96.2	10 µM	98.4
	2 µM	97.0	50 µM	102.8
	5 µM	101.1	100 µM	102.1
	10 µM	102.0	250 µM	102.1
	20 µM	102.3	500 µM	100.6
	40 µM	101.3	2000 µM	101.3

Lower limit of detection (LLOD) and Lower limit of quantitation (LLOQ) were estimated by Calibration Approach as described in methods. Upper limit of quantitation (ULOQ) = greater than highest standard investigated. Accuracy = ratio of the TMA and TMAO concentrations measured to the TMA and TMAO concentrations added to water followed by the same procedure as the urine samples for the TMAO and TMA measurement, respectively.

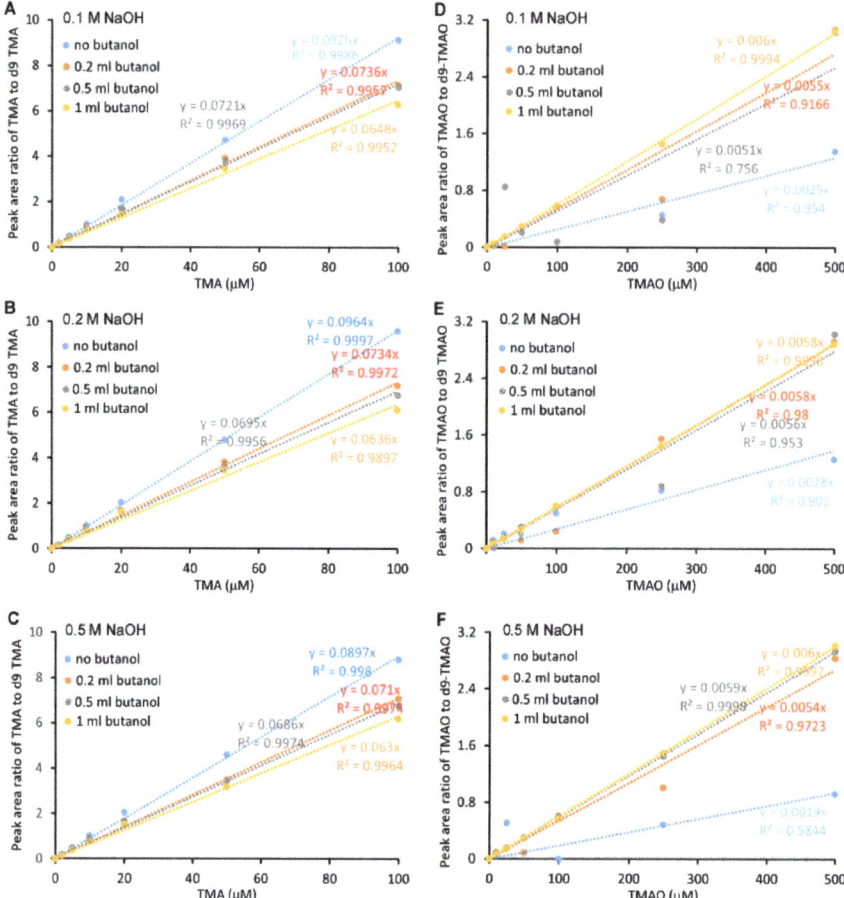

Figure 3. Standard curves for the Liquid Chromatography Electrospray Ionization Tandem Mass Spectrometric (LC/ESI/MS/MS) analysis of TMA and TMAO. Here, 500 µL of varying concentrations of TMA and TMAO were mixed with 20 µL of 1 mM (d9-TMA + d9-TMAO) and extracted with 2 mL hexane and different volumes of butanol in the presence of 1 mL NaOH at varying concentrations ((**A**, **D**), 0.1 M; (**B**, **E**) 0.2 M; (**C**, **F**), 0.5 M). Then, analytes were transferred to the aqueous phase by adding 0.2 mL of 0.2 M formic acid. Curves were plotted as the peak area ratio of TMA to d9-TMA (**A**–**C**) and TMAO to d9-TMAO (**D**–**F**) versus concentration. Data were acquired in a QTRAP5500 mass spectrometer with a Shimadzu auto-sampler and a LC pump system.

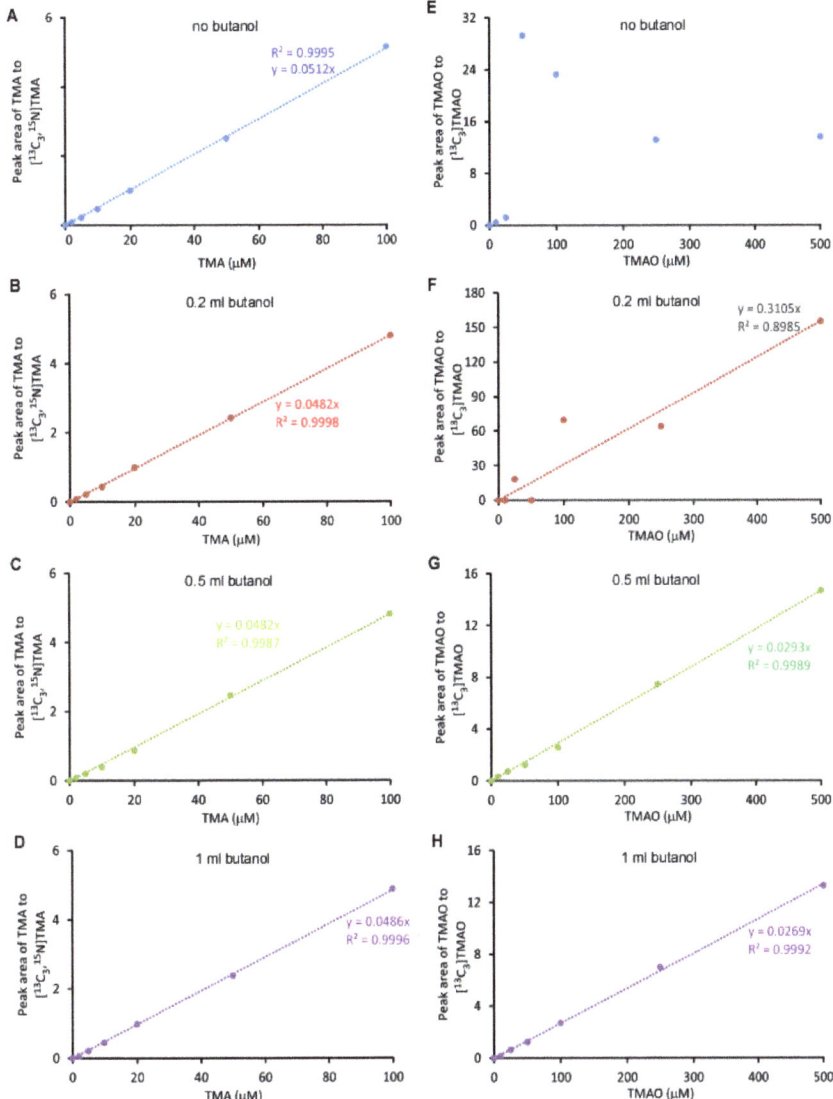

Figure 4. Standard curves for the LC/ESI/MS/MS analysis of TMA and TMAO with [^{13}C$_3$,^{15}N]TMA and [^{13}C$_3$]TMAO as internal standards, respectively. Here, 500 µL of varying concentrations of TMA and TMAO were mixed with 20 µL of 1 mM [^{13}C$_3$, ^{15}N]TMA and 1 mM [^{13}C$_3$]TMAO and extracted with 2 mL hexane and varying volumes of butanol (no butanol (**A,E**), 0.2 mL butanol (**B,F**), 0.5 mL butanol (**C,G**), 1 mL butanol (**D,H**)) in the presence of 1 mL 0.5 M NaOH. Then, analytes were transferred to the aqueous phase by adding 0.2 mL of 0.2 M formic acid. The analysis was performed using electrospray ionization in positive-ion mode with multiple reaction monitoring of precursor and characteristic product ions. The transitions monitored were mass-to-charge ratio (m/z): m/z 60 → 44, m/z 64 → 47, m/z 76 → 58 and m/z 79 → 61 for TMA, [^{13}C$_3$, ^{15}N]TMA, TMAO and [^{13}C$_3$]TMAO, respectively. Curves were plotted as the peak area ratio of TMA to [^{13}C$_3$, ^{15}N]TMA (**A–D**) and TMAO to [^{13}C] TMAO (**E–H**) versus concentration. Data were acquired in a Thermo Quantiva interfaced with a Vanquish LC system with 0.5 µL injection to column.

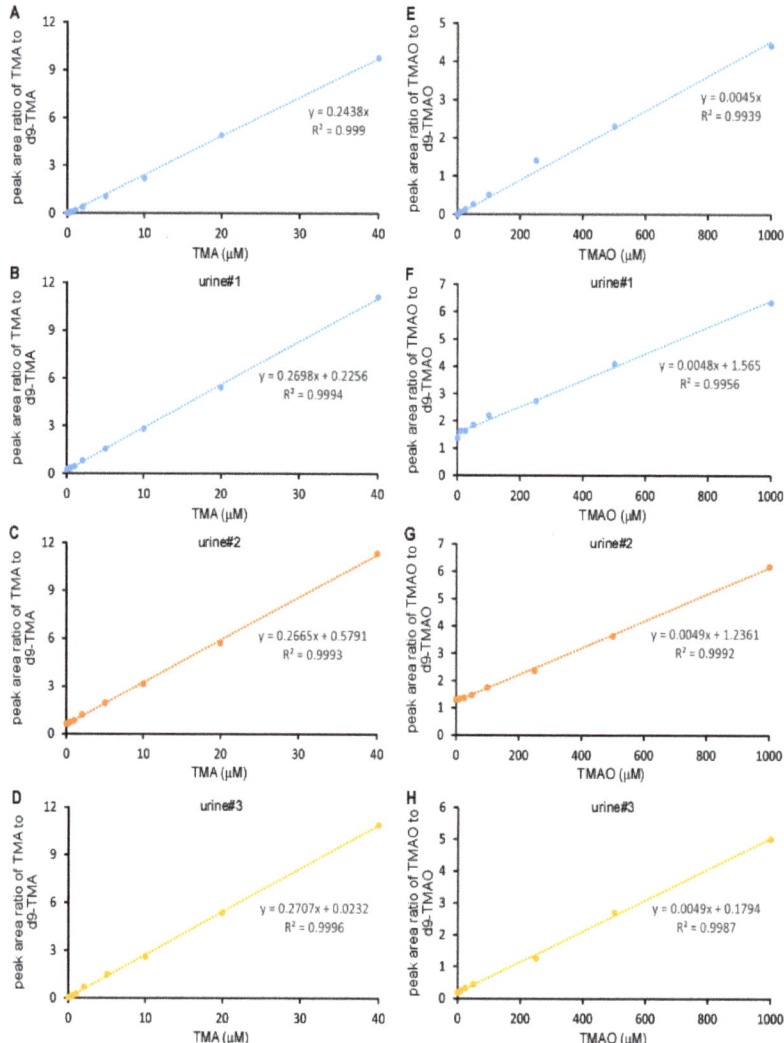

Figure 5. Standard curves for the LC/ESI/MS/MS analysis of the TMA and TMAO spiked into urine or non-urine control with d9-TMA and d9-TMAO as internal standards, respectively. Varying concentrations of TMA and TMAO standards were spiked into 3 urine samples or water. Then, 500 µL each was mixed with 20 µL of 1 mM d9-TMA and 1 mM d9-TMAO, extracted with 2 mL hexane and 1 mL butanol in the presence of 1 mL 0.5 N NaOH, and transferred to the aqueous phase by adding 0.2 mL 0.2 N formic acid. Curves were plotted as the peak area ratio of TMA to d9-TMA (**A–D**) and TMAO to d9-TMAO (**E–H**) versus concentration. Data were acquired in a Shimadzu 8050 LC/MS/MS with 2 µL sample injected into column.

To examine assay precision, urine TMA and TMAO were measured in six different human urine samples over a span of more than 10 days. Shown in Table 2, we can see that the intra-day CV%s for TMA and TMAO are less than 4.0% and 9.0%, respectively, and the inter-day CV%s are higher than intra-day CV%s. Only the urine sample with a relatively low TMA concentration of 0.99 µM gave a CV% greater than 15% while all the other samples gave CV%s less than 15% for both TMA and TMAO.

Table 2. Precision of the TMA and the TMAO concentrations (in μM) measured in 6 urine samples with different levels over a span of more than 10 days.

Urine	TMA			TMAO		
	Mean ± SD	Intraday CV%	Interday CV%	Mean ± SD	Intraday CV%	Interday CV%
1	0.99 ± 0.09	3.4	17.6	305.6 ± 22.7	5.2	12.4
2	2.41 ± 0.10	2.2	8.6	236.8 ± 15.9	5.1	10.5
3	2.28 ± 0.14	2.1	12.3	303.3 ± 11.8	2.9	6.1
4	4.60 ± 0.18	2.5	7.0	393.3 ± 30.3	8.5	2.3
5	19.54 ± 0.65	2.1	6.2	825.2 ± 68.1	7.8	10.1
6	9.96 ± 0.34	2.5	5.9	1215.2 ± 47.3	2.6	7.3

The calculated mean, SD, intraday CV% and interday CV% are given. For each of the 6 different urine samples, a cluster of 4 determinations is presented as they were run on four separate days. CV, coefficient of variance.

Compounds with a TMA structural moiety can serve as substrates for TMA production by gut microbial enzymatic cleavage [2,10,23,24]. In order to test whether these substrates can be spontaneously degraded to form TMA during sample processing in our method, we spiked 100 μM d9-choline, d9-betaine, d9-TMAO, d9-carnitine, d9-γ-butyrobetaine, and d9-crotonobetaine into five randomly selected human urine samples (500 μL each), separately, followed by the addition of 2 mL hexane/1 mL butanol and 1 mL 0.5 M NaOH. After being vortexed and spun down, the hexane layer was acidified with 0.2 mL 0.2 M formic acid to collect the aqueous phase. By LC/MS/MS, we failed to detect any d9-TMA, suggesting that no artificial production of TMA occurred during sample processing.

2.2. Normal Range of Urinary TMA and TMAO in Healthy Subjects

To determine the normal range of urinary TMA and TMAO levels, apparently healthy human subjects (n = 29) undergoing routine healthy screens in the community were examined. Table 3 shows the distribution of urinary TMA and TMAO. Urinary TMAO has a relatively larger concentration range compared to TMA, with a ratio of maximum to minimum concentration of TMAO at 22 and 6 for TMA. The minimum TMA and TMAO concentrations were 0.70 and 52.0 μM, respectively, which were above the LLOQs. Shown here, we can see that urine TMAO concentrations were much greater than TMA, with half of the human samples containing TMAO at a concentration >200 fold higher than TMA. Furthermore, urinary TMAO and TMA were highly correlated to each other (r^2 = 0.27, p = 0.004).

Table 3. Quantile distribution of urinary TMA and TMAO.

	0%	25%	50%	75%	100%
TMA (μM)	0.70	1.37	2.07	2.62	4.39
TMAO (μM)	52.0	243.7	379.0	648.1	1141.0
TMAO/TMA (mol/mol)	66	132	208	315	506
TMA/creatinine (mmol/mol)	0.05	0.09	0.13	0.23	1.43
TMAO/creatinine (mmol/mol)	11.3	22.5	33.8	45.6	106.4

Twenty-nine healthy human subjects' urine samples were collected to measure TMA and TMAO. Urine creatinine was measured by LC/MS following the protocol as reported previously [10].

2.3. Comparison of the Measured TMAO Results by Two Different Methods

Urine TMAO can be directly measured by mixing with the internal standard dissolved in methanol. After precipitation and removal of protein by centrifugation, the supernatant can be injected into an LC/MS system as described above. We compared this method with the hexane/butanol extraction method for quantitation of urine TMAO. Results are shown in Figure 6, where we observed a linearity between the concentration of urine TMAO acquired by the two different methods (r^2 = 0.948, slope = 1.094, CV%=14.3). These data suggested that the results determined by either method are both comparable and reliable.

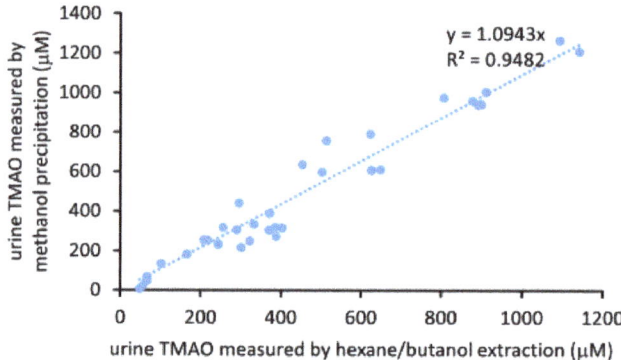

Figure 6. Correlation between the two different methods for the measurement of urine TMAO. Hexane/butanol extraction performed as described in Figure 5. For the methanol precipitation method, urine was diluted 20 times first, then mixed with 4 volumes of methanol containing 10 µM d9-TMAO as the internal standard following the method for the determination of the plasma TMAO [25].

2.4. Stability of Urine TMA

Urine is a collection of metabolic waste containing commensal genitourinary tract microbes responsible for maintaining the health of the urinary tract in vivo [26]. Given that the molecular TMA precursors, choline, betaine, carnitine, γ-butyrobetaine and crotonobetaine are abundant in urine [10], TMA concentrations may increase over time due to metabolism by microbial enzymes. We selected two urine samples and monitored the concentration of TMA during storage at different time points and temperatures. We reported urinary TMA concentration to be stable after one week of storage at −80 °C, −20 °C and 4 °C since the concentration differed from baseline by less than 15%. However, after prolonged storage up to five years, a dramatic increase in urine TMA was observed (Figure 7). The storage at room temperature, 37 °C and 60 °C will lead to a rapid increase in urine TMA concentration.

Additionally, due to its gaseous properties, TMA may be volatile during storage. In order to confirm this, we tested the stability of a TMA stock solution in sterile water with concentrations ranging from 10 to 40 mM during storage at room temperature while maintaining a pH close to 7.0. After two and three months of storage, there was no observable difference in the TMA concentration as evidenced by a deviation of less than 15% from baseline and a CV% of 12.0% across three different timepoints for the seven stock solutions. These data suggested that TMA in sterile water is stable and the gaseous properties of TMA do not affect its stability in water during storage at a neutral pH (Figure 8). However, the stock solution shown here has a concentration ~1000 times higher than that what is found in human urine samples. Consequently, the possibility persists that the volatility of TMA may still contribute to instability at physiological concentrations. In some papers, the acidification of urine was suggested for storage for TMA quantification [21]. In our study, we compared the urine TMA concentration changes between the samples with 60 mM HCl added versus samples with no HCl added. We determined that HCl can significantly attenuate the increase in urine TMA ($p < 0.05$) during storage for three months or five years either at −80 °C or room temperature (Figure 9). Granted, compared to baseline, urine with HCl added still showed an increase in TMA concentration after 5 years, suggesting that urinary TMA is continually generated during storage but acidification can delay this spontaneous production. The observed increase in urinary TMA during storage may be related to the presence of genitourinary bacteria. In order to test this, we sterilized urine using a syringe filter (0.22 µm, Millipore) and incubated the sterilized urine with deuterium-labeled TMA-containing compounds. Here we reported that d9-TMAO can gradually be metabolized to d9-TMA in non-sterilized urine and that sterilized urine is less conducive to the catabolism of TMAO to TMA (Figure 10). These data suggested that the bacterial TMAO reductases may contribute to the increase in TMA observed during storage.

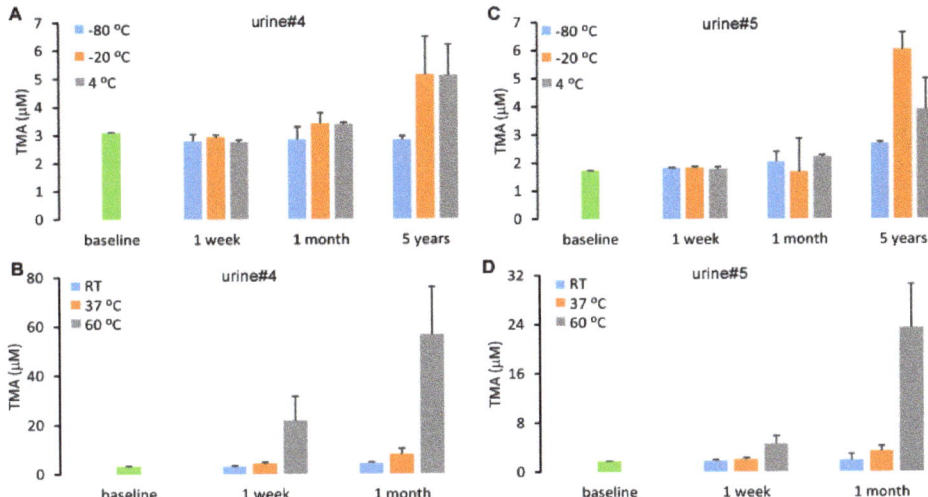

Figure 7. Urine TMA concentration changes during storage. Two urine samples were stored at different temperatures, −80 °C, −20 °C 4 °C (**A**,**C**), room temperature (RT), 37 °C and 60 °C for varying times (**B**,**D**). Urine TMA was extracted with hexane/butanol under alkaline pH with d9-TMA as an internal standard and quantified by LC/MS/MS. Data were presented as mean ± SD from two replicates.

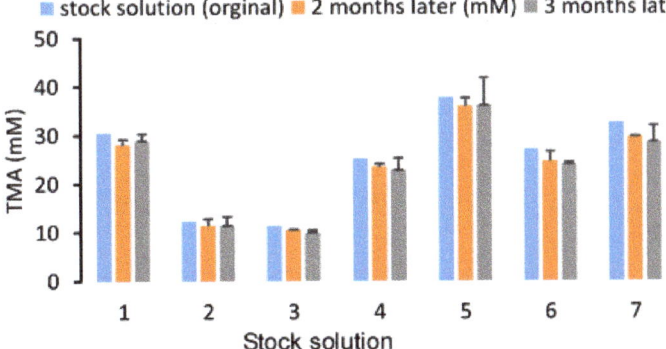

Figure 8. Stability of the TMA stock solution during storage at room temperature. Varying concentrations of TMA standards in sterile water were stored in glass vials and kept at room temperature. The concentrations after 2 and 3 months of storage were measured by LC/MS/MS. Each stock solution was diluted to 10–50 µM and calibrated with standard curves after spiking-in a fixed amount of d9-TMA as an internal standard. Data were presented as mean±SD from three replicates.

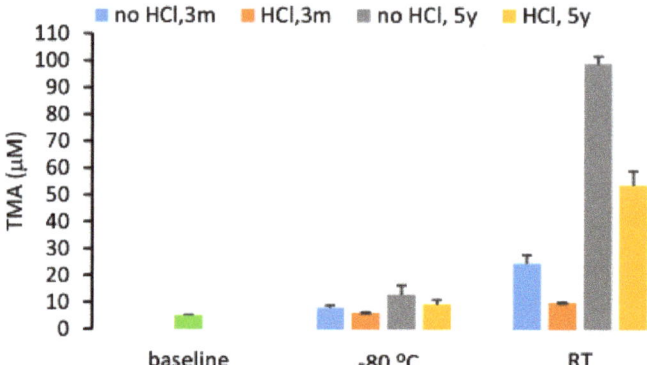

Figure 9. Urine TMA concentration changes during storage after acidification at −80 °C or at room temperature (RT). The HCl added to urine has a final concentration of 60 mM. Data were presented as mean±SD from two replicates.

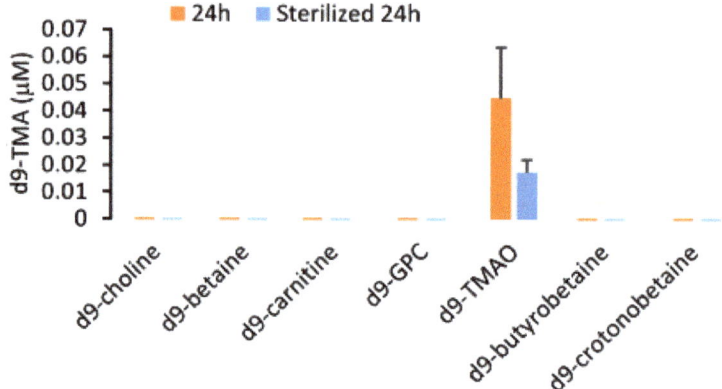

Figure 10. The d9-TMA monitoring in urine after incubation with different stable isotope-labeled compounds containing a TMA (d9-TMA) moiety. Here, 100 μM d9-choline, d9-betaine, d9-carnitine, d9-glycerophosphocholine (GPC), d9-TMAO, d9-butyrobetaine and d9-crotonobetaine were separately spiked into a pooled urine sample randomly selected from 5 human subjects. Urine filtered through a 0.22 μm filter was used as sterilized urine. Then, 500 μL of the non-sterilized urine or sterilized urine were put in glass tube and incubated at 37 °C for 24 h for determination of d9-TMA. In addition, 20 μL of 100 μM [$^{13}C_3,^{15}N$]TMA was added to the urine sample as the internal standard followed by the addition of 2 mL hexane/1 mL butanol and 1 mL of 0.5 N NaOH for extraction. The extracted TMA was transferred to the aqueous phase by adding 0.2 mL of 0.2 M formic acid. Serial dilution of the d9-TMA in 500 μL water mixed with 20 μL of 100 μM [$^{13}C_3,^{15}N$]TMA followed by the same procedure that was used to determine the concentration of TMA in urine was used to prepare a calibration curve by LC/MS/MS. The precursor-to-product transitions in positive MRM mode monitoring by mass spectrometry were m/z 69 → 49 and m/z 64 → 47 for d9-TMA and [$^{13}C_3,^{15}N$]TMA, respectively. Data were presented as mean±SD from two replicates.

3. Discussion

Herein we presented an LC/MS/MS method for the simultaneous measurement of TMA and TMAO in urine with high accuracy and precision. Hexane/butanol extraction under alkaline condition and subsequent transfer to the aqueous phase after acidification resulted in a sample suitable for mass spectrometric analysis. Furthermore, this method of sample preparation ensured the MS source free of contamination and shortened the data acquisition window due to the removal of compounds with a structural formula containing a TMA moiety. The different recovery rates of TMAO and TMA by sample preparation offset the concentration difference while running LC/MS/MS and the linearity of urinary TMAO measurement can be extended to more than 10 mM from 5 µM. Moreover, it was observed that the accuracy of urine TMA and TMAO measurement was largely affected by the volume of butanol added to the extraction system. This is a high-throughput method, where prepared samples can be injected using an automated sampling system and the LC/MS can be done in 9.5 min per injection. Using a two LC column system as we reported previously [25], a single injection can be completed in 5 min.

Urinary TMA is not stable during storage and the concentration will increase with time and storage temperature. The acidification of urine was expected to decrease the volatility of TMA but meanwhile it was determined to play a role in the inhibition of TMA production from other TMA containing compounds, such as TMAO. TMAO may spontaneously be metabolized to TMA and urogenital bacterial flora may speed up this conversion process. Compared with TMA, urinary TMAO concentrations are much higher, thus the slow conversion will not lead to a significant change in urine TMAO concentration.

The highly positive correlation between urine TMAO and plasma TMAO [10], and the clinical relevance of TMAO [1,5,6,27], confer the practical use of urine TMAO measurement. Monitoring urine TMAO can serve as a biomarker for diet–gut microbiota interactions.

The demonstration of the involvement of TMAO in cardiovascular disease [3,5] suggests that inhibition in this metaorganismal pathway can be beneficial for cardiovascular health [28,29]. From dietary TMA precursors to TMAO, there are two steps: the first step is gut microbial enzymatic cleavage to produce TMA, which is the rate-limiting step. As such, the inhibition of this first step becomes a therapeutic strategy to attenuate atherosclerosis and thrombosis. Using the present method for measuring urine TMA, we can generate a high-throughput screen of inhibitors to specific TMA-containing compounds by incubation with purified bacterial enzymes, bacterial lysate, intact bacterial cells, or human fecal samples. The hexane/butanol extraction under alkaline pH can also be used to measure the TMA content in fecal samples after homogenization in cold water.

4. Materials and Methods

4.1. Reagents

Deuterated trimethylamine-N-oxide (d9-TMAO, DLM-4779-1) and TMA (d9-TMA, DLM-1817-5) were purchased from Cambridge Isotope Laboratories (Tewksbury, MA, USA). [$^{13}C_3$]Trimethylamine N-oxide was purchased from IsoSciences (Ambler, PA, USA). All other reagents were purchased from either Sigma-Aldrich (St. Louis, MO, USA) or Fisher Scientific Chemicals (Waltham, MA, USA) unless otherwise stated.

4.2. Research Subjects

Urine samples were collected from subjects undergoing community health screens. All subjects gave written informed consent and the Institutional Review Board of the Cleveland Clinic approved all study protocols.

4.3. Sample Processing

500 µL urine was aliquoted in a 13 × 100 mm glass tube, and 20 µL isotope labeled TMA and TMAO, d9-TMA and d9-TMAO, were added with fixed concentrations. Then, the urine TMA and TMAO were extracted in the organic phase by adding 2 mL hexane and different volumes of butanol and 1 mL NaOH of varying concentrations. After being vortexed for 1 min and spun down at 2500 g, 4 °C for 10 min, the hexane layer was collected and transferred to a 12 × 75 mm glass tube with 0.2 mL 0.2 N formic acid pre-loaded. After being vortexed for 1 min and spun down at 2500 g, 4 °C for 5 min,

the aqueous phase was collected in a mass spec vial (Agilent Technologies, Santa Clara, CA, USA, Part: 5182-0714) with plastic insert (P. J. Cobert Association, St. Louis, MO, USA, Cat. No. 95172) and a Teflon cap (Agilent Technologies, Part: 5182-0717) for liquid chromatography–tandem mass spectrometry (LC/MS/MS) analysis. Different concentrations of TMA and TMAO replacing urine or spiked to urine samples were used to prepare calibration curves in parallel.

For comparison, an alternative method for urine TMAO quantitation is to dilute urine 20 times with water, then aliquot 20 µL diluted urine to a 1.5 mL Eppendorf tube, mixed with 80 µL 10 µM d9-TMAO in methanol. Then, vortex for 1 min and spin down at 20,000 g, 4 °C for 10 min and transfer the supernatant to a mass spectrometry (MS) vial for LC/MS/MS analysis. Then, 20 µL aliquots of varying concentrations (0–100 µM) of TMAO were mixed with 80 µL 10 µM d9-TMAO in methanol to prepare a calibration curve.

4.4. LC/MS/MS

Prepared MS samples were analyzed by injection onto a silica column (2 × 150 mm, 5 mm Luna silica; Cat. No: 00F-4274-B0, Phenomenex, Torrance, CA, USA) at a flow rate of 0.3 mL.min^{-1} using a 2 LC-20AD Shimadzu pump system, a SIL-HTC autosampler interfaced with a QTRAP5500 mass spectrometer (AB SCIEX, Framingham, MA, USA) or a Shimadzu 8050 mass spectrometer or a Vanquish LC system interfaced with a Thermo Quantiva mass spectrometer. A discontinuous gradient was generated to resolve the analytes by mixing solvent A (0.2% formic acid in water) with solvent B (0.2% formic acid in methanol) at different ratios starting at 0% B for 2 min then linearly to 15% within 4 min, then linearly to 100% B within 0.5 min, then held for 3 min and then back to 0% B. TMA, TMAO, d9-TMA, d9-TMAO, [$^{13}C_3$,^{15}N] TMA, [$^{13}C_3$]TMAO were monitored using electrospray ionization (ESI) in positive-ion mode with multiple reaction monitoring (MRM) of precursor and characteristic product ion transitions of m/z 60 → 44 amu, 76 → 58 amu, 69 → 49 amu, 85 → 66 amu, 64 → 47 amu and 79 → 61 amu, respectively. The parameters for the ion monitoring were optimized in individual mass spectrometers. The internal standards d$_9$-TMA (or [$^{13}C_3$,^{15}N] TMA), d$_9$-TMAO (or [$^{13}C_3$] TMAO) were used for quantification of TMA and TMAO, respectively.

4.5. Precision, Accuracy, Limit of Quantitation and Linearity

Four replicates were performed on a single day to establish the intra-day coefficient of variation (CV) for 6 different urine samples with different concentrations of TMA and TMAO. The inter-day CV was determined by assaying aliquots of these samples daily over a span of more than 10 days. Carry-over between injections was not observed. Accuracy was expressed as the ratio of the TMA, the TMAO concentration measured to the TMA, the TMAO concentration added to water undergoing the hexane/butanol extraction. The lower limit of detection (LLOD) was calculated via calibration approach using Equation C [30]. The lower limit of quantification (LLOQ) was estimated from calibration standards as the lowest concentration with a calculated deviation ≤ 30% based on the calibration curve [30]. To determine assay linearity, a standard curve over the 0.01–100 µM, 10–3000 µM concentration range for TMA and TMAO, respectively, was checked for linearity by linear regression fit. The linear range was defined as the region of the standard curve where the difference between the calculated TMA or TMAO concentration and the standard TMA or TMAO concentration was less than 15%.

Author Contributions: X.J. performed mass spectrometry studies. L.J.O. assisted in drafting of the manuscript. Z.W. conceived of the idea, designed experiments, assisted in data analyses, drafting and critical review of the manuscript, and provided funding for the study. All authors have read and agreed to the published version of the manuscript.

Funding: This research was funded by the National Institutes of Health (NIH) and Office of Dietary Supplement grant R01 HL130819 (Z.W.) and R01 HL144651 (Z.W.) and the APC was funded by R01 HL130819 (Z.W.)

Conflicts of Interest: Wang reports being listed as co-inventor on pending and issued patents held by the Cleveland Clinic relating to cardiovascular diagnostics and therapeutics. Wang reports having the right to receive royalty payments for inventions or discoveries related to cardiovascular diagnostics from Cleveland Heart Lab and Procter & Gamble.

References

1. Wang, Z.; Klipfell, E.; Bennett, B.J.; Koeth, R.; Levison, B.S.; DuGar, B.; Feldstein, A.E.; Britt, E.B.; Fu, X.; Chung, Y.M.; et al. Gut flora metabolism of phosphatidylcholine promotes cardiovascular disease. *Nature* **2011**, *472*, 57–63. [CrossRef] [PubMed]
2. Koeth, R.A.; Wang, Z.; Levison, B.S.; Buffa, J.A.; Org, E.; Sheehy, B.T.; Britt, E.B.; Fu, X.; Wu, Y.; Li, L.; et al. Intestinal microbiota metabolism of L-carnitine, a nutrient in red meat, promotes atherosclerosis. *Nat. Med.* **2013**, *19*, 576–585. [CrossRef] [PubMed]
3. Zhu, W.; Gregory, J.C.; Org, E.; Buffa, J.A.; Gupta, N.; Wang, Z.; Li, L.; Fu, X.; Wu, Y.; Mehrabian, M.; et al. Gut Microbial Metabolite TMAO Enhances Platelet Hyperreactivity and Thrombosis Risk. *Cell* **2016**, *165*, 111–124. [CrossRef] [PubMed]
4. Seldin, M.M.; Meng, Y.; Qi, H.; Zhu, W.; Wang, Z.; Hazen, S.L.; Lusis, A.J.; Shih, D.M. Trimethylamine N-Oxide Promotes Vascular Inflammation Through Signaling of Mitogen-Activated Protein Kinase and Nuclear Factor-kappa B. *J. Am. Heart Assoc.* **2016**, *5*, e002767. [CrossRef]
5. Tang, W.W.; Wang, Z.; Levison, B.S.; Koeth, R.A.; Britt, E.B.; Fu, X.; Wu, Y.; Hazen, S.L. Intestinal Microbial Metabolism of Phosphatidylcholine and Cardiovascular Risk. *N. Engl. J. Med.* **2013**, *368*, 1575–1584. [CrossRef]
6. Tang, W.W.; Wang, Z.; Kennedy, D.J.; Wu, Y.; Buffa, J.A.; Agatisa-Boyle, B.; Li, X.S.; Levison, B.S.; Hazen, S.L. Gut Microbiota-Dependent Trimethylamine N-Oxide (TMAO) Pathway Contributes to Both Development of Renal Insufficiency and Mortality Risk in Chronic Kidney Disease. *Circ. Res.* **2015**, *116*, 448–455. [CrossRef]
7. Shan, Z.; Sun, T.; Huang, H.; Chen, S.; Chen, L.; Luo, C.; Yang, W.; Yang, X.; Yao, P.; Cheng, J.; et al. Association between microbiota-dependent metabolite trimethylamine-N-oxide and type 2 diabetes. *Am. J. Clin. Nutr.* **2017**, *106*, 888–894. [CrossRef]
8. Lever, M.; George, P.M.; Slow, S.; Bellamy, D.; Young, J.M.; Ho, M.; McEntyre, C.J.; Elmslie, J.L.; Atkinson, W.; Molyneux, S.L.; et al. Betaine and Trimethylamine-N-Oxide as Predictors of Cardiovascular Outcomes Show Different Patterns in Diabetes Mellitus: An Observational Study. *PLoS ONE* **2014**, *9*, e114969. [CrossRef]
9. Barrea, L.; Annunziata, G.; Muscogiuri, G.; Di Somma, C.; Laudisio, D.; Maisto, M.; De Alteriis, G.; Tenore, G.C.; Colao, A.; Savastano, S. Trimethylamine-N-oxide (TMAO) as Novel Potential Biomarker of Early Predictors of Metabolic Syndrome. *Nutrients* **2018**, *10*, 1971. [CrossRef]
10. Wang, Z.; Bergeron, N.; Levison, B.S.; Li, X.S.; Chiu, S.; Jia, X.; Koeth, R.A.; Li, L.; Wu, Y.; Tang, W.W.; et al. Impact of chronic dietary red meat, white meat, or non-meat protein on trimethylamine N-oxide metabolism and renal excretion in healthy men and women. *Eur. Heart J.* **2019**, *40*, 583–594. [CrossRef]
11. Dolphin, C.T.; Riley, J.H.; Smith, R.L.; Shephard, E.A.; Phillips, I.R. Structural organization of the human flavin-containing monooxygenase 3 gene (FMO3), the favored candidate for fish-odor syndrome, determined directly from genomic DNA. *Genomics* **1997**, *46*, 260–267. [CrossRef] [PubMed]
12. Al-Waiz, M.; Ayesh, R.; Mitchell, S.C.; Idle, J.R.; Smith, R.L. Trimethylaminuria ('fish-odour syndrome'): A study of an affected family. *Clin. Sci.* **1988**, *74*, 231–236. [CrossRef]
13. al-Waiz, M.; Ayesh, R.; Mitchell, S.C.; Idle, J.R.; Smith, R.L. Trimethylaminuria: The detection of carriers using a trimethylamine load test. *J. Inherit. Metab. Dis.* **1989**, *12*, 80–85. [CrossRef] [PubMed]
14. Johnson, D.W. A flow injection electrospray ionization tandem mass spectrometric method for the simultaneous measurement of trimethylamine and trimethylamine N-oxide in urine. *J. Mass Spectrom.* **2008**, *43*, 495–499. [CrossRef] [PubMed]
15. Schlesinger, P. An improved gas–liquid chromatographic method for analysis of trimethylamine in urine. *Anal. Biochem.* **1979**, *94*, 358–359. [CrossRef]
16. Matsushita, K.; Kato, K.; Ohsaka, A.; Kanazawa, M.; Aizawa, K. A simple and rapid method for detecting trimethylamine in human urine by proton NMR. *Physiol. Chem. Phys. Med. NMR* **1989**, *21*, 3–4.
17. Zhang, A.Q.; Mitchell, S.C.; Ayesh, R.; Smith, R.L. Determination of trimethylamine and related aliphatic amines in human urine by head-space gas chromatography. *J. Chromatogr.* **1992**, *584*, 141–145. [CrossRef]

18. Maschke, S.; Wahl, A.; Azaroual, N.; Boulet, O.; Crunelle, V.; Imbenotte, M.; Foulard, M.; Vermeersch, G.; Lhermitte, M. 1H-NMR analysis of trimethylamine in urine for the diagnosis of fish-odour syndrome. *Clin. Chim. Acta* **1997**, *263*, 139–146. [CrossRef]
19. Mamer, O.A.; Choiniere, L.; Treacy, E.P. Measurement of trimethylamine and trimethylamine N-oxide independently in urine by fast atom bombardment mass spectrometry. *Anal. Biochem.* **1999**, *276*, 144–149. [CrossRef]
20. Mills, G.A.; Walker, V.; Mughal, H. Quantitative determination of trimethylamine in urine by solid-phase microextraction and gas chromatography-mass spectrometry. *J. Chromatogr. B Biomed. Sci. Appl.* **1999**, *723*, 281–285. [CrossRef]
21. Zhao, X.; Zeisel, S.H.; Zhang, S. Rapid LC-MRM-MS assay for simultaneous quantification of choline, betaine, trimethylamine, trimethylamine N-oxide, and creatinine in human plasma and urine. *Electrophoresis* **2015**, *36*, 2207–2214. [CrossRef] [PubMed]
22. Awwad, H.M.; Geisel, J.; Obeid, R. Determination of trimethylamine, trimethylamine N-oxide, and taurine in human plasma and urine by UHPLC-MS/MS technique. *J. Chromatogr.* **2016**, *1038*, 12–18. [CrossRef] [PubMed]
23. Koeth, R.A.; Levison, B.S.; Culley, M.K.; Buffa, J.A.; Wang, Z.; Gregory, J.C.; Org, E.; Wu, Y.; Li, L.; Smith, J.D.; et al. Gamma-Butyrobetaine is a proatherogenic intermediate in gut microbial metabolism of L-carnitine to TMAO. *Cell Metab.* **2014**, *20*, 799–812. [CrossRef] [PubMed]
24. Li, X.S.; Wang, Z.; Cajka, T.; Buffa, J.A.; Nemet, I.; Hurd, A.G.; Gu, X.; Skye, S.M.; Roberts, A.B.; Wu, Y.; et al. Untargeted metabolomics identifies trimethyllysine, a TMAO-producing nutrient precursor, as a predictor of incident cardiovascular disease risk. *JCI Insight* **2018**, *3*. [CrossRef] [PubMed]
25. Wang, Z.; Levison, B.S.; Hazen, J.E.; Donahue, L.; Li, X.M.; Hazen, S.L. Measurement of trimethylamine-N-oxide by stable isotope dilution liquid chromatography tandem mass spectrometry. *Anal. Biochem.* **2014**, *455*, 35–40. [CrossRef]
26. Whiteside, S.A.; Razvi, H.; Dave, S.; Reid, G.; Burton, J.P. The microbiome of the urinary tract–a role beyond infection. *Nat. Rev. Urol.* **2015**, *12*, 81–90. [CrossRef]
27. Tang, W.W.; Wang, Z.; Li, X.S.; Fan, Y.; Li, D.S.; Wu, Y.; Hazen, S.L. Increased Trimethylamine N-Oxide Portends High Mortality Risk Independent of Glycemic Control in Patients with Type 2 Diabetes Mellitus. *Clin. Chem.* **2017**, *63*, 297–306. [CrossRef]
28. Wang, Z.; Roberts, A.B.; Buffa, J.A.; Levison, B.S.; Zhu, W.; Org, E.; Gu, X.; Huang, Y.; Zamanian-Daryoush, M.; Culley, M.K.; et al. Non-lethal Inhibition of Gut Microbial Trimethylamine Production for the Treatment of Atherosclerosis. *Cell* **2015**, *163*, 1585–1595. [CrossRef]
29. Roberts, A.B.; Gu, X.; Buffa, J.A.; Hurd, A.G.; Wang, Z.; Zhu, W.; Gupta, N.; Skye, S.M.; Cody, D.B.; Levison, B.S.; et al. Development of a gut microbe-targeted nonlethal therapeutic to inhibit thrombosis potential. *Nat. Med.* **2018**, *24*, 1407–1417. [CrossRef]
30. Wenzl, T.; Haedrich, J.; Schaechtele, A.; Robouch, P.; Stroka, J. *Guidance Document on the Estimation of LOD and LOQ for Measurements in the Field of Contaminants in Feed and Food*; Publications Office of the European Union: Luxemburg, 2016.

Sample Availability: Samples of the compounds are not available from the authors.

© 2020 by the authors. Licensee MDPI, Basel, Switzerland. This article is an open access article distributed under the terms and conditions of the Creative Commons Attribution (CC BY) license (http://creativecommons.org/licenses/by/4.0/).

Article

Significantly Elevated Levels of Plasma Nicotinamide, Pyridoxal, and Pyridoxamine Phosphate Levels in Obese Emirati Population: A Cross-Sectional Study

Ghada Rashad Ibrahim [1], Iltaf Shah [1], Salah Gariballa [2], Javed Yasin [2], James Barker [3] and Syed Salman Ashraf [4,*]

1. Department of Chemistry, College of Science, UAE University, P.O. Box 15551, Al Ain, UAE; 201570102@uaeu.ac.ae (G.R.I.); altafshah@uaeu.ac.ae (I.S.)
2. Department of Internal Medicine, College of Medicine, UAE University, P.O. Box 15551, Al Ain, UAE; s.gariballa@uaeu.ac.ae (S.G.); javed.yasin@uaeu.ac.ae (J.Y.)
3. Department of Chemical and Pharmaceutical Sciences, Kingston University, Penrhyn Road, Kingston upon Thames, Surrey KT1 2EE, UK; j.barker@kingston.ac.uk
4. Department of Chemistry, College of Arts and Sciences, Khalifa University, P.O. Box 127788, Abu Dhabi, UAE
* Correspondence: syed.ashraf@ku.ac.ae; Tel.: +971-2501-8483

Academic Editors: Gavino Sanna and Stefan Leonidov Tsakovski
Received: 28 July 2020; Accepted: 15 August 2020; Published: 28 August 2020

Abstract: Water-soluble vitamins like B3 (nicotinamide), B6 (pyridoxine), and B9 (folic acid) are of utmost importance in human health and disease, as they are involved in numerous critical metabolic reactions. Not surprisingly, deficiencies of these vitamins have been linked to various disease states. Unfortunately, not much is known about the physiological levels of B6 vitamers and vitamin B3 in an ethnically isolated group (such as an Emirati population), as well as their relationship with obesity. The aim of the present study was to quantify various B6 vitamers, as well as B3, in the plasma of obese and healthy Emirati populations and to examine their correlation with obesity. A sensitive and robust HPLC-MS/MS-based method was developed for the simultaneous quantitation of five physiologically relevant forms of vitamin B6, namely pyridoxal, pyridoxine, pyridoxamine, pyridoxamine phosphate, and pyridoxal phosphate, as well as nicotinamide, in human plasma. This method was used to quantify the concentrations of these vitamers in the plasma of 57 healthy and 57 obese Emirati volunteers. Our analysis showed that the plasma concentrations of nicotinamide, pyridoxal, and pyridoxamine phosphate in the obese Emirati population were significantly higher than those in healthy volunteers ($p < 0.0001$, $p = 0.0006$, and $p = 0.002$, respectively). No significant differences were observed for the plasma concentrations of pyridoxine and pyridoxal phosphate. Furthermore, the concentrations of some of these vitamers in healthy Emirati volunteers were significantly different than those published in the literature for Western populations, such as American and European volunteers. This initial study underscores the need to quantify micronutrients in distinct ethnic groups, as well as people suffering from chronic metabolic disorders.

Keywords: vitamins; obesity; Emirati population; bioanalytical quantification; serum

1. Introduction

It is well-established that vitamins and their metabolites are critical for cellular homeostasis, as well as cellular metabolism, mainly as coenzymes. For example, the phosphorylated forms of thiamine (vitamin B1) play a key role in the Krebs cycle [1,2], whereas riboflavin (vitamin B2), nicotinamide (vitamin B3), pantothenic acid (vitamin B5), pyridoxal 5-phosphate (circulating form of vitamin B6), 5-methyl tetrahydrofolate (circulating form of vitamin B9), and biotin (vitamin B8) are involved in oxidation/reduction reactions, fatty acid and neurotransmitter synthesis and metabolism [3–6]. Because all these vitamers are directly involved in various important cellular functions, their deficiency has a direct impact on human health. Supplementation is therefore highly recommended for targeted populations such as pregnant women, lactating women, infants, the elderly, and athletes to prevent various diseases, such as cardiovascular risk [7], anemia, cognitive impairment [8], and neural tube defects in newborns [9,10]. Vitamin B6 has also been shown to be important for normal cognitive function and in lowering the incidence of coronary heart disease among the elderly [11–13]. Supplementation also improves the physical performance of the same targeted population [14]. For example, vitamin B6 supplementation has been shown to reduce diabetic complications and incidences of neurodegenerative diseases in varying degrees [15]. The vitamin B6 group includes pyridoxal, pyridoxine, pyridoxamine, and their metabolites. The phosphate ester derivative pyridoxal 5′-phosphate (PLP) is the biologically active form of this vitamin [16] and reflects long-term body storage [17]. Studies have shown that low plasma PLP concentrations are associated with an increased risk of cardiovascular disease (CVD) [18,19]. Recent data have shown that plasma PLP is adversely associated with inflammatory markers, which include C-reactive protein, fibrinogen, and blood cell count [19–22] Additionally, low vitamin B6 concentrations are commonly present in diseases with a strong inflammatory basis, such as diabetes [23], rheumatoid arthritis [24], and inflammatory bowel disease [25]. Current evidence highlights the notion that inflammation may represent another link between vitamin B6 and CVD. However, the relationship of vitamin B6 status with inflammation and other CVD risk factors has not yet been extensively investigated in a population at high risk of CVD [26].

Niacin, also known as vitamin B3, is the precursor of the redox mediator coenzymes nicotinamide adenine dinucleotide (NAD) and nicotinamide adenine dinucleotide phosphate (NADP) [27]. Niacin has also been shown to decrease low-density lipoprotein cholesterol (LDL), very low-density lipoprotein cholesterol (VLDL-C), and triglycerides (TG), as well as to increase high-density lipoprotein cholesterol (HDL) levels [28]. Niacin alone [29] or in combination with other lipid-lowering agents such as statin [30] or ezetimibe [31] has been shown to significantly reduce the risk of cardiovascular disease and atherosclerosis progression [32]. There is some evidence that niacin might also help in lowering the risk of Alzheimer's disease, cataracts, osteoarthritis, and type 1 diabetes [33]. Though the role of vitamin B6 in reducing complications associated with diabetes, aging, and neurodegenerative diseases has been widely reported [11,12,34,35], most of the published work on vitamin B6 is limited to general clinical observations and case studies.

In the past ten years, LC-MS/MS has been demonstrated to be uniquely suitable for the analysis of many water-soluble vitamins. However, only a few methods have been published for the multi-analyte quantification of water-soluble vitamins in complex biological matrices such as human milk [36,37], human plasma [38], and urine [39]. For instance, in 2012, Hampel et al. quantified four different water-soluble vitamins represented by five analytes in human milk by LC-MS/MS [40]. Thus far, an analytical approach for performing the simultaneous quantification of vitamins B1, B2, B3, B5, B6, B8, B9, and their main circulating forms in human plasma has not been published. This can be of great advantage when vitamer profiling is required in large longitudinal studies [41]. The objectives of the present work were two-fold: (1) to develop a sensitive robust and easy LC-MS/MS-based assay for measuring vitamins B6 and B3 in human plasma and (2) to use this method to measure plasma vitamins B6 and B3 in healthy and obese Emirati populations.

2. Results and Discussion

The most widely used variation of the LC-MS method are the tandem mass-spectrometry and "multiple reaction monitoring" (MRM) methods. This approach basically uses tandem mass-spectrometers to detect a specific product ion that is generated from a precursor ion (the parent compound) under a given set of fragmentation conditions. The monitoring of precursor-to-product ion allows for the specific and accurate determination of given analytes, even if they are not chromatographically resolved in the liquid chromatography part of the LC-MS/MS method. Initial experiments were conducted with pure analytical standards to identify the precursor and product ions for specific analytes. The vitamers, their structures, abbreviations, the mass-to-charge ratio (m/z) for the precursor and products ions, and the fragmentor voltage and collision energies values determined for the various vitamers are summarized in Table 1.

These parameters were then used to develop a short, 10-min, LC-MS/MS-based chromatographic method (described more under Materials and Methods) that was able to quantify the five different B6 vitamers, nicotinamide, and an internal standard (pyridoxine hydrochloride (PN)-D3).

Since the final application of our assay was the analysis of vitamins B6 and B3 in human plasma, we also wanted to determine the best method for the extraction of these vitamins from plasma. The choice of precipitating agent [42,43] and incubation conditions [44] are known to greatly influence the effectiveness of protein precipitation and analyte extraction. TCA has been reported to be effective in precipitating proteins in human plasma [42,43]. However, it is also known that some vitamers, such as PLP, are strongly bound to plasma proteins [45,46] and may require vigorous vortexing as well as incubation conditions at 50 °C for their complete release. Therefore, experiments were conducted to determine which temperature incubation of the TCA-mixed samples would result in the highest recovery of our analytes from spiked simulated plasma. Three different conditions were chosen to be tested with 5 and 60 min incubations: 0 °C, room temperature, and 50 °C. Our results showed that 50 °C incubation of the TCA-precipitated samples for 5 min gave similar results to those incubated at room temperature or ice for one hour (Supplementary information, Table S1). Therefore, all the experiments shown here were performed using a 0.3 N TCA precipitation of each plasma sample at 50 °C for 5 min, followed by 1 min of vortexing and then centrifugation at 15,000× g for 10 min to obtain the supernatant that was then used for LC-MS/MS analysis, as shown in Figure 1. Table 2 shows the typical recovery of spiked vitamers obtained using our optimized TCA-based extraction in simulated plasma. Several mixtures of vitamers of different concentrations were then analyzed using the optimized LC-MS/MS method to establish standard curves for the five B6 vitamers as well as nicotinamide (B3). Table 3 shows the lower limit of detection (LLOD) and the lower limit of quantification (LLOQ) determined for each of the vitamers of interest under our optimized conditions. Further statistics of the partially validated method including precision, accuracy, and linear range are shown in Supplementary Table S2.

Table 1. Names, structures, precursor, and product ions of all the analytes used in this study.

Name	Vitamer Structure	Mass (g/mol)	Precursor Ion [M + H]+ (m/z)	Product Ion [M + H]+ (m/z)	Fragmentor Voltage (V)	Collision Energy (eV)
Pyridoxal-5′-phosphate (PLP)	$C_8H_{10}NO_6P$	247	248	149.7	45	15
Pyridoxal hydrochloride (PL)	$C_8H_9NO_3 \cdot HCl$	203.63 167.06 (-HCl)	168	149.9	94	10
Pyridoxamine dihydrochloride (PM)	$C_8H_{12}N_2O_2 \cdot 2HCl$	241.11 168.09 (-2HCl)	169	152	45	10
Pyridoxamine-5′-phosphate (PMP)	$C_8H_{13}N_2O_5P$	248	249	232.1	94	10
Pyridoxine hydrochloride (PN)	$C_8H_{11}NO_3 \cdot HCl$	205.64 169.07 (-HCl)	170	151.9	94	10
Nicotinamide	$C_6H_6N_2O$	122.12	123	80.2	94	20
Pyridoxine-(methyl-d3) hydrochloride	$C_8D_3H_8NO_3 \cdot HCl$	208.6	173	155	94	10

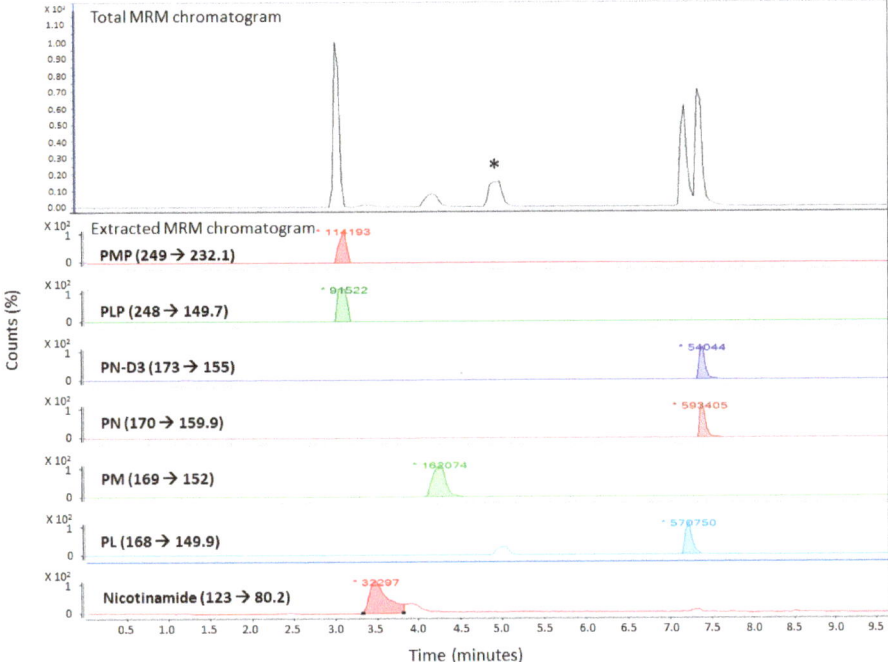

Figure 1. LC-MS/MS chromatogram of all analytes in the simulated plasma solution in the optimized LC-MS/MS method. (* refers to a peak which most likely is an isomer of PL as it is only seen in the extracted MRM chromatogram of PL (MRM 168→149.9).).

Table 2. Percentage recovery of all analytes in simulated plasma using trichloroacetic acid (TCA) and 50 °C (5 min) incubation.

Analyte	Recovery (%)	Standard Deviation
PLP	121.3	19.2
PM	82.5	1.8
PL	118.0	25.6
PN	102.8	2.6
PMP	100.8	34.1
Nicotinamide	83.6	9.2
PN-D3	95.2	2.6

Table 3. The lower limit of detection (LLOD) and the lower limit of quantification (LLOQ) of each analyte.

Analyte	LLOD (pg)	LLOQ (pg)
PLP	0.66	56.0
PL	6.0	18.4
PMP	6.0	18.4
PM	18.0	56.0
PN	2.0	6.0
Nicotinamide	4400	4400

This optimized, robust, and relatively simple method was then used to screen plasma samples from 57 healthy and 57 obese Emirati volunteers. Table 4 shows the concentrations of the vitamers that were detected in the human plasma samples of healthy patients. Surprisingly, we were not able to detect pyridoxamine (PM) in any of the samples, while the other five vitamers were detected with varying concentrations. Table 4 (and Supplementary Figure S1a) also shows that the average concentration of the pyridoxamine-5′-phosphate (PMP) analyte in all 57 samples was around 30 nM, and the patient samples showed an even distribution around this average. The PLP analyte was only detected in the plasma samples of 14 patients, with an average concentration of 36 nM (Supplementary Figure S1b). On the other hand, the PN and pyridoxal hydrochloride (PL) analytes were detected in the plasma samples of all 57 patients, with average concentrations of 21 and 45 nM, respectively, (as shown in Figure S1c,d). Finally, the nicotinamide analyte was only detected in 54 samples, with an average concentration of 850 nM.

Table 4. Quantification of B3 and B6 vitamers in 57 healthy plasma samples.

Sample	PMP	PLP	PN	PM	PL	Nicotinamide
			Concentration (nM)			
1	42.9	n.d. *	25.6	n.d.	20.3	1365.0
2	36.5	35.3	25.0	n.d.	84.3	544.9
3	21.0	n.d.	18.1	n.d.	66.8	2600.6
4	35.9	30.6	13.3	n.d.	53.8	1035.9
5	29.7	n.d.	19.2	n.d.	24.5	2492.7
6	21.4	23.5	14.9	n.d.	33.6	1338.1
7	30.7	17.6	12.7	n.d.	34.3	776.9
8	20.0	n.d.	12.3	n.d.	32.6	949.6
9	18.6	18.8	13.3	n.d.	23.8	2881.2
10	36.3	n.d.	11.3	n.d.	27.5	1138.4
11	21.6	n.d.	9.9	n.d.	31.1	3032.2
12	21.4	n.d.	15.7	n.d.	21.6	588.1
13	13.8	n.d.	17.0	n.d.	22.6	1143.8
14	17.1	38.4	9.5	n.d.	44.5	393.9
15	18.1	n.d.	22.4	n.d.	13.8	884.8
16	28.2	n.d.	23.2	n.d.	51.4	491.0
17	23.7	n.d.	18.4	n.d.	53.3	841.7
18	34.2	n.d.	22.7	n.d.	50.2	286.0
19	33.6	n.d.	16.3	n.d.	44.6	701.4
20	35.0	20.0	20.9	n.d.	58.6	1251.7
21	23.3	n.d.	30.8	n.d.	57.7	458.6
22	30.7	90.5	28.7	n.d.	130.1	n.d.
23	30.9	n.d.	21.6	n.d.	46.7	669.0
24	27.2	30.6	21.6	n.d.	53.6	1381.2
25	38.3	n.d.	19.8	n.d.	42.7	415.4
26	29.3	n.d.	19.4	n.d.	43.9	577.3
27	45.1	n.d.	19.9	n.d.	44.0	863.3
28	19.0	n.d.	21.9	n.d.	11.3	604.3
29	52.2	n.d.	18.3	n.d.	53.0	750.0
30	56.9	n.d.	13.1	n.d.	32.7	1084.5
31	82.3	38.0	17.5	n.d.	60.1	825.5
32	47.4	n.d.	21.0	n.d.	38.5	1165.4
33	51.7	n.d.	19.5	n.d.	32.2	1764.3
34	21.4	n.d.	22.0	n.d.	46.5	2390.2
35	26.6	18.8	19.4	n.d.	37.7	6005.1
36	35.9	68.9	19.1	n.d.	95.9	3620.3
37	41.2	n.d.	19.1	n.d.	32.5	830.9
38	27.4	n.d.	22.2	n.d.	42.0	965.8
39	13.8	n.d.	19.9	n.d.	42.0	1705.0
40	11.3	n.d.	22.8	n.d.	39.7	690.6
41	13.8	n.d.	20.3	n.d.	41.2	572.7
42	10.9	n.d.	21.7	n.d.	47.9	793.1
43	20.2	n.d.	21.0	n.d.	32.9	2303.8
44	26.6	n.d.	17.5	n.d.	48.9	n.d.
45	14.6	n.d.	19.3	n.d.	50.7	566.5

Table 4. Cont.

Sample	PMP	PLP	PN	PM	PL	Nicotinamide
			Concentration (nM)			
46	19.2	n.d.	22.1	n.d.	52.1	1348.9
47	35.9	37.6	24.3	n.d.	88.8	1980.1
48	15.9	n.d.	26.6	n.d.	43.3	n.d.
49	63.3	n.d.	27.1	n.d.	41.2	588.1
50	16.7	n.d.	28.2	n.d.	52.3	372.3
51	28.4	n.d.	26.0	n.d.	49.5	755.4
52	22.7	n.d.	30.1	n.d.	52.2	1429.8
53	21.4	n.d.	31.8	n.d.	49.0	302.1
54	55.9	35.6	32.4	n.d.	58.2	388.5
55	29.9	n.d.	33.2	n.d.	48.4	275.2
56	43.1	n.d.	28.5	n.d.	36.0	825.5
40	11.3	n.d.	22.8	n.d.	39.7	690.6
Average	30.2	36.0	21.0	n.d	45.8	1206.5
Max	82.3	90.5	33.2	n.d	130.1	6005.1
Min	10.9	n.d.	9.5	n.d	11.3	275.2

* n.d.: not detected.

Similarly, a total of 57 human plasma samples of obese patients were also analyzed for the concentrations of the analytes in these samples. Table 5 (and Supplementary Figure S2) shows the analysis results of five analytes (PMP, PLP, PN, PL, and nicotinamide), while PM (just like in the healthy population) was not detected in any of the obese plasma samples. All plasma samples showed variable concentrations of PMP, with an average value of 50 nM, as shown in Figure S2a. On the other hand, the concentration of PLP was only detected in the plasma samples of eight patients with an average value of 37 nM (Figure S2b). The concentrations of PN and PL were detected in the plasma samples of all obese patients, with average concentrations of 21 and 61 nM, respectively (Figure S2c,d). Finally, the concentration of nicotinamide was detected in a total of 45 samples, with an average value of 3700 nM.

Table 5. Quantification of B3 and B6 vitamers in 57 obese plasma samples.

Sample	PMP	PLP	PN	PM	PL	Nicotinamide
			Concentration (nM)			
1	12.0	n.d. *	24.7	n.d.	51.5	5471.0
2	34.4	n.d.	20.8	n.d.	n.d.	1823.7
3	13.6	n.d.	21.7	n.d.	48.4	1348.9
4	n.d.	n.d.	39.5	n.d.	n.d.	n.d.
5	64.3	26.2	20.9	n.d.	114.9	2314.6
6	27.8	n.d.	20.1	n.d.	58.3	1246.3
7	69.1	n.d.	21.0	n.d.	46.4	3118.6
8	14.2	n.d.	26.3	n.d.	62.2	1516.1
9	79.2	29.8	25.8	n.d.	74.7	2514.3
10	37.1	n.d.	23.8	n.d.	57.0	n.d.
11	36.1	30.2	27.2	n.d.	41.7	n.d.
12	28.7	n.d.	24.5	n.d.	45.4	n.d.
13	46.4	n.d.	23.4	n.d.	57.9	n.d.
14	15.0	n.d.	23.6	n.d.	n.d.	2044.9
15	14.4	n.d.	30.8	n.d.	n.d.	2217.5
16	24.1	n.d.	25.8	n.d.	n.d.	2854.2
17	31.3	36.8	21.2	n.d.	70.0	3177.9
18	77.9	n.d.	35.3	n.d.	63.1	2082.6
19	15.9	69.3	28.7	n.d.	95.1	3679.7
20	33.2	n.d.	24.4	n.d.	41.2	3323.6

Table 5. Cont.

Sample	PMP	PLP	PN	PM	PL	Nicotinamide
			Concentration (nM)			
21	20.6	n.d.	26.1	n.d.	57.6	n.d.
22	24.3	61.5	21.3	n.d.	97.3	1818.3
23	22.5	n.d.	20.3	n.d.	41.4	5956.5
24	18.6	n.d.	21.1	n.d.	41.1	6167.0
25	59.8	n.d.	28.0	n.d.	40.7	2271.5
26	42.9	n.d.	29.1	n.d.	86.2	1645.6
27	18.3	n.d.	22.4	n.d.	41.5	4526.8
28	33.2	n.d.	25.0	n.d.	40.0	3275.0
29	21.2	n.d.	23.4	n.d.	70.1	4402.7
30	16.1	n.d.	20.6	n.d.	54.3	3258.8
31	77.3	n.d.	12.6	n.d.	56.7	5287.5
32	32.4	22.7	11.8	n.d.	62.6	8810.7
33	94.8	n.d.	14.3	n.d.	59.0	2401.0
34	86.0	n.d.	14.8	n.d.	35.8	1494.5
35	38.8	n.d.	24.3	n.d.	93.9	5406.2
36	243.9	n.d.	18.0	n.d.	63.6	9625.4
37	138.3	n.d.	20.7	n.d.	65.3	11,303.4
38	44.3	n.d.	12.1	n.d.	76.3	5109.5
39	100.6	n.d.	12.0	n.d.	58.3	4337.9
40	95.4	n.d.	6.7	n.d.	75.6	3345.2
41	101.0	n.d.	n.d.	n.d.	58.3	4105.9
42	47.8	n.d.	20.7	n.d.	n.d.	n.d.
43	25.6	n.d.	17.6	n.d.	n.d.	3998.0
44	34.8	n.d.	21.8	n.d.	n.d.	3064.6
45	90.9	n.d.	16.4	n.d.	n.d.	3933.3
46	168.2	27.8	21.3	n.d.	60.6	1386.6
47	17.5	n.d.	19.2	n.d.	n.d.	7181.3
48	27.2	n.d.	12.7	n.d.	25.2	10,693.7
49	191.9	n.d.	7.7	n.d.	n.d.	372.3
50	65.3	n.d.	13.5	n.d.	n.d.	2832.6
51	53.8	n.d.	15.4	n.d.	n.d.	2498.1
52	19.8	n.d.	14.0	n.d.	n.d.	3884.7
53	39.0	n.d.	7.8	n.d.	n.d.	3965.6
54	87.2	n.d.	12.9	n.d.	n.d.	933.4
55	26.8	n.d.	9.5	n.d.	n.d.	2476.5
56	54.6	n.d.	11.4	n.d.	n.d.	2633.0
57	23.7	n.d.	8.8	n.d.	n.d.	3539.4
Average	53.2	38.0	20.1	n.d.	60.2	3733.5
Max	243.9	69.3	39.5	n.d.	114.9	11,303.4
Min	12.0	n.d.	6.7	n.d.	n.d	372.3

* n.d.: not detected.

A comparative analysis was also carried out between the concentrations of the analytes found in the plasma samples of the healthy and obese patients in order to see if there were any significant differences between the two Emirati populations. As can be seen in Table 6 and Figure 2, obese Emirati patients showed significantly higher average concentrations of PMP ($p = 0.002$), PL ($p = 0.0006$), and nicotinamide ($p < 0.0001$) than healthy patients, as judged by a Student's t-test analysis. No significant differences in the plasma concentrations of PLP and PN were observed between the healthy and obese Emirati populations. The exact implications and reasons behind these significant differences are not clear. However, it is well known that ever since the late 1930s, when the fortification of foods and supplements by synthetic vitamins started, vitamin intake has significantly increased. The vitamin paradox in obesity may reflect excess vitamin intake rather than vitamin deficiency given that there is a correlation between high vitamin intake and increased obesity [47]. These interesting differences could also be related to various factors, such as vitamin supplementation, higher caloric and food intake, other metabolic imbalances, and single nucleotide polymorphisms or genetic differences. The exact nature and cause of these findings will be carefully examined in future studies involving larger population sizes. Additionally, these data could also be correlated with gender, body weight, and other plasma parameters, such as complete blood count, HbA1c, interleukins, and liver function enzymes. Nevertheless, this initial pilot study points to the need and value of studying the plasma concentration of vitamers in metabolically distinct populations.

Table 6. Concentration of various B6 vitamers and nicotinamide in healthy and obese Emirati populations.

Analyte (nM)	Healthy Plasma			Obese Plasma		
	Average	Max	Min	Average	Max	Min
PMP	30.2	82.3	10.9	53.2	243.9	12.0
PLP	36.0	90.5	n.d. *	38.0	69.3	n.d.
PN	21.0	33.2	9.5	20.1	39.5	6.7
PM			n.d.			
PL	45.8	130.1	11.3	60.2	114.9	n.d.
Nicotinamide	1206.5	6005.1	275.2	3733.5	11303.4	372.3

* n.d.: not detected.

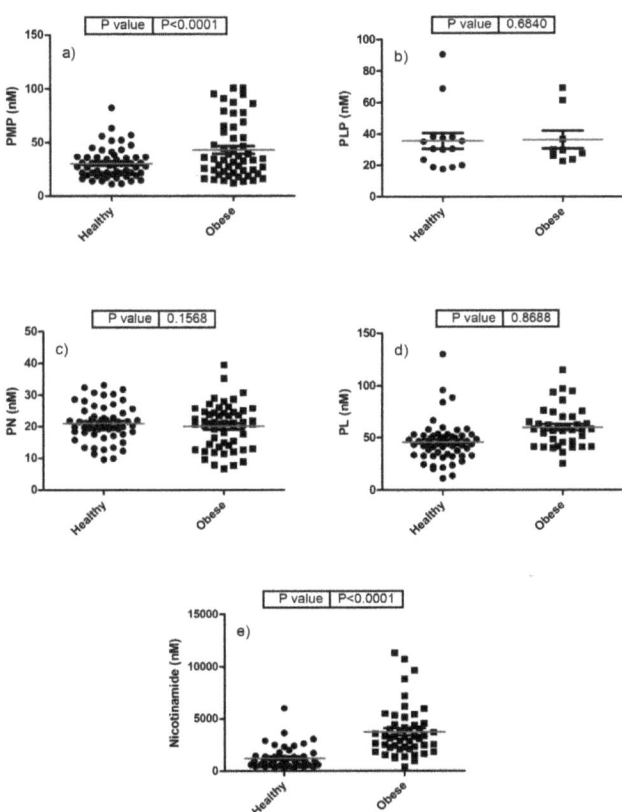

Figure 2. Comparative analysis of various B6 vitamers and B3 (nicotinamide) in healthy and obese Emirati populations (panels (a–e)) refer to PMP, PLP, PN, PL, and nicotinamide, respectively.

We also compared our data of the healthy population with two representative data sets of vitamin concentrations in healthy human plasma previously published by Redeuil et al. in 2015 [41] and Midttun et al. in 2005 [48] for US and European populations, respectively (Table 7). Compared with an average normal concentration of 4.9 nM PMP for the healthy US population [41] and below the detection limit for the European population [48], we found higher PMP concentrations in healthy and obese Emirati populations, with values of 30.2 and 53.2 nM of PMP, respectively. A similar pattern was also observed for nicotinamide, where our values in the healthy and obese plasma samples (1206.5 and 3733.5 nM, respectively) were higher than the published average value of 274.4 nM for the US samples (Tables 6 and 7). On the other hand, the average concentrations of PLP (36 and 38 nM in the healthy and obese Emirati populations, respectively) were lower than the US population (average 92 nM). However, the European population had plasma PLP concentrations (34.4 nM) that were very similar to the UAE population. Similarly, PN, and PL were found to be lower in the plasma samples of our healthy population, as compared with those in the US population that had higher levels than the Europeans. It is interesting to note that only two of the vitamin B6 vitamers (namely PLP and PL) were detected in the plasma samples of the European healthy population, as analyzed by Midttun et al. [48]. Again, the reasons for these interesting and significant differences are most-likely multi-factorial and involve polymorphisms, dietary, and lifestyle factors. Future studies with larger cohorts of patients will need to be undertaken to establish the physiological ranges of these and other vitamers for different population groups.

Table 7. Comparison between the concentrations of all analytes in the plasma samples of a healthy population and those mentioned in the literature.

Analyte (nM)	Healthy Emirati Plasma			US Population [41]			European Population [48]		
	Average	Max	Min	Average	Max	Min	Average	Max	Min
PMP	30.2	82.3	10.9	4.9	7.6	2.1	Not detected		
PLP	36.0	90.5	n.d.	92.1	163.3	20.9	34.4	102.3	17.0
PN	21.0	33.2	9.5	142.8	285.4	0.2	Not detected		
PM	Not detected			4.1	7.7	0.4	Not detected		
PL	45.8	130.1	11.3	118.4	233.5	3.2	9.9	28.2	5.7
Nicotinamide	1206.5	6005.1	275.2	274.4	479.6	69.1	Not included		

3. Materials and Methods

3.1. Materials

Vitamin standards and other reagents were purchased from different suppliers as follows: pyridoxal-5′-phosphate hydrate, pyridoxal hydrochloride, pyridoxamine-5′-phosphate, nicotinamide, phosphate-buffered saline, Tris(2-carboxyethyl) phosphate hydrochloride, heptafluorobutyric acid, trichloroacetic acid, formic acid, and LC-MS-grade water were purchased from Sigma-Aldrich, St. Louis, MO, USA. Pyridoxine hydrochloride was purchased from Supelco, Sigma-Aldrich, St. Louis, MO, USA and pyridoxamine dihydrochloride was purchased from Fluka, Fisher Scientific, Waltham, MA, USA. HPLC-grade acetonitrile was purchased from Merck, Sigma-Aldrich, St. Louis, MO, USA.

3.2. Preparation of Standard Solutions

Individual stock solutions of B6 and B3 vitamers, as well as the internal standard, were prepared at 1000 ppm (1 µg/mL) in deionized water. These stock solutions were kept in Eppendorf tubes and stored at −80 °C to avoid degradation. Working solutions of vitamins standards were prepared daily by mixing and diluting individual stock solutions in deionized water to desired concentrations (three-fold serially diluted starting from 360 ng/mL). Preparation steps were protected from light during laboratory handling by using amber tubes to prevent vitamins from degradation.

3.3. Plasma Sample Extraction Method

Plasma samples from the test subjects were stored at −80 °C were thawed right before the analysis. An aliquot of 300 µL was taken into an Eppendorf tube and spiked with 10 µL of the internal standard (100 ppm), and then the mixture was vortexed for 2 min. The proteins were precipitated by adding an equal volume of 0.6 N trichloroacetic acid (TCA) to produce a final TCA concentration of 0.3 N. The samples were vortexed for 2 min and then incubated for 5 min at 50 °C. The samples were then centrifuged at 11,000 rpm for 10 min at 4 °C. The resulting supernatant was filtered using a CA (Cellulose Acetate) filter (0.22 µm), transferred into HPLC amber vials, and then placed in an autosampler where the samples were kept at 4 °C and protected from light. Normally, 8 µL of each sample extract was injected into the LC-MS/MS system.

3.4. Liquid Chromatography and Mass Spectrometry

The LC separation of vitamins was achieved with an Agilent 1260 HPLC system on a reversed-phase column Poroshell 120 EC-C18 (Agilent Technology, Santa Clara, CA, USA) with a particle size of 2.7 µm, an inner diameter of 3.0 mm, and a length of 100 mm. The column was maintained at 35 °C and a constant flow rate of 0.4 mL/min. Two mobile phases were used: A was LC-MS-grade water containing 0.1% formic acid, and 0.1% heptafluorobutyric acid; B, which was acetonitrile containing 0.1% formic acid. The LC method was set as follows: 3 min of 100% A, followed by a 0–100% gradient of B for 3–5 min, then 100% of B for 5–8:5 min, finally 100% A for 8:6–10 min, and finally by 100% A for 5 min as a post run. The mass spectrometry analysis was performed on an Agilent 6420 Triple Quadrupole MS system in positive electrospray ionization (ESI$^+$) mode. The electrospray voltage was set at 4 kV, the ion source gas 1 (a desolvation gas consisting of nitrogen 99.9%) pressure was set at 20 psi, the ion source gas 2 (a nebulizer gas consisting of nitrogen) was set at 45 psi, and the drying gas (N_2) flow was 8 L/min at 325 °C. Table 6 shows the precursor and product ions, along with their collision energies.

3.5. Study Design and Sample Collection

Study participants (57 healthy and 57 obese Emiratis) were recruited from the local Tawam Hospital in Al Ain (UAE), and signed consent forms were obtained from all the volunteers, as per the UAE University ethical approval protocol number (UAEU Ref# 09/70). The demographics and BMI (Body Mass Index) values of the volunteers were as follows: the healthy group was comprised 53 females and 4 males with an average age of 33 years (minimum age = 18; maximum age = 55), with an average BMI = 30.9 ± 0.8; the obese group was comprised of 56 females and 1 male with an average age of 35 years (minimum age = 18; maximum age = 65), with an average BMI = 33.9 ± 0.3). Plasma was prepared immediately from 10 mL of blood drawn from fasting volunteers and then stored in −80 °C. For the purpose of this preliminary study, a margin of error of 11% and a confidence level of 90% were chosen, which corresponded to an ideal sample size of 56 volunteers for each group (Raosoft sample size calculator).

4. Conclusions

In summary, the results presented here summarize the development of a rapid, sensitive, and robust LC-MS/MS-based assay for the simultaneous quantification of six different vitamers in human plasma. The method involves the simple, single step precipitation-based extraction of vitamins from human plasma for the subsequent analysis by an MRM-based LC-MS/MS method. This technique was subsequently used to analyze plasma samples taken from 57 healthy and 57 obese Emirati patients from a local hospital. We observed significant differences in the plasma vitamin B6 and B3 concentrations between the healthy and obese Emirati samples. Additionally, our results showed that B6 vitamers, as well as nicotinamide concentrations in the healthy Emirati population, were significantly different than those published in the literature for Western populations. The reasons behind this interesting finding will be the focus of future studies. It will also be interesting to see if the increased levels of

B6 vitamers and vitamin B3 are correlated with any physiological imbalances or disease states in obese patients.

Supplementary Materials: The following are available online. Figure S1: Concentrations of PMP, PL, PN, PL and Nicotinamide in healthy Emirati population (n = 57), Figure S2: Concentrations of PMP, PL, PN, PL and Nicotinamide in obese Emirati population (n = 57), Table S1: Effect of incubation temperature on TCA-precipitated release of vitamers from spiked simulated plasma, Table S2: Intra-day and inter-day accuracy, precision and linear range of the partially validated LC-MSMS method.

Author Contributions: S.S.A., I.S., and S.G. conceived and designed the experiments. G.R.I., S.S.A., and I.S. did the analytical work, whereas S.G. and J.Y. were in charge of patient recruitment and sample collection. S.S.A. and G.R.I. wrote the first draft of the manuscript. J.B., I.S., and S.G. helped with the data analysis and finalizing the manuscript. All authors have read and agreed to the published version of the manuscript.

Funding: The authors acknowledge funding from UAEU College of Graduate Studies which partially supported the work reported here.

Conflicts of Interest: All the authors declare no conflicts of interest.

Abbreviations

PL	Pyridoxal
PLP	Pyridoxal 5′-phosphate
PM	Pyridoxamine
PMP	Pyridoxamine 5′-phosphate
PN	Pyridoxine; PN-d3: Pyridoxine - (methyl-d3)
TCA	Trichloroacetic acid
n.d.	Not detected
LC-MS/MS	Liquid Chromatography tandem Mass Spectrometry
CVD	Cardiovascular disease
BMI	Body Mass Index

References

1. Furdui, C.; Ragsdale, S.W. The role of pyruvate ferredoxin oxidoreductase in pyruvate synthesis during autotrophic growth by the Wood-Ljungdahl path-way. *J. Biol. Chem.* **2000**, *275*, 28494–28499. [CrossRef] [PubMed]
2. Manzetti, S.; Zhang, J.; Van Der Spoel, D. Thiamin function, metabolism, uptake, and transport. *Biochemistry* **2014**, *53*, 821–835. [PubMed]
3. Kotloski, N.J.; Gralnick, J.A. Flavin electron shuttles dominate extracellular electron transfer by Shewanella oneidensis. *mBio* **2013**, *4*, e00553-12. [CrossRef]
4. Velasquez-Orta, S.B.; Head, I.M.; Curtis, T.; Scott, K.; Lloyd, J.R.; Von Canstein, H. The effect of flavin electron shuttles in microbial fuel cells current production. *Appl. Microbiol. Biotechnol.* **2009**, *85*, 1373–1381. [CrossRef] [PubMed]
5. Selhub, J. Folate, vitamin B12 and vitamin B6 and one carbon metabolism. *J. Nutr. Heal. Aging* **2002**, *6*, 39–42.
6. Clayton, P.T. B6-responsive disorders: A model of vitamin dependency. *J. Inherit. Metab. Dis.* **2006**, *29*, 317–326. [CrossRef]
7. Wang, L.; Li, H.; Zhou, Y.; Jin, L.; Liu, J. Low-dose B vitamins supplementation ameliorates cardiovascular risk: A double-blind randomized controlled trial in healthy Chinese elderly. *Eur. J. Nutr.* **2014**, *54*, 455–464. [CrossRef]
8. Selhub, J.; Morris, M.S.; Jacques, P.F.; Rosenberg, I.H. Folate-vitamin B-12 inter-action in relation to cognitive impairment, anemia, and biochemical indicators of vitamin B-12 deficiency. *Am. J. Clin. Nutr.* **2009**, *89*, 702S–706S. [CrossRef]
9. Czeizel, A.E.; Dudás, I.; Vereczkey, A.; Bánhidy, F. Folate deficiency and folic acid supplementation: The prevention of neural-tube defects and congenital heart defects. *Nutrients* **2013**, *5*, 4760–4775. [CrossRef]
10. Heseker, H. Folic acid and other potential measures in the prevention of neural tube defects. *Ann. Nutr. Metab.* **2011**, *59*, 41–45. [CrossRef]

11. Bryan, J.; Calvaresi, E.; Hughes, D. Short-Term Folate, Vitamin B-12 or Vitamin B-6 supplementation slightly affects memory performance but not mood in women of various ages. *J. Nutr.* **2002**, *132*, 1345–1356. [CrossRef] [PubMed]
12. Calvaresi, E.; Bryan, J. B vitamins, cognition, and aging: A review. *J. Gerontol. Ser. B* **2001**, *56*, P327–P339. [CrossRef] [PubMed]
13. Fletcher, R.H.; Fairfield, K.M. Vitamins for chronic disease prevention in adults, clinical applications. *JAMA* **2002**, *287*, 3127–3129. [CrossRef] [PubMed]
14. Dunn-Lewis, C.; Kraemer, W.J.; Kupchak, B.R.; Kelly, N.A.; Creighton, B.A.; Luk, H.-Y.; Ballard, K.D.; Comstock, B.A.; Szivak, T.K.; Hooper, D.R.; et al. A multi-nutrient supplement reduced markers of inflammation and improved physical performance in active individuals of middle to older age: A randomized, double-blind, placebo-controlled study. *Nutr. J.* **2011**, *10*, 90. [CrossRef]
15. Fairfield, K.M.; Fletcher, R.H. Vitamins for chronic disease prevention in adults. *JAMA* **2002**, *287*, 3116–3126. [CrossRef]
16. Tully, D.B.; Allgood, V.E.; Cidlowski, J.A. Modulation of steroid receptor-mediated gene expression by vitamin B6. *FASEB J.* **1994**, *8*, 343–349. [CrossRef]
17. Morris, M.S.; Picciano, M.F.; Jacques, P.F.; Selhub, J. Plasma pyridoxal 5'-phosphate in the US population: The National Health and Nutrition Examination Survey, 2003–2004. *Am. J. Clin. Nutr.* **2008**, *87*, 1446–1454. [CrossRef]
18. Friso, S.; Girelli, D.; Martinelli, N.; Olivieri, O.; Lotto, V.; Bozzini, C.; Pizzolo, F.; Faccini, G.; Beltrame, F.; Corrocher, R. Low plasma vitamin B-6 concentrations and modulation of coronary artery disease risk. *Am. J. Clin. Nutr.* **2004**, *79*, 992–998. [CrossRef]
19. Rimm, E.B.; Willett, W.C.; Hu, F.B.; Sampson, L.; Colditz, G.A.; E Manson, J.; Hennekens, C.; Stampfer, M.J. Folate and vitamin B6 from diet and supplements in relation to risk of coronary heart disease among women. *JAMA* **1998**, *279*, 359–364. [CrossRef]
20. Folsom, A.; Desvarieux, M.; Nieto, F.J.; Boland, L.L.; Ballantyne, C.M.; Chambless, L.E. B vitamin status and inflammatory markers. *Atherosclerosis* **2003**, *169*, 169–174. [CrossRef]
21. Friso, S.; Jacques, P.F.; Wilson, P.W.; Rosenberg, I.H.; Selhub, J. Low circulating vitamin B(6) is associated with elevation of the inflammation marker C-reactive protein independently of plasma homocysteine levels. *Circulation* **2001**, *103*, 2788–2791. [CrossRef] [PubMed]
22. James, S.; Vorster, H.H.; Venter, C.S.; Kruger, H.S.; Nell, T.A.; Veldman, F.J.; Ubbink, J.B. Nutritional status influences plasma fibrinogen concentration: Evidence from the THUSA survey. *Thromb. Res.* **2000**, *98*, 383–394. [CrossRef]
23. Okada, M.; Shibuya, M.; Yamamoto, E.; Murakami, Y. Effect of diabetes on vitamin B6 requirement in experimental animals. *Diabetes Obes. Metab.* **1999**, *1*, 221–225. [CrossRef] [PubMed]
24. Roubenoff, R.; Roubenoff, R.A.; Selhub, J.; Nadeau, M.R.; Cannon, J.G.; Freeman, L.M.; Dinarello, C.A.; Rosenberg, I.H. Abnormal vitamin b6status in rheumatoid cachexia association with spontaneous tumor necrosis factor α production and markers of inflammation. *Arthritis Rheum.* **1995**, *38*, 105–109. [CrossRef] [PubMed]
25. Saibeni, S.; Cattaneo, M.; Vecchi, M.; Zighetti, M.L.; Lecchi, A.; Lombardi, R.; Meucci, G.; Spina, L.; de Franchis, R. Low vitamin B(6) plasma levels, a risk factor for thrombosis, in inflammatory bowel disease: Role of inflammation and correlation with acute phase reactants. *Am. J. Gastroenterol.* **2003**, *98*, 112–117. [CrossRef]
26. Shen, J.; Lai, C.-Q.; Mattei, J.; Ordovás, J.M.; Tucker, K.L. Association of vitamin B-6 status with inflammation, oxidative stress, and chronic inflammatory conditions: The Boston Puerto Rican Health Study. *Am. J. Clin. Nutr.* **2009**, *91*, 337–342. [CrossRef]
27. Wan, P.; Moat, S.; Anstey, A. Pellagra: A review with emphasis on photosensitivity. *Br. J. Dermatol.* **2011**, *164*, 1188–1200. [CrossRef]
28. Villines, T.C.; Kim, A.S.; Gore, R.S.; Taylor, A.J. Niacin: The evidence, clinical use, and future directions. *Curr. Atheroscler. Rep.* **2011**, *14*, 49–59. [CrossRef]
29. Bruckert, E.; Labreuche, J.; Amarenco, P. Meta-analysis of the effect of nicotinic acid alone or in combination on cardiovascular events and atherosclerosis. *Atherosclerosis* **2010**, *210*, 353–361. [CrossRef]
30. Taylor, A.J.; Lee, H.J.; Sullenberger, L.E. The effect of 24 months of combination statin and extended-release niacin on carotid intima-media thickness: ARBITER 3. *Curr. Med. Res. Opin.* **2006**, *22*, 2243–2250. [CrossRef]

31. Taylor, A.J.; Villines, T.C.; Stanek, E.J.; Devine, P.J.; Griffen, L.; Miller, M.; Weissman, N.J.; Turco, M. Extended-release niacin or ezetimibe and carotid intima–media thickness. *N. Engl. J. Med.* **2009**, *361*, 2113–2122. [CrossRef] [PubMed]
32. Lukasova, M.; Hanson, J.; Tunaru, S.; Offermanns, S. Nicotinic acid (niacin): New lipid-independent mechanisms of action and therapeutic potentials. *Trends Pharmacol. Sci.* **2011**, *32*, 700–707. [CrossRef] [PubMed]
33. Roy, B.; Singh, B.; Rizal, A.; Malik, C.P. Bioanalytical method development and validation of niacin and nicotinuric acid in human plasma by LC-MS/MS. *Int. J. Pharm. Clin. Res.* **2014**, *6*, 206–213.
34. Cohen, K.; Gorecki, G.; Silverstein, S.; Ebersole, J.; Solomon, L. Effect of pyridoxine (vitamin B6) on diabetic patients with peripheral neuropathy. *J. Am. Podiatr. Med. Assoc.* **1984**, *74*, 394–397. [CrossRef] [PubMed]
35. Ellis, J.M.; Folkers, K.; Minadeo, M.; VanBuskirk, R.; Xia, L.-J.; Tamagawa, H. A deficiency of vitamin B6 is a plausible molecular basis of the retinopathy of patients with diabetes mellitus. *Biochem. Biophys. Res. Commun.* **1991**, *179*, 615–619. [CrossRef]
36. Hamaker, B.R.; Kirksey, A.; Borschel, M.W. Distribution of B-6 vitamers in human milk during a 24-h period after oral supplementation with different amounts of pyridoxine. *Am. J. Clin. Nutr.* **1990**, *51*, 1062–1066. [CrossRef]
37. Taguchi, K.; Fukusaki, E.; Bamba, T. Determination of niacin and its metabolites using supercritical fluid chromatography coupled to tandem mass spectrometry. *Mass Spectrom.* **2014**, *3*, A0029. [CrossRef]
38. Hamaker, B.; Kirksey, A.; Ekanayake, A.; Borschel, M. Analysis of B-6 vitamers in human milk by reverse-phase liquid chromatography. *Am. J. Clin. Nutr.* **1985**, *42*, 650–655. [CrossRef]
39. Heydari, R.; Elyasi, N.S. Ion-pair cloud-point extraction: A new method for the determination of water-soluble vitamins in plasma and urine. *J. Sep. Sci.* **2014**, *37*, 2724–2731. [CrossRef]
40. Hampel, D.; York, Y.R.; Allen, L.H. Ultra-performance liquid chromatography tandem mass-spectrometry (UPLC-MS/MS) for the rapid, simultaneous analysis of thiamin, riboflavin, flavin adenine dinucleotide, nicotinamide and pyridoxal in human milk. *J. Chromatogr. B* **2012**, *903*, 7–13. [CrossRef]
41. Redeuil, K.M.; Redeuil, K.; Bénet, S.; Munari, C.; Giménez, E.C. Simultaneous quantification of 21 water soluble vitamin circulating forms in human plasma by liquid chromatography-mass spectrometry. *J. Chromatogr. A* **2015**, *1422*, 89–98. [CrossRef] [PubMed]
42. Blanchard, J. Evaluation of the relative efficacy of various techniques for deprotenizing plasma samples prior to high performance liquid chromatographic analysis. *J. Chromatogr. B* **1981**, *226*, 455–460. [CrossRef]
43. Polson, C.; Sarkar, P.; Incledon, B.; Raguvaran, V.; Grant, R. Optimization of protein precipitation based upon effectiveness of protein removal and ionization effect in liquid chromatography-tandem mass spectrometry. *J. Chromatogr. B* **2003**, *785*, 263–275. [CrossRef]
44. Ubbink, J.B.; Serfontein, W.J.; de Villiers, L.S. Analytical Recovery of protein-bound pyridoxal-5′-phosphate in plasma analysis. *J. Chromatogr. B* **1986**, *375*, 399–404. [CrossRef]
45. Lumeng, L.; Brashear, R.E.; Li, T.K. Pyridoxal 5′-phosphate in plasma: Source, protein-binding, and cellular transport. *J. Lab. Clin. Med.* **1974**, *84*, 334–343.
46. Bates, C.J.; Pentieva, K.D.; Matthews, N.; Macdonald, A. A simple, sensitive and reproducible assay for pyridoxal 5′-phospate and 4-pyridoxic acid in human plasma. *Clin. Chem. Acta* **1999**, *280*, 101–111. [CrossRef]
47. Zhou, S.-S.; Li, D.; Chen, N.-N.; Zhou, Y. Vitamin paradox in obesity: Deficiency or excess? *World J. Diabetes* **2015**, *6*, 1158–1167. [CrossRef]
48. Midttun, Ø.; Hustad, S.; Solheim, E.; Schneede, J.; Ueland, P.M. Multianalyte Quantification of Vitamin B6 and B2 Species in the nanomolar range in human plasma by liquid chromatography–tandem mass spectrometry. *Clin. Chem.* **2005**, *51*, 1206–1216. [CrossRef]

Sample Availability: Samples and compounds mentioned in the manuscript are available from the authors.

© 2020 by the authors. Licensee MDPI, Basel, Switzerland. This article is an open access article distributed under the terms and conditions of the Creative Commons Attribution (CC BY) license (http://creativecommons.org/licenses/by/4.0/).

Article

Determination of Vitamin B3 Vitamer (Nicotinamide) and Vitamin B6 Vitamers in Human Hair Using LC-MS/MS

Sundus M. Sallabi, Aishah Alhmoudi, Manal Alshekaili and Iltaf Shah *

Department of Chemistry, College of Science, UAE University, Al Ain 15551, United Arab Emirates; 201640052@uaeu.ac.ae (S.M.S.); 201303508@uaeu.ac.ae (A.A.); 201250724@uaeu.ac.ae (M.A.)
* Correspondence: altafshah@uaeu.ac.ae

Abstract: Water-soluble B vitamins participate in numerous crucial metabolic reactions and are critical for maintaining our health. Vitamin B deficiencies cause many different types of diseases, such as dementia, anaemia, cardiovascular disease, neural tube defects, Crohn's disease, celiac disease, and HIV. Vitamin B3 deficiency is linked to pellagra and cancer, while niacin (or nicotinic acid) lowers low-density lipoprotein (LDL) and triglycerides in the blood and increases high-density lipoprotein (HDL). A highly sensitive and robust liquid chromatography–tandem mass spectroscopy (LC/MS-MS) method was developed to detect and quantify a vitamin B3 vitamer (nicotinamide) and vitamin B6 vitamers (pyridoxial 5′-phosphate (PLP), pyridoxal hydrochloride (PL), pyridoxamine dihydrochloride (PM), pridoxamine-5′-phosphate (PMP), and pyridoxine hydrochloride (PN)) in human hair samples of the UAE population. Forty students' volunteers took part in the study and donated their hair samples. The analytes were extracted and then separated using a reversed-phase Poroshell EC-C18 column, eluted using two mobile phases, and quantified using LC/MS-MS system. The method was validated in human hair using parameters such as linearity, intra- and inter-day accuracy, and precision and recovery. The method was then used to detect vitamin B3 and B6 vitamers in the human hair samples. Of all the vitamin B3 and B6 vitamers tested, only nicotinamide was detected and quantified in human hair. Of the 40 samples analysed, 12 were in the range 100–200 pg/mg, 15 in the range 200–500 pg/mg, 9 in the range of 500–4000 pg/mg. The LC/MS-MS method is effective, sensitive, and robust for the detection of vitamin B3 and its vitamer nicotinamide in human hair samples. This developed hair test can be used in clinical examination to complement blood and urine tests for the long-term deficiency, detection, and quantification of nicotinamide.

Keywords: nicotinamide; vitamin B3; vitamin B6; hair analysis; vitamin; vitamers; LC-MS/MS

Citation: Sallabi, S.M.; Alhmoudi, A.; Alshekaili, M.; Shah, I. Determination of Vitamin B3 Vitamer (Nicotinamide) and Vitamin B6 Vitamers in Human Hair Using LC-MS/MS. *Molecules* **2021**, *26*, 4487. https://doi.org/10.3390/molecules26154487

Academic Editors: Gavino Sanna and Stefan Leonidov Tsakovski

Received: 16 June 2021
Accepted: 23 July 2021
Published: 25 July 2021

Publisher's Note: MDPI stays neutral with regard to jurisdictional claims in published maps and institutional affiliations.

Copyright: © 2021 by the authors. Licensee MDPI, Basel, Switzerland. This article is an open access article distributed under the terms and conditions of the Creative Commons Attribution (CC BY) license (https://creativecommons.org/licenses/by/4.0/).

1. Introduction

Vitamin B and its vitamers (metabolites) are water-soluble nutritional elements critical for maintaining cellular metabolism, and cellular homeostasis, mainly as coenzymes. For example, nicotinamide (a vitamer of vitamin B3), pyridoxal 5′-phosphate (a vitamer of vitamin B6), pantothenic acid (vitamin B5), riboflavin (vitamin B2), biotin (vitamin B8), and 5-methyl tetrahydrofolate (a vitamer of vitamin B9) are all involved in neurotransmission and fatty acid synthesis, oxidation/reduction reactions, or one-carbon metabolism [1–4]. Vitamin B is usually obtained from foods such as almonds, whole grains, meat, fish, and leafy greens. However, a deficiency of vitamin B and its vitamers has negative effects on human health. Many diseases have been linked to the deficiency of vitamin B and its vitamers, such as cognitive impairment, anaemia [5,6], cardiovascular disease [7], neural tube defects [8,9], neuropsychiatric disorders [10], and thromboembolic processes [11].

Niacin, or nicotinic acid (a vitamer of vitamin B3), is absorbed by the body when dissolved in water and taken orally. It is converted to the major vitamer niacinamide in the body, along with other minor vitamers such as nicotinamide N-oxide and nicotinuric acid [12]. Niacin and nicotinamide are crucial for all living cells. Vitamin B3 is converted biosynthetically to nicotinamide adenine dinucleotide (NAD+) and nicotinamide adenine dinucleotide phosphate (NADP), oxidising agents that accept an electron from a reducing agent to change to the reduced form, NAD(P)H and are involved in DNA repair, calcium mobilisation, and deacetylation [12,13]. Moreover, NAD can be derived biosynthetically from the amino acid tryptophan through the kynurenine pathway. Niacin is also synthesised endogenously through the kynurenine pathway (see Figure 1) [14–18]. NAD+/NADH mediates redox reactions for energy metabolism and cellular biochemistry. NAD is also required as a coenzyme for the catalysis of oxidoreductases (dehydrogenases) [19]. Furthermore, NAD+/NADH is essential for mediating the electron transport chain, which fuels oxidative phosphorylation in mitochondria. Therefore, energy metabolism in cells is mainly mediated by cofactors derived from vitamin B3 and is involved in the majority of anabolic and catabolic pathways. Diseases caused by vitamin B3 deficiency are pellagra (dermatitis, depression, and diarrhoea) and cancer. Vitamin B3 (niacin form) is also known to lower total cholesterol, bad cholesterol (such as low-density lipoprotein (LDL)), triglycerides, and lipoprotein in the blood [12].

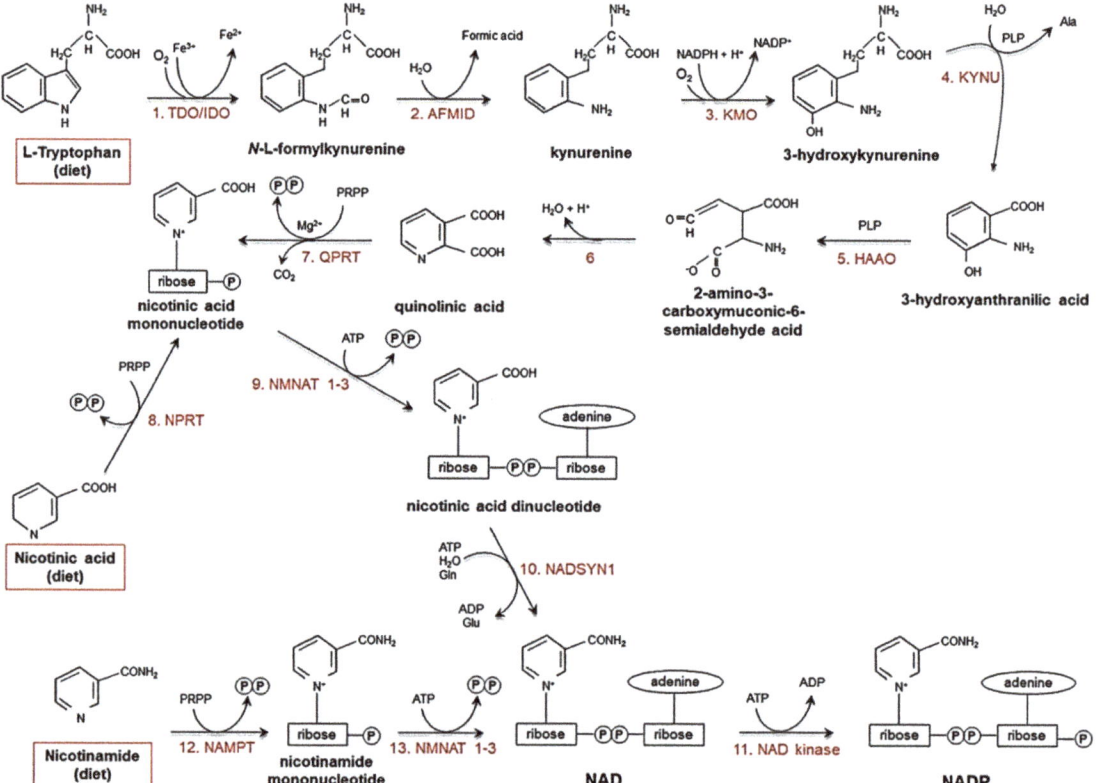

Figure 1. Pathways that biosynthesise NAD(P)H and niacin [16]. Ala: alanine; Glu: glutamate; Gln: glutamine; PRPP: 5-phosphoribosyl-1-pyrophosphate; PLP: pyridoxal phosphate. (Reproduced with permission from [16]).

Vitamin B6, administered as pyridoxine hydrochloride, is used therapeutically to treat pyridoxine-responsive inherited disorders and some types of seizures in neonates and infants and to improve immune function in vitamin-B6-deficient individuals [20]. Pyridoxine has a hydroxymethyl group at four positions; pyridoxamine, an aminomethyl group; and pyridoxal, an aldehyde. Each vitamer may be phosphorylated at its 5-substituent to form an active coenzyme [21]. Pyridoxal 5′-phosphate (PLP) is the most common of these.

It is the active form of vitamin B6 in humans and functions as a cofactor for more than 140 distinct enzyme-catalysed reactions [22].

Recently, many studies have been published on the analysis of vitamins B vitamers in human plasma, serum, and blood using LC/MS-MS. Asante et al. (2018) investigated the relationship between the deficiency of vitamin B vitamers involved in the folate-mediated one-carbon metabolism (FOCM) cycle and the pathogenesis of colorectal cancer, but this study was limited to plasma [23]. Zhang et al. (2019) examined the status of vitamins B1, B2, and B6 using dried blood spots (DBS) collected from Chinese children, but this assay was limited to only blood spots [24]. Roy et al. (2014) successfully quantified niacin and nicotinuric acid in humans, but the application was limited to human plasma [25]. Vitamin B3 vitamers, such as nicotinamide and nicotinuric acid, were also found in humans by Sutherland et al. but only in plasma [26]. Several studies have analysed nicotinamide in different matrices such as food, animal and human blood, serum, and plasma to quantify nicotinamide by LC/MS-MS, but there is no literature on the analysis and quantification of vitamin B vitamers in human hair using LC/MS-MS [27–30].

Zgaga et al. (2019) [31] and another study published by our lab recently [32] analysed and quantified 25-hydroxyvitamin D concentrations in human hair using LC/MS-MS. These are the first two studies to use human hair and beard samples for LC/MS-MS to quantify 25(OH)D3 and to show the existence of 25(OH)D3 in human hair samples. Zgaga et al. (2019) [31] compared the 25(OH)D3 levels in human serum and hair samples. Hair samples showed 25(OH)D3 hormone levels for an extended period, depending on the hair length, whereas serum samples showed short-lived 25(OH)D3 levels. Furthermore, there was a large and small variation in 25(OH)D3 levels between and within-subjects, respectively, as opposed to slightly similar concentrations in serum samples. Therefore, hair testing should be applied as a complementary test and not as a replacement for blood testing for 25(OH)D3 levels for clinical examination [31].

Hair samples could be easily collected (non-invasive), stored and transported without any fear of infection, while blood collection required trained professionals, and storage requires freezing at very low temperatures and special care. Hair analysis could be useful in verifying self-reported histories of nicotinamide deficiency. It has been well documented in literature that hair analysis has a wider window of detection and could show the presence of analytes in a person's body from a few weeks up to a year, while blood and urine tests could detect the presence of analyte from 2–4 days. This could confirm the long-term deficiency of analyte or could show if the deficiency was more recent. Lastly the hair test could be used as a complementary test to blood and urinalysis as the studies revealed [31,33,34].

Given the advantages of hair testing, the objective of this study was to develop and validate an innovative LC/MS-MS method for the accurate detection and quantification of vitamin B3 and B6 vitamer in human hair samples collected from the UAE population. The secondary aim was to use the validated LC/MS-MS assay for accurate measurement of vitamin B3 and B6 vitamer levels in the hair samples collected.

2. Results

2.1. Development of an LC-MS/MS-Based Method

The method of extracting vitamin B3 and B6 vitamers, along with a robust and sensitive LC-MS/MS for vitamer analysis, was developed after investigating many extraction techniques. Next, we will discuss the series of experimental methods that were performed. First, we identified the product ions of each precursor ion of each analyte. Subsequently, the multiplier voltage (Delta EMV) was subjected to variation to check its effect on the peak intensity produced. The effect of different column types on the resolution of the peaks obtained was examined. Lastly, the method reproducibility was tested and tried, in addition to analysing the recovery of all the analytes. Based on the in-depth study of the parameters mentioned above, the method optimisation state was reached and evaluated.

2.2. Multiple Reaction Monitoring (MRM) (Precursors and Product Ion Identification)

Compounds with different structures behave differently to various fragmentor voltage settings. The injection of individual standards enabled the detection of the ideal fragmentation voltage for every single component. The fragmentor was set to positive electrospray ionisation mode. The positively charged molecules of analytes entered the mass spectrometer. Precursor ions were generated, which, by collision-induced dissociation, formed product ions. Both precursor and product ions were measured together in multiple reaction monitoring (MRM) mode. The daughter ions were found based on the mass of the standards at varying collision energies. The values of nicotinamide, pyridoxamine 5′-phosphate (PMP), pyridoxamine (PM) [35], and pyridoxal 5′-phosphate (PLP) [17] matched those in the literature, while the values of pyridoxal (PL) and pyridoxine (PN) were close to those in the literature [35].

PLP showed a precursor ion of m/z 248 with complete fragmentation and produced an intensity signal of the quantifier product ion of m/z 149.7 and qualifier product ion of 94 with collision energies 15 eV. The precursor ion of m/z 168 for PL was completely fragmented and produced an intensity signal of the quantifier product ion of m/z 149.9 and qualifier product ion of 94 with collision energies 10 eV. The PMP quantifier product ion of m/z 232.1 and qualifier ion of 134.1 was detected at collision energies 10 eV. PM showed a quantifier product ion of m/z 152 and qualifier product ion of 134.1 at collision energies 10 eV, produced from the intense precursor ion of m/z 169. The quantifier product ion of m/z 151.9 of PM was the strongest at collision energy 10 eV and it was used for quantitation while the qualifier product ion was 134.1 at 10 eV collision energy. Nicotinamide had a fully fragmented precursor ion of m/z 123 with a high-intensity quantifier product ion of m/z 80.2 and a qualifier product ion of 96 at collision energy 20 eV. The internal standard showed a daughter quantifier ion at 155.1 and a qualifier ion at m/z 137.1 at collision energy 10 eV. Table 1 summarises the precursor and product ion, fragmentor voltage, and collision energy values found for every analyte.

Table 1. Fragmentation parameters identified for all analytes examined in this study. Transitions marked with * were used for quantitation.

Analyte	Precursor Ion (m/z) [M + H]+	Product Ion (m/z)	Fragmentor Voltage (V)	Collision Energy (eV)
PLP (Quantifier)	248 *	149.7	45	15
(Qualifier)	248	94	62	15
PL	168 *	149.9	94	10
	168	94	98	10
PMP	249 *	232.1	45	10
	249	134.1	60	10
PM	169 *	152	94	10
	169	134.1	98	10
PN	170 *	151.9	94	10
	170	134.1	97	10
Nicotinamide	123 *	80.2	94	20
	123	96	90	20
PN-d3	173 *	155	94	10
	173	137.1	90	10

Two columns with distinct dimensions and similar packing materials were used to test the chromatographic separation method. Eclipse plus C-18 was first column tested with a 1.8 μm particle size, 50 mm length, and inner diameter 2.1 mm, while the second column was Poroshell 120 EC-C18 with a 2.7 μm particle size, 100 mm length, and 3.0 mm inner diameter. The investigation was based on testing which column would give the highest resolution and intensity. Both columns showed an almost similar number of peaks, but differences were observed in elution profiles. However, the Poroshell column obtained a chromatographic profile with a separated peak with high intensity compared with the Eclipse plus C-18 column, whose chromatographic profile showed separated, broad peaks [36].

The human hair samples were analysed for estimating and detecting seven different vitamin B3 and B6 vitamers: nicotinamide (vitamin B3 vitamer); PMP, PM, PLP, PN, and PL (vitamin B6 vitamers); and the IS. However, only nicotinamide was detected and quantified in the human hair samples. This suggests that nicotinamide is the only vitamin B3 form that can be quantified in human hair. Figure 2 below shows the chromatographic peaks and MRM transitions of vitamin B3 vitamer, nicotinamide, and vitamin B6 vitamers PMP, PM, PLP, PN, and PL along with the internal standard PN-d3.

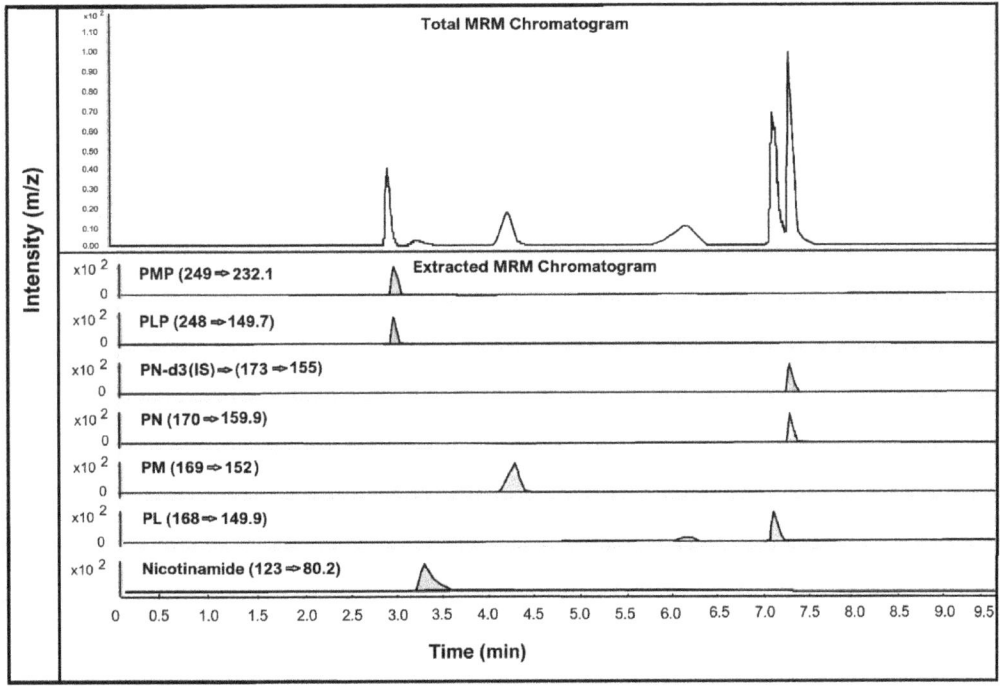

Figure 2. LC-MS chromatogram of all vitamin B3 and vitamin B6 vitamers analysed in human hair. All analytes spiked at a concentration of 150 pg/mg of hair.

The chromatogram in Figure 3 below shows the chromatographic peaks and MRM transitions for nicotinamide in human hair samples, along with the internal standard.

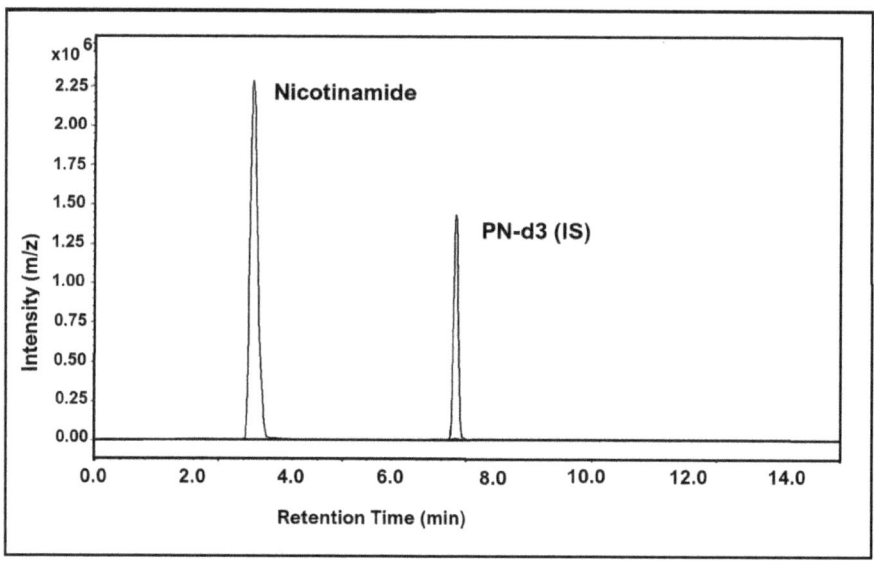

Figure 3. LC-MS/MS chromatogram showing nicotinamide (vitamin B3 vitamer) found in human hair at a concentration of 758.8 pg/mg (sample 27), along with the internal standard peak.

3. Method Validation

The LC-MS/MS validation method was evaluated using the following parameters: linear range, the intra- and inter-day accuracy and precision, recovery LOD and LOQ. The validation results are presented in Table 2. Table 2 summarises the intra- and inter-day precision and accuracy of three QC samples. The intra-day precision and accuracy values were all within the acceptable range. The percentage recovery of the low, medium, and high QC of vitamin B3 and B6 vitamers is also given in the Table 2 below. The recovery ranged between 73% and 89%. The LOD was 10 pg/mg, and LOQ was 50 pg/mg with a linear range from 50 to 4000 pg/mg.

Table 2. LC-MS/MS validation parameters for intra- and inter-day accuracy, precision, and recovery.

Analytes	Conc. QC's (pg/mg)	% Recovery	Intraday (n = 6)		Interday (n = 6)	
			Precision, % CV	Accuracy, %	Precision, % CV	Accuracy, %
Nicotinamide	200	73	11.6	94.2	9.2	91.1
	400	78	6.3	88.2	11.5	86.5
	1200	87	7.5	87.0	8.5	85.6
PLP	200	81	12.5	88.9	12.2	89.8
	400	89	14.3	89.4	13.3	88.9
	1200	91	12.9	89.9	12.3	87.8
PL	200	85	13.5	112.2	14.7	94.5
	400	86	9.1	102.3	10.9	91.6
	1200	78	5.3	100.1	8.9	93.9

Table 2. Cont.

Analytes	Conc. QC's (pg/mg)	% Recovery	Intraday (n = 6)		Interday (n = 6)	
			Precision, % CV	Accuracy, %	Precision, % CV	Accuracy, %
PMP	200	86	14.9	92.8	13.2	100.5
	400	82	13.9	85.6	12.7	86.9
	1200	89	7.2	89.9	13.2	94.3
PM	200	78	11.2	98.3	9.3	85.2
	400	89	8.3	102.6	8.1	87.3
	1200	81	6.1	100.9	13.6	87.1
PN	200	82	13.8	93.3	4.3	101.2
	400	84	7.5	98.7	12.1	89.8
	1200	87	8.2	101.2	13.2	99.9

Analysis of Human Hair Samples

The analysis results for the 40 hair samples collected are given in Table A1. The nicotinamide concentration was 104.9 to 2706.5 pg/mg of hair in the female participants and 106.9 to 3349.9 pg/mg of hair in the male participants.

The minimum nicotinamide concentration was 106.9 pg/mg in males and 104.9 pg/mg in females (Figure 4). The maximum nicotinamide concentration was 3349.9 pg/mg in males (sample 14) and 2706.5 pg/mg in females (sample 4). Quartile 1 values for male and female participants were 171.5 and 187.1 pg/mg, respectively, and quartile 3 values were 443.4 and 905.9 pg/mg, respectively. Mean nicotinamide concentrations for male and female participants were 573.7 and 726.6 pg/mg, respectively. The interquartile range (IQR) was −236.4 to 851.42 pg/mg. The average nicotinamide concentrations in male and female participants are shown in Table A1. In Figure A1, the results are represented as a histogram detailing the nicotinamide ranges for the male and female participants. The graph depicts the sample numbers vs. the nicotinamide concentrations quantified in each sample.

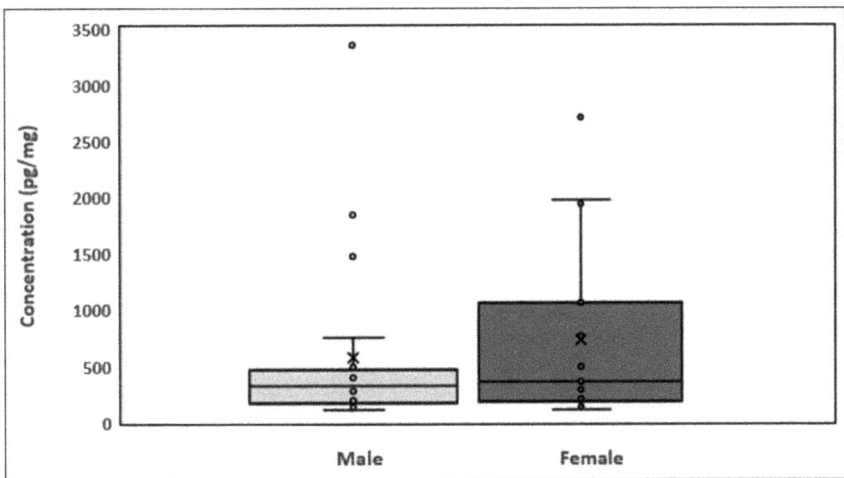

Figure 4. Box plots showing the distribution of nicotinamide in hair and data skewness through data quartiles, "×" shows the mean nicotinamide concentrations and "•" shows the outliers.

4. Discussion

Nicotinamide was found in all hair samples, except one male and three female samples that had no nicotinamide. The exact reason behind these results is unknown. The results between the two genders were compared, and no major differences were found. These findings prove that nicotinamide, a vitamin B3 vitamer, can be easily detected in human hair. Further research is required to detect other vitamin B vitamers compared to blood and urine tests. The results also show the prevalence and distribution of nicotinamide in hair. Of 40 samples, 12 were in the range 100–200 pg/mg, 15 in the range of 200 to 500 pg/mg, 9 in the range of 500–4000 pg/mg, and 4 samples with no nicotinamide.

In this study, we detected nicotinamide, a vitamin B3 vitamer, in human hair. As there are no previous studies on vitamin B vitamers and their levels in human hair, we have just correlated our results with what is available in the literature: the nicotinamide quantified in human hair samples in our study was compared with the average levels of nicotinamide quantified in human plasma. Redeuil et al. (2015) quantified 21 water-soluble vitamin Bs in human plasma samples of the American population. Since the only vitamin B3 vitamer found and quantified in human hair samples was nicotinamide, we wanted to find the average levels of nicotinamide in human blood samples and it seemed rational to compare it with the results of Redeuil et al. [35]. A comparison of our human hair results of average nicotinamide (104.9 to 3349.9 pg/mg) with average levels of nicotinamide in human plasma (69.1–479.6 nmol/L, or 8.5–59 ng/mL) shows good correlation [35].

Hair testing is also non-invasive compared with blood testing. Hair samples can be easily collected and transported, with no fear of infection. Moreover, hair is easily storable, whereas blood, urine, and saliva storage requires special care [37]. Hair testing has also demonstrated that the concentration of parent drug ions is higher than product ions compared with urinalysis [37]. Segmental hair analysis can confirm nicotinamide deficiency information of the past few weeks to many months, depending on the length of the hair analysed, while in urine and blood analytes, nicotinamide can be detected up to 2 to 4 days [33,34]. The practical major advantage of hair analysis over blood and urine tests for nicotinamide deficiency is its wider window of detection. The presence of a vitamin or its vitamer in hair can confirm whether the person has a long-term deficiency of nicotinamide or has become recently deficient due to a disease or dietary constraints. In addition, hair analysis is useful when a history of nicotinamide deficiency is difficult to obtain (e.g., in psychiatric patients). Due to these characteristics, hair analysis is extremely valuable, especially when the other biological matrices are unsuitable. In hair analysis, however, the interpretation of results is still debatable because of some unanswered questions, such as the effect of cosmetic treatment, genetic differences, and external contamination [38]. In addition, the hair matrix should not be considered an ultimate substitute of urine and/or blood matrices in evaluating analyte deficiency but rather as a source of important and complementary information. Due to these benefits, hair testing will continue to develop further, and new applications will be discovered with time.

The nicotinamide concentrations in hair are always low as a small amount of hair is analysed and a small amount of nicotinamide usually reaches and gets stored in the hair shaft. This also depends on the biochemical composition of the hair and blood cells, interaction with other cells, lifestyle, polymorphism, and diet. The next step is to compare the human blood and hair vitamin B3 levels and also perform a segmental analysis of hair samples to depict the long-term deficiency of nicotinamide and other vitamin B vitamers in individuals, which will be a good predictor of obesity and high cholesterol and many other diseases.

5. Materials and Methods

5.1. Sample Collection

For this study, 40 healthy male and female students (aged 18–35 years) from the UAE University were recruited in December 2019. All the participants read the participant information sheet and signed consent forms as per the UAE University ethical approval protocol (UAEU Ref# SNA/fa/19-15). All participants were in good health and free from any known diseases and vitamin deficiencies. About 500 mg of hair was collected from the crown. A rubber band was used to tie hair strands. Hairs were cut close to the surface of skin and collected from different spots on the crown to avoid creating a bald spot. The samples were stored in labeled plastic envelopes in a cold, dry, and dark place before processing. The length of the hair samples was approx. 6 cm. Six different vitamin B3 (nicotinamide) and B6 vitamers (pyridoxial 5′-phosphate (PLP), pyridoxal hydrochloride (PL), pyridoxamine dihydrochloride (PM), pridoxamine-5′-phosphate (PMP), and pyridoxine hydrochloride (PN)) were analysed in the hair samples; pyridoxine-(methyl-d3) hydrochloride was used as the internal standard.

5.2. Chemical and Reagents

PN, PMP, PM, PLP, PL, nicotinamide, heptafluorobutyric acid, and LC-MS-grade water were purchased from Sigma-Aldrich, St. Louis, MO, USA. Formic acid, isopropanol, methanol, ethyl acetate, acetonitrile, deionised water, ammonium hydroxide, ammonium formate, and pentane were obtained from LABCO LLC (Dubai, UAE).

5.3. Preparation of Standard Solutions

Individual stock solutions of all the vitamin B3 and B6 vitamers and the internal standard were prepared at a concentration of 1 ppm (1000 µg/mL) in formic acid solution. Stock solution standards were mixed to form a working solution with a specific composition. Calibration was performed by the working solution. The stock solutions were used to prepare quality control (QC) samples. Each stock solution was diluted with DI water to the desired concentration to prepare working solutions of vitamin standards and internal standard. The solutions were prepared using amber glass tubes to prevent the degradation of vitamins by the light of the laboratory. Standard mixture solutions were prepared by diluting individual stock solutions. Hair powder containing no vitamin B3 and B6 vitamers were spiked with different concentrations to prepare calibrants and quality control samples. All solutions were stored at $-20\ °C$ in amber glass tubes until further processing.

5.4. Extraction Method

The 500 mg hair samples were decontaminated with isopropanol to remove contaminants, sweat, sebum, and coloring. The hair strands were dried under a gentle stream of air. The decontaminated hair was ground into a fine powder using a Fritsch Mini-ball mill (Fritsch and Gerhardt UK Ltd., Brackley, UK). Internal standard solution was added to all samples calibrants and QC's except blank hair sample. The powder was transferred into a test tube, 2 mL of methanol was added, and the mixture was vortexed for a few minutes. The mixture was further subjected to sonication for 60 min in a water bath at $35\ °C$ to release hair-bound vitamin B3 and B6 vitamers from the hair matrix. For extraction, we tried water alone and then a water/methanol mixture, but extraction with methanol was found to give the best release. The hair mixture was centrifuged at $1500\times g$ for 5 min to obtain the supernatant (expected vitamin B3 and B6 vitamers). The supernatant was collected and filtered through a syringe filter having a PTFE membrane (0.45 µm). The filtrate was dried in a sample concentrator using a gentle stream of N_2 at $40\ °C$ and reconstituted with 50 µL of a methanol:water (50:50, v/v) mixture, and 20 µL was injected into the LC-MS/MS system.

5.5. Liquid Chromatography–Tandem Mass Spectrometry (LC-MS/MS)

The separation and quantification of analytes were performed using 8060-LCMS coupled with Nexera ultra-high-pressure liquid chromatography (UHPLC), Shimadzu Corporation (Kyoto, Japan). The quantification of each analyte was performed using the signature multiple reaction monitoring (MRM). The UHPLC system is equipped with a reversed-phase Poroshell EC-C18 column with a particle size of 2.7 µm, a length of 100 mm, and a diameter 3.0 mm. The column was kept at $35\ °C$ with a constant flow rate of 0.4 mL/min. The elution of analytes was carried out with two mobile phases: A comprising LC-MS-grade water consisting of 0.1% formic acid and 0.1% heptafluorobutyric acid and B comprising acetonitrile with 0.1% formic acid. LC started with 100% A for 3 min, followed by 0–100% B for 3–5 min, then 100% B from 5–8.50 min, and declined to 100% A from 8.60–10 min. Then, the column was recovered to the initial conditions in the post-run for 5 min by eluting with 100% of the mobile phase. Heptafluorobutyric acid has been used and added to the mobile phase in the past for optimal retention and separation, as an ion-pair reagent in mass spectrometry. It has also been used in the analysis of highly polar compounds and for robust and sensitive detection of metabolites related to the tryptophan-kynurenine pathway and quantitative profiling of biomarkers related to B-vitamin status [39].

For the optimisation of the parameters for LC-MS/MS, first, a series of steps was performed to liberate the analytes from the hair. Then, a suitable separation method was performed, which gave well-separated peaks with high resolution. As soon as the optimised conditions were determined, they were used to investigate analyte recovery, similar to the analysis of spiked human plasma samples performed in our laboratory previously [36].

A Shimadzu 8060 Triple Quadrupole MS system in positive electrospray ionisation (ESI+) mode was used to perform mass spectroscopy analysis. The electrospray voltage was set at 4000 V, and the optimal conditions were follows: ion source gas (1) set at 20 psi, ion source (2) set at 45 psi, and the flow of the drying gas (N_2) set at 8 L/min at a source temperature of 300 °C. The names, structures, and molecular masses of B6 and B3 vitamers are shown in Table 3 below.

Table 3. Names, structures, and masses of B6 and B3 vitamers.

Name	Vitamer Structure	Molecular Mass (g/mol)
Pyridoxal-5'-phosphate (PLP) (vitamin B6 vitamer)	$C_8H_{10}NO_6P$	247.1
Pyridoxal hydrochloride (PL) (vitamin B6 vitamer)	$C_8H_9NO_3 \cdot HCl$	203.6 167.1 (− Cl)
Pyridoxamine dihydrochloride (PM) (vitamin B6 vitamer)	$C_8H_{12}N_2O_2 \cdot 2HCl$	241.1 168.1 (− 2HCl)
Pyridoxamine-5'-phosphate (PMP) (vitamin B6 vitamer)	$C_8H_{13}N_2O_5P$	248.1

Table 3. Cont.

Name	Vitamer Structure	Molecular Mass (g/mol)
Pyridoxine hidrochloride (PN) (vitamin B6 vitamer)	$C_8H_{11}NO_3 \cdot HCl$	205.6 169.1 (− HCl)
Nicotinamide (vitamin B3 vitamer)	$C_6H_6N_2O$	122.1
Pyridoxine-(methyl-d3) hydrochloride	$C_8H_9NO_3 \cdot HCl$	208.6

5.6. Method Validation

The LC-MS/MS method was validated by examining the following parameters: linear range, lower limit of detection (LOD), lower limit of quantitation (LOQ), intra- and inter-day accuracy and precision and recovery. The US Food and Drug Administration (US-FDA) guidelines were used for bioanalytical method development and validation [40]. Intra- and inter-day accuracy and precision were determined for three quality control (QC) samples. The evaluation of intra- and inter-day accuracy and precision was performed at high, low, and medium QC sample concentrations by investigating the replicates at every level. The inter-day precision accuracy was accepted if the experimental concentrations were within 15% of the actual concentrations and the lower limit of quantification (LOQ) was accepted within the 20% limit range.

The inter/intra-day precision (% CV) data were calculated from quality control data obtained after analysis using the equation:

$$\left[\% \ CV = \left(\frac{Standard \ deviation}{mean}\right) \times 100\right] \quad (1)$$

The assay % inter/intra-day accuracy was obtained from the quality control data using the equation:

$$\left[\% \ Accuracy = \left(\frac{Mean \ value}{Nominal \ value}\right) \times 100\right] \quad (2)$$

The % absolute recoveries were calculated using the equation:

$$\left[\% \ Recovery = \left(\frac{Mean \ unextracted \ QC \ values}{Mean \ extracted \ QC \ values}\right) \times 100\right] \quad (3)$$

The lower limit of detection (LOD) was established by comparing the instrument signal to noise ratio (S/N) to the analytes lowest concentration. The analytes lowest concentrations were determined by decreasing the analyte concentrations until an LC-MS/MS detector response equal to 3 times the instrument background noise level was observed.

Quality control samples were prepared at the following concentrations: QCH at 1200 pg/mg, QCM at 400 pg/mg, and QCL at 200 pg/mg. All QC and calibrant samples were prepared with human hair containing no vitamin B3 and B6 vitamers or no interfering or matrix peaks at required retention times. For analyte recovery, QC samples at the above three concentrations were prepared in methanol, dried, reconstituted in the mobile phase, and injected into the LC-MS/MS system and the area under the curve for all QC samples was calculated. Then, QC samples at the same concentrations were spiked in blank human hair samples, extracted using the normal protocol, dried, reconstituted, and injected into the LC-MS/MS system, and the area under the curve was then calculated. The percentage recovery was calculated by taking the ratio of extracted vs. non-extracted vitamin B3 and B6 vitamers and multiplying it with 100. The specificity of the method was determined by running 6 blank hair samples and finding no interfering peaks at the retention times of vitamin B3 and B6 vitamers.

From the 40 human hair samples, six different lots of presumed vitamin B3- and B6-free samples were ball-milled, and one representative sample from each lot was selected. The selected samples were extracted and analysed in triplicate using LC-MS/MS. No interferences from the matrix and no vitamer peaks were detected in few of the selected samples. One of the six samples was chosen as a blank; this sample was the cleanest of all, with no co-eluting and interfering peaks detected at retention time of the hair matrix and no vitamin B vitamer peaks. This blank hair sample was used for further preparation of quality controls and calibrants. Lastly, under the given analytical conditions employed, there were no major matrix interferences associated with blank hair that affected the analysis of vitamin B vitamers."

6. Conclusions

To the best of our knowledge, this is the first pilot study to detect vitamin B3 and B6 vitamers in human hair. Our sensitive, rapid, and robust LC-MS/MS-based assay was able to quantify seven different vitamin B3 and B6 vitamers in human hair. Further work is warranted to develop and validate the procedure by including more vitamin B vitamers and to produce a gold-standard test for human hair analyses. This might be possible by using advanced sample concentration steps, using larger hair samples, and exploring more advanced sensitive commercially available instrumentation. Future investigative work should also include a full spectrum of ethnicity-based hair types, different population groups, age groups, genders, seasonal variations, and color combinations. Hair from other parts of the human body and animals could also be investigated. This innovation will help explore many serious diseases related to vitamin B deficiency.

In this study, human hair was chosen for the detection and quantification of vitamin B3 and vitamin B6 vitamers because hair is tamper-resistant and easily storable, with no risk of infection, and the method is non-invasive as opposed to human plasma testing. The method is free from interferences resulting from instrumental means and matrices. The LC-MS/MS method was validated with multiple parameters: linearity, reliability, robustness, specificity, rapidness, and accuracy. It will be possible to detect vitamin B3 coenzymes in human hair and prevent serious diseases. Therefore, the novel LC-MS/MS method is significant for future clinical trials and long-term evaluation of the vitamin B3 status in humans.

Author Contributions: I.S. conceived and designed the research project. S.M.S., A.A. and M.A. carried out the lab work with help from I.S. S.M.S. wrote the first draft of the manuscript with help from I.S. All authors helped with data generation, analysis, and manuscript finalisation. All authors have read and agreed to the final version of the manuscript.

Funding: The authors acknowledge funding from the UAEU start-up grant no. 31S213, which partially supported the work reported here.

Institutional Review Board Statement: The study was conducted according to the guidelines of the Declaration of Helsinki and approved by the Human Research Ethics Committee of the United Arab Emirates University (protocol code UAEU Ref# SNA/fa/19-15 and 31 December 2019).

Informed Consent Statement: Informed consent was obtained from all subjects involved in the study.

Data Availability Statement: All relevant data are included in the paper or its Supplementary Information.

Conflicts of Interest: The authors declare that they have no conflict of interest.

Sample Availability: Samples of the compounds are not available from the authors.

Appendix A

Table A1. The average concentrations of nicotinamide in male and female volunteers. (here, SD means standard deviation and "ND" means "not detected").

Male		Female	
Hair Sample	Mean ± SD (pg/mg)	Hair Sample	Mean ± SD (pg/mg)
Sample-01	131.1 ± 6.72	Sample-23	284.4 ± 7.11
Sample-02	1468.4 ± 3.9	Sample-24	200.8 ± 3.1
Sample-03	316.8 ± 14.2	Sample-25	243.1 ± 2.1
Sample-04	195.6 ± 0.62	Sample-26	ND
Sample-05	269.6 ± 3.29	Sample-27	758.8 ± 1.49
Sample-06	169.2 ± 17.2	Sample-28	173.3 ± 2.5
Sample-07	107.0 ± 2.3	Sample-29	132.1 ± 4.2
Sample-08	493.1 ± 3.23	Sample-30	1972.1 ± 1.1
Sample-09	135.6 ± 8.2	Sample-31	1939.5 ± 3.1
Sample-10	147.7 ± 1.59	Sample-32	2706.0 ± 4.4
Sample-11	422.1 ± 5.62	Sample-33	356.1 ± 0.5
Sample-12	171.5 ± 5.11	Sample-34	1052.9 ± 0.9
Sample-13	186.7 ± 9.93	Sample-35	483.7 ± 3.1
Sample-14	ND	Sample-36	358.5 ± 8.1
Sample-15	224.9 ± 11.2	Sample-37	105.0 ± 0.9
Sample-16	443.4 ± 4.7	Sample-38	131.6 ± 3.2
Sample-17	747.5 ± 3.9	Sample-39	ND
Sample-18	400.2 ± 2.1	Sample-40	ND
Sample-19	430.8 ± 3.5		
Sample-20	3350 ± 0.71		
Sample-21	1837.1 ± 1.2		
Sample-22	397.9 ± 1.1		

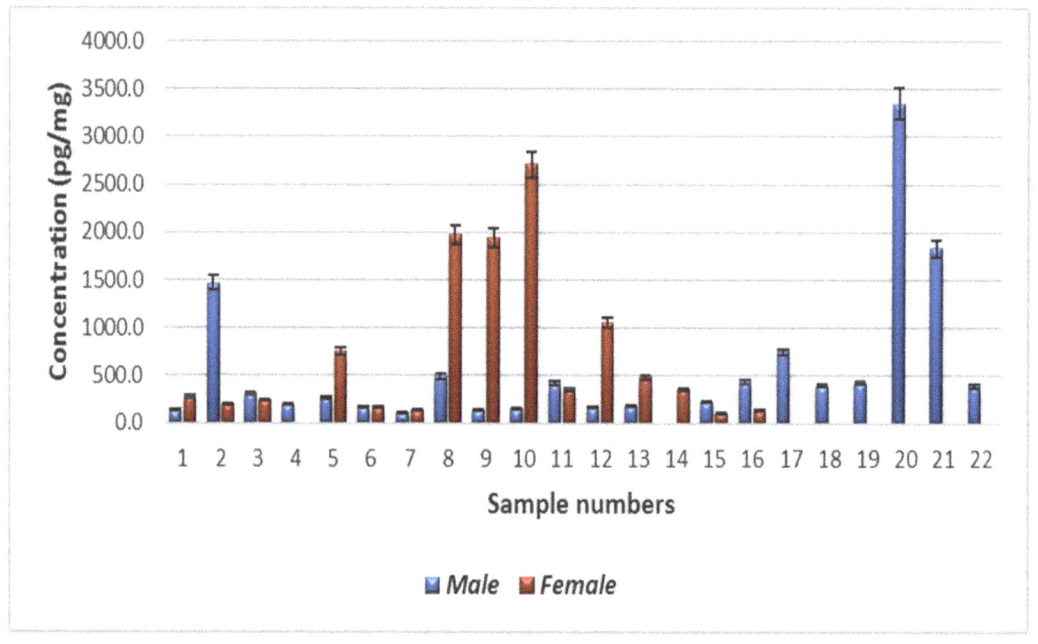

Figure A1. Levels of nicotinamide in the human hair samples. Error bars above shows the standard error of the mean.

References

1. Clayton, P.T. B 6-responsive disorders: A model of vitamin dependency. *J. Inherit. Metab. Dis.* **2006**, *29*, 317–326. [CrossRef] [PubMed]
2. Kotloski, N.J.; Gralnick, J.A. Flavin electron shuttles dominate extracellular electron transfer by Shewanella oneidensis. *MBio* **2013**, *4*, e00553-12. [CrossRef] [PubMed]
3. Selhub, J. Folate, vitamin B12 and vitamin B6 and one carbon metabolism. *J. Nutr. Health Aging* **2002**, *6*, 39–42. [PubMed]
4. Velasquez-Orta, S.B.; Head, I.M.; Curtis, T.P.; Scott, K.; Lloyd, J.R.; von Canstein, H. The effect of flavin electron shuttles in microbial fuel cells current production. *Appl. Microbiol. Biotechnol.* **2010**, *85*, 1373–1381. [CrossRef]
5. Bryan, J.; Calvaresi, E.; Hughes, D. Short-term folate, vitamin B-12 or vitamin B-6 supplementation slightly affects memory performance but not mood in women of various ages. *J. Nutr.* **2002**, *132*, 1345–1356. [CrossRef]
6. Selhub, J.; Morris, M.S.; Jacques, P.F.; Rosenberg, I.H. Folate–vitamin B-12 interaction in relation to cognitive impairment, anemia, and biochemical indicators of vitamin B-12 deficiency. *Am. J. Clin. Nutr.* **2009**, *89*, 702S–706S. [CrossRef]
7. Wang, L.; Li, H.; Zhou, Y.; Jin, L.; Liu, J. Low-dose B vitamins supplementation ameliorates cardiovascular risk: A double-blind randomized controlled trial in healthy Chinese elderly. *Eur. J. Nutr.* **2015**, *54*, 455–464. [CrossRef]
8. Czeizel, A.E.; Dudás, I.; Vereczkey, A.; Bánhidy, F. Folate deficiency and folic acid supplementation: The prevention of neural-tube defects and congenital heart defects. *Nutrients* **2013**, *5*, 4760–4775. [CrossRef]
9. Heseker, H. Folic acid and other potential measures in the prevention of neural tube defects. *Ann. Nutr. Metab.* **2011**, *59*, 41–45. [CrossRef]
10. Seshadri, S.; Beiser, A.; Selhub, J.; Jacques, P.F.; Rosenberg, I.H.; D'Agostino, R.B.; Wilson, P.W.; Wolf, P.A. Plasma homocysteine as a risk factor for dementia and Alzheimer's disease. *N. Engl. J. Med.* **2002**, *346*, 476–483. [CrossRef]
11. Ray, J.G. Meta-analysis of hyperhomocysteinemia as a risk factor for venous thromboembolic disease. *Arch. Intern. Med.* **1998**, *158*, 2101–2106. [CrossRef] [PubMed]
12. Sauve, A.A. NAD+ and vitamin B3: From metabolism to therapies. *J. Pharmacol. Exp. Ther.* **2008**, *324*, 883–893. [CrossRef] [PubMed]
13. Preedy, V.R. *B Vitamins and Folate: Chemistry, Analysis, Function and Effects*; Royal Society of Chemistry, Thomas Graham House, Science Park: Cambridge, UK, 2012.
14. Badawy, A.A. Kynurenine pathway of tryptophan metabolism: Regulatory and functional aspects. *Int. J. Tryptophan Res.* **2017**, *10*, 1178646917691938. [CrossRef] [PubMed]
15. Badawy, A.A.-B. Tryptophan metabolism in alcoholism. *Nutr. Res. Rev.* **2002**, *15*, 123–152. [CrossRef] [PubMed]
16. Gasperi, V.; Sibilano, M.; Savini, I.; Catani, M.V. Niacin in the central nervous system: An update of biological aspects and clinical applications. *Int. J. Mol. Sci.* **2019**, *20*, 974. [CrossRef]

17. Roelofsen-de Beer, R.; van Zelst, B.; Wardle, R.; Kooij, P.; de Rijke, Y. Simultaneous measurement of whole blood vitamin B1 and vitamin B6 using LC-ESI–MS/MS. *J. Chromatogr. B* **2017**, *1063*, 67–73. [CrossRef]
18. Shibata, K.; Kobayashi, R.; Fukuwatari, T. Vitamin B1 deficiency inhibits the increased conversion of tryptophan to nicotinamide in severe food-restricted rats. *Biosci. Biotechnol. Biochem.* **2015**, *79*, 103–108. [CrossRef]
19. Kaneko, J.J.; Harvey, J.W.; Bruss, M.L. *Clinical Biochemistry of Domestic Animals*; Academic Press: Burlington, MA, USA, 2008.
20. Ueland, P.M.; McCann, A.; Midttun, Ø.; Ulvik, A. Inflammation, vitamin B6 and related pathways. *Mol. Asp. Med.* **2017**, *53*, 10–27. [CrossRef]
21. Erdman, J.W., Jr.; Macdonald, I.A.; Zeisel, S.H. *Present Knowledge in Nutrition*; John Wiley & Sons: Oxford, UK, 2012.
22. Kushnir, M.M.; Song, B.; Yang, E.; Frank, E.L. Development and Clinical Evaluation of a High-Throughput LC–MS/MS Assay for Vitamin B6 in Human Plasma and Serum. *J. Appl. Lab. Med.* **2021**, *6*, 702–714. [CrossRef]
23. Asante, I.; Pei, H.; Zhou, E.; Liu, S.; Chui, D.; Yoo, E.; Louie, S.G. Simultaneous quantitation of folates, flavins and B6 metabolites in human plasma by LC–MS/MS assay: Applications in colorectal cancer. *J. Pharm. Biomed. Anal.* **2018**, *158*, 66–73. [CrossRef]
24. Zhang, M.; Liu, H.; Huang, X.; Shao, L.; Xie, X.; Wang, F.; Yang, J.; Pei, P.; Zhang, Z.; Zhai, Y. A novel LC-MS/MS assay for vitamin B1, B2 and B6 determination in dried blood spots and its application in children. *J. Chromatogr. B* **2019**, *1112*, 33–40. [CrossRef]
25. Roy, B.; Singh, B.; Rizal, A.; Malik, C. Bioanalytical method development and validation of niacin and nicotinuric acid in human plasma by LC-MS/MS. *Int. J. Pharm. Clin. Res* **2014**, *6*, 206–213.
26. Sutherland, K.D.; Gibbs, A.J. Method Development for Separation and Quantification of Niacin and Its Metabolites in Human Blood Plasma. Presented at the 18th Annual Phi Kappa Phi Student Research and Fine Arts Conference: Posters, Department of Chemistry and Physics: Student Research and Presentations, Augusta University, Augusta, GA, USA, March 2017. Available online: http://hdl.handle.net/10675.2/621328 (accessed on 24 July 2021).
27. Lang, R.; Yagar, E.F.; Eggers, R.; Hofmann, T. Quantitative investigation of trigonelline, nicotinic acid, and nicotinamide in foods, urine, and plasma by means of LC-MS/MS and stable isotope dilution analysis. *J. Agric. Food Chem.* **2008**, *56*, 11114–11121. [CrossRef] [PubMed]
28. Catz, P.; Shinn, W.; Kapetanovic, I.M.; Kim, H.; Kim, M.; Jacobson, E.L.; Jacobson, M.K.; Green, C.E. Simultaneous determination of myristyl nicotinate, nicotinic acid, and nicotinamide in rabbit plasma by liquid chromatography–tandem mass spectrometry using methyl ethyl ketone as a deproteinization solvent. *J. Chromatogr. B* **2005**, *829*, 123–135. [CrossRef] [PubMed]
29. Redeuil, K.; Vulcano, J.; Prencipe, F.P.; Bénet, S.; Campos-Giménez, E.; Meschiari, M. First Quantification of Nicotinamide Riboside with B 3 Vitamers and Coenzymes Secreted in Human Milk by Liquid Chromatography-Tandem-Mass Spectrometry. *J. Chromatogr. B Anal. Technol. Biomed. Life Sci.* **2019**, *1110*, 74–80. [CrossRef] [PubMed]
30. Xue, Y.; Redeuil, K.M.; Giménez, E.C.; Vinyes-Pares, G.; Zhao, A.; He, T.; Yang, X.; Zheng, Y.; Zhang, Y.; Wang, P. Regional, socioeconomic, and dietary factors influencing B-vitamins in human milk of urban Chinese lactating women at different lactation stages. *BMC Nutr.* **2017**, *3*, 22. [CrossRef] [PubMed]
31. Zgaga, L.; Laird, E.; Healy, M. 25-Hydroxyvitamin D measurement in human hair: Results from a proof-of-concept study. *Nutrients* **2019**, *11*, 423. [CrossRef] [PubMed]
32. Shah, I.; Mansour, M.; Jobe, S.; Salih, E.; Naughton, D.; Salman Ashraf, S. A Non-Invasive Hair Test to Determine Vitamin D3 Levels. *Molecules* **2021**, *26*, 3269. [CrossRef]
33. Manson, P.; Zlotkin, S. Hair analysis—A critical review. *Can. Med Assoc. J.* **1985**, *133*, 186. [PubMed]
34. Harkins, D.K.; Susten, A.S. Hair analysis: Exploring the state of the science. *Environ. Health Perspect.* **2003**, *111*, 576–578. [CrossRef] [PubMed]
35. Redeuil, K.M.; Longet, K.; Bénet, S.; Munari, C.; Campos-Giménez, E. Simultaneous quantification of 21 water soluble vitamin circulating forms in human plasma by liquid chromatography-mass spectrometry. *J. Chromatogr. A* **2015**, *1422*, 89–98. [CrossRef]
36. Ibrahim, G.R.; Shah, I.; Gariballa, S.; Yasin, J.; Barker, J.; Salman Ashraf, S. Significantly elevated levels of plasma nicotinamide, pyridoxal, and pyridoxamine phosphate levels in obese Emirati population: A cross-sectional study. *Molecules* **2020**, *25*, 3932. [CrossRef] [PubMed]
37. Shah, I.; Petroczi, A.; Uvacsek, M.; Ránky, M.; Naughton, D.P. Hair-based rapid analyses for multiple drugs in forensics and doping: Application of dynamic multiple reaction monitoring with LC-MS/MS. *Chem. Cent. J.* **2014**, *8*, 1–10. [CrossRef] [PubMed]
38. Tsanaclis, L.; Andraus, M.; Wicks, J. Hair analysis when external contamination is in question: A review of practical approach for the interpretation of results. *Forensic Sci. Int.* **2018**, *285*, 105–110. [CrossRef] [PubMed]
39. Midttun, Ø.; Hustad, S.; Ueland, P.M. Quantitative profiling of biomarkers related to B-vitamin status, tryptophan metabolism and inflammation in human plasma by liquid chromatography/tandem mass spectrometry. *Rapid Commun. Mass Spectrom.* **2009**, *23*, 1371–1379. [CrossRef] [PubMed]
40. Zimmer, D. New US FDA draft guidance on bioanalytical method validation versus current FDA and EMA guidelines: Chromatographic methods and ISR. *Bioanalysis* **2014**, *6*, 13–19. [CrossRef]

MDPI
St. Alban-Anlage 66
4052 Basel
Switzerland
Tel. +41 61 683 77 34
Fax +41 61 302 89 18
www.mdpi.com

Molecules Editorial Office
E-mail: molecules@mdpi.com
www.mdpi.com/journal/molecules

www.ingramcontent.com/pod-product-compliance
Lightning Source LLC
LaVergne TN
LVHW070155100526
838202LV00015B/1951